Studies in Computational Intelligence

Volume 902

The series "Studies in Computational Intelligence" (SCI) publishes new developments and advances in the various areas of computational intelligence—quickly and with a high quality. The intent is to cover the theory, applications, and design methods of computational intelligence, as embedded in the fields of engineering, computer science, physics and life sciences, as well as the methodologies behind them. The series contains monographs, lecture notes and edited volumes in computational intelligence spanning the areas of neural networks, connectionist systems, genetic algorithms, evolutionary computation, artificial intelligence, cellular automata, self-organizing systems, soft computing, fuzzy systems, and hybrid intelligent systems. Of particular value to both the contributors and the readership are the short publication timeframe and the world-wide distribution, which enable both wide and rapid dissemination of research output.

The books of this series are submitted to indexing to Web of Science, EI-Compendex, DBLP, SCOPUS, Google Scholar and Springerlink.

More information about this series at http://www.springer.com/series/7092

Ivan Dimov · Stefka Fidanova
Editors

Advances in High Performance Computing

Results of the International Conference on "High Performance Computing" Borovets, Bulgaria, 2019

Springer

Editors
Ivan Dimov
Institute of Information
and Communication Technology
Bulgarian Academy of Sciences
Sofia, Bulgaria

Stefka Fidanova
Institute of Information
and Communication Technology
Bulgarian Academy of Sciences
Sofia, Bulgaria

ISSN 1860-949X ISSN 1860-9503 (electronic)
Studies in Computational Intelligence
ISBN 978-3-030-55346-3 ISBN 978-3-030-55347-0 (eBook)
https://doi.org/10.1007/978-3-030-55347-0

This Springer imprint is published by the registered company Springer Nature Switzerland AG
The registered company address is: Gewerbestrasse 11, 6330 Cham, Switzerland

Preface

Many challenging real-world problems arising in engineering, economics, medicine and other areas can be formulated as large-scale computational tasks. Every day, we need to solve large problems for which supercomputers are needed. High performance computing (HPC) is a paradigm that allows to efficiently implement large-scale computational tasks on powerful supercomputers. We face many challenges in the area of HPC. Supercomputers are not only tools of paramount importance to research & development projects in all areas of science, but also an instrument that is likely to have a positive impact on people's quality of life in the long run. Without such an instrument progress is impossible. But what is the distribution of these supercomputers around the world? Among the top 500 most powerful supercomputers, there are: China—219 (43%); USA—116; EU—88; and Japan—29 (June 2019). So, the European Union is not the first player, not even the second player in the world. Obviously, at least here in Europe, we are going to have to play catch-up (to the rest). An important part of offsetting this discrepancy in raw computational power is the development of clever new software tools, methods and parallel algorithms capable of increasing the efficiency of our systems. Maybe this vital part of the development has been neglected in recent years, but now is the time to revisit it.

This volume is a result of very vivid and fruitful discussions held during the High Performance Computing Conference held in Borovets, Bulgaria, from September 2 to 6, 2019. The participants have agreed that the relevance of the conference topic and quality of the contributions have clearly suggested that a more comprehensive collection of extended contributions devoted to the area would be very welcome and would certainly contribute to a wider exposure and proliferation of the field and ideas. The topics of interest included into this volume are: HP software tools, parallel algorithms and scalability, HPC in big data analytics, modeling, simulation & optimization in a data-rich environment, advanced numerical methods for HPC, hybrid parallel or distributed algorithms. The volume is focused on important large-scale applications like environmental and climate modeling, computational chemistry and heuristic algorithms.

November 2019

Ivan Dimov
Stefka Fidanova
Editor of Advances in HPC'2019

Organization

High Performance Computing Conference was held in Borovets, Bulgaria, from September 2 to 6, 2019.

Conference Co-chairs

Ivan Dimov	IICT-BAS, Bulgaria
Stefka Fidanova	IICT-BAS, Bulgaria

Program Committee

Angelova, Galia	IICT-BAS, Bulgaria
Denkov, Nikolay	Sofia University, Bulgaria
Dimov, Ivan	IICT-BAS, Bulgaria
Dongarra, Jack	University of Tennessee, USA
Drensky, Vesselin	IMI-BAS, Bulgaria
Fidanova, Stefka	IICT-BAS, Bulgaria
Ilieva, Nevena	IICT-BAS, Bulgaria
Karaivanova, Aneta	IICT-BAS, Bulgaria
Lazarov, Raytcho	Texas A& M University, USA
Margenov, Svetozar	IICT-BAS, Bulgaria
Markov, Stoyan	NCSA, Bulgaria
Schilders, Wim	TU Eindhoven, Netherlands
Lippert, Tomas	Julich Supercomputer Center, Germany
Zlatev, Zahari	Aarhus University, Denmark
Ostromsky, Tzvetan	IICT, Bulgaria
Asenov, Asen	University of Glasgow, UK
Farago, Istvan	Eötvös Loránd University, Hungary

Skordas, Thomas	EU Commission, Belgium
Georgiev, Krassimir	IICT-BAS, Bulgaria
Verwaerde, Daniel	Teratec, France
Schulthess, Thomas	ETH Zurich, Switzerland

About the Book

Many challenging real world problems arising in engineering, economics, medicine and other areas can be formulated as large-scale computational tasks. Every day we need to solve large problems for which supercomputers are needed. High performance computing (HPC) is a paradigm that allows to efficiently implement large-scale computational tasks on powerful supercomputers. We face many challenges in the area of HPC. Supercomputers are not only tools of paramount importance to research & development projects in all areas of science, but also an instrument that is likely to have a positive impact on people's quality of life in the long run.

The volume shows how to develop algorithms for them based on new intelligent methods and algorithms capable of increasing the efficiency of our systems. This research demonstrates how some real-world problems arising in engineering, economics and other domains can be solved in efficient way.

The book can be useful for both researchers and practitioners.

Contents

Contributed Papers

Application of Artificial Intelligence in Optimization and Modeling

Modeling and Simulation of Low Power Wireless Sensor Networks Based on Generalized Nets

A. Alexandrov(✉), R. Andreev, S. Ilchev, A. Boneva, S. Ivanov, and J. Doshev

Institute of Information and Communication Technologies,
Bulgarian Academy of Sciences, Sofia, Bulgaria
akalexandrov@iit.bas.bg

Abstract. The Wireless Sensor Networks (WSNs) in now days are facing the additional requirement to provide the functionality to collect, analyze and integrate the sensor data, i.e. to acts as intelligent sensor networks, in addition to the typical transmission and routing functions. The energy minimization at a fixed level performance is very important in wireless sensor nodes which are mainly battery powered. Therefore the understanding of the energy consumption characteristics of each sensor node is critical for the design of energy saving monitoring strategies. The Low Power Wireless Sensor Networks (LPWSN) which are a matter of the current research are applied in many domains, such as ecological and environmental monitoring, traffic management, military applications and etc. The right and realistic model and an adequate simulation of these networks is a key step in the design of a reliable sensor networks architecture and can reduce sensitively the development cost and time. This paper develops a model of LPWSN based on Generalized Nets (GNs) to evaluate the energy consumption of Wireless Sensor Nodes. The proposed model factors important components of a typical sensor node, including SoC (System on Chip) microcontrollers with energy-saving features, wireless front-end components, and low power sensor modules. A Markov chain model of the same network with the same input data was realized as a benchmark. Both models were simulated in WSNet simulator and the data after that compared to real low powered wireless sensor node in lab environment. The output experimental results show that the GN model is more flexible and accurate to the real sensor device than the Markov Chain model and provides a scalable simulation platform to study energy-saving strategies in WSNs.

1 Introduction - Low Power WSNs

The battery-powered sensor nodes data transmission process has been identified as one of the main causes of the lifetime limitation of the WSNs. The sensor node's battery replacements during a regular time period are impractical in large wireless networks or may even be impossible in some environments. Although,

I. Dimov and S. Fidanova (Eds.): HPC 2019, SCI 902, pp. 3–14, 2021.
https://doi.org/10.1007/978-3-030-55347-0_1

the module's energy consumption and optimization have a direct impact on a network's lifetime. Therefore, the energy consumption has to be analyzed [1,2] as a first level towards achieving maximum shelf life. An energy model is required to capture all energy consumption sources in all communication layers stack and who covers various scenarios that occur during the inter-node communication process. One of the key parameters to achieving a longer lifetime of a WSN is the hardware design customization of the wireless sensor nodes which are key factors and can minimize the overall WSN power consumption. One of the ways to reduce overall WSN's power consumption is the possibility of the low power wireless sensor networks to control the active time of every sensor and limit the current draw when sensors are in no active mode. Because of that was created a new class of WSNs focused on minimization of the energy consumption and extension of the sensor's lifetime.

The Low-power Wireless Sensor Networks (LPWSN) are optimal for monitoring of for example climate, public safety, transport monitoring and etc.

In environmental monitoring, a common function of a wireless sensor node is to track relatively slow changing parameters, such as temperature, humidity, and barometric pressure. To maximize the battery capacity, the sensor node states in low-power "sleep" mode, waking up periodically according to a fixed time period. During the active period, the node collects sensor data and sends it by radio channel to a central controller. Then it goes back to "sleep" mode to save energy. The block hardware diagram of typical low powered sensor node is shown in Fig. 1.

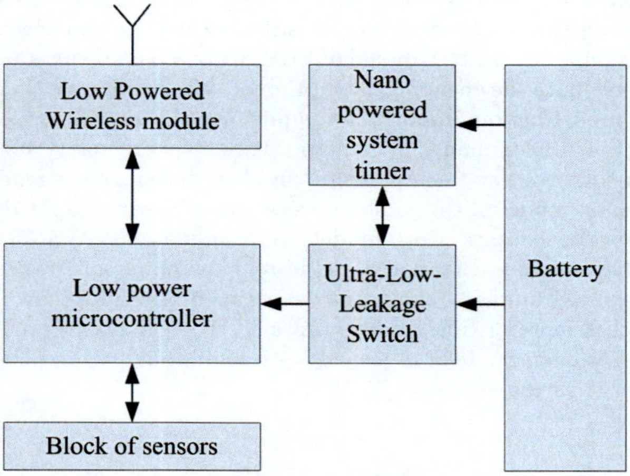

Fig. 1. Low Power Wireless sensor node - block diagram

As is shown on the diagram above the main difference between WSN and the typical LPWSN based hardware is the availability of a new component - nano-powered system timer with a combination of ultra-low current leakage switch who is responsible to total switch off and on of the overall sensor hardware during fixed time periods.

2 Related Work

Several types of research in WSN have looked at various ways of saving energy. Much research has been done on the physical layer energy modeling for WSNs [3–6]. However, these papers ignored an important source of energy consumption, such as physical layer overhead and channel error, which made these models not accurate.

In [7], the authors give general average power consumption formula and mathematical model for different MAC protocols which include the effect of duty cycling. In the other hand, the authors ignore the effect of overhead and error probability.

In [8], the authors present energy models based on the energy consumption due to transmitting, receiving and overhearing but they again ignore the important source of energy consumption, such as the effect of duty cycling and overhead.

The work proposed by [9] is an estimation model based on the Hardware Management Console (HMC) to predict the energy level of a sensor node. The proposed process contains two main parts: a first part to train the protocol-specific HMC via the Baum-Welch algorithm and a second part to predict energy levels via the Viterbi algorithm. This approach suffers from several limits. It does not predict the value of power remaining in the battery of the sensor. Also, the algorithms used are very complex and resources consuming. Additionally, it does not focus on optimizing the process of the QoS values transmission.

The research in [10] proposes a Markov chain model for the standard IEEE 802.15.4 MAC protocol considering a simple M/G/1/K system queue model (the finite population of M/G/1 with maximum K jobs) and super-frame with both active/only and active/inactive duty-cycle periods for a one-hop star topology. Expressions for the access delay, probability distribution of the packet service time as well as probability distribution of the queue length are presented. The limitation of this model is that all results were obtained for 1- transmission where a device sends a packet to a coordinator and waits for the acknowledgment. Even considering an M/G/1/K queue system for the first node, taking into account that the output distribution of an M/G/1/K is not Markov, it is not possible to extend the proposed model for multi-hop transmissions by chaining M/G/1/K queue system.

Thus, while a significant amount of research has been done on energy consumption based modeling of WSNs, it does not provide an accurate energy model of WSNs taken into account the physical layer and MAC layer parameters.

2.1 Model of Low Power Wireless Sensor Node Based on Continuous Time Markov Chains

For the Continuous Time Markov chains as described in details in [11,12], the used in the proposed model transition probability function for $t > 0$ can be described as:

$$P_{ij}(t) = P(X(t+u) = j|X(u) = i) \tag{1}$$

and is independent of $u \geq 0$. In the equation above $P(X(t+u) = j|X(u) = i)$ is a function of t and describes a time homogeneous transition function for this process. In our case to construct a discrete time Markov process, we need just to specify a transition matrix together with a distribution function. The Birth-Death process [10] is a variation of the Continuous Time Markov process. In this case all the transitions are limited to birth and death. When we have a birth, the process goes from state i to state $i+1$. By the same way, when death happens, the current process goes from state i to state $i-1$. It is mandatory to be assumed that the birth and death events are independent.

The general description of the Birth-and-Death process [10] can be as follows: after the process enters state i, it states temporary in the given state for a random amount of time, exponentially distributed with parameter $(\lambda_i + \mu_i)$. In the next step the process i, enters either $i+1$ with probability

$$\frac{\lambda_i}{\lambda_i + \mu_i} \tag{2}$$

or $i-1$ with probability

$$\frac{\mu_i}{\lambda_i + \mu_i} \tag{3}$$

If the next state is $i+1$, then the process stays temporary in this state and then chooses the next state etc. For example let's consider two exponentially distributed random variables $A(i)$ and $B(i)$ with parameters λ_i and μ_i respectively. These variables describe the birth and death time of i, respectively. The population of i increases by one if the birth occurs prior to death and decreases by one otherwise. The variables $A(i)$, $B(i)$ be default are independent exponentially distributed random variables, then their minimum is exponentially distributed with parameter $(\lambda_i + \mu_i)$. Thus the transition from i to $i+1$ is made if $A(i) < B(i)$, which occurs with probability

$$P[A(i) < B(i)] = \frac{\lambda_i}{\lambda_i + \mu_i} \tag{4}$$

In Fig. 2 the sensor node goes to "active" mode P_{active} from "sleep" mode P_{sleep} when the nano-powered system timer activates it by the Ultra-Low-Leakage Switch as is shown on Fig. 1. The Markov model describes the various increasing states ($P_{i,1}, P_{i,2}, P_{i,n}$, etc.) and the sensor node enters as the number of jobs increase under a given job arrival rate λ. The sensor node services the jobs at rate μ and moves the node to "sleep" states P_{sleep} and eventually to the "active" state P_{active}. If the node remains in the active state for some time interval greater than some threshold, the embedded in the node's microcontroller watchdog timer moves back the sensor module to the "sleep" P_{sleep} state.

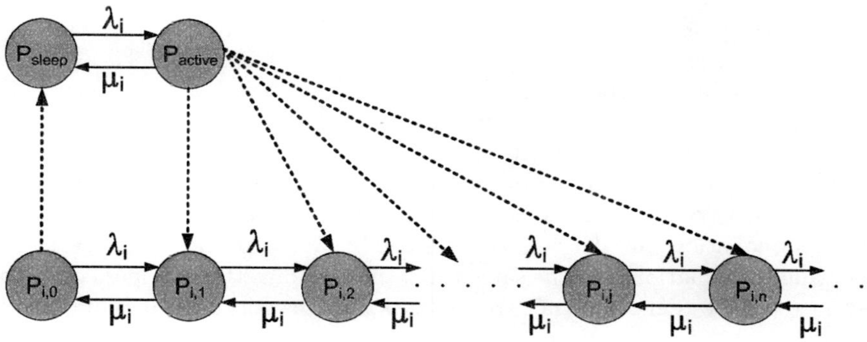

Fig. 2. LPWSN sensor node Markov chain model

In the proposed model, the following assumptions are made:

1) The request arrivals follow a standard Markov chain related Birth-Death process, described in [10]
2) The service time is exponentially distributed.
3) The sensor node enters the sleep mode if there are no more jobs to be processed for a fixed time interval.
4) The activation process P_{active} takes a fixed time T_{active}.

The proposed Markov chains based LPWSN sensor node model consists of a mix of exponential and deterministic transitions. All the exponential time followed transitions are shown as solid lines in Fig. 2. At the same time, the deterministic transitions are shown as dashed lines. The sensor node enters the "sleep" state after a fixed time T_{sleep}. Therefore, as it is visible from the diagram, the power down transition depends on its previous state only.

The stable state probabilities of P_{sleep} and P_{active} are given by the following equations:

$$P_{sleep} = \frac{1 - \frac{\lambda_i}{\mu_i}}{e^{\lambda_i T_{sleep}} + (1 - \frac{\lambda_i}{\mu_i})(1 - e^{-\lambda_i B}) + \frac{\lambda_i}{\mu_i}\lambda} \tag{5}$$

$$P_{active} = \frac{(1 - \frac{\lambda_i}{\mu_i})(1 - e^{-\lambda_i A})}{e^{\lambda_i T_{active}} + (1 - \frac{\lambda_i}{\mu_i})(1 - e^{-\lambda_i A}) + \frac{\lambda_i}{\mu_i}\lambda} \tag{6}$$

and the total energy consumption is:

$$E = (P_{i,n} P_{sleep} + P_{i,n} P_{active} + G_i P_{active} \frac{N + L(n)^2}{\lambda_i}) \tag{7}$$

where $G_i(t) = e^{-t(\lambda_i = \mu_i)}$, n is the total number of the sensor processes, P_{sleep}, and P_{active} are the power consumption rates in the sensor node's sleep, and active states respectively.

The described WSN Markov chain model was simulated by NS3 simulator using the embedded in the simulator GaussMarkovMobilityModel Class Reference. Markov chains can be used for WSN modeling, however, they are very restrictive in the type of cases that can be modeled. A Markov model is a discrete-time stochastic process composed of event chains with the Markov property. It's mean that the next state in a Markov model is only dependent upon the current state instead of previous ones. At the same time, the Markov chains cannot be used to model transitions between states that require fixed deterministic intervals. The benefit of using Markov chains for modeling sensor systems is that once the proper equations are derived, the average behavior can be easily reached by evaluating the equations. At the same time, the task of getting proper equations, relevant to the system can be not an easy task, and in some cases it is impossible.

2.2 Model of Low Power Wireless Sensor Networks Based on GNs

The Generalized Nets (GNs) [13–15] are radically new approach in WSNs modeling. The main benefits of the proposed GNs based model are that they are extremely flexible and reliable, especially in the area of the parallel processes modeling and analysis. A new GN model of the sensor data clustered parallel integration process is presented in Fig. 3. The proposed new model is related to WSN's with star topology. In the model the input $l_{i,0}$ represents the start point of power up and the initialization process of the sensor node.

In the model the token α enters GN in place $l_{i,0}$ with characteristic "A_i - power-up and initialization where $i \in N$". N is the number of the sensor nodes in the modeled WSN with star topology.

$$Z_0 = <\{l_{i,0}, l_{i,2}\}, \{l_{i,1}, l_{i,2}\}, r_0 \vee (l_{i,0}, l_{i,2})> \tag{8}$$

$$r_0 = \begin{array}{c|cc} & l_{i,1} & l_{i,2} \\ \hline l_{i.0} & false & true \\ l_{i,2} & W_{i2,i1} & true \end{array} \tag{9}$$

Where $W_{i2,i1}$ is "finished time of sensor node initialization".

The transition Z_1 in Fig. 3 represents the "sleep" mode of the wireless node and the activation of the ultra low powered system timer.

$$Z_1 = <\{l_{i,1}, l_{i,5}, l_{i,4}\}, \{l_{i,3}, l_{i,4}\}, r_1 \vee (l_{i,1}, l_{i,5}, l_{i,4})> \tag{10}$$

$$r_1 = \begin{array}{c|cc} & l_{i,3} & l_{i,4} \\ \hline l_{i.1} & false & true \\ l_{i.5} & false & true \\ l_{i,4} & W_{i4,i3} & true \end{array} \tag{11}$$

Where $W_{i4,i3}$ is "finished sleep mode cycle" and activation of the process of sensor data collection.

The transition Z_2 represents the "active mode" of the sensor process of the sensor data collection and transmission i, $i \in N$

$$Z_2 = <\{l_{i,3}, l_{i,6}\}, \{l_{i,5}, l_{i,6}\}, r_2 \vee (l_{i,3}, l_{i,6})> \tag{12}$$

$$r_2 = \begin{array}{c|cc} & l_{i,5} & l_{i,6} \\ \hline l_{i.3} & false & true \\ l_{i,6} & W_{i6,i5} & true \end{array} \tag{13}$$

Where $W_{i6,i5}$ is "finished active mode cycle" and initiation of the sleep mode process.

As is shown on Fig. 3 the combination of the transitions Z_1 and Z_2 and the output $l_{i,5}$ represents the duty cycle process of the wireless battery powered sensor node.

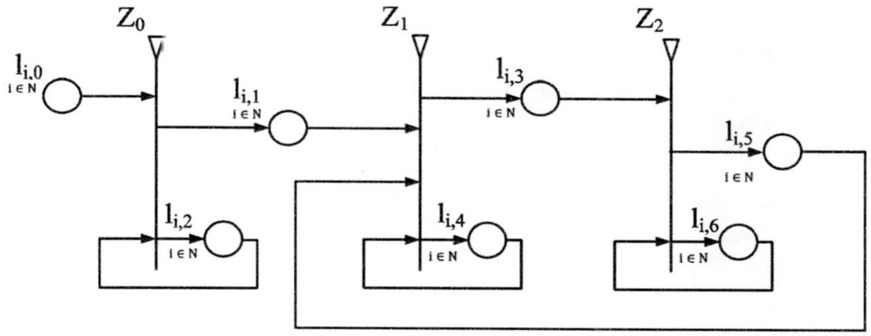

Fig. 3. GN model of parallel sensor data integration in clustered WSNs

The new developed GNs based model was simulated in WSNet simulator ver. 9.07, using the embedded in the simulator support for energy consumption simulation. In general, the processes modeled by the Generalized Nets can be defined easily and detailed in death if needed. At the same time, they have not the limitations typical for the Markov chain based models.

2.3 Experimental and Simulation Results

In order to compare results from real hardware experiment and the simulation results based on NS3 and WSNet simulators, we suppose a real WSN node that consists of a ARM core Texas Instrument (TI) MSP430F449 Ultra-Low-Power SoC microcontroller, TI TPL5010 Nano-Power Watchdog Timer, TI ZigBee CC2531 RF transceiver module and TI HDC1080 humidity and temperature sensor shown in Fig. 4.

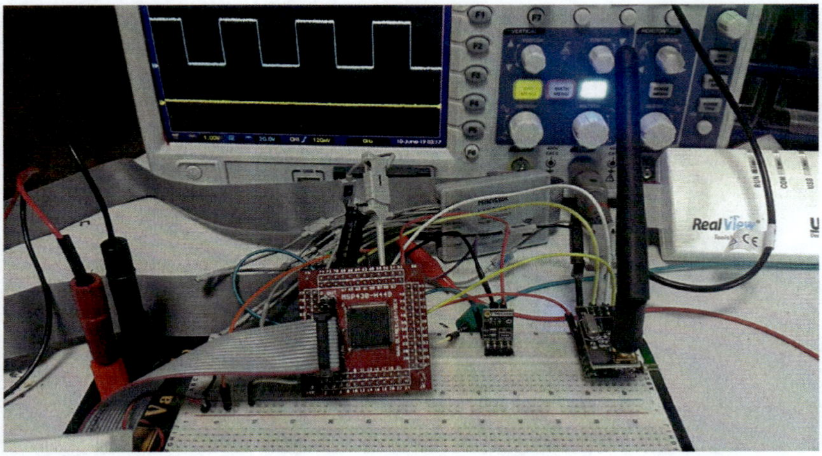

Fig. 4. WSN node experimental lab

Table 1 presents the power consumption in power up, sleep mode and active modes of the sensor node based on 3 V power supply with the hardware configuration described above.

Table 1. Measured real sensor node energy consumption - average rates

Sensor device state	Average power consumption mW	Average current consumption
Active mode sensor data collection	1.5 mW	0.5 mA
Active Sensor Rx mode	72 mW	24 mA
Active Sensor Tx mode	87 mW	29 mA
Sleep mode	0.105 µW	35 nA

In this experiment, the sensor node uses an embedded AODV (RFC3561) routing protocol [16]. The maximum simulation time length is set to 600 s. The data is transmitted randomly in series in predefined time windows. The range of wireless communication is 50 m.

2.3.1 Comparison Between Hardware Lab Wireless Sensor Node, Markov Chain Simulated Model and Generalized Nets Simulated Model

The WSNet simulator is used to define the timing emulation of state transitions of the sensor node. The Eq. (7) is used to calculate the total energy for the WSNet simulator as well. We will compare the predicted stable state probabilities and the calculated energy estimates of the Markov based model and the Generalized Nets model while the T_{sleep} interval is varied in 50 ms–500 ms range. The total energy spend by the sensor node is:

$$P_{total} = (P_{powerUp} + P_{sleep} + P_{active})T_{work} \tag{14}$$

The Table 2 shows some of the experimental and simulation data related to sensor "sleep" mode only.

Table 2. Current consumption average estimates in sleep mode measured in nA

Time sample	10 ms	20 ms	40 ms	60 ms	80 ms	100 ms
Real sensor device	69.2	62.4	41.2	34.5	35.8	34.9
Markov chain based estimation	67.2	60.1	50.2	46.4	44.2	45.1
Generalized Nets based estimation	67.6	60.8	38.7	33.9	36.2	39.9

The Fig. 5 illustrates the current consumption experimental and simulation results for P_{sleep} = 100 ms compared to a real sensor node duty cycle for each of the models - Markov chains and GNs.

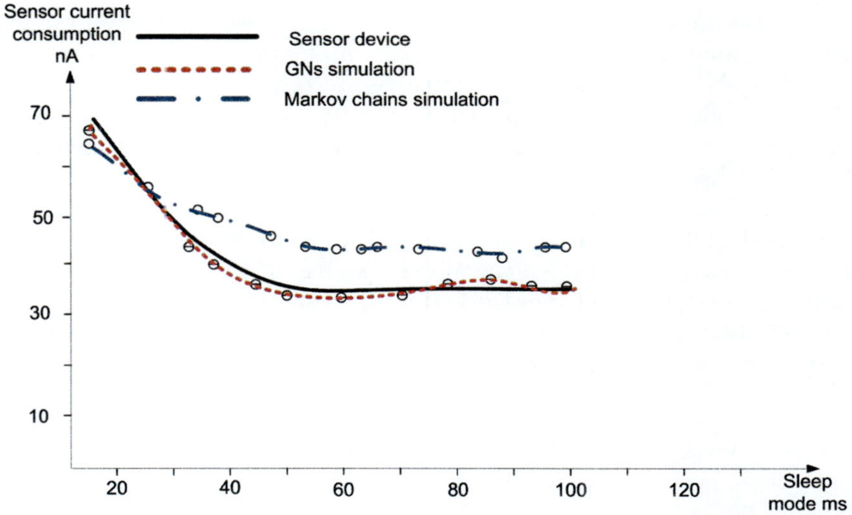

Fig. 5. The current simulation results in nA for sleep mode, compared to real sensor node consumption

The energy estimates for P_{active} = 20 ms compared to real sensor node duty cycle are shown on Fig. 6.

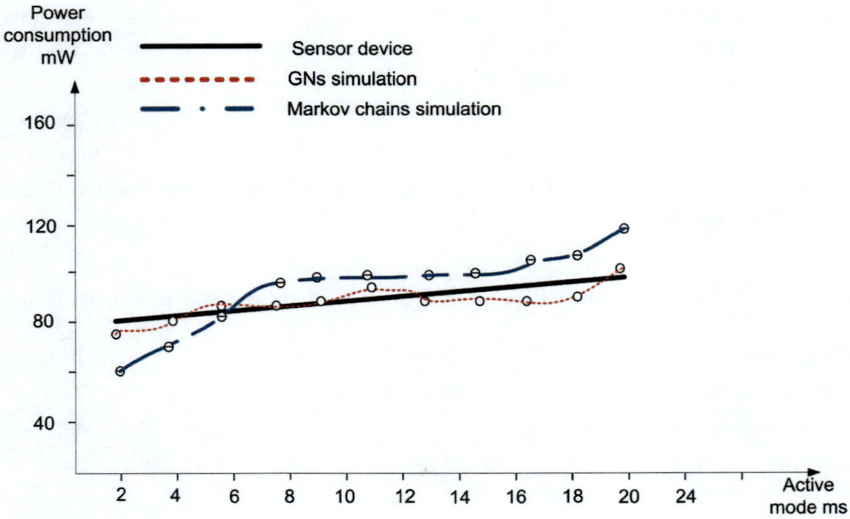

Fig. 6. Sensor device Active Tx mode energy estimates comparison in mW

It is interesting to note that the average difference between the Markov model based energy estimates calculated by the WSNet simulator and the real node measurements are sensitively bigger to the average difference between the Generalized Nets simulation results and the real sensor's measurements. The experimental data shows that the power switching process between the "active" and the "sleep" mode has an accompanying transition associated with the battery power switching parasitic capacitance. The transition process was included in both simulations.

Figure 6 depicts the behavior of the Sensor node when the active mode P_{active} is 20 ms. Although, the Markov chains model seems to over-estimate the power consumption compared to the Generalized nets based simulation results and it tends to be a better indicator of the system than the Markov model.

It is interesting to note that again the Generalized nets energy consumption estimates are now closer to the real sensor device results than the Markov model.

3 Conclusion

This paper proposes a new model for a Low Powered Wireless Sensor Node by using the Generalized Nets approach. The experimental results indicate that the Generalized nets model of a sensor node is more accurate than one based on Markov chains. This is due to the fact that a Markov model requires the modeled system to have a memoryless state. A standard wireless sensor node that relies on time to dynamically change its power state does not satisfy the Markov chain's memoryless requirements and will be not so easy to be modeled.

During the simulations in WSNet, the Generalized Nets model in Fig. 3 took around 10 min to stabilize. Depending on the desired accuracy, the simulation time can be longer. The disadvantage of Generalized Nets models is the relatively long simulation time to achieve stable state probabilities in high complicated systems. Nevertheless, in general, the Generalized Nets model is more flexible and reliable than the Markov chains model and can be modified easily.

We also illustrated that the energy estimates in a Generalized Nets were sensitively closer to the actual energy needs of a ZigBee wireless network system triggered by random events. The possibility of the dynamical identification of the time periods in the duty cycling technique can result in larger power savings.

Through this example, we were able to show that GNs method of modeling is more accurate than the Markov chain models and provides a valuable platform for energy optimization in wireless sensor networks.

Acknowledgements. This paper is supported by the National Scientific Program "Information and Communication Technologies for a Single Digital Market in Science, Education and Security (ICTinSES)", financed by the Ministry of Education and Science.

References

1. Atanasova, T.: Modelling of complex objects in distance learning systems. In: Proceedings of the First International Conference - "Innovative Teaching Methodology", Tbilisi, Georgia, 25–26 October 2014, pp. 180–190 (2014). ISBN 978-9941-9348-7-2
2. Balabanov, T., Zankinski, I., Barova, M.: Strategy for individuals distribution by incident nodes participation in star topology of distributed evolutionary algorithms. Cybern. Inf. Technol. **16**(1), 80–88 (2016)
3. Akyildiz, I.F., Su, W., Sankarasubramaniam, Y., Cayirci, E.: Wireless sensor networks: a survey. Comput. Netw. **38**(4), 393–422 (2002)
4. Estrin, D., Girod, L., Pottie, G., Srivastava, M.: Incrementing the world with wireless sensor networks. In: IEEE ICASSP 2001, vol. 4, pp. 2033–2036 (2001)
5. Feng, J., Koushanfar, F., Potkonjak, M.: Sensor network architecture. In: Mahgoub, I., Ilyas, M. (eds.) Handbook of Sensor Networks, Section III, no. 12. CRC Press (2004)
6. Sachdeva, G., Doemer, R., Chou, P.: System modeling: a case study on a wireless sensor network. Technical Report CECS-TR-05-12, University of California, 15 June 2005
7. Obaidat, M.S., Green, D.B.: On Modeling Networks of Wireless Microsensors, pp. 115–153. Kluwer Academic Publishers (2003)
8. Karl, H., Willig, A.: Protocols and Architectures for Wireless Sensor Networks. Wiley, Hoboken (2005)
9. Hu, P., Zhou, Z., Liu, Q., Li, F.: The HMM-based modeling for the energy level prediction in wireless sensor networks? In: Proceedings of the 2007 2nd IEEE Conference on Industrial Electronics and Applications (ICIEA 2007), pp. 2253–2258, May 2007
10. Misic, J., Misic, V.: Wireless Personal Area Networks: Performance, Interconnection, and Security with IEEE 802.15.4. Wiley (2008)

11. Mitrofanova, A.: NYU, Department of Computer Science, 18 December 2007. Lecture 3. https://cs.nyu.edu/mishra/COURSES/09.HPGP/scribe3.pdf
12. Cox, D.R.: The analysis of non-markovian stochastic processes by the inclusion of NN variables. Proc. Cambridge Philoso. Soc. **51**(3), 433–441 (1955)
13. Atanassov, K.: Generalized Nets. World Scientific, Singapore (1991)
14. Fidanova, S., Atanasov, K., Marinov, P.: Generalized Nets and Ant Colony Optimization. Academic Publishing House (2011). Sofia, ISBN 978-954-322-473-9
15. Doukovska, L., Atanassova, V., Shahpazov, G., Sotirov, E.: Generalized net model of the creditworthiness financial support mechanism for the SMEs. Int. J. Comput. Inform. (2016). ISSN 1335-9150
16. Bobade, N.P., Mhala, N.N.: Performance evaluation of AODV and DSR on-demand routing protocols with varying manet size. Int. J. Wirel. Mob. Netw. (IJWMN) **4**(1), 183–196 (2012)

WSN-Based Prediction Model of Microclimate in a City Urbanized Areas Based on Extreme Learning and Kalman Filter

A. Alexandrov(✉), R. Andreev, S. Ilchev, A. Boneva, S. Ivanov, and J. Doshev

Institute of Information and Communication Technologies,
Bulgarian Academy of Sciences, Sofia, Bulgaria
akalexandrov@iit.bas.bg

Abstract. The monitoring and microclimate controlling in urbanized areas become one of the research hotspots in the field of air quality, where the application of Wireless Sensor Networks (WSN) recently attracts more attention due to its features of self-adaption, resilience, and cost-effectiveness. Present microclimate monitoring and control systems achieve their prediction by manipulating captured environmental factors and traditional neural network algorithms. However, these systems have a problem to solve the challenges of the quick prediction (e.g. hourly and even minutely) when the WSN network is deployed. In this paper, a novel prediction method based on a combination of Extended Kalman Filter (EKF) and an Extreme Learning Machine (ELM) algorithm is proposed to predict the key microclimate parameters like temperature, humidity, and barometric pressure level in a city urbanized areas. The outdoor air temperature, humidity, and barometric pressure in the air are measured as data samples via WSN based clusters managed by custom design operator stations. The results of the realized model simulation show that the processing speed rate of the proposed prediction model is significantly higher than other ANN-based models at a relatively good level of precision.

1 Introduction

Due to accelerated urbanization, more than half of the Earth's population currently live in urbanized areas and this tendency may continue to increase in the future [1]. This means that the number of people affected by the hazards of their housing and working environments such as air and light pollution and noise and thermal load (extremal levels of heat or cold) is growing rapidly. In addition to urbanization, the trends of climate change, particularly the increasing frequency and intensification of extreme heat waves, increase the importance of urban climatological research.

In urban environments, climate parameters are modified faster compared to the rural areas and even micro-scale climatic conditions change rapidly during

I. Dimov and S. Fidanova (Eds.): HPC 2019, SCI 902, pp. 15–26, 2021.
https://doi.org/10.1007/978-3-030-55347-0_2

the last years which have a great influence on the thermal sensations and thermal stress levels of the residents and visitors. Some microclimatic aspects substantially determine the behavioral reactions as well as the decisions of the citizens related to the living area design, and ultimately area usage.

Application of microclimate models in urban planning is suggested together with adequate onsite measurements reflecting the real microclimatic conditions in order to validate the outcomes of the simulations. There is a need for a generic microclimate model that takes the effects of all factors influencing thermal conditions (vegetation, buildings, and surface cover) into consideration.

The urbanized area's microclimate is an extremely complicated system, influenced by many environmental factors as the structures of the urbanized areas (industrial or living), type of the buildings, specific geographic relief climate anomalies and etc.

2 Related Work

The temperature difference between urban and its surrounding geographic areas creates Urban Heat Island (UHI). This phenomenon can be researched in detail by using remote sensing technology. This temperature discrepancy can be influenced by a lot of factors. In [2] are used artificial neural networks to investigate the correlation between the Land Surface Temperature(LST) and factors like wind, land-cover, geographic relief, population density, energy consumption and etc. in the urban area.

A number of researches on urban modeling have been realized based on cellular automata (CA) [3–5]. In CA based models there is a set of cells, each in one has a finite number of states. The goal of CA is to create a model of how these states of cells change in discrete time steps.

In [6], to find the CA model's transfer rules, Artificial Neural Networks (ANN) and Support Vector Regression (SVR) have been processed separately and after that, the accuracy of these methods have been compared.

Artificial neural networks (ANNs) have been used in some of the prediction simulations that involve parameters as air temperature, wind speed, and relative humidity.

In [7] is described e ANN based weather prediction model which is implemented in Saskatchewan, Canada for 24 h weather forecasting. The results show that combining the outputs of a standard Feed-Forward ANN, a recurrent ANN, a radial basis function network, and a Hopfield network into a combination can lead to more accurate predictions of wind speed, relative humidity, and air temperature compared to any of the individual component networks.

In [8] is presented a multi-objective genetic algorithm to develop a radial basis function ANN model for the prediction of air temperature in a secondary school building in Portugal.

A back-propagation ANN with batch learning scheme is presented in [9] for 24-h prediction in ambient temperature on a coastal location in Saudi Arabia.

The results show that temperature can be predicted even with only one input with good accuracy.

Additionally, in [10] a number of predictions for UHI, based on ANN are realized. A custom designed machine learning architecture was developed to predict the UHI intensity in Athens, Greece, during both day and night.

In [11] is using historical data for a two-year time period. In [12] also is used input data from meteorological stations as well as historic measured air temperatures within the city to predict the UHI intensity in London based on neural network architecture.

3 WSN Based Prediction Model of Microclimate Based on Extreme Learning and Kalman Filter

3.1 Meteorological Wireless Sensor Network

The hardware architecture-specific design is based on the widespread concept of using a Low Power single-chip microcontroller with built-in memory, external non-volatile memory, and data buses for which all peripheral devices, such as sensors, communication modules, and the power supplies. The presented intelligent wireless sensor module is made up of four main components:

Sensor Block - A set of 3 main sensors that measure air temperature, humidity, and barometric pressure. From a functional point of view, the group of 3 main sensors includes the GPS unit as it has sensor functions in terms of measuring the current GPS coordinates of the module.

Processor Block - SOC (System on Chip) single-chip microcontroller, external operational and external non-volatile memory. This block manages the entire process of collecting and processing data from the sensors. The algorithms governing the overall operation of the sensor module, depending on the environment and other sensor modules, are also set in the processor block.

Additionally the processor block is responsible for the sensor data collection and processing, self-test and self-checking procedures, validation and integration of the sensor data and other functionalities.

Communication Block - An electronics module combined with embedded software with data transmission functionality and functionality of receiving of commands related to the sensor module operations.

Power Supply - The power of the developed sensor module is autonomous and is based on a LiON rechargeable battery in combination with a solar panel built into the housing. A combination of electronic components and software is provided to control one or more power modes, necessary for the normal operation of the sensor module. The hardware consist of Li-ion Battery, DC-DC converter, photovoltaic panel, and Ultra Low powered System Timer which manages a nano-powered power supply switch. In the proposed design, the power supply unit operates under a special adaptive energy saving duty cycle algorithm which manage the Ultra Low powered System Timer.

The functional block diagram of the sensor module for monitoring meteorological data is shown in Fig. 1.

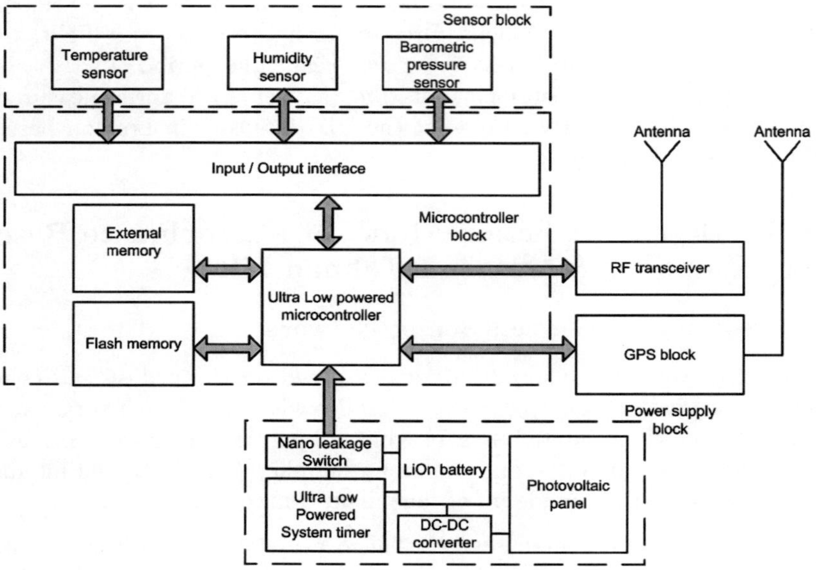

Fig. 1. Low Power Wireless Sensor Node - block diagram

The developed wireless sensor modules are able to exchange data on distance up to 500 m. Based on the designed wireless sensor module, a sensor network consisting of 10 wireless sensor devices is built to allow the measurement and control of the environmental parameters in the sensitivity zone of the sensors.

In this case, all sensor modules of the system perform the functions of coordinators and have the task of ensuring the synchronization of data (in this case, temperature, humidity, barometric pressure and GPS data in the NMEA format) to other sensors and/or network coordinators.

The real measurements of temperature, humidity, and barometric pressure are used in the current research as a benchmark of the results calculated by the proposed ELKF algorithm.

3.2 Prediction Model Based on Extreme Learning and Kalman Filtering

The Extreme Learning Machine (ELM) algorithm [13–15] and [16] has the advantages of faster training speed, global optimal solution, and good scalability. In addition, the Kalman Filtering approach has been applied. In the current research, the Extreme Learning approach in combination with Kalman filtering is used to establish a reliable and fast model-based microclimate simulation of a city urbanized areas.

3.2.1 Extreme Learning Approach

Unlike the conventional learning algorithms for neural networks such as Back Propagation (BP) algorithms, which may face some difficulties to find local minimums, ELM is fully automatically implemented without iterative tuning, and in theory, no additional intervention is required. At the same time, the learning speed of ELM is much faster compared to other neural networks based methods. In the Extreme Learning approach, the learning parameters of the hidden nodes (input weights and biases) are randomly assigned in most cases, and the output weights of neural networks can be determined by the simple inverse operation.

During the training stage, a number of fixed nonlinear transformations are used without a time-consuming learning process. One of the benefits of the ELM algorithm is that is can be achieved a good generalization performance and scalability.

As mentioned in [17], the key idea of ELM is the random initialization of the hidden layer weights and the subsequent training consists of computing the least-squares solution to the linear system defined by the hidden layer outputs and targets. An overview of the structure of an ELM is given in Fig. 2 and the algorithm for training this network is described in [14] and can be summarized as follows:

Consider a set of N unique samples (x_i, y_i) with $x_i \in \mathbb{R}^d$ and $y_i \in \mathbb{R}$. Then, the output of an Single-Layer Feedforward Networks (SLFNs) [3] with K hidden neurons can be written as:

$$\hat{y_i} = \sum_{i=1}^{M} \beta_i f(w_k, x_i + b_i), i \in [1, N] \tag{1}$$

where $\hat{y_i}$ is its approximation to y_i, f is the activation function, w_k the input weight vector, b_i the hidden layer bias and β_i the output weight corresponding to the ith neuron in the hidden layer. In case the SLFN would perfectly approximate the data (meaning the error between the output $\hat{y_i}$ and the actual value y_i is zero), the relation would be:

$$\sum_{i=1}^{M} \beta_i f(w_k, x_i + b_i) = y_i, i \in [1, N] \tag{2}$$

which can be written compactly as $H\beta = Y$ where H is the hidden layer output matrix defined as

$$H = \begin{pmatrix} w_1, x_1 + b_1 & \cdots & w_K, x_1 + b_K \\ \vdots & \ddots & \vdots \\ w_1, x_N + b_1 & \cdots & w_K, x_i + b_K \end{pmatrix} \tag{3}$$

and $\beta = (\beta_1 \ldots \beta K)^T$ and $Y = (y_1 \ldots y_N)^T$.

The benefit of the Extreme Learning Machine is the fact that there is no need to iterate the randomly initialized network weights, which makes it faster compared to other neural network algorithms. At the same time, nevertheless

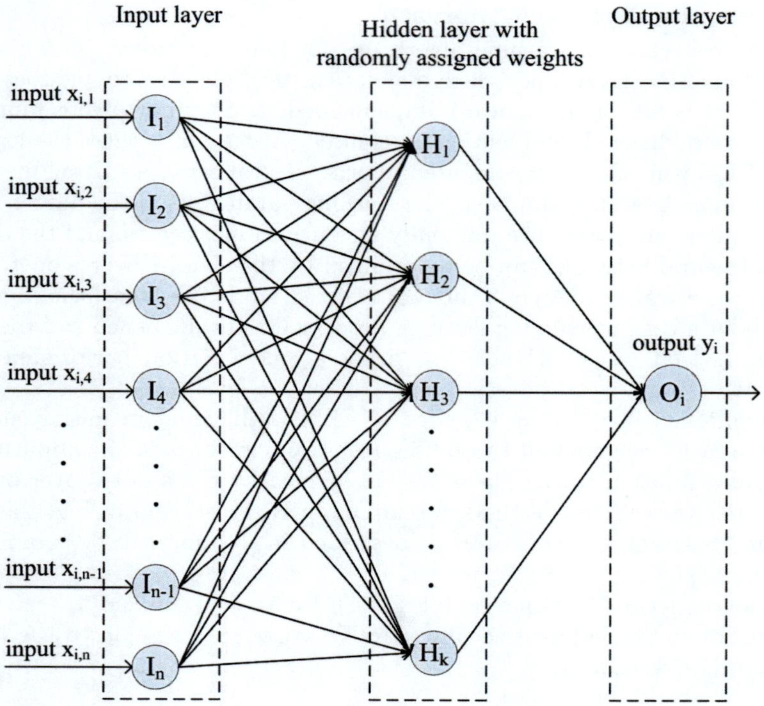

Fig. 2. Extreme Learning Machine overview

that the hidden neurons not being adjusted, relatively accurate network can be obtained.

3.2.2 Kalman Filtering Approach

In general, the equations for the Kalman filter [18] can be divided into two categories: time update equations and measurement update equations. The equations for time update are used to predict the current state and error estimates are used to obtain the a priori estimates for the next time step.

The measurement focused update equations are used for the feedback creation. For example to incorporate a new measurement into the a priori estimate to obtain an improved a posteriori estimate. The time update equations very often are presented as predictor equations, at the same time the measurement update equations can be presented as corrector equations (see Fig. 3).

The prediction stage starts with initial estimation of $x_{\widehat{k-1}}$ and covariance vector P_{k-1} and proceeds with

$$\hat{x_k} = Ax_{\widehat{k-1}} + Bu_k + Cw_k \tag{4}$$

$$z_k = H_k x_k + D_k v_k \tag{5}$$

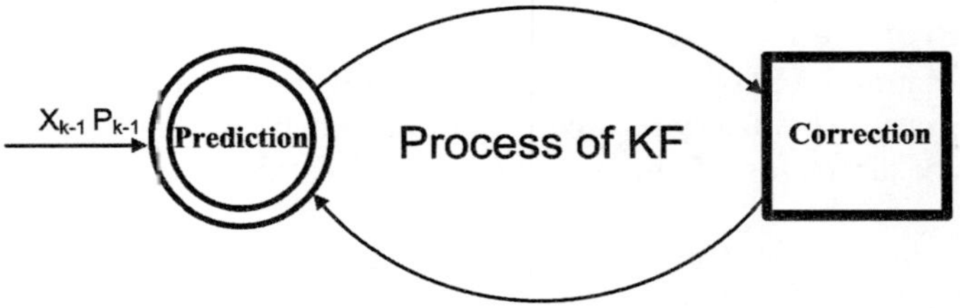

Fig. 3. Kalman filtering overview

where: $\hat{x_k}$ is an estimated value, A is the transition state matrix of the process, B is the input matrix, C is the noise state transition matrix H_k is the known input value, w_k is the noise, z_k is the observation vector, v_k is a variable describing observation noise. H_k is the matrix of the observed value z_k, and D_k is a matrices describing the contribution of noise to the observation.

The measurement correction tunes the projected estimate by an actual measurement at that time. In the current development, we will be focused mainly on the measurement update algorithm. The current paper doesn't focus on details on the mathematical side of the Kalman filter measured updated equations. This is realized in detail in [18].

Based on the papers above we accept that the final extended Kalman filter measurement update equations are formulated as follows:

$$G_k = (P_k H_k)/(H_k P_k H_k^t + R_k) \tag{6}$$

$$\hat{x_k} = \hat{x_k} + G_k(z_k - H_k\hat{x_k}) \tag{7}$$

$$P_k = (1 - G_k H_k)P_k \tag{8}$$

where G_k is the Kalman gain, P_k is error covariance vector, H_k is as described above the matrix of the observed value vector z_k and R_k is the covariance matrix. The initial task during the measurement update is to calculate the Kalman gain G_k (6). The next phase is to measure the process, and then to calculate an a posteriori state estimate by adding the correction based on the measurement and estimation as in (7). The final step is to obtain an a posteriori error covariance estimate via (8). After each measurement cycle update pair, the process is repeated with the previous a posteriori estimates used to project. This recursive principle is one of the key important benefits of the Kalman filter - the software implementation is easier and faster. In general, the Kalman filter algorithm generates recursively a condition of the current estimate based on past measurements. In the current research, we consider a prediction model [19,20] based on the combination of Extreme Learning approach and the Kalman Filter algorithm (ELKF).

3.2.3 The Proposed ELKF Algorithm

The relatively fast and reliable prediction of temperature, humidity and barometric pressure in city urbanized areas can be improved by using enough amount of historical measurements of these parameters to establish precise prediction models.

In general, the proposed method uses a Kalman filter iterations to reduce the fluctuations in the inputs of the Extreme Learning Machine [19,20] and to increase the data reliability. The block diagram of the ELKF algorithm is shown on Fig. 8.

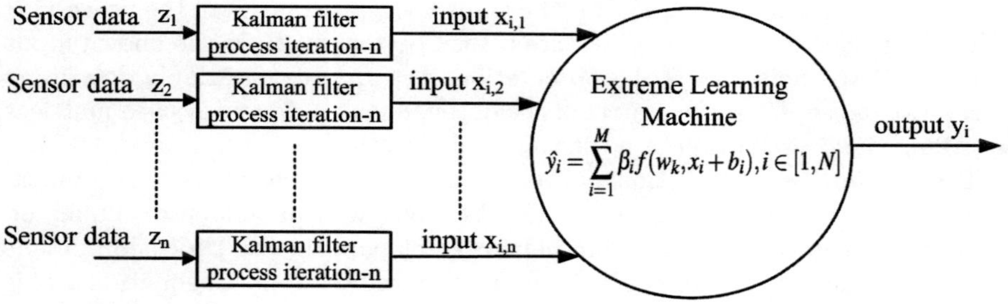

Fig. 4. ELKF block diagram

To establish the barometric pressure model, the training parameters of ELM were as follows: the number of neurons in the hidden layer is 64 and the activation function of neurons in the hidden layer is a sin.

Kalman Filter algorithm - sample pseudo code

Input: $\hat{x_{k-1}}, P_{k-1}, u_k, z_k$
Output: $\hat{x_k}, P_k$
Predicted estimate
1. Calculate predicted state estimate $\hat{x_k} = A\hat{x_{k-1}} + Bu_k + Cw_k$
2. Calculate predicted error covariance $z_k = H_k x_k + D_k v_k$

Update
3. Calculate Kalman gain $G_k = (P_k H_k)/(H_k P_k H_k^t + R_k)$ -
4. Update state estimate $\hat{x_k} = \hat{x_k} + G_k(z_k - H_k \hat{x_k})$
5. Update error covariance $P_k = (1 - G_k H_k) P_k$ -
6. Return $\hat{x_k}, P_k$

ELM algorithm

Given a set (x_i, y_i) with $x_i \in \mathbb{R}^d$ and $y_i \in \mathbb{R}$ a probability distribution from which to draw random weights, an activation function f and M the number of hidden nodes:

1: Randomly assign input weights w_i and biases b_i, $i \in [1, K]$;
2: Calculate the hidden layer output matrix H;
3: Calculate output weights matrix $\beta = HyY$;

The main benefit of the proposed algorithm is that he is extremal fast, and needs a very small amount of memory and processor power because of the used recursive functions. Therefore, the software implementation of the algorithm can be easily embedded in battery powered sensor devices.

3.3 Results of Prediction Using ELKF Based Model

A meteorological monitoring system is implemented based on a Wireless Sensor Network with a cluster topology. The designed wireless sensor nodes use 802.15.4 based communication platform with implemented AODV (RFC3561) routing protocol. The hardware block diagram is shown in Fig. 1 and the patent-pending prototype design is shown in Fig. 4.

The real sensor devices measurement experiment started on the 4th of May 2019 and continued till the 10th of May 2019 in Sofia, Bulgaria. Approximately, at the same time were realized and the modeling processes.

To establish the temperature, humidity, and barometric pressure model simulation was used the open source Python Extreme Learning Machine software library *elm*0.1.3 (https://pypi.org/project/elm/0.1.3/) and the Python-based Kalman filter library *filterPy*1.4.4. For calculation speed test was executed ELM algorithm and after that, the ELKF algorithm was executed with the same input data.

The parameters of the ELM processing were as follows: the number of neurons in the hidden layer is 64 and the activation function of neurons in the hidden layer is a sin.

In order to meet the requirements of the prediction ELM based model, all input data have been normalized. Establishing the humidity model uses the same

Fig. 5. Meteorological sensor prototype

method with the temperature model, the difference is in the sensor data set, so establishing the humidity model was not described in detail (Fig. 5).

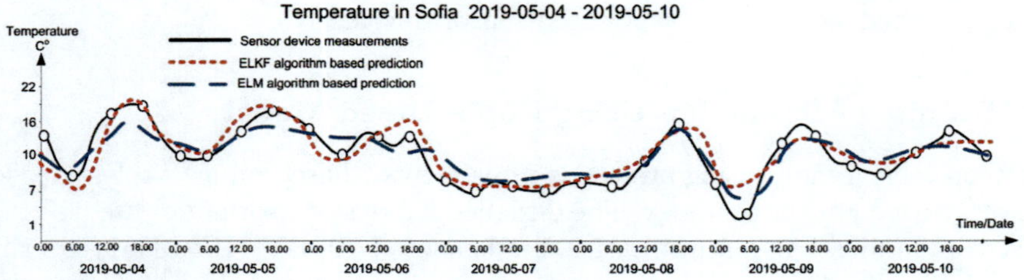

Fig. 6. Air Temperature prediction compared to real sensor data

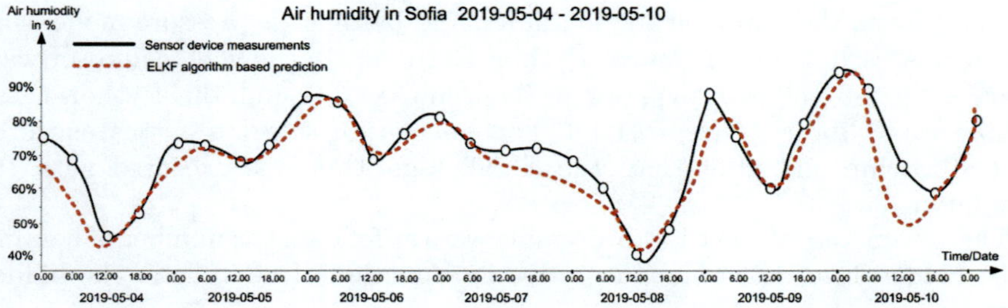

Fig. 7. Air humidity prediction compared to real sensor data

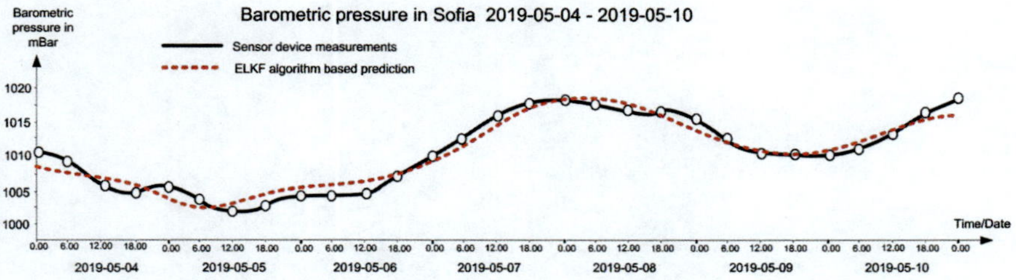

Fig. 8. Barometric pressure prediction compared to real sensor data

As is seen in diagrams on Fig. 6, 7 and 8 the proposed ELKF algorithm provide relatively good prediction level in slow changing processes as barometric pressure changes for example.

When the input data has a bigger deviation the proposed ELKF algorithm needs additional tunning.

4 Conclusion

This paper proposes a new ELKF algorithm to predict specific environmental factors like temperature, air humidity, and barometric pressure in the urbanized areas. Unlike the traditional learning algorithms, the proposed approach uses as input data, the sensor data preliminary processed by a Kalman filter algorithm. In the next stage, the ELKF algorithm generates input weights and thresholds and sets the weight coefficients of the hidden layer neurons.

As a result in high probability, we can obtain the unique global optimal solution. The algorithm is simple, extremal fast and with relatively good simulation precision. Compared with the prediction model based on the ELM only, which is tested in temperature prediction only, the ELKF reaches better results on accuracy but at the same time, the ELM had a better execution time which in some calculations reaches 30%. The proposed algorithm requires less training time but at the same time shows stronger fitness and more stable performance.

During future research, the model will be improved further. Building a real-time model of microclimate in specific city urbanized areas is a necessary task for prediction of future fast climate and environmental changes and disasters as floods and fires.

Acknowledgements. This paper is supported by the National Scientific Program "Information and Communication Technologies for a Single Digital Market in Science, Education and Security (ICTinSES)", financed by the Ministry of Education and Science.

References

1. UNFPA: 7e state of world population YZDD. Report of the United Nations Population Fund, New York, NY, USA (2011)
2. Ng, E., Chen, L., Wang, Y., Yuan, C.: A study on the cooling effects of greening in a high-density city: an experience from Hong Kong. Build. Environ. **47**, 256–271 (2012)
3. White, R., Engelen, G., Uijee, I.: The use of constrained cellular automata for high-resolution modelling of urban land-use dynamics. Environ. Plan. **24**, 323–343 (1997)
4. Gar-On Yeh, A., Li, X.: Urban simulation using neural networks and cellular automata for land use planning. Int. J. Geogr. Inf. Sci. **16**, 323–343 (2002)
5. Alkheder, S., Shan, J.: Cellular Automata Urban Growth Simulation and Evaluation-A Case Study of Indianapolis, Geomatics Engineering, School of Civil Engineering, Purdue University (2005). http://www.geocomputation.org/2005/Alkheder.pdf (14) (PDF) Investigation of Urban Sprawl using Remote Sensing and GIS: A case of Onitsha and its Environs. Available from: https://www.researchgate.net/publication/259391154Investigation of Urban Sprawl using Remote Sensing and GIS A case of Onitsha and its Environs
6. Menard, A., Marceau, D.J.: Exploration of spatial scale sensitivity in geographic cellular automata. Environ. Plan. **32**, 693–714 (2005)
7. Maqsood, I., Khan, M.R., Abraham, A.: An ensemble of neural networks for weather forecasting. Neural Comput. Appl. **13**, 112–122 (2004)

8. Ruano, A.E., Crispim, E.M., Conceicão, E.Z.E., Lucio, M.M.J.R.: Prediction of building's temperature using neural networks models. Energy Build. **38**, 682–694 (2006)
9. Tasadduq, I., Rehman, S., Bubshait, K.: Application of neural networks for the prediction of hourly mean surface temperatures in Saudi Arabia. Renew. Energy **36**, 545–554 (2002)
10. Santamouris, M., Paraponiaris, K., Mihalakakou, G.: Estimating the ecological footprint of the heat Island effect over Athens. Greece Clim. Change **80**, 265–276 (2007)
11. Mihalakakou, M., Santamouris, N., Papanikolaou, C.: Cartalis simulation of the urban heat Island phenomenon in Mediterranean climates. Pure Appl. Geophys. **161**, 429–451 (2004)
12. Kolokotroni, M., Davies, M., Croxford, B., Bhuiyan, S., Mavrogianni, A.: A validated methodology for the prediction of heating and cooling energy demand for buildings within the urban heat island: case-study of London. Solar Energy **84**(12), 2246–2255 (2010)
13. Heeswijk, M.: Advances in Extreme Learning Machines, Aalto University publication series, 43/2015, pp. 13–17. ISBN 978-952-60-6148-1
14. Huang, B., Zhu, Q.Y., Siew, C.-K.: Extreme learning machine: a new learning scheme of feedforward neural networks. In: Proceedings of the 2004 IEEE International Joint Conference on Neural Networks, vol. 2, pp. 985–990 (2004)
15. Gu, B., Sheng, V.S., Tay, K.Y., Romano, W., Li, S.: Incremental support vector learning for ordinal regression. IEEE Trans. Neural Netw. Learn. Syst. **26**(7), 1403–1416 (2015)
16. Gu, B., Sheng, V.S., Wang, Z., Ho, D., Osman, S., Li, S.: Incremental learning for v-support vector regression. Neural Netw. **67**, 140–150 (2015)
17. Huang, G.-B., Zhu, Q.-Y., Siew, C.-K.: Extreme learning machine: a new learning scheme of feedforward neural networks. In: IEEE Proceedings of International Joint Conference on Neural Networks, pp. 985–990 (2004)
18. Kalman, R.E.: A new approach to linear filtering and prediction problems. J. Basic Eng. **82**(1), 35–45 (1960)
19. Patil, S.L., Tantau, H.J., Salokhe, V.M.: Modelling of tropical greenhouse temperature by auto regressive and neural network models. Biosyst. Eng. **99**(3), 423–431 (2008)
20. Shao, B., Zhang, M., Mi, Q., Xiang, N.: Prediction and visualization for urban heat island simulation. In: Transactions on Edutainment VI, pp. 1–11. Springer (2011)

Four Operations over Extended Intuitionistic Fuzzy Index Matrices and Some of Their Applications

Krassimir Atanassov[1,2] and Veselina Bureva[2](✉)

[1] Department of Bioinformatics and Mathematical Modelling,
Institute of Biophysics and Biomedical Engineering, Bulgarian Academy of Sciences,
Acad. G. Bonchev Street, Bl. 105, 1113 Sofia, Bulgaria
krat@bas.bg

[2] Intelligent Systems Laboratory, Prof. Asen Zlatarov University,
8000 Bourgas, Bulgaria
vesito_ka@abv.bg

Abstract. In this paper, four new operations are introduced over extended intuitionistic fuzzy index matrices and over their simpler cases, such as intuitionistic fuzzy index matrices, extended index matrices and index matrices. Some of the properties of the four new operators are discussed and their applications in intercriteria analysis are shown. A theorem for stratification of an index matrix is formulated and proved.

1 Introduction and Preliminary Remarks

Here, as a continuation of [3,5–9], we discuss the concepts of an Extended Intuitionistic Fuzzy Index Matrix (EIFIM), that generalizes the concepts of an Intuitionistic Fuzzy Index Matrix (IFIM), an Extended Index Matrix (EIM) and Index Matrix (IM). All these concepts are described in detail in [9]. There, the operation substitution was introduced over standard IMs, IFIMs, EIFIMs and other types of IMs, as EIMs, temporal IMs. It changes only the identifiers of IM's rows and/or columns with other symbols.

Here, we introduce two new operations, that change the places of two rows and of two columns of a given EIFIM and show their activity for the EIFIM's simpler cases, as IFIM, EIM, IM. Two other auxiliary operations are introduced.

Firstly, we give some remarks on Intuitionistic Fuzzy Sets (IFSs, see, e.g., [4,7]) and especially, of their partial case, Intuitionistic Fuzzy Pairs (IFPs; see [11]). The IFP is an object in the form $\langle a, b\rangle$, where $a, b \in [0, 1]$ and $a + b \leq 1$, that is used as an evaluation of some object or process and which conponents (a and b) are interpreted as degrees of membership and non-membership, or degrees of validity and non-validity, or degree of correctness and non-correctness, etc.

Let us have two IFPs $x = \langle a, b\rangle$ and $y = \langle c, d\rangle$. Then, following, e.g., [9], we define

$$x \leq y \text{ if and only if } a \leq c \text{ and } b \geq d,$$

I. Dimov and S. Fidanova (Eds.): HPC 2019, SCI 902, pp. 27–39, 2021.
https://doi.org/10.1007/978-3-030-55347-0_3

$$x = y \text{ if and only if } a = c \text{ and } b = d.$$

Second, following [9], the definition of EIFIM is given:

$$A = [K^*, L^*, \{\langle \mu_{k_i,l_j}, \nu_{k_i,l_j} \rangle\}]$$

$$\equiv \begin{array}{c|ccccc} & l_1, \langle \alpha_1^l, \beta_1^l \rangle & \dots & l_j, \langle \alpha_j^l, \beta_j^l \rangle & \dots & l_n, \langle \alpha_n^l, \beta_n^l \rangle \\ \hline k_1, \langle \alpha_1^k, \beta_1^k \rangle & \langle \mu_{k_1,l_1}, \nu_{k_1,l_1} \rangle & \dots & \langle \mu_{k_1,l_j}, \nu_{k_1,l_j} \rangle & \dots & \langle \mu_{k_1,l_n}, \nu_{k_1,l_n} \rangle \\ \vdots & \vdots & \ddots & \vdots & \ddots & \vdots \\ k_i, \langle \alpha_i^k, \beta_i^k \rangle & \langle \mu_{k_i,l_1}, \nu_{k_i,l_1} \rangle & \dots & \langle \mu_{k_i,l_j}, \nu_{k_i,l_j} \rangle & \dots & \langle \mu_{k_i,l_n}, \nu_{k_i,l_n} \rangle \\ \vdots & \vdots & \ddots & \vdots & \ddots & \vdots \\ k_m, \langle \alpha_m^k, \beta_m^k \rangle & \langle \mu_{k_m,l_1}, \nu_{k_m,l_1} \rangle & \dots & \langle \mu_{k_m,l_j}, \nu_{k_m,l_j} \rangle & \dots & \langle \mu_{k_m,l_n}, \nu_{k_m,l_n} \rangle \end{array},$$

where for every $1 \leq i \leq m, 1 \leq j \leq n$:

$$0 \leq \mu_{k_i,l_j}, \nu_{k_i,l_j}, \mu_{k_i,l_j} + \nu_{k_i,l_j} \in [0,1],$$

$$\alpha_i^k, \beta_i^k, \alpha_i^k + \beta_i^k \in [0,1],$$

$$\alpha_j^l, \beta_j^l, \alpha_j^l + \beta_j^l \in [0,1],$$

and here and below

$$K^* = \{\langle k_i, \alpha_i^k, \beta_i^k \rangle | k_i \in K\} = \{\langle k_i, \alpha_i^k, \beta_i^k \rangle | 1 \leq i \leq m\},$$

$$L^* = \{\langle l_j, \alpha_j^l, \beta_j^l \rangle | l_j \in L\} = \{\langle l_j, \alpha_j^l, \beta_j^l \rangle | 1 \leq j \leq n\},$$

where $\{k_i | 1 \leq i \leq m\} \cup \{l_j | 1 \leq j \leq n\} \subseteq \mathscr{I}$ - an index set.

When the evaluation IFPs of the indices k_i and l_j are omitted, we obtain an IFIM. When the IFPs $\langle \mu_{k_i,l_j}, \nu_{k_i,l_j} \rangle$ of an IFIM are changed with arbitrary other objects (numbers, propositions or predicates, whole IMs, etc.) we obtain an EIM. In the partial case, when these elements are numbers (natural, real, etc), we obtain a standard IM.

Let the two EIFIMs $A = [K^*, L^*, \{\langle a_{k,l}, b_{k,l} \rangle\}]$ and $B = [P^*, Q^*, \{\langle c_{p,q}, d_{p,q} \rangle\}]$ be given. Following and extending [9], we shall introduce the following definitions:

Definition 1. The strict relation "inclusion about dimension" is

$$A \subset_d B \text{ iff } (((K^* \subset P^*) \,\&\, (L^* \subset Q^*)) \vee ((K^* \subseteq P^*) \,\&\, (L^* \subset Q^*))$$

$$\vee((K^* \subset P^*) \,\&\, (L^* \subseteq Q^*))) \,\&\, (\forall k \in K)(\forall l \in L)(\langle a_{k,l}, b_{k,l} \rangle = \langle c_{k,l}, d_{k,l} \rangle).$$

Definition 2. The non-strict relation "inclusion about dimension" is

$$A \subseteq_d B \text{ iff } (K^* \subseteq P^*) \,\&\, (L^* \subseteq Q^*) \,\&\, (\forall k \in K)(\forall l \in L)$$

$$(\langle a_{k,l}, b_{k,l} \rangle = \langle c_{k,l}, d_{k,l} \rangle).$$

Definition 3. The strict relation "inclusion about value" is

$$A \subset_v B \text{ iff } (K^* = P^*) \ \& \ (L^* = Q^*) \ \& \ (\forall k \in K)(\forall l \in L)$$
$$(\langle a_{k,l}, b_{k,l}\rangle < \langle c_{k,l}, d_{k,l}\rangle).$$

Definition 4. The non-strict relation "inclusion about value" is

$$A \subseteq_v B \text{ iff } (K^* = P^*) \ \& \ (L^* = Q^*) \ \& \ (\forall k \in K)(\forall l \in L)$$
$$(\langle a_{k,l}, b_{k,l}\rangle \le \langle c_{k,l}, d_{k,l}\rangle).$$

Definition 5. The strict relation "inclusion" is

$$A \subset B \text{ iff } (((K^* \subset P^*) \ \& \ (L^* \subset Q^*)) \ \vee \ ((K^* \subseteq P^*) \ \& \ (L^* \subset Q^*))$$
$$\vee \ ((K^* \subset P^*) \ \& \ (L^* \subseteq Q^*))) \ \& \ (\forall k \in K)(\forall l \in L)(\langle a_{k,l}, b_{k,l}\rangle < \langle c_{k,l}, d_{k,l}\rangle).$$

Definition 6. The non-strict relation "inclusion" is

$$A \subseteq B \text{ iff } (K^* \subseteq P^*) \ \& \ (L^* \subseteq Q^*) \ \& \ (\forall k \in K)(\forall l \in L)$$
$$(\langle a_{k,l}, b_{k,l}\rangle \le \langle c_{k,l}, d_{k,l}\rangle).$$

Definition 7. The relation "equality" is

$$A = B \text{ iff } (K^* = P^*) \ \& \ (L^* = Q^*) \ \& \ (\forall k \in K)(\forall l \in L)$$
$$(\langle a_{k,l}, b_{k,l}\rangle = \langle c_{k,l}, d_{k,l}\rangle).$$

From the last definition it is clear that, e.g., the following relation "equality"

$$\begin{array}{c|cc} & l_1, \langle \alpha_1^l, \beta_1^l\rangle & l_2, \langle \alpha_2^l, \beta_2^l\rangle \\ \hline k_1, \langle \alpha_1^k, \beta_1^k\rangle & \langle \mu_{k_1,l_1}, \nu_{k_1,l_1}\rangle & \langle \mu_{k_1,l_2}, \nu_{k_1,l_2}\rangle \\ k_2, \langle \alpha_2^k, \beta_2^k\rangle & \langle \mu_{k_2,l_1}, \nu_{k_2,l_1}\rangle & \langle \mu_{k_2,l_2}, \nu_{k_2,l_2}\rangle \end{array} = \begin{array}{c|cc} & l_2, \langle \alpha_2^l, \beta_2^l\rangle & l_1, \langle \alpha_1^l, \beta_1^l\rangle \\ \hline k_2, \langle \alpha_2^k, \beta_2^k\rangle & \langle \mu_{k_2,l_2}, \nu_{k_2,l_2}\rangle & \langle \mu_{k_2,l_1}, \nu_{k_2,l_1}\rangle \\ k_1, \langle \alpha_1^k, \beta_1^k\rangle & \langle \mu_{k_1,l_2}, \nu_{k_1,l_2}\rangle & \langle \mu_{k_1,l_1}, \nu_{k_1,l_1}\rangle \end{array}$$

is valid.

2 Definitions and Properties of Four Operations over Extended Intuitionistic Fuzzy Index Matrices

In [9], for the IM $A = [K, L, \{a_{k,l}\}]$, the local substitution is defined for the pairs of indices (p, k) and/or (q, l), respectively, by

$$\left[\frac{p}{k}; \perp\right] A = [(K - \{k\}) \cup \{p\}, L, \{a_{k,l}\}],$$

$$\left[\perp; \frac{q}{l}\right] A = [K, (L - \{l\}) \cup \{q\}, \{a_{k,l}\}].$$

On the next step, this operation is extended for a set of row and/or column indices, but the values of the matrix elements that are labeled by these indices are not changed.

Now, for the EIFIM

$$A = [K^*, L^*, \{\langle \mu_{k_i,l_j}, \nu_{k_i,l_j} \rangle\}]$$

$$= \begin{array}{c|cccccc} & \dots & l_j, \langle \alpha_j^l, \beta_j^l \rangle & \dots & l_q, \langle \alpha_q^l, \beta_q^l \rangle & \dots \\ \hline \vdots & \vdots & \vdots & \ddots & \vdots & \ddots \\ k_i, \langle \alpha_i^k, \beta_i^k \rangle & \dots & \langle \mu_{k_i,l_j}, \nu_{k_i,l_j} \rangle & \dots & \langle \mu_{k_i,l_q}, \nu_{k_i,l_q} \rangle \\ \vdots & \vdots & \vdots & \ddots & \vdots & \ddots \\ k_p, \langle \alpha_p^k, \beta_p^k \rangle & \dots & \langle \mu_{k_p,l_j}, \nu_{k_p,l_j} \rangle & \dots & \langle \mu_{k_p,l_q}, \nu_{k_p,l_q} \rangle \\ \vdots & \vdots & \vdots & \ddots & \vdots & \ddots \end{array}$$

we, introduce the two new operations by

$$\left\{ \frac{k_p}{k_i} \right\}_\rho A = \begin{array}{c|cccccc} & \dots & l_j, \langle \alpha_j^l, \beta_j^l \rangle & \dots & l_q, \langle \alpha_q^l, \beta_q^l \rangle & \dots \\ \hline \vdots & \vdots & \vdots & \ddots & \vdots & \ddots \\ k_p, \langle \alpha_p^k, \beta_p^k \rangle & \dots & \langle \mu_{k_p,l_j}, \nu_{k_p,l_j} \rangle & \dots & \langle \mu_{k_p,l_q}, \nu_{k_p,l_q} \rangle \\ \vdots & \vdots & \vdots & \ddots & \vdots & \ddots \\ k_i, \langle \alpha_i^k, \beta_i^k \rangle & \dots & \langle \mu_{k_i,l_j}, \nu_{k_i,l_j} \rangle & \dots & \langle \mu_{k_i,l_q}, \nu_{k_i,l_q} \rangle \\ \vdots & \vdots & \vdots & \ddots & \vdots & \ddots \end{array}$$

and

$$\left\{ \frac{l_q}{l_j} \right\}_\sigma A = \begin{array}{c|cccccc} & \dots & l_q, \langle \alpha_q^l, \beta_q^l \rangle & \dots & l_j, \langle \alpha_j^l, \beta_j^l \rangle & \dots \\ \hline \vdots & \vdots & \vdots & \ddots & \vdots & \ddots \\ k_i, \langle \alpha_i^k, \beta_i^k \rangle & \dots & \langle \mu_{k_i,l_q}, \nu_{k_i,l_q} \rangle & \dots & \langle \mu_{k_i,l_j}, \nu_{k_i,l_j} \rangle \\ \vdots & \vdots & \vdots & \ddots & \vdots & \ddots \\ k_p, \langle \alpha_p^k, \beta_p^k \rangle & \dots & \langle \mu_{k_p,l_q}, \nu_{k_p,l_q} \rangle & \dots & \langle \mu_{k_p,l_j}, \nu_{k_p,l_j} \rangle \\ \vdots & \vdots & \vdots & \ddots & \vdots & \ddots \end{array}$$

The two definitions keep their forms for the cases of IFIMs, of EIMs and of IMs.

Both operations can be extended in the forms

$$\left\{ \frac{k_{p_1}}{k_{i_1}}, ..., \frac{k_{p_{s-1}}}{k_{i_{s-1}}}, \frac{k_{p_s}}{k_{i_s}} \right\}_\rho A = \left\{ \frac{k_{p_s}}{k_{i_s}} \right\}_\rho \left(\left\{ \frac{k_{p_1}}{k_{i_1}}, ..., \frac{k_{p_{s-1}}}{k_{i_{s-1}}} \right\}_\rho A \right)$$

and

$$\left\{ \frac{l_{q_1}}{l_{j_1}}, ..., \frac{l_{q_{s-1}}}{l_{j_{s-1}}}, \frac{l_{q_s}}{l_{j_s}} \right\}_\sigma A = \left\{ \frac{l_{q_s}}{l_{j_s}} \right\}_\sigma \left(\left\{ \frac{l_{q_1}}{l_{j_1}}, ..., \frac{l_{q_{s-1}}}{l_{j_{s-1}}} \right\}_\sigma A \right),$$

for the natural number $s \geq 2$.

The validity of the following two assertions is checked by the definitions.

Theorem 1. *For every two natural numbers u and v such that* $1 \leq u, v \leq s$ *and* $s \geq 2$*:*

$$\left\{\frac{k_{p_1}}{k_{i_1}}, ..., \frac{k_{p_u}}{k_{i_u}}, ..., \frac{k_{p_v}}{k_{i_v}}, ... \frac{k_{p_s}}{k_{i_s}}\right\}_\rho A = \left\{\frac{k_{p_1}}{k_{i_1}}, ..., \frac{k_{p_v}}{k_{i_v}}, ..., \frac{k_{p_u}}{k_{i_u}}, ... \frac{k_{p_s}}{k_{i_s}}\right\}_\rho A,$$

$$\left\{\frac{l_{q_1}}{l_{j_1}}, ..., \frac{l_{q_u}}{l_{j_u}}, ..., \frac{l_{q_v}}{l_{j_v}}, ... \frac{l_{q_s}}{l_{j_s}}\right\}_\sigma A = \left\{\frac{l_{q_1}}{l_{j_1}}, ..., \frac{l_{q_v}}{l_{j_v}}, ..., \frac{l_{q_u}}{l_{j_u}}, ... \frac{l_{q_s}}{l_{j_s}}\right\}_\sigma A.$$

Theorem 2. *For every two natural numbers* $s, t \geq 1$*:*

$$\left\{\frac{k_{p_1}}{k_{i_1}}, ..., \frac{k_{p_s}}{k_{i_s}}\right\}_\rho \left(\left\{\frac{l_{q_1}}{l_{j_1}}, ..., \frac{l_{q_s}}{l_{j_s}}\right\}_\sigma A\right) = \left\{\frac{l_{q_1}}{l_{j_1}}, ..., \frac{l_{q_s}}{l_{j_s}}\right\}_\sigma \left(\left\{\frac{k_{p_1}}{k_{i_1}}, ..., \frac{k_{p_s}}{k_{i_s}}\right\}_\rho A\right).$$

In [9], for the EIFIM A and for the fixed indices k_0 and l_0 some aggregation operations are defined. Here, a new operation is introduced with the aim to determine the indices of the A-element with an extremal value.

Let ext $\in \{\min, \max\}$.

$$\mathscr{E}_{\text{ext}} A = \langle\{\langle k_{r_1}, l_{s_1}\rangle, ..., \langle k_{r_t}, l_{s_t}\rangle\}; \langle\mu, \nu\rangle\rangle,$$

where $1 \leq t \leq \max(m, n)$,

$$\langle\mu, \nu\rangle = \langle\mu_{k_{r_1}, l_{s_1}}, \nu_{k_{r_1}, l_{s_1}},\rangle = ... = \langle\mu_{k_{r_t}, l_{s_t}}, \nu_{k_{r_t}, l_{s_t}}\rangle$$

and IFP $\langle\mu, \nu\rangle$ has an extremal (maximal or minimal in respect of the index ext of $\mathscr{E}$) value among all A-elements. The set $\mathscr{E}_{\text{ext}} A$ can be single (if there is only one extremal element of A) or with a grater number of elements, but smaller or equal to mn. For the case of EIFIMs, the order of the IFPs is determined by the relation $\leq$ from Sect. 1.

If we like to have the indices of exactly one extremal A-element, then using the relation $\leq$ from Sect. 1 we determine the IFP with extremal values among elements of set $\{\langle k_{r_1}, l_{s_1}\rangle, ..., \langle k_{r_t}, l_{s_t}\rangle\}$, having in mind their IF estimations $\langle\alpha_i^k, \beta_i^k\rangle$ and $\langle\alpha_j^l, \beta_j^l\rangle$, respectively.

Now, by analogy with operation "substitution" of indices, described in [9], for the IM A from Sect. 1, we will introduce the operation "substitution" of values as follows:

$$\mathscr{V}(A; k_i, l_j; \langle\phi, \psi\rangle)$$

$$= \begin{array}{c|ccccc} & l_1, \langle\alpha_1^l, \beta_1^l\rangle & \dots & l_j, \langle\alpha_j^l, \beta_j^l\rangle & \dots & l_n, \langle\alpha_n^l, \beta_n^l\rangle \\ \hline k_1, \langle\alpha_1^k, \beta_1^k\rangle & \langle\mu_{k_1,l_1}, \nu_{k_1,l_1}\rangle & \dots & \langle\mu_{k_1,l_j}, \nu_{k_1,l_j}\rangle & \dots & \langle\mu_{k_1,l_n}, \nu_{k_1,l_n}\rangle \\ \vdots & \vdots & \ddots & \vdots & \ddots & \vdots \\ k_i, \langle\alpha_i^k, \beta_i^k\rangle & \langle\mu_{k_i,l_1}, \nu_{k_i,l_1}\rangle & \dots & \langle\phi, \psi\rangle & \dots & \langle\mu_{k_i,l_n}, \nu_{k_i,l_n}\rangle \\ \vdots & \vdots & \ddots & \vdots & \ddots & \vdots \\ k_m, \langle\alpha_m^k, \beta_m^k\rangle & \langle\mu_{k_m,l_1}, \nu_{k_m,l_1}\rangle & \dots & \langle\mu_{k_m,l_j}, \nu_{k_m,l_j}\rangle & \dots & \langle\mu_{k_m,l_n}, \nu_{k_m,l_n}\rangle \end{array}.$$

It can be directly seen that

$$\mathscr{V}(\mathscr{V}(A; k_i, l_j; \langle \phi, \psi \rangle); k_i, l_j; \langle \mu_{k_i,l_j}, \nu_{k_i,l_j} \rangle) = A$$

and

$$\mathscr{V}(\mathscr{V}(A; k_i, l_j; \langle \phi, \psi \rangle); k_p, l_q; \langle \chi, \omega \rangle) = \mathscr{V}(\mathscr{V}(A; k_p, l_q; \langle \chi, \omega \rangle); k_i, l_j; \langle \phi, \psi \rangle).$$

3 Applications of the New Operations in Intercriteria Analysis

In this Section, we will construct a multi-dimensional EIFIM (see [10]) with the form

$$A = [K_1, K_2, ..., K_n, \{a_{k_{1,s_1}, k_{2,s_2}, ..., k_{n,s_n}}\}]$$

$$= \left\{ \begin{array}{c|ccc} \langle k_{3,s_3}, ..., k_{n,s_n} \rangle & k_{2,1} & \dots & k_{2,m_2} \\ \hline k_{1,1} & a_{k_{1,1},k_{2,1},k_{3,s_3},...,k_{n,s_n}} & \dots & a_{k_{1,1},k_{2,m_2},k_{3,s_3},...,k_{n,s_n}} \\ \vdots & \vdots & \ddots & \vdots \\ k_{1,i} & a_{k_{1,i},k_{2,1},k_{3,s_3},...,k_{n,s_n}} & \dots & a_{k_{1,i},k_{2,m_2},k_{3,s_3},...,k_{n,s_n}} \\ \vdots & \vdots & \ddots & \vdots \\ k_{1,m_1} & a_{k_{1,m_1},k_{2,1},k_{3,s_3},...,k_{n,s_n}} & \dots & a_{k_{1,m_1},k_{2,m_2},k_{3,s_3},...,k_{n,s_n}} \end{array} \right.$$

$$|\langle k_{3,s_3}, ..., k_{n,s_n} \rangle \in K_3 \times ... \times K_n\},$$

where $K_1, K_2, ..., K_n \subseteq \mathscr{I}$, $K_i = \{k_{i,1}, k_{i,2}, ..., k_{i,m_i}\}$, $m_i \geq 1$ and $a_{k_{1,s_1}, k_{2,s_2}, ..., k_{n,s_n}} \in \mathscr{X}$, a fixed set for $1 \leq i \leq n$ and $1 \leq s_i \leq m_i$.

Here, we describe a procedure for ordering of the A-elements following their maximality in the upper left angle. In practice, this procedure is a "procedure for stratification" of EIFIM A.

Let us have the EIFIM A from Sect. 1. Now, we will describe an algorithm for stratification of A. It has the following steps. We will discuss the case when the extremum is maximum, but the case of minimum is analogous.

Step 1. Construct an EIFIM $B = A$.
Step 2. Construct the set $\mathscr{E}_{\text{ext}}B$.
Step 3. Construct the EIFIM C_1 with dimension $m \times n$, whose elements are $\langle 0, 1 \rangle$. In the case, when the extremum is minimum, these elements will be $\langle 1, 0 \rangle$.
Step 4. Using the relation $\leq$ over IFPs, we determine the IFP of $\mathscr{E}_{\text{ext}}B$ with maximal values $\langle \alpha_i^k, \beta_i^k \rangle$ and $\langle \alpha_j^l, \beta_j^l \rangle$ among elements of set $\{\langle k_{r_1}, l_{s_1} \rangle, ..., \langle k_{r_t}, l_{s_t} \rangle\}$. If there is no IFP with two maximal IFP-values, we determine the one with maximal IFP $\langle \alpha_i^k, \beta_i^k \rangle$ or even with maximal value α_i^k. Now, using the local substitutions, we put this element with its indices and values on the upper - left place in EIFIM C_1. If there are other elements of $\mathscr{E}_{\text{ext}}B$ with the same first index as the maximal element, we put them as the second, third, etc. elements of the first row of C_1, ordering them by relation $\leq$ about their second indices. Also, if

there are other elements of $\mathscr{E}_{\text{ext}}B$ with the same second index as the maximal element, we put them as the second, third, etc. elements of the first column of C_1, ordering them by relation $\leq$ about their first indices. The element of $\mathscr{E}_{\text{ext}}B$ that has the largest values among the non-putting in C_1 elements is put in C_1 on the next (lower) row and next (right) column and the procedure is repeated by the putting of all elements of set $\mathscr{E}_{\text{ext}}B$.
Step 5. Using substitution $\mathscr{V}$, we change the values of B-elements from set $\mathscr{E}_{\text{ext}}B$ with value $\langle 0, 1\rangle$. In the case, when the extremum is minimum, these elements will be $\langle 1, 0\rangle$.
Step 6. If all elements of B are already equal to $\langle 1, 0\rangle$, the process stops, else, we return to point **2**.

Obviously, if all A-elements are equal, the C-matrix will be only one, while if all A-elements are different, we must have mn C-matrices with only one element in the respective corner. These C-matrices generate one 3-dimensional EIFIM. For it, the following Theorem is valid.

Theorem 3. *For each EIFIM A that has s in number C-matrices:*

$$C_1 \oplus_{\langle \max,\min\rangle} C_2 \oplus_{\langle \max,\min\rangle} \cdots \oplus_{\langle \max,\min\rangle} C_s = A,$$

$$C_1 \oplus_{\langle \min,\max\rangle} C_2 \oplus_{\langle \min,\max\rangle} \cdots \oplus_{\langle \min,\max\rangle} C_s = A.$$

Proof. First, we will modify some steps from the above algorithm. It is oriented to the software realization, while the new form will give possibility for a shorter proof.

Now, **Step 1.** is "Construct for $i = 1$ an EIFIM $B_i = A$" and **Step 2.** "Construct the set $\mathscr{E}_{\text{ext}}B_i$. In **Step 4**, for each element $\langle k, l\rangle; \langle \mu, \nu\rangle \in \mathscr{E}_{\text{ext}}B_i$ we construct the EIFIM $\begin{array}{c|c} & l \\ \hline k & \langle \mu, \nu\rangle \end{array}$ and

$$C_i = (\sum_{\langle\langle k,l\rangle;\langle \mu,\nu\rangle\rangle \in \mathscr{E}_{\text{ext}}B_i})_{\langle \max,\min\rangle} \begin{array}{c|c} & l \\ \hline k & \langle \mu, \nu\rangle \end{array},$$

where for $\circ \in \{\langle \max,\min\rangle, \langle \min,\max\rangle\}$:

$$(\sum)_\circ X_i = X_1 \oplus_\circ X_2 \oplus_\circ \ldots \oplus_\circ$$

and for two arbitrary EIFIMs

$$A = [K^*, L^*, \{\langle \mu_{k_i,l_j}, \nu_{k_i,l_j}\rangle\}], B = [P^*, Q^*, \{\langle \rho_{p_r,q_s}, \sigma_{p_r,q_s}\rangle\}],$$

operation "Addition-$(\text{ext}_1, \text{ext}_2)$" has the form

$$A \oplus_{(\text{ext}_1,\text{ext}_2)} B = [T^*, V^*, \{\langle \varphi_{t_u,v_w}, \psi_{t_u,v_w}\rangle\}],$$

where

$$T^* = K^* \cup P^* = \{\langle t_u, \alpha_u^t, \beta_u^t\rangle | t_u \in K \cup P\},$$

$$V^* = L^* \cup Q^* = \{\langle v_w, \alpha_w^v, \beta_w^v\rangle | v_w \in L \cup Q\},$$

$$\alpha_u^t = \begin{cases} \alpha_i^k, & \text{if } t_u \in K - P \\ \alpha_r^p, & \text{if } t_u \in P - K \\ \text{ext}_1(\alpha_i^k, \alpha_r^p), & \text{if } t_u \in K \cap P \end{cases},$$

$$\beta_w^v = \begin{cases} \beta_j^l, & \text{if } v_w \in L - Q \\ \beta_s^q, & \text{if } t_w \in Q - L \\ \text{ext}_2(\beta_j^l, \beta_s^q), & \text{if } t_w \in L \cap Q \end{cases},$$

and

$$\langle \varphi_{t_u,v_w}, \psi_{t_u,v_w}\rangle = \begin{cases} \langle \mu_{k_i,l_j}, \nu_{k_i,l_j}\rangle, & \text{if } t_u = k_i \in K \\ & \quad \text{and } v_w = l_j \in L - Q \\ & \quad \text{or } t_u = k_i \in K - P \\ & \quad \text{and } v_w = l_j \in L; \\ \langle \rho_{p_r,q_s}, \sigma_{p_r,q_s}\rangle, & \text{if } t_u = p_r \in P \\ & \quad \text{and } v_w = q_s \in Q - L \\ & \quad \text{or } t_u = p_r \in P - K \\ & \quad \text{and } v_w = q_s \in Q; \\ \langle \text{ext}_1(\mu_{k_i,l_j}, \rho_{p_r,q_s}), & \text{if } t_u = k_i = p_r \in K \cap P \\ \quad \text{ext}_2(\nu_{k_i,l_j}, \sigma_{p_r,q_s})\rangle, & \quad \text{and } v_w = l_j = q_s \in L \cap Q \\ \langle 0, 1\rangle, & \text{otherwise} \end{cases},$$

where $\langle \text{ext}_1, \text{ext}_2\rangle \in \{\langle \max, \min\rangle, \langle \min, \max\rangle\}$.

Therefore, C_i is an EIFIM containing all elements with their indices of $\mathscr{E}_{\text{ext}} B_i$ and hence the EIFIM $C_1 \oplus_{\langle \max,\min\rangle} C_2 \oplus_{\langle \max,\min\rangle} \cdots \oplus_{\langle \max,\min\rangle} C_s$ will contain all elements with their indices of EIFIM A. Since the pair of used operations is $\langle \max, \min\rangle$, the values $\langle 0, 1\rangle$ do not change the values of the elements with other forms.

For the second equality, the proof is similar, but now we will work with the pair of operations $\langle \min, \max\rangle$ and pair of neutral elements $\langle 1, 0\rangle$ that will again not influence the results of operation with the A-elements.

4 Algorithm Implementation Remarks

The realization of the algorithm is performed using C# programming language on Windows 64 bits operating system in using Visual Studio 2017 Community environment. Two-dimensional array A with $m \times n$ size is declared. The elements are initialized manually (or can be generated automatically by random values). A jagged array is defined to store the maximum elements of $ExtB$ set. It contains the elements in the range of an array (jagged array is an array of arrays) to multiple arrays, but less than mn. Each sub-array contains the following list of elements: "index-key for rows, index-key for column, value (IFP)", i.e. $[0, 0, 25]$ (multidimensional array concept is also possible). Thereafter a jagged array of two-dimensional arrays with size $m \times n$ is declared. It represents the matrix C_1.

In the next step a searching procedure over the two-dimensional array A is performed. It finds the maximal element/s using the conditional statement. In the conditional statement comparing operator and definition of an initial maximal element are included. The selected maximal elements with corresponding values for indexes of row and column are assigned to the jagged array B. If the maximal elements (arrays in the jagged array) are more than one it is compared by their indexes for rows and columns. The comparing procedure is made by selecting the elements with determined positions (indices). For example the index for row of first maximal element is compared with the index of the row for the second array. The step is repeated for the column indexes also. The whole procedure is repeated until all elements are compared. Then the maximal elements are assigned to the multidimensional array C_1 according to the specified order, starting from the cell with indexes $(0, 0)$ (the indexes started from the number 0). Therefore, the elements are initialized by row- and column-major order. The searching procedure is repeated until all the elements of matrix A are passed. The maximal elements in the each pass are assigned to the jagged array $ExtB$ and then to the two-dimensional arrays $C_1, ..., C_s$. In the last step a new two-dimensional array $D(C)$ with size $m \times n$ is defined. We compare the indexes of C-matrices to determine the element with minimum indexes of rows and columns. The C-matrices are compared by conditional statement having initial minimal values for minimal indexes for row and column. When the C_i-matrix containing the element with indexes $(0, 0)$ is found its value is selected and assigned to the element with indexes $(0, 0)$ of D-matrix (or C-matrix). The elements are initialized by row- and column-major order.

In the presented procedure the elements of jagged arrays are selected by choosing the element of the jagged array plus selecting the sub-index of the sub-array, i.e. *name of jagged array*[*number of the array*][*number of the element of the selected array*].

5 Conclusion

The defined here operators can be combined with the rest research over IMs and with some IM applications, described, in [1,2,9,12–59].

Some modifications and extensions of the new operators will be discussed in future. Also, in a next research, the new operators will be extended for the case of 3- and n-dimensional extended intuitionistic fuzzy index matrices and over their simpler cases and their properties will be studied.

Acknowledgments. The first author is grateful for the support under Grant Ref. No. DN 17/6 "A New Approach, Based on an Intercriteria Data Analysis, to Support Decision Making in *in silico* Studies of Complex Biomolecular Systems" of the Bulgarian National Science Fund.

References

1. Andreev, N., Sotirova, E., Ribagin, S.: Intercriteria analysis of data from data from the centers for transfusion haematology in Bulgaria. Comptes Rendus de l'Academie Bulgare des Science **72**(7), 982–990 (2019)
2. Angelova, M., Pencheva, T.: InterCriteria analysis approach for comparison of simple and multi-population genetic algorithms performance. In: Fidanova, S. (ed.) Recent Advances in Computational Optimization. Studies in Computational Intelligence, vol. 795, pp. 117–130 (2019)
3. Atanassov, K.: Generalized index matrices. Comptes rendus de l'Academie Bulgare des Sciences **40**(11), 15–18 (1987)
4. Atanassov, K.: Intuitionistic Fuzzy Sets. Springer, Heidelberg (1999)
5. Atanassov, K.: On index matrices. Part 1: Standard cases. Adv. Stud. Contemp. Math. **20**(2), 291–302 (2010)
6. Atanassov, K.: On index matrices. Part 2: Intuitionistic fuzzy case. Proc. Jangjeon Math. Soc. **13**(2), 121–126 (2010)
7. Atanassov, K.: On Intuitionistic Fuzzy Sets Theory. Springer, Berlin (2012)
8. Atanassov, K.: On extended intuitionistic fuzzy index matrices. Notes Intuitionistic Fuzzy Sets **19**(4), 27–41 (2013)
9. Atanassov, K.: Index Matrices: Towards an Augmented Matrix Calculus. Springer, Cham (2014)
10. Atanassov, K.: n-Dimensional extended index matrices. Adv. Stud. Contemp. Math. **28**(2), 245–259 (2018)
11. Atanassov, K., Szmidt, E., Kacprzyk, J.: On intuitionistic fuzzy pairs. Notes Intuitionistic Fuzzy Sets **19**(3), 1–13 (2013)
12. Atanassov, K., Szmidt, E., Kacprzyk, J., Bureva, V.: Two examples for the use of 3-dimensional intuitionistic fuzzy index matrices. Notes Intuitionistic Fuzzy Sets, **20**(2), 52–59 (2014)
13. Atanassova, V.: Interpretation in the intuitionistic fuzzy triangle of the results, obtained by the InterCriteria analysis. In: 16th World Congress of the International Fuzzy Systems Association (IFSA), 9th Conference of the European Society for Fuzzy Logic and Technology (EUSFLAT), Gijon, Spain, 30 June–03 July 2015, pp. 1369–1374 (2015). https://doi.org/10.2991/ifsa-eusflat-15.2015.193
14. Atanassova, V.: New modified level operator N over intuitionistic fuzzy sets. In: Christiansen, H., Jaudoin, H., Chountas, P., Andreasen, T., Larsen, H.L. (eds.) Proceedings of 12th International Conference on Flexible Query Answering Systems (FQAS 2017), London, UK, 21–22 June 2017. LNAI, vol. 10333, pp. 209–214. Springer (2017). https://doi.org/10.1007/978-3-319-59692-1_18. ISBN 978-3-319-59691-4. ISSN 0302-9743
15. Atanassova, V., Doukovska, L., Michalikova, A., Radeva, I.: Intercriteria analysis: from pairs to triples. Notes Intuitionistic Fuzzy Sets **22**(5), 98–110 (2016). Print ISSN 1310-4926. Online ISSN 2367-8283
16. Atanassova, V., Vardeva, I.: Sum- and average-based approach to criteria shortlisting in the InterCriteria Analysis. Int. J. Notes Intuitionistic Fuzzy Sets **20**(4), 41–46 (2014)
17. Bureva, V.: Application of the index matrix operations over relational databases. Adv. Stud. Contemp. Math. **26**(4), 669–679 (2016)

18. Bureva, V., Sotirova, E., Moskova, M., Szmidt, S.: Solving the problem of appointments using index matrices. In: 13th International Workshop on Generalized Nets London, 29 October 2012, pp. 43–48 (2012)
19. Fidanova, S., Roeva, O.: InterCriteria analysis of different metaheuristics applied to E. coli cultivation process. Numerical Methods for Scientific Computations and Advanced Applications, pp. 21–25 (2016)
20. Fidanova, S., Roeva, O.: Comparison of different metaheuristic algorithms based on InterCriteria analysis. J. Comput. Appl. Math. **7** (2017). https://doi.org/10.1016/j.cam.2017.07.028
21. Fidanova, S., Roeva, O., Mucherino, A., Kapanova, K.: InterCriteria analysis of ant algorithm with environment change for GPS surveying problem. Lecture Notes in Computer Science, vol. 9883, pp. 271–278. Springer (2016)
22. Fidanova, S., Roeva, O., Paprzycki, M.: InterCriteria analysis of ant colony optimization application to GPS surveying problems. Issues Intuitionistic Fuzzy Sets Generalized Nets **12**, 20–38 (2015/2016)
23. Georgieva, V., Angelova, N., Roeva, O., Pencheva, T.: InterCriteria analysis of wastewater treatment quality. J. Int. Sci. Publ. Ecol. Saf. **10**, 365–376 (2016)
24. Georgieva, V., Angelova, N., Roeva, O., Pencheva, T.: Simulation of parallel processes in wastewater treatment plant using generalized net integrated development environment. Comptes rendus de l'Académie bulgare des Sciences **69**(11), 1493–1502 (2016). ISI IF:0.233
25. Ikonomov, N., Vassilev, P., Roeva, O.: ICrAData – software for InterCriteria analysis. Int. J. Bioautomation **22**(1), 1–10 (2018)
26. Ilkova, T., Petrov, M.: Application of InterCriteria analysis to the Mesta river pollution modelling. Notes Intuitionistic Fuzzy Sets **21**(2), 118–125 (2015)
27. Ilkova, T., Roeva, O., Vassilev, P., Petrov, M.: InterCriteria analysis in structural and parameter identification of L-lysine production model. Issues Intuitionistic Fuzzy Sets Generalized Nets **12**, 39–52 (2015/2016)
28. Krawczak, M., Bureva, V., Sotirova, E., Szmidt, E.: Application of the InterCriteria decision making method to universities ranking. In: Atanassov, K., et al. (eds.) Novel Developments in Uncertainty Representation and Processing. Advances in Intelligent Systems and Computing, vol. 401, pp. 365–372. Springer, Cham (2016)
29. Krumova, S., Todinova, D., Mavrov, P., Marinov, V., Atanassova, K., Atanassov, S., Taneva, S.: Intercriteria analysis of calorimetric data of blood serum proteome. Biochimica et Biophysica Acta - Gener. Subjects **2017**(2), 409–417 (1861). https://doi.org/10.1016/j.bbagen.2016.10.012
30. Marinov, P., Fidanova, S.: Intercriteria and correlation analyses: similarities, differences and simultaneous use. Ann. "Inf." Sect. Union Sci. Bulgaria, **8**, 45–53 (2015/2016)
31. Mucherino, A., Fidanova, S., Ganzha, M.: Ant colony optimization with environment changes: an application to GPS surveying. In: Proceedings of the Federated Conference on Computer Science and Information Systems, ACSIS, vol. 5, pp. 495–500. https://doi.org/10.15439/2015F33
32. Pencheva, T., Angelova, M., Atanassova, V., Roeva, O.: InterCriteria analysis of genetic algorithm parameters in parameter identification. Notes Intuitionistic Fuzzy Sets **21**(2), 99–110 (2015)
33. Pencheva, T., Roeva, O., Angelova, M.: Investigation of genetic algorithm performance based on different algorithms for InterCriteria relations calculation. In: Lirkov, I., Margenov, S. (eds.) Large-Scale Scientific Computing. Lecture Notes in Computer Science, vol. 10665, pp. 390–398 (2018)

34. Petrov, M., Ilkova, T.: Application of the intercriteria analysis for selection of growth rate models for cultivation of strain Kluyveromyces marxianus var. lactis MC 5. Notes Intuitionistic Fuzzy Sets **21**(5), 49–60 (2015)
35. Petrov, M., Ilkova, T.: Intercriteria decision analysis for choice of growth rate models of batch cultivation by strain Kluyveromyces marxianus var. lactis MC 5. J. Int. Sci. Publ. Mater. Methods Technol. **10**, 468–486 (2016)
36. Roeva, O., Fidanova, S., Paprzycki, M.: InterCriteria analysis of ACO and GA hybrid algorithms. Studies in Computational Intelligence, vol. 610, pp. 107–126. Springer (2016). https://doi.org/10.1007/978-3-319-21133-6_7
37. Roeva, O., Vassilev, P.: InterCriteria analysis of generation gap influence on genetic algorithms performance. In: Atanassov, K.T., et al. (eds.) Novel Developments in Uncertainty Representation and Processing, Part V. Advances in Intelligent Systems and Computing, vol. 401, pp. 301–313 (2016)
38. Roeva, O., Vassilev, P., Angelova, M., Su, J., Pencheva, T.: Comparison of different algorithms for InterCriteria relations calculation. In: IEEE 8th International Conference on Intelligent Systems, pp. 567–572 (2016)
39. Roeva, O., Vassilev, P., Angelova, M., Pencheva, T.: InterCriteria analysis of parameters relations in fermentation processes models. Lecture Notes Computer Science, vol. 9330, pp. 171–181 (2015)
40. Roeva, O., Vassilev, P., Ikonomov, N., Angelova, M., Su, J., Pencheva, T.: On different algorithms for InterCriteria relations calculation. In: Hadjiski, M., Atanassov, K.T. (eds.) Intuitionistic Fuzziness and Other Intelligent Theories and Their Applications. Studies in Computational Intelligence, vol. 757, pp. 143–160 (2019)
41. Roeva, O., Zoteva, D.: Knowledge discovery from data: InterCriteria analysis of mutation rate influence. Notes Intuitionistic Fuzzy Sets **24**(1), 120–130 (2018)
42. Shannon, A., Atanassova, V., Atanassov, K.: Operation, "concatenation" over intuitionistic fuzzy index matrices. Notes Intuitionistic Fuzzy Sets **22**(3), 106–111 (2016)
43. Sotirov, S., Atanassova, V., Sotirova, E., Bureva, V., Mavrov, D.: Application of the intuitionistic fuzzy InterCriteria analysis method to a neural network preprocessing procedure. In: 9th Conference of the European Society for Fuzzy Logic and Technology (EUSFLAT), Gijon, Spain, 30 June–03 July 2015, pp. 1559–1564 (2015). https://doi.org/10.2991/ifsa-eusflat-15.2015.222
44. Sotirov, S., Atanassova, V., Sotirova, E., Doukovska, L., Bureva, V., Mavrov, D., Tomov, J.: Application of the intuitionistic fuzzy InterCriteria analysis method with triples to a neural network preprocessing procedure. In: Computational Intelligence and Neuroscience, Hindawi 2017, Article ID 2157852, p. 9 (2017). https://doi.org/10.1155/2017/2157852
45. Stratiev, D., Sotirov, S., Shishkova, I., Nedelchev, A., Sharafutdinov, I., Vely, A., Mitkova, M., Yordanov, D., Sotirova, E., Atanassova, V., Atanassov, K., Stratiev, D.D., Rudnev, N., Ribagin, S.: Investigation of relationships between bulk properties and fraction properties of crude oils by application of the intercriteria analysis. Petrol. Sci. Technol. **34**(13), 1113–1120 (2016)
46. Todorova, L., Vassilev, P., Surchev, J.: Using Phi coefficient to interpret results obtained by InterCriteria analysis. In: Novel Developments in Uncertainty Representation and Processing. Advances in Intelligent Systems and Computing, vol. 401, pp. 231–239. Springer (2106). https://doi.org/10.1007/978-3-319-26211-6_20
47. Traneva, V.: Internal operations over 3-dimensional extended index matrices. In: Proceedings of the Jangjeon Mathematical Society, vol. 18, no. 4, pp. 547–569 (2015)

48. Traneva, V.: One application of the index matrices for a solution of a transportation problem. Adv. Stud. Contemp. Math. **26**(4), 703–715 (2016)
49. Traneva, V., Atanassova, V., Tranev, S.: Index matrices as a decision-making tool for job appointment. In: Nikolov, G., et al. (eds.) NMA 2018, pp. 1–9. Springer, Cham (2019). https://doi.org/10.1007/978-3-030-10692-8_18
50. Traneva, V., Marinov, P., Atanassov, K.: Index matrix interpretations of a new transportation-type problem. Comptes Rendus de L'Academie Bulgare des Sciences **69**(10), 1275–1282 (2016)
51. Traneva, V., Marinov, P., Atanassov, K.: Transportation-type problems and their index matrix interpretations. Adv. Stud. Contemp. Math. **26**(4), 587–594 (2016)
52. Traneva, V., Sotirova, E., Bureva, V., Atanassov, K.: Aggregation operations over 3-dimensional extended index matrices. Adv. Stud. Contemp. Math. **25**(3), 407–416 (2015)
53. Traneva, V., Tranev, S.: Existence of solution of the problem of optimal control of mines for minerals. In: Proceedings of the Jangjeon Mathematical Society, vol. 21, no. 3, pp. 443–478 (2018)
54. Traneva, V., Tranev, S., Atanassova, V.: An intuitionistic fuzzy approach to the Hungarian algorithm. In: Nikolov, G., Kolkovska, N., Georgiev, K. (eds.) Numerical Methods and Applications. NMA 2018. Lecture Notes in Computer Science, vol. 11189, pp. 167–175. Springer, Cham (2019). https://doi.org/10.1007/978-3-030-10692-8_19
55. Traneva, V., Tranev, S., Szmidt, E., Atanassov, K.: Three dimensional intercriteria analysis over intuitionistic fuzzy data. In: Kacprzyk, J., et al. (eds.) Advances in Fuzzy Logic and Technology 2017. Advances in Intelligent Systems and Computing, vol. 641, pp. 442–449. Springer International Publishing AG (2018). https://doi.org/10.1007/978-3-319-66827-7_40
56. Vassilev, P., Ribagin, S.: Todorova, on an aggregation of expert value assignments using index matrices. Notes Intuitionistic Fuzzy Sets **23**(4), 75–78 (2017)
57. Vankova, D., Sotirova, E., Bureva, V.: An application of the InterCriteria method approach to health-related quality of life. Notes Intuitionistic Fuzzy Sets **21**(5), 40–48 (2015)
58. Vassilev, P., Todorova, L., Andonov, V.: An auxiliary technique for InterCriteria analysis via a three dimensional index matrix. Notes Intuitionistic Fuzzy Sets **21**(2), 71–76 (2015)
59. Zoteva, D., Roeva, O.: InterCriteria analysis results based on different number of objects. Notes Intuitionistic Fuzzy Sets **24**(1), 110–119 (2018)

Genetic Algorithm Based Formula Generation for Curve Fitting in Time Series Forecasting Implemented as Mobile Distributed Computing

Rumen Ketipov, Georgi Kostadinov, Plamen Petrov, Iliyan Zankinski, and Todor Balabanov(✉)

Institute of Information and Communication Technologies, Bulgarian Academy of Sciences, Acad. Georgi Bonchev Street, Block 2, 1113 Sofia, Bulgaria
{rketipov,g.kostadinov,p.petrov}@iit.bas.bg, iliyan@hsi.iccs.bas.bg, todorb@iinf.bas.bg

Abstract. Times series forecasting has many important real life applications. Such forecasting is widely used in statistics, signal processing, pattern recognition, econometrics, mathematical finance, weather forecasting, earthquake prediction, electroencephalography, control engineering, astronomy, communications engineering and in any applied mathematics field where temporal measurements are done.

In last few decades time series forecasting receives a lot of attention from the researchers in the machine learning domain. Many different forecasting models are developed with the usage of different prediction approaches. Artificial neural networks are a bright example of such forecasting technique. The main goal is learning of data dependency between past and future values of the time series when artificial neural network is used. If the weights of the artificial neural networks are taken as coefficients of a complex polynomial the forecasting can be presented as curve fitting problem.

This research proposes forecasting approach a little bit different than the approach used in the artificial neural networks. Set of mathematical formulas are presented as expression trees in a genetic algorithm population. The goal in this genetic algorithm based optimization is searching of a mathematical expression which can provide the best curve fitting formula according time series values. Because of the genetic algorithms' extremely high degree of parallelism possibilities calculations in this research are organized as distributed computing solutions on a mobile devices with Android operating system.

1 Introduction

This research focuses on the capabilities distributed genetic algorithms to be used for mathematical formulas generation as tool for time series curve fitting. Each of the components used in this research will be introduced in the following subsections.

I. Dimov and S. Fidanova (Eds.): HPC 2019, SCI 902, pp. 40–47, 2021.
https://doi.org/10.1007/978-3-030-55347-0_4

1.1 Time Series

Financial time series forecasting is an area of high researchers interest [1] from many decades. Having an accurate forecast in the financial world is crucial for many important decision makings [2]. Time series are ordered measurements of particular variable done in a temporal manner [3]. In the most cases values are measured on equal intervals, but it is not a mandatory condition. Time series analysis is applicable in processes with clear repetition pattern [4]. Measurements done in a temporal order are presented as points in a two-dimensional space. For visualization purposes in many cases these points are connected with straight lines, which is a simplified form of linear interpolation for the values between two neighboring measurements. With such organization of points in two-dimensional space many different curves can be proposed for some generalization of the data nature (curve fitting problem). Linear regression for example can be used for trend estimation. In this case parameters of line equation are calculated such as that this line is the closest line for the given points. Another much more complicated generalization can be the Lagrange polynomial. In this case the parameters of a polynomial of the lowest degree that assumes at each value from X the corresponding value on Y (curve coincide at each point) are estimated. Theoretically infinite number of curves can satisfy the condition to pass across finite number of points in two-dimensional space. In the same context if calculations done in an artificial neural network are evaluated as mathematical expression it will give rough description of an equation in two-dimensional space. The general advantage of artificial neural networks is that they are capable to self-adjust coefficients [5] in this equation. In some cases the optimization inside the artificial neural network is done by genetic algorithm [6,7]. For this study it is much more interesting how genetic algorithms can be used for curve fitting as self-adaptive system similar to the artificial neural networks.

1.2 Genetic Algorithms

Genetic algorithms are metaheuristic for global optimization inspired by the natural processes in the biological evolution. Genetic algorithms are population based [8] and they are subset of a larger class algorithms called evolutionary algorithms [9]. Searching for a solution with genetic algorithms is based on three general operations - selection, crossover and mutation applied over population with individuals. Each individual is a vector into the solutions space.

Solving a problem starts with proper encoding of the individuals. There are many different encoding possibilities like binary encoding, integer numbers encoding, real numbers encoding, permutation encoding and etc. One less popular encoding is the expression tree encoding. In this case the individuals are represented with tree as data structure. Each node of the tree is a mathematical operation, on each leaf of the tree is an expression operand [6,8].

The process of the selection is related to pickup of two parents, which will be recombined for the creation of the new generation. There are many ways for the selection to be done. The most popular are - roulette wheel selection, rank selection, steady-state selection and etc. The process of parents selection is important, because it is expected that better parents have better chances to produce better children. During the selection process individuals with higher survival capabilities (fitness value) have better chances to mate. In the same time individuals with worse survival capabilities should also have some chances to mate. In some implementations of genetic algorithms elitism rule is applied. This means that the best found solutions are not removed from the population and this individuals have much higher chances and options to mate.

Crossover operation gives to the algorithm exploration capabilities and it is very dependent of the individuals encoding. In most of the cases crossover is done by single cut, two points cut or uniform merge. The main idea for the crossover is that two parents are exchanging different parts in order new individual to be created. When single cut point is applied parents are divided in two parts and the second half of the parents is swapped. The place of the cut point is chosen randomly. For expression tree encoding single cut crossover is almost intuitive choice. Branches (sub-trees) in both parents are randomly selected and swapped.

Mutation operation gives to the algorithm exploitation capabilities and it is also very dependent of the individuals encoding. During mutation only small part of the individual is changed. In the simplest form only a single bit is inverted. With integer or real numbers small delta is added or subtracted. When permutations are encoded two elements are exchanged for achieving mutation. With expression tree encoding few mutations are possible - random change of mathematical operation in the randomly selected node, random change of the operand in randomly selected leaf, random swap of nodes and etc. [5,6].

After creation of the new generation each individual is evaluated and a fitness value is assigned. The individuals from the old generation and the new generation are sorted according their fitness values and the second half of the list is removed from the population. The fitness function used for individuals evaluation is extremely problem dependent.

Genetic algorithms have a group of parameters like population size (usually estimated experimentally), crossover rate (in most cases 0.95 is advisable), mutation rate (in most cases 0.1 or 0.05 is advisable), elitism flag and etc. There is no general theory about population size and because of this the only applicable way for estimation of this parameters is by trying different possibilities. The crossover rate shows the chance a child to be produced by exchanging parts of the parents. When this parameter is less than one it means that there would be chances some parents to skip crossover procedure and to bypass directly to mutation procedure. Mutation rate shows the chance of every single element in the particular individual to be changed. When the mutation rate is very low few neighboring solutions are checked and the chances to approach global optimum are lower [7,10].

The organization of genetic algorithms computation by presence of population and operators applied over this population makes genetic algorithms perfect for parallel computations. In distributed computing system genetic algorithm global population is used (client-server architecture) and many local sub-populations are evolved without real need for information exchange between territorial separated computers.

1.3 Distributed Computing

Distributed computing is a computational organization in a system where components are located on different networked computers [11]. These computers communicates and coordinates their activities by passing messages between each other. Distributed computing is remarkable with: concurrency of components, lack of a global clock, and independent failure of components. In distributed computing, a problem is divided into many tasks, each of which is solved by one or more computers [12], which communicate with each other via message passing [13]. Distributed computing is generally used for solving a large computational problems.

In this study for distributed computing will be accepted the definition which states that each processor has its own memory and it is embedded in a separate device. Under these conditions calculating instructions run in parallel, which is the only relation with the parallel computing. The system is tolerant to a single component failure. The exact structure of the system is not known in advance and it changes with the time passing, because some nodes get in when some nodes get out [14]. The whole system consists of heterogeneous hardware and variety types of network connectivity.

When end users donate their computational resources it is called donated distributed computing and in most cases computations are organized as screensaver applications or low consumption demon processes. Many devices around the world have periods of idle usage. In such cases their owners are capable to donate this calculating power. Donated calculating power is mainly used for low budget projects usually in society help. Many scientific projects are organized around donated distributed computing and the most popular of them is SETI@home, which analyzes radio signals from the deep space [12].

Last decade was marked with a huge progress of mobile devices and communications, which gave to the scientist new capabilities for implementation of distributed computing software implementation. For example modern mobile devices are capable to provide active wallpapers instead of screensavers. Another option are the widgets supported in the modern mobile operating systems. But the best advantage of the mobile distributed computing is that the most of the devices are operating 24/7, which is not the case with the desktop computers and working stations.

2 Proposed Solution

There are different ways in which arithmetic expressions can be presented into the computer memory. One of the most used representations is an expression tree. Such representation is convenient in many cases, but for the current research arithmetic expressions are encoded as simple text strings. The crossover operator in this case is simple single cut and swap. The mutation is organized as random change of a mathematical operator, random change of a mathematical function and random change of a operand.

Listing 1. Genetic algorithm pseudo-code.

```
begin
  t = 0;
  initialize P(t);
  evaluate structures in P(t);
  while termination condition not satisfied do
  begin
    t = t + 1;
    select_repro C(t) from P(t-1);
    recombine and mutate structures in C(t)
    forming C'(t);
    evaluate structures in C'(t);
    select_replace P(t) from C'(t) and P(t-1);
  end
end
```

For the local genetic algorithm a classical implementation is proposed (Listing 1) as described in [15]. Each new individual in the local genetic algorithm population is tested by calculation of the proposed formula and mean square of the error with the real life values. If the calculated values of the formula matches perfectly with the real life values the total mean square value would be zero. This means that better individuals have lower values for their fitness.

The best found local individuals are reported to the centralized server and become part of a global genetic algorithm population. When there is a connection of a new mobile device a subset of the global genetic population is sent to the device in order new local population to be established and evolved.

3 Experiments

Experiments are done as Android mobile application. On the client side Apache Commons Genetic Algorithms Framework [16] is used with the parameters listed in Table 1. Mathematical expressions are presented and evaluated with Java build of mXparser [17] software library. On the server side PHP/MySQL based solution [18] is responsible for synchronization and financial time series data supply.

Publicly available data are used for EUR/USD rate for two months in 2019 year (Fig. 1). Bars in the chart shows the real world price. The line above the bars is the line of the formula generated by the proposed genetic algorithm. The approximation of the price values follows the shape of the bars.

Date	Price	Formula
30 Apr 2019	1.1217	1.1301
29 Apr 2019	1.1186	1.1343
26 Apr 2019	1.1149	1.1143
25 Apr 2019	1.1132	1.1181
24 Apr 2019	1.1154	1.1275
23 Apr 2019	1.1227	1.1340
22 Apr 2019	1.1258	1.1315
19 Apr 2019	1.1248	1.1161
18 Apr 2019	1.1231	1.1247
17 Apr 2019	1.1297	1.1329
16 Apr 2019	1.1281	1.1331
15 Apr 2019	1.1309	1.1251
12 Apr 2019	1.1302	1.1218
11 Apr 2019	1.1253	1.1311
10 Apr 2019	1.1274	1.1339
09 Apr 2019	1.1263	1.1278
08 Apr 2019	1.1261	1.1183
05 Apr 2019	1.1216	1.1287
04 Apr 2019	1.1221	1.1340
03 Apr 2019	1.1235	1.1302
02 Apr 2019	1.1204	1.1207
01 Apr 2019	1.1213	1.1143
29 Mar 2019	1.1218	1.1333
28 Mar 2019	1.1222	1.1321
27 Mar 2019	1.1247	1.1234
26 Mar 2019	1.1266	1.1153
25 Mar 2019	1.1313	1.1151
22 Mar 2019	1.1314	1.1334
21 Mar 2019	1.1375	1.1262
20 Mar 2019	1.1414	1.1169
19 Mar 2019	1.1352	1.1141
18 Mar 2019	1.1337	1.1203
15 Mar 2019	1.1325	1.1288
14 Mar 2019	1.1304	1.1192
13 Mar 2019	1.1327	1.1139
12 Mar 2019	1.1288	1.1178
11 Mar 2019	1.1249	1.1273
08 Mar 2019	1.124	1.1217
07 Mar 2019	1.1194	1.1145
06 Mar 2019	1.1307	1.1157
05 Mar 2019	1.1308	1.1244
04 Mar 2019	1.134	1.1325
01 Mar 2019	1.1377	1.1158

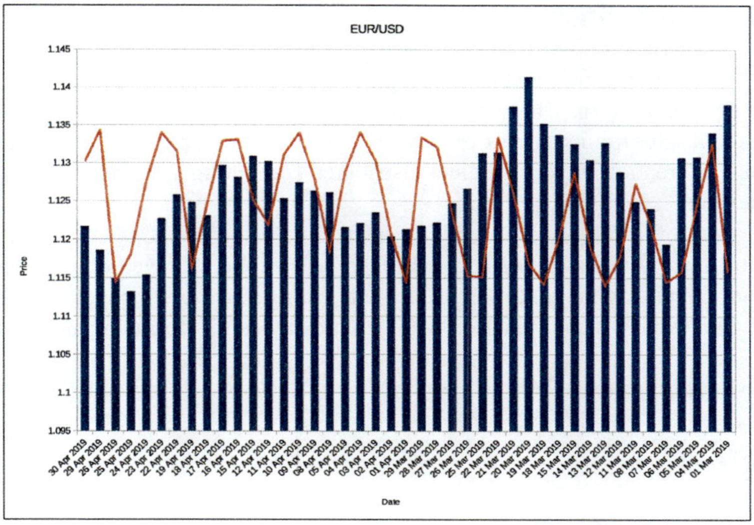

Fig. 1. EUR/USD rate for two months.

Table 1. Genetic algorithm parameters.

Parameter	Value
Generation gap	0.97
Crossover rate	0.95
Mutation rate	0.03
Maximum generations	100
Number of individuals	37
Number of variables	Floating
Inserted rate	100 %

In 100 generations the genetic algorithm proposes a solution presented with the formula in Eq. 1.

$$price = 0.485 + 0.01 * sin(day - 2.3) + log(day/10000) \quad (1)$$

The equation has few terms with sine and logarithmic functions used. The input of the equation is time value (number of days). The output of the equation is a prediction of the EUR/USD rate.

Even that initial results are promising much more extensive operation of the entire distributed computing system should be executed in order much better results to be achieved.

4 Conclusions

This research shows that genetic algorithms are a promising approach for financial time series forecasting, when they are used for formula generated predictors. The nature of genetic algorithms allows extremely high degree of parallel computations. This fact allows creation of very efficient distributed computing systems even on modern mobile devices.

As a future research it will be interesting to be investigated different communication capabilities as described in [19] in order to be achieved better genetic algorithms individuals exchange. It will be also interesting to be tested generalized artificial neural networks [20] in combination with genetic algorithms formula generation.

Acknowledgement. This work was funded by Velbazhd Software LLC.

References

1. Nava, N., Di Matteo, T., Aste, T.: Financial time series forecasting using empirical mode decomposition and support vector regression. Risks **6**(1), Article no. 7 (2018)
2. Catania, L., Grassi, S., Ravazzolo, F.: Predicting the volatility of cryptocurrency time-series. In: Mathematical and Statistical Methods for Actuarial Sciences and Finance, pp. 203–207. Springer, Cham (2018)
3. Chen, J., Boccelli, D.L.: Real-time forecasting and visualization toolkit for multi-seasonal time series. Environ. Model. Softw. **105**, 244–256 (2018)
4. Mueen, A., Keogh, E., Zhu, Q., Cash, S., Westover, B.: Exact discovery of time series motifs. In: Proceedings of the SIAM International Conference on Data Mining, pp. 473–484 (2009)
5. Aljarah, I., Faris, H., Mirjalili, S.: Optimizing connection weights in neural networks using the whale optimization algorithm. Soft Comput. **22**(1), 1–15 (2016)
6. Zhang, R., Tao, J.: A nonlinear fuzzy neural network modeling approach using an improved genetic algorithm. IEEE Trans. Ind. Electron. **65**(7), 5882–5892 (2018)
7. Kapanova, K., Dimov, I., Sellier, J.M.: A genetic approach to automatic neural network architecture optimization. Neural Comput. Appl. **29**(5), 1481–1492 (2016)
8. Mirjalili, S.: Genetic algorithm. In: Evolutionary Algorithms and Neural Networks. Studies in Computational Intelligence, vol. **780**, pp. 43–55. Springer, Cham (2019)
9. Liu, J., Abbass, H.A., Tan, K.C.: Evolutionary computation. In: Evolutionary Computation and Complex Networks, pp. 3–22. Springer, Cham (2019)
10. Cheng, J.R., Gen, M.: Accelerating genetic algorithms with GPU computing: a selective overview. Comput. Ind. Eng. **128**, 514–525 (2019)
11. Balabanov, T., Genova, K.: Distributed system for artificial neural networks training based on mobile devices. In: Proceedings of International Conference AUTOMATICS AND INFORMATICS, Federation of the Scientific Engineering Unions John Atanasoff Society of Automatics and Informatics, pp. 49–52 (2016)
12. Godfrey, B.: A primer on distributed computing (2006). http://billpg.com/bacchae-co-uk/docs/dist.html. Accessed 01 May 2019
13. Andrews, G.: Foundations of Multithreaded, Parallel, and Distributed Programming, pp. 291–292. Addison-Wesley (2000)
14. Balabanov, T., Zankinski, I., Barova, M.: Strategy for individuals distribution by incident nodes participation in star topology of distributed evolutionary algorithms. Cybern. Inf. Technol. **16**(1), 80–88 (2016)

15. Baeck, T., Fogel, D.B., Michalewicz, Z.: Evolutionary Computation 1 - Basic Algorithms and Operators, pp. 65–68. Institute of Physics Publishing, Bristol and Philadelphia (2000)
16. Apache Commons Genetic Algorithms Framework (2016). http://commons.apache.org/proper/commons-math/userguide/genetics.html. Accessed 12 May 2019
17. Gromada, M.: mXparser - Math Expression Evaluator/Parser - Library. http://www.mathparser.org/. Accessed 12 May 2019
18. Balabanov, T.: Distributed System for Time Series Prediction with Evolutionary Algorithms and Artificial Neural Networks. Abstracts of Dissertations, Institute of Information and Communication Technologies, Bulgarian Academy of Sciences, vol. 6 (2017)
19. Alexandrov, A.: Comparative analysis of IEEE 802.15.4 based communication protocols used in wireless intelligent sensor systems. In: Proceedings of the International conference RAM, pp. 51–54 (2014)
20. Tashev, T., Hristov, H.: Modeling of synthesis of information processes with generalized nets. Cybern. Inf. Technol. **3**(2), 92–104 (2003)

Generalized Net Model Simulation of Cluster Analysis Using CLIQUE: Clustering in Quest

Veselina Bureva[1], Velichka Traneva[1](✉), Dafina Zoteva[2], and Stoyan Tranev[1]

[1] "Prof. Asen Zlatarov" University, "Prof. Yakimov" Blvd, 8000 Bourgas, Bulgaria
{vesito_ka,tranev}@abv.bg, veleka13@gmail.com
[2] Institute of Biophysics and Biomedical Engineering, Bulgarian Academy of Sciences, Sofia, Bulgaria
dafy.zoteva@gmail.com

Abstract. Cluster analysis searches for similarities between data objects according to their characteristics and groups the similar objects into clusters. One of the techniques which combines subspace grid-based clustering and density-based cluster analysis, namely Clustering In Quest (CLIQUE), is studied in the present research. The main steps performed in the process of detecting groups of objects with similar behaviour are: dividing the data space into a finite number of cells, forming a grid-based structure, detecting groups of similar objects and defining the clusters.

Generalized Nets (GNs) have been introduced by Atanassov as an extension of the ordinary Petry nets and other their extensions and modifications. They are a powerful tool for modelling real processes.

A GN-model of the CLIQUE real-time data clustering process is constructed here and a simulation of the model is performed using a platform independent software, called GN Integrated Development Environment (GN IDE). An open-source version of the RapidMiner software is used for performing the cluster analysis on real datasets.

Keywords: Cluster analysis · Clique · Generalized nets · Modelling · Simulation

1 Introduction

Cluster analysis is used in Data Mining to organize objects with similar characteristics into groups, called clusters. The objects in the same cluster have more similarities to each other than to the objects in other clusters. In general, the concept of cluster analysis is not referred to one single algorithm, rather to the general problem itself. The results are achieved by different in their nature clustering algorithms.

The variety of clustering algorithms [1, 15] can be categorized, based on the clustering method used in the process, into: partitioning clustering, hierarchical clustering, density-based clustering, grid-based clustering, etc. These techniques are often combined to create new clustering algorithms.

I. Dimov and S. Fidanova (Eds.): HPC 2019, SCI 902, pp. 48–60, 2021.
https://doi.org/10.1007/978-3-030-55347-0_5

Partitioning clustering constructs a number of partitions, the potential clusters, based on a specific objective function. Each object of the input dataset have to be assigned to one cluster only. Each cluster consists of at least one object. Examples of such algorithms are K-Mean algorithm [23], K-Medoid algorithm [21], K-median algorithms [18].

Hierarchical clustering builds a hierarchy of clusters, determined by a series of partitions [15]. Which of the clusters should be combined and which of them split is determined by an appropriate metric, a measure of dissimilarity, and a linkage criterion to specify the distance between the subsets. Examples of such algorithms are CURE(Clustering using REpresentatives) [17], SLINK [29], etc.

The density-based clusters are considered as areas of connected, dense objects. They are separated by sparser areas, treated as noise, with density lower than the density in any of the clusters [1]. Some of the popular density-based clustering algorithms are DBSCAN [14], OPTICS [3], etc.

Grid-based clustering algorithms divide the data space into a finite number of cells in order to form a grid structure. The clusters are then formed in this grid structure. Clusters here are the more dense areas than the surroundings [1]. These algorithms are usually applied on large multidimensional datasets. Some of the commonly used grid-based clustering algorithms are STING [31], CLIQUE [2], etc.

CLIQUE: CLustering In QUEst which successfully combines the principles of the density-based and grid-based clustering is the object of research considered here. The workflow of the process of the cluster analysis with CLIQUE is studied in terms of Generalized Nets.

Generalized Nets (GNs) have been introduced as an extension of Petri nets and all other their extensions and modifications [6,7,9]. GNs are more complex in nature than any other Petri nets' extension or modification. Just like the other types of nets, GNs compose of places, transitions and tokens which are transferred from place to place during the functioning of the net. During this time, however, GN-tokens accumulate "history" as a result of the characteristic functions applied in the output places. GN-transitions are more complex in their nature since the number of the related input and output places might be more than one. The capacities of the arcs, as well as the conditions for tokens' transfer between the input and the output places of a transition, are described by Index Matrices (IMs) [5,8]. Finally, a global time scale is associated with GNs. Time in the GN's definition is discrete and increments with discrete steps.

Several GN-models have been developed so far to describe different processes related to cluster analysis. Hierarchical GN-model of the process of clustering have been introduced in [10]. The process of selecting a method for clustering have been modelled in terms of GNs in [11]. The GN-model of the STING clustering algorithm have been presented in [12].

The focus of the present research is a simulation of a GN-model of the CLIQUE clustering algorithm in GN Integrated Development Environment (GN IDE) [4]. GN IDE is a platform independent software for constructing GN-models and performing simulations of such. For the purpose of the simulation, the CLIQUE clustering procedure is applied on "User Knowledge Modelling Data Set" from the Machine Learning

Repository [22]. The RapidMiner software [27] with OpenSubspace plugin [30] is used to verify the results of the simulation.

The paper is organized in the following way. A short description of the simulated GN-model of the CLIQUE clustering algorithm is introduced in Sect. 2. Specifics of the simulation and the obtained results are discussed in Sect. 3. Section 4 contains details of the application of the CLIQUE algorithm in the RapidMiner environment on the same set of data used for the GN-model simulation. The concluding remarks are given in Sect. 5.

2 Generalized Net Model of the Process of CLIQUE

A detailed GN-model of CLIQUE clustering algorithm [2] has already been introduced in [13]. The model describes the workflow of the CLIQUE algorithm in 5 transitions and 17 places.

A modified version of this model is considered here. In order to simplify the simulation process of the GN-model in GN IDE, the preprocessing of the input data is omitted, i.e. the second transition Z_2 of the GN-model in [13] is removed.

The simplified GN-model of the CLIQUE algorithm, used here for simulation of the process, comprises of 4 transitions and 12 places (Fig. 1).

The transitions represent the following steps of the algorithm:

- Z_1 – Loading the input multidimensional database;
- Z_2 – Partitioning the data space and finding the number of points inside each cell of the partition;
- Z_3 – Identifying the subspaces that contain clusters using the Apriori principle;
- Z_4 – Generating minimal description for the clusters.

Initially, there is a δ-token in place L_3 which stays there during the entire GN's functioning. Its initial characteristic is the entire *"Database with multidimensional information"*.

An α-token enters the net through place L_1 with initial characteristics *"Query for selecting the initial dataset for the clustering procedure"*.

The transition Z_1 has the following form:

$$Z_1 = \langle \{L_1, L_3, L_9\}, \{L_2, L_3\}, R_1, \vee(\wedge(L_1, L_3), L_9) \rangle, \text{where}$$

$$R_1 = \begin{array}{c|cc} & L_2 & L_3 \\ \hline L_1 & false & true \\ L_3 & W_{3,2} & true \\ L_9 & false & true \end{array},$$

$W_{3,2}$ = "There are selected multidimensional data for the clustering procedure".

The tokens which enter place L_3 do not receive new characteristics. The α_1-token which enters the output place L_2 when the predicate $W_{3,2}$ is evaluated as *true* receives the characteristic *"Selected multidimensional data"*.

A β-token enters the net through places L_4 with initial characteristics *"Threshold for separating the cells"*.

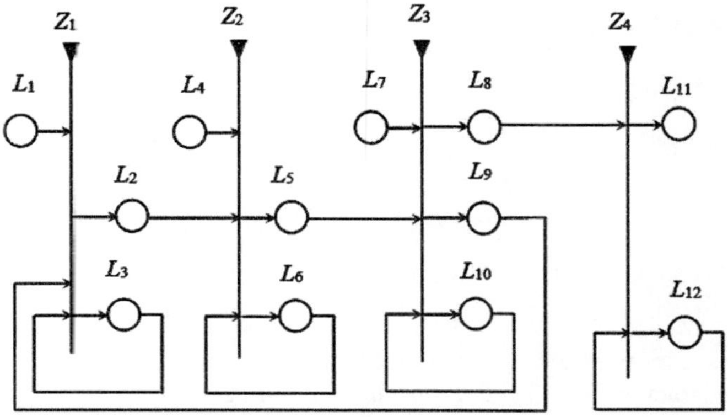

Fig. 1. Generalized net of the cluster analysis process using CLIQUE

The transition Z_2 has the form:

$$Z_2 = \langle\{L_2, L_4, L_6\}, \{L_5, L_6\}, R_2, \vee(\wedge(L_2, L_4), L_6)\rangle, \text{ where}$$

$$R_2 = \begin{array}{c|cc} & L_5 & L_6 \\ \hline L_2 & false & true \\ L_4 & false & true \\ L_6 & W_{6,5} & W_{6,6} \end{array},$$

- $W_{6,5}$ = "There is spatial data divided into cells";
- $W_{6,6} = \neg W_{6,5}$.

The tokens which enter the output place L_5 receive a new characteristic *"Spatial data separated into cells"*.

A γ-token enters the net through the input place L_7. The token has initial characteristic *"Frequency"*.

The transition Z_3 has the following form:

$$Z_3 = \langle\{L_5, L_7, L_{10}\}, \{L_8, L_9, L_{10}\}, R_3, \vee(\wedge(L_5, L_7), L_{10})\rangle, \text{ where}$$

$$R_3 = \begin{array}{c|ccc} & L_8 & L_9 & L_{10} \\ \hline L_5 & false & false & true \\ L_7 & false & false & true \\ L_{10} & W_{10,8} & W_{10,9} & W_{10,10} \end{array},$$

- $W_{10,8}$ = "There are identified clusters after applying the Apriori principle";
- $W_{10,9}$ = "There are cells which do not satisfy the frequency threshold";
- $W_{10,10} = \neg(W_{10,8} \wedge W_{10,9})$.

The tokens which enter place L_{10} do not receive new characteristics. The tokens entering places L_8 and L_9 obtain the characteristics *"Identified clusters after applying the Apriori principle"* and *"Cells which do not satisfy the frequency threshold"*, respectively.

The transition Z_4 has the following form:

$$Z_4 = \langle\{L_8, L_{12}\}, \{L_{11}, L_{12}\}, R_4, \vee(L_8, L_{12})\rangle, \text{ where}$$

$$R_4 = \begin{array}{c|cc} & L_{11} & L_{12} \\ \hline L_8 & false & true \\ L_{12} & W_{12,11} & W_{12,12} \end{array},$$

- $W_{12,11}$ = "There are clusters with a minimal description";
- $W_{12,12} = \neg W_{12,11}$

The tokens entering place L_{12} do not receive new characteristics. Tokens enter place L_{11} with the characteristic *"Result of the clustering process"*.

3 Simulation of the GN-Model of the CLIQUE in GN IDE

The independent software tool GN IDE [4], which implements the algorithms of transitions' and GN's functioning, has been employed here to simulate the considered GN-model.

The GN-model in Sect. 2 is slightly changed for the purpose of the simulation. It allows the clustering algorithm to be applied on an entire multidimensional dataset or on a part of it, while the choice of the input dataset in the simulated GN-model is left to the user.

The simulated GN-model does not require the selection query and the entire multidimensional database separately. The input dataset is loaded directly as an initial characteristic of the α-token in input place L_1. Therefore, the presence of the δ-token in place L_3 holding the entire multidimensional database is unnecessary.

The input dataset used for the simulated cluster analysis is "User Knowledge Modelling Data Set" [22]. There are 258 records in the training dataset and other 145 records in the testing dataset. The datasets are multidimensional with five attributes, which hold the following meaning:

- STG: The degree of study time for goal object materials;
- SCG: The degree of repetition number of user for goal object materials;
- STR: The degree of study time of user for related objects with goal object;
- LPR: The exam performance of user for related objects with goal object;
- PEG: The exam performance of user for goal objects.

Positive real numbers are used as estimations of the attributes in the two datasets. The two datasets, therefore, can be considered as real matrices with 258 rows and 5 columns for the training dataset, and 145 rows and 5 columns for the testing dataset.

A preview of the simulated GN-model in GN IDE is shown in Fig. 2.

The simulation is performed on Windows 64 bits operating system with Java 8 installed.

The transition Z_1 is responsible for loading this particular part of the data on which the analysis will be applied. In this particular case this is the testing dataset of the "User Knowledge Modelling Data Set".

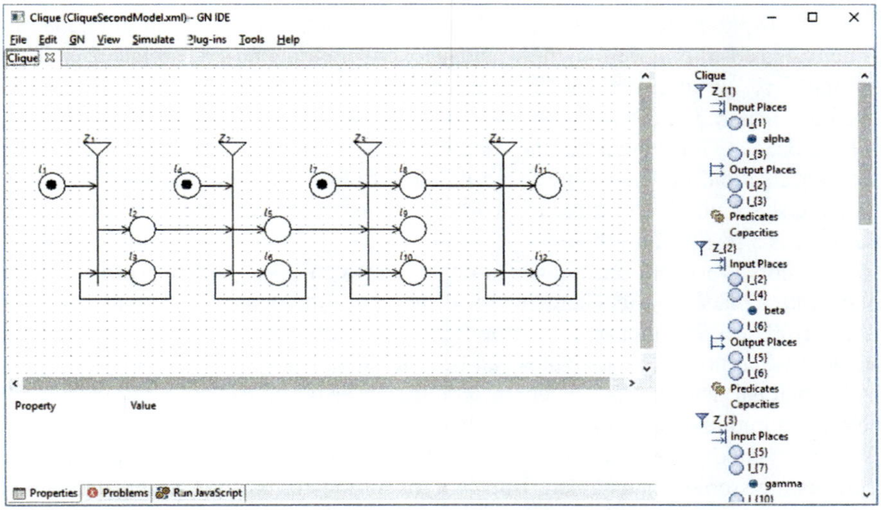

Fig. 2. GN model preview in GN IDE

The input dataset is stored as a characteristic of the α-token which enters the GN through the input place L_1, along with the number of dimensions of the input dataset. When the transition Z_1 is activated for the first time, this token is transferred to place L_3. The next time the transition Z_1 is activated, a new α_1-token with a copy of the input dataset as a characteristic is generated and transferred to output place L_2 for further processing. The original α-token stays in L_3 till the end of the GN's functioning.

The transition Z_2 is responsible for partitioning each and every dimension of the dataset into ξ intervals with equal length. The required input arguments for the partitioning are the input dataset and the desired number of intervals in each dimension. The input dataset is a characteristic of the α_1-token and the parameter ξ is stored in the β-token which enters the GN through the input place L_4. Therefore, the transition Z_2 can be activated for the first time only when both of the tokens are in the input places L_2 and L_4, respectively. The value set for the characteristic ξ is 2.

The two tokens α_1 and β are merged in place L_6. The resulting κ-token is transferred to L_5 place when all dimensions of the input dataset are processed. In order to track the count of the currently processed dimensions, the κ-token receives a new characteristic *"Current number of processed dimensions"* in L_6 place. It has an initial value of 0, which is incremented by 1 each time the transition Z_2 is activated, till it reaches the number of dimensions of the input dataset. The κ-token receives new characteristics when enters the place L_5. It stores the resulting intervals from the partitioning of each dimension. They can be interpreted as 1–dimensional candidate dense units. The other characteristic is the current number of processed dimensions, which in this particular case has a value of 1.

The transition Z_3 is activated only when there is a token in place L_5 and a γ-token in the L_7 input place. The γ-token has the frequency parameter τ as an initial characteristic,

$\tau = 0.04$. This parameter is required to determine the dense units. When the transition Z_3 is activated for the first time, the κ and γ tokens are transferred to place L_{10}. They merge there, resulting in ρ-token. The selectivity of the generated intervals inside the ρ-token is compared to the frequency parameter τ. Only the dense ones are kept.

The procedure for generating the candidate dense units is performed in place L_{10} of the Z_3 transition, i.e. each time the ρ-token is transferred from L_{10} back to L_{10}. If D_{k-1} is the set of all $(k-1)$-dimensional dense units, the next generation of candidate dense units is a result of a self-join operation over D_{k-1}. Then those candidates which are dense units are kept, while the non-dense ones can be discarded. Each time the ρ-token enters place L_{10}, its characteristics (the list of the dense units and the current number of processed dimensions) are updated. The candidate generation procedure is terminated when no more candidates can be generated. Then the ρ-token with the list of k-dimensional dense units as a characteristic is transferred to the output place L_8. k is the maximum number of dimensions where dense units can be found.

In the GN-model presented in Sect. 2, each time candidates are generated, the non–dense ones are stored as a characteristic of a token transferred to place L_9. This token is later passed to place L_3 of the Z_1 transition and stored there along with the α-token with the original dataset. Since the non–dense units are to be discarded, the model can be simplified by removing the cyclic arc from L_9 place to Z_1 transition. There is no particular need to store these units for now. They can be simply removed from the corresponding set of candidate dense units. Therefore, the L_9 place may turn out to be unnecessary.

The transition Z_4 is responsible for the identification of clusters, based on the k-dimensional dense units generated during the functioning of the transition Z_3, and generating their minimal description.

The number of the clusters generated by the simulated GN-model of the CLIQUE clustering algorithm is 31. The ρ-token receives as final characteristics in place L_{11} the number of the clusters, as well as the clusters in the form of nested lists.

The GN-model of the CLIQUE algorithm can be improved with further research. The most important part of the algorithm is to determine the values of the input parameters ξ and τ, which correspond to the number of intervals for the partitioning of the dimensions and the frequency threshold, respectively. The GN-model could be modified so that the GN itself can make decisions for the values of these parameters, based on the results of the first or the current iterations.

An interesting place of the GN-model is the place L_9, where the discarded candidates are stored. The selectivity of these units is known, so as the frequency threshold, stored in the γ-token. Their difference could be evaluated and the value of the frequency threshold could be changed accordingly, so that a new parallel iteration of the clustering process to be run on the same dataset with improved values of the parameters. This can be interpreted as a type of self-learning GN.

4 Realization of the GN Model of the CLIQUE

An open-source version of the RapidMiner [27] software has been chosen for the verification of the results of the simulated clustering procedure. The additional library

"OpenSubspace" [24,25] has been used to develop high-dimensional subspace clustering process using CLIQUE algorithm [30].

The multivariate input dataset used for the cluster analysis in the RapidMiner environment is again "User Knowledge Modelling Data Set" [19,20,22] with 403 records in total, 258 records (Fig. 3, 4) in the training subset and 145 records in the testing one.

ExampleSet (258 examples, 0 special attributes, 5 regular attributes)

Row No.	STG	SCG	STR	LPR	PEG
1	0	0	0	0	0
2	0.080	0.080	0.100	0.240	0.900
3	0.060	0.060	0.050	0.250	0.330
4	0.100	0.100	0.150	0.650	0.300
5	0.080	0.080	0.080	0.980	0.240

Fig. 3. Part of the input data

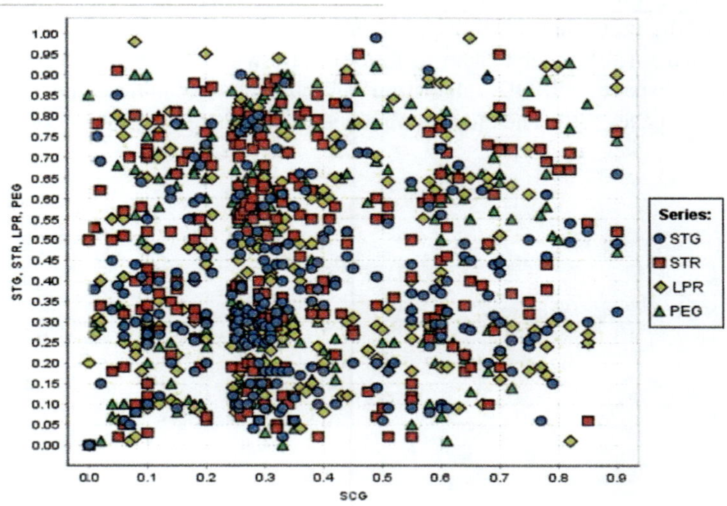

Fig. 4. The input data visualisation

The clustering process of these high-dimensional data using the CLIQUE algorithm in RapidMiner environment is presented in Fig. 5. The main purpose is to classify the knowledge level of the user according to the considered data.

The training set is loaded in the *Read Excel*-operator. The testing set is loaded in the *Read Excel (2)*-operator. The *Multiply operators* are used to copy the input data.

Several copies are obtained as an output without any processing step. The operators *Clique clustering algorithm* apply the clustering procedure on the input datasets. The cluster model operator is used to display the clusters visually. *Coda* and *MCExplorer visualization* are also used to present Interactive Exploration of Multiple (Subspace) Clustering Solutions. The operator *E4SC Measure* is added to calculate the performance of the clustering procedure.

The input parameters of the clustering procedure *xi* and *tau* represent respectively the number of grid-cells of the subspace and the frequency of the number of objects in each unit. A natural number can be used as a value of the parameter *xi*. The value of the parameter *tau* should be in the range [0; 1]. After multiple tests, the input parameters of the clustering procedure *xi* and *tau* are set to 2 and 0.04, respectively (Fig. 6).

The choice of appropriate values for these parameters is challenging for developing a clustering procedure with good results. Hybrid meta-heuristic algorithms for parameters identification problem [16,28] could be used in future studies for this purpose.

The clusters, results of the clustering procedure applied on the training set are visualized in Fig. 7, Fig. 8. The *Cluster model visualization* operator presents the clusters as circles. The size of the subspace cluster is represented as a diameter of the circle. The dimensionality of the clusters is encoded in the colour of the circle. Therefore, similar subspace clusters, clusters of similar dimensionality, or of similar size could be identified.

There are 5 dimensions and 31 clusters representing the user knowledge level according to the input dataset. Subspace clustering algorithms localize the search for relevant dimensions, allowing them to find clusters which exist in multiple, possibly overlapping subspaces. CLIQUE algorithm can found clusters in the same, overlapping, or disjoint subspaces [26]. The clusters obtained by the current clustering procedure are overlapping.

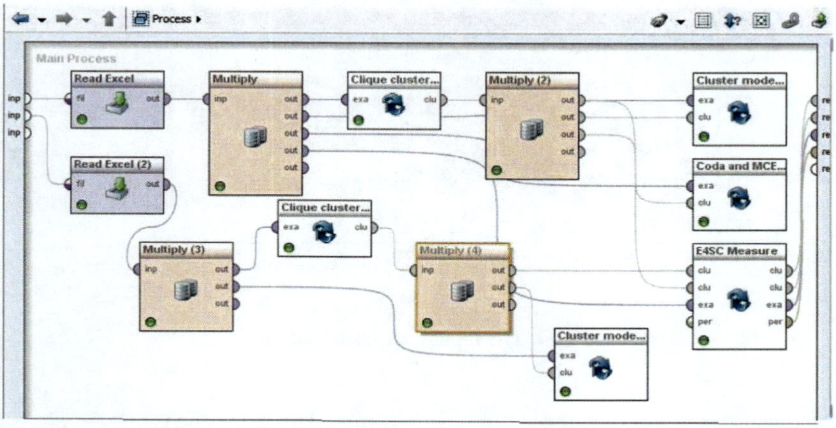

Fig. 5. High-dimensional subspace clustering with CLIQUE algorithm in RapidMiner

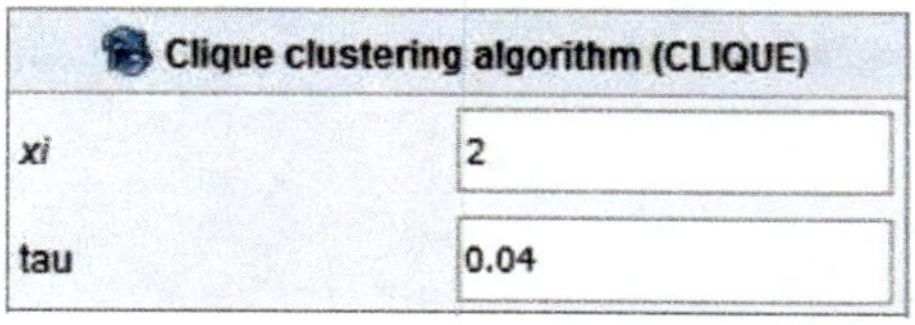

Fig. 6. Parameters of the clustering procedure

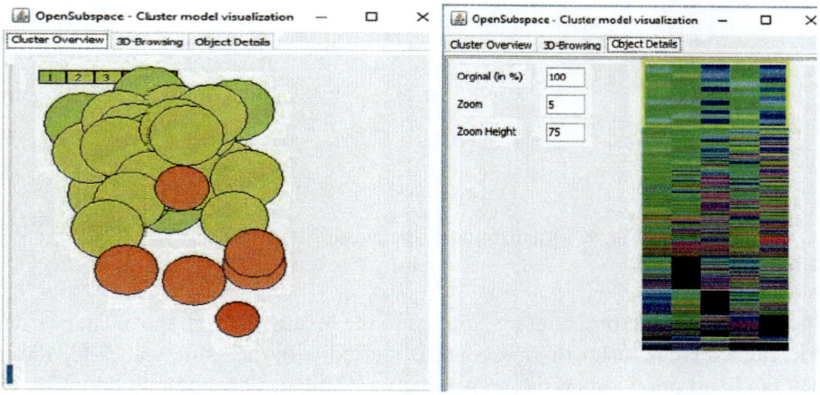

Fig. 7. Visualization of the clusters (training set)

Detailed view of the objects is shown in Fig. 8. Each object of the clusters can be presented with a degree of interestingness and additional information.

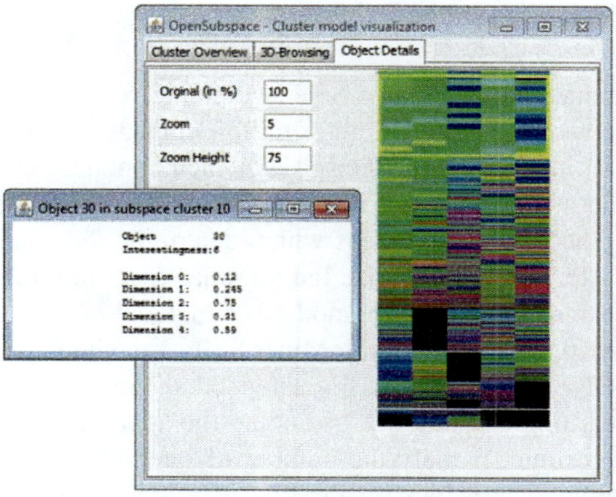

Fig. 8. Detailed view of the objects

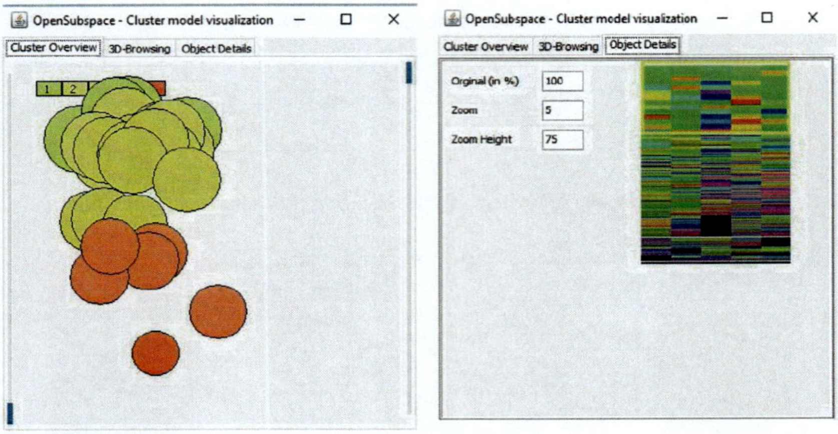

Fig. 9. Visualization of the clusters (testing set)

When the clustering procedure is applied to the testing dataset, the clusters are visualized in Fig. 9. This clustering procedure is started with the same values for the input parameters *xi* and *tau*, 2 and 0.04, respectively. 31 clusters are found. Obviously, overlapping clusters exist again. Each cluster presents the particular degree of success of the users in the educational process.

The performance of the current clustering procedure is measured by the operator *E4SC Measure*. Usually, the performance of such procedures is estimated in the range [0; 1]. Therefore, the value 0.726 is considered a good result. However, there is some space for improvement in future studies.

5 Conclusion

The GN-model presented and simulated here is a simple version, designed to describe and to follow the steps of the CLIQUE algorithm. It emphasizes on the first two major step of the algorithm, identifying the subspaces which contain clusters and determining the dense units in all subspaces of interests. These steps are described using three transitions and ten places. The next steps which determine the connected dense units in all subspaces of interests and generate the minimal description for the clusters, are described in the last transition. The GN-model is simulated in GN IDE. The results of the simulation are verified using the RapidMiner Software with OpenSubspace plugin.

This simulated GN-model can be easily extended and improved in future. It can be designed as a self-learning GN in terms of finding the appropriate values for the input arguments of the algorithm, namely the number of intervals for the partitioning and the frequency threshold. The values for the first iteration of the algorithm have to be specified, but later they can be evaluated based on the results of the previous iteration. That way the results of several parallel iterations of the algorithm can be observed during the GN simulation.

Future research will be related also to the development and the simulation of GN-models of other clustering methods.

Acknowledgements. The work on Sect. 3 and Sect. 4 is supported by the Bulgarian National Science Fund under Ref. No. DN-02-10 "New Instruments for Knowledge Discovery from Data, and their Modelling". The GN-model in Sect. 2 is supported by the "Asen Zlatarov" University project under Ref. No. NIX-423/2019 "Innovative methods for extracting knowledge management".

References

1. Aggarwal, C.C., Reddy, C.K.: Data Clustering: Algorithms and Applications. Chapman and Hall/CRC (2013)
2. Agrawal, R., Gehrke, J., Gunopulos, D., Raghavan, P.: Automatic subspace clustering of high dimensional data for data mining applications, vol. 27, no. 2, pp. 94–105. ACM (1998)
3. Ankerst, M., Breunig, M.M., Kriegel, H., Sander, J.: OPTICS: ordering points to identify the clustering structure. In: ACM SIGMOD International Conference on Management of Data, pp. 49–60. ACM Press (1999)
4. Angelova, N., Todorova, M., Atanassov, K.: GN IDE: implementation, improvements and algorithms. Comptes rendus de l'Académie bulgare des Sciences **69**(4), 411–420 (2016)
5. Atanassov, K.: Generalized index matrices. Comptes rendus de l'Academie Bulgare des Sciences **40**(11), 15–18 (1987)
6. Atanassov, K.: Generalized Nets. World Scientific, Singapore (1991)
7. Atanassov, K.: On Generalized Nets Theory. "Prof. M. Drinov" Academic Publ. House, Sofia (2007)
8. Atanassov, K.: Index Matrices: Towards an Augmented Matrix Calculus. Studies in Computational Intelligence, vol. 573. Springer, Cham (2014). https://doi.org/10.1007/978-3-319-10945-9
9. Atanassov, K., Sotirova, E.: Generalized Nets. Prof. M. Drinov Academic Publishing House, Sofia (2017). (in Bulgarian)
10. Bureva, V., Sotirova, E., Atanassov, K.: Hierarchical generalized net model of the process of clustering. In: Issues in Intuitionistic Fuzzy Sets and Generalized Nets. School of Information Technology, vol. 1, pp. 73–80 (2014)
11. Bureva, V., Sotirova, E., Atanassov, K.: Hierarchical generalized net model of the process of selecting a method for clustering. In: 15th International Workshop on Generalized Nets, Burgas, pp. 39–48 (2014)
12. Bureva, V., Sotirova, E., Popov, S., Mavrov, D., Traneva, V.: Generalized net of cluster analysis process using STING: a statistical information grid approach to spatial data mining. In: Christiansen, H., Jaudoin, H., Chountas, P., Andreasen, T., Legind Larsen, H. (eds.) Flexible Query Answering Systems. FQAS 2017. Lecture Notes in Computer Science, vol. 10333, pp. 239–248 (2017)
13. Bureva, V., Popov, S., Traneva, V., Tranev, S.: Clustering in QUEst. Int. J. Bioautomation **23**(2), 131–138 (2019)
14. Ester, M., Kriegel, H., Sander, J., Xu, X.: A density-based algorithm for discovering clusters in large spatial databases with noise. In: Proceedings of the Second International Conference on Knowledge Discovery and Data Mining (KDD 1996), pp. 226–231. AAAI Press (1996)
15. Everitt, B., Landau, S., Leese, M., Stahl, D.: Cluster Analysis, 5th edn. Wiley, Hoboken (2011)

16. Fidanova, S., Paprzycki, M., Roeva, O.: Hybrid GA-ACO algorithm for a model parameters identification problem. In: Proceedings of the Federated Conference on Computer Science and Information Systems (FedCSIS), WCO, Poland, pp. 413–420 (2014). https://doi.org/10.15439/2014F373
17. Guha, S., Rastogi, R., Shim, K.: Cure: an efficient clustering algorithm for large databases. ACM SIGMOD Rec. **27**, 73–84 (1998)
18. Jain, A., Dubes, R.: Algorithms for Clustering Data. Prentice Hall, Upper Saddle River (1988)
19. Kahraman, H., Sagiroglu, S., Colak, I.: Developing intuitive knowledge classifier and modeling of users' domain dependent data in web. Knowl. Based Syst. **37**, 283–295 (2013)
20. Kahraman, H.: Designing and application of web-based adaptive intelligent education system. Gazi University Ph.D. thesis, Turkey, pp. 1–156 (2009)
21. Kaufman, L., Rousseeuw, P.J.: Clustering by means of Medoids. Statistical Data Analysis Based on the L_1 – Norm and Related Methods. Edited by Y. Dodge, pp. 405–416. North-Holland (1987)
22. Machine Learning Repository. http://archive.ics.uci.edu/ml/datasets/User+Knowledge+Modeling. Accessed 15 May 2019
23. MacQueen, J.B.: Some methods for classification and analysis of multivariate observations. In: Proceedings of 5th Berkeley Symposium on Mathematical Statistics and Probability, pp. 281–297. University of California Press (1967)
24. Müller, E., Assent, I., Günnemann, S., Jansen, T., Seidl, T.: OpenSubspace: an open source framework for evaluation and exploration of subspace clustering algorithms in WEKA. In: Proceedings of the 1st Open Source in Data Mining Workshop (OSDM 2009) in Conjunction with 13th Pacific-Asia Conference on Knowledge Discovery and Data Mining (PAKDD 2009), Bangkok, Thailand, pp. 2–13 (2009)
25. Müller, E., Günnemann, S., Assent, I., Seidl, T.: Evaluating clustering in subspace projections of high dimensional data. In: Proceedings of the 35th International Conference on Very Large Data Bases (VLDB), Lyon, France, pp. 1–12 (2009). http://dme.rwth-aachen.de/OpenSubspace/. Accessed 15 May 2019
26. Parsons, L., Haque, E., Liu, H.: Subspace clustering for high dimensional data: a review. SIGKDD Explor. **6**(1), 90–105 (2004)
27. RapidMiner: Data Science Platform. https://rapidminer.com/. Accessed 15 May 2019
28. Roeva, O., Fidanova, O., Vassilev, P., Gepner, P.: InterCriteria analysis of a model parameters identification using genetic algorithm. In: Proceedings of the Federated Conference on Computer Science and Information Systems. Annals of Computer Science and Information Systems, vol. 5, pp. 501–506 (2015). https://doi.org/10.15439/2015F223
29. Sibson, R.: SLINK: an optimally efficient algorithm for the single-link cluster method. Comput. J. **16**(1), 30–34 (1973)
30. Subspace Clustering Extensions for Knowledge Discovery Frameworks. http://dme.rwth-aachen.de/en/KDD-SC. Accessed 15 May 2019
31. Wang, W., Yang, J., Muntz, R.R.: STING: a statistical information grid approach to spatial data mining. In: Proceedings of 23rd International Conference on Very Large Data Bases, VLDB 1997, Greece, Athens, pp. 186–195 (1997)

Representation of Initial Temperature as a Function in Simulated Annealing Approach for Metal Nanoparticle Structures Modeling

Vladimir Myasnichenko[1], Stefka Fidanova[2], Rossen Mikhov[2], Leoneed Kirilov[2](✉), and Nickolay Sdobnyakov[1]

[1] Tver State University, Tver, Russia
viplabs@yandex.ru
[2] Institute of Information and Communication Technologies, Bulgarian Academy of Sciences, Sofia, Bulgaria
stefka@parallel.bas.bg, l_kirilov_8@abv.bg

Abstract. Very important in the study of the thermodynamic characteristics of nanostructures (melting/crystallization) is the structure of nanoclusters. The properties of the nanoparticles can be predicted if we know the mechanism of the formation and the dynamic of changes in the internal structure. The problem to find the stable structure of nanoparticle is NP-hard and it needs development of special methods coming from Artificial Intelligence to be solved.

In this paper we apply Simulated Annealing Method to find approximate solution. The proposed algorithm is designed for metal nanoparticle structures optimization. This problem has an exceptional importance in studying the properties of nanomaterials. The problem is represented as a global optimization problem. The most important algorithm parameter is the temperature. The main focus in this paper is on representation of the initial temperature as a function. Thus the algorithm parameters will be closely related with the input data.

The experiments are performed with real data as follows. One set of mono metal clusters is chosen for investigation: Silver (Ag) where the size of clusters for Ag varies from Ag150 (atoms) to Ag3000 (atoms). Several dependencies are derived between the number and configuration of atoms in the cluster on one hand, and temperature representation and stopping rule on the other hand.

Keywords: Simulated annealing · Metal nanoparticle modeling · Metaheuristics

1 Introduction

In the past years, both theoretical and experimental research is conducted for studying nanostructures very extensively. One of the reasons is that they possess

I. Dimov and S. Fidanova (Eds.): HPC 2019, SCI 902, pp. 61–72, 2021.
https://doi.org/10.1007/978-3-030-55347-0_6

very interesting and in some cases very different properties in comparison with usual materials. The diversity of properties becomes even greater when nanostructures of two elements are considered. In particular, metal nanostructures are very important for their applications in areas such as electronics, medicine, biology.

Further, it is proven nowadays that the structure of nanoclusters has significant influence in defining the physico-chemical properties of the considered clusters. Understanding the mechanisms of formation and dynamics of changes in the internal structure of nanoclusters will allow predicting of these properties. For this purpose theoretical models have to be developed and solved. The methods for solving can be classified into two groups. *Ab initio* methods exploit the molecular Schrödinger equation in the form of the Born–Oppenheimer approximation. These methods are suitable for nanoparticles with a small number of atoms. For nanoparticles of sizes from several hundred to several thousand atoms, semi-empirical modeling methods have to be applied [8,16,19].

The potential energy surface (PES) is a reflection of the structure, dynamics, and thermodynamics of any system within the Born–Oppenheimer approximation for a particular electronic state [22]. In this way the problem for finding stable configuration(s) of nanoparticles is converted into a Global Minimization (GM) problem for the potential energy function. In computational complexity theory, global optimization problems are NP-hard, meaning that they cannot be solved in polynomial time. Therefore developing of new efficient approaches for minimizing PES is an important task.

In this paper we present a study of the computational behavior of a Simulated Annealing approach for metal nanoparticle structures modeling developed by the authors [14,15]. The initial temperature, which is a main parameter, is presented as a function of the initial cluster configuration. In this way the temperature is tightly bounded with the initial solution. Also there is no need to guess/predict what starting temperature is to be set.

2 Literature Review

In [11], a method TOP (topological energy expression) for global optimization of chemical ordering in bimetallic nanoparticles is proposed. The method is based on DFT (density functional theory). The TOP method allows effective identification of the global minimum of PES. The authors claim the method is efficient for nanoparticles with thousands of atoms. In [10] and [20], the method is applied for studying Pd-Au and Pd-Rh nanoparticles with different sizes. The study of the surface/landscape of potential energy (PES) for a given set of atoms [21] underlies the popular methods for predicting cluster structures today. This surface is a graph of potential energy as a function of atomic coordinates. The greatest difficulty in the global search is related to the complex topography of the PES [1], consisting of many funnels. Therefore, even relatively small clusters and molecules have a large number of configurations corresponding to local minima. However, the global minimum belongs to a subset of local minima. Therefore, global optimization algorithms most often search not across the entire PES, but only among the phase points corresponding to local minima. A series of optimization problems in the physics of materials are characterized by a "curse of dimension"—they have a very high computational complexity. In particular, multiple funnels of the energy landscape significantly impede the global optimization of any nanoscale system of atoms where there is not a single strongly dominating structural type [3]. The application of classical deterministic optimization methods is also difficult due to the multi-criteria and non-linear constraints.

In the study of structural transformations in nanoparticles, their specific features as compared with the corresponding bulk phases are of interest from both scientific and applied points of view, including the possibility of creating new metals with predefined specific properties. The basic step of learning metal clusters is to locate their stable geometrical structures. In the case of metallic nanoparticles, the reliability of the simulation results is primarily determined by the interatomic interaction potential used. The analytical form of potential energy is created in such a way that it contains a certain set of independent parameters. Using these parameters, the potentials are fitted to the experimental data or to *ab initio* data of the calculations. When describing metal nanoclusters, it is legitimate to use the multiparticle form of the potential energy function, which takes into account the band nature of the metal bonds. The main advantage of the multiparticle interaction potential, as compared with the simple pair potential, is its ability to better reproduce some of the main features of metallic systems. A relatively simple way to describe the atomic and electronic structure is tight-binding (TB) [2,7], in which the ion-ion interaction is described taking into account the band nature of the bond and is supplemented by a short-range pair repulsion potential. The method is based on the fact that a significant group of properties of transition metals can be completely determined on the basis of the density of states of external d-electrons. Despite the simple functional form, the strong-binding model describes well enough the elastic properties, defect characteristics and melting for a wide range of fcc and hcp metals. In our opinion, this scheme is one of the most suitable for computer analysis of small metal particles consisting of several hundred or thousands of atoms. There exist different approaches: the Gupta potential function has been widely applied to the study of face-centered-cubic structures [2], hexagonal closed packed [12], transition metals and their alloy clusters [5]. The stable geometrical structure of six metallic clusters is optimized by a dynamic lattice search method [23].

In nanocluster science, silver clusters have become a hot topic because of their application in catalysis, photography, electronic materials and production of metal materials [4]. Silver nanoparticles can be prepared by physical, chemical, and biological synthesis [6]. The structure of silver clusters needs to be optimized according the potential function which gives the stable structure of the metal.

3 Simulated Annealing Method and Temperature Function

Simulated Annealing (SA) is a nature inspired method for finding approximate solution of hard optimization problems. It is a flexible method based on local search and is successfully applied on various optimization problems. The fundamental idea is to allow moves resulting in solutions of worse quality than the current solution in order to escape from local optimums. First it is proposed by Kirkpatrick [9].

SA is a stochastic metaheuristic method which explores the solution space using a stochastic hill-climbing process. SA is inspired by the Metropolis scheme [13]. The method imitates the annealing process in metals and glass. An initial state of a thermodynamic system is chosen with energy E and temperature T, the initial configuration is perturbed and the change of energy dE is computed. The current state of the thermodynamic system is analogous to the current solution to the combinatorial problem, the energy equation for the thermodynamic system is analogous to the objective function, and the ground state is analogous to the global minimum. The algorithm starts by generating an initial solution and by initializing the so-called temperature parameter T. The temperature is decreased during the search process, thus at the beginning of the search the probability of accepting uphill moves is high and it gradually decreases.

In order to implement SA a number of decisions have to be made. The first ones, are the definition of the solution space Q and neighborhood structure I, the form of the objective function $C(V)$ and the way in which a starting solution V is obtained. Secondly, the generic choices which govern the working of the algorithm itself, they are mainly concerned with the components of the cooling parameters: control parameter T (temperature) and its initial starting value, the cooling rate F and the temperature update function, the Markov's chain (number of iterations between decreases) L and the condition under which the system will be terminated. The performance of the achieved result is highly dependent on the right choice of both specific and generic decisions. The structure of the SA algorithm is shown in Fig. 1. In our case $C(V)$ is the total potential energy of the system E.

The algorithm starts by generating an initial solution and by initializing the temperature parameter T. We generate the initial solution in a random way. We apply SA method developed by the authors in their previous work [14].

Interaction between atoms is calculated using the multi-particle tight-binding potential of Gupta - Cleri & Rosato [2]. The total potential energy of the system is defined as follows:

$$E = \sum_i \left(\sum_{j \neq i} E_{ij}(a,b) - \sqrt{\sum_{j \neq i} B_{ij}(a,b)} \right) \tag{1}$$

$$E_{ij}(a,b) = A_{ab} exp \left(-p_{ab} \left(\frac{r_{ij}}{r_{0,ab}} - 1 \right) \right) \tag{2}$$

```
Simulated Annealing

t:=0;
Initialize(V);
Initialize(T);
Initialize(L);
Initialize(F);
while not end_condition(t,V) do
      while not cooling_condition(t)
            V' := Choose_neighbor(V);
            Δ = C(V')-C(V);
            if Δ < 0 then
               V := V';
            end if
            else
               Generate_random number Θ
               if e^(-Δ/T) > Θ
                      V:=V';
               end if
            end else
            t := t+1;
      end while
      Cooldown(T);
end while
```

Fig. 1. Pseudocode for SA

$$B_{ij}(a,b) = \xi_{ab}^2 exp\left(-2q_{ab}\left(\frac{r_{ij}}{r_{0,ab}} - 1\right)\right), \tag{3}$$

where i ranges over all atoms; j ranges over all atoms other than i but within distance R_{cut} from i; a and b represent the species of the atoms i and j; $E_{ij}(a,b)$ is the repulsive component of the potential due to the atoms i and j; $B_{ij}(a,b)$ is the binding component of the potential due to the atoms i and j; r_{ij} is the distance between the atoms; $r_{0,ab}$, A_{ab}, p_{ab}, ξ_{ab}, q_{ab} are parameters that depend only on the species of the atoms. R_{cut} is the maximum distance beyond which the interaction is assumed to be zero.

The algorithm takes as input a list of nodes, given by their Cartesian coordinates. Each node can either be empty or contain an atom. The distance between adjacent nodes in the input data may slightly vary (within 15%). We keep the nodes in two arrays N and A, where N gives the index of a node into A and A gives the index of a node into N. N is sorted in input order, while A is sorted to begin with the atoms and end with the holes. This allows us to query information about nodes and atoms, select atoms at random, add, remove and move atoms around, all in constant time. Before starting the SA algorithm we do some preprocessing which is described in details in [14].

The right parameter settings play crucial role for good algorithm performance. The main SA parameters are initial temperature T_0 and cooling rate. Most authors use fixed value for T_0, which is not related to the solved problem [17,18]. Others start from very high temperature close to ∞ and decrease it very quickly. Their goal is to speedup the search process.

Our main idea is the initial temperature to be related with the input data. We set the initial temperature T_0 to be a linear function of the cost of the initial solution, $T_0 = K(C(V_0) + A)$, where K and A are parameters. We decided T_0 to be a linear function of $C(V_0)$, because Δ is a linear function of $C(V)$—see (4). The acceptance probability depends on the difference Δ between the costs of the current and candidate solutions. Thus, if T is large, the probability of accepting the candidate solution is higher. The main problem is the meaning of large. For example: Let $C(V) = 100$ and $T = 20$ thus T is 20% of the value of $C(V)$ and T is large with respect to $C(V)$. Let $C(V) = 100\,000$ and $T = 20$, thus T is 0.02% of the value of $C(V)$ and T is not large with respect to $C(V)$. For most problems the expected cost of the optimal solution is unknown, so our proposal is to automatically set the initial temperature to be proportional to the initial solution cost. Thus the value of the initial temperature will be related with the input data:

$$\frac{\Delta}{T} = \frac{C(V') - C(V)}{T} \tag{4}$$

4 Computational Results

The performance and successes of the SA highly depends on proper choice of the initial temperature. It needs to be high enough to allow the system to explore different areas and configurations in a search space, without falling into a local minimum too early. If it is set too high, the algorithm might need too many iterations to find good solution. On the other hand, a lower value of T_0 will allow running more trials. Thus finding appropriate starting temperature related with the input data is not trivial.

In order to find initial temperature in an automatic way and it to be closely related with the input data, we propose the initial temperature to be a linear function of the value of the initial solution. For our tests we chose icosahedral lattice in two different sizes, 1415 nodes and 10170 nodes. On each of those lattices, we investigated mono-metal clusters of silver (Ag). From statistical point of view we ran 30 trials for each variant, taking the average. We tested on lattices with 1415 nodes and 10179 nodes. For the smaller lattice we used from 150 to

Table 1. Algorithm and problem parameters

Lattices	Small variant 1415; Large variant 10179
Nanoclusters	17 variants of silver: Ag150 to Ag310 with step 10; 10 larger variants of silver: Ag620, Ag940, Ag1100, Ag1260, Ag1420, Ag1740, Ag2060, Ag2380, Ag2700, Ag3000 (placed only on the large lattice)
R_{cut}	$\sqrt{5.5} \times r_0$
Cooling speed	-7×10^-5
Temperature change	Once every 10,000 iterations

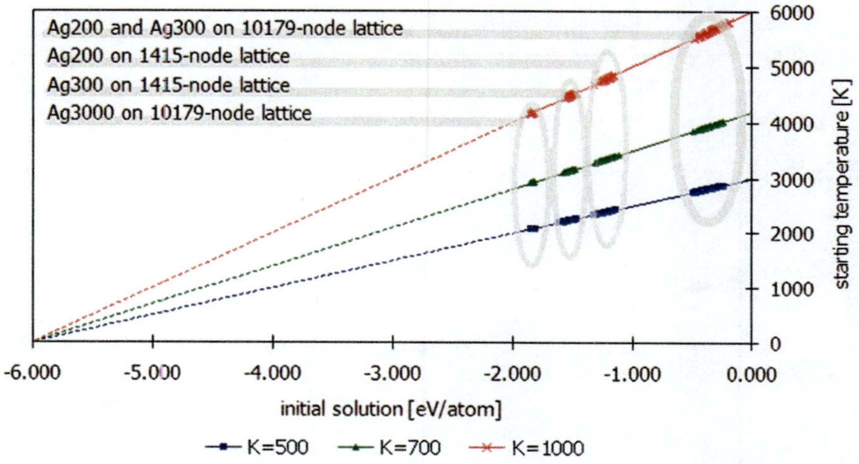

Fig. 2. Actual T_0 values for the trials with Ag200, Ag300 and Ag3000 (total 150 data points plotted)

300 atoms and for the larger lattice we used up to 3000 atoms. The formula for the initial temperature is as follows:

$$T_0 = K(C(V_0) + A) \tag{5}$$

We tested with the following values for the coefficients: $K \in \{500, 700, 1000\}$ and $A = 6$. The initial temperatures determined in this way for some of the examples are illustrated on Fig. 2. It can be seen that smaller initial temperatures are selected for initial solutions with lower energy.

All results are summarized on Figs. 3, 4 and 5. On the horizontal axis is the number of used atoms and on the vertical axis is the potential energy of the system. The achieved energy when $K = 500$ is with blue color, when $K = 700$ is with green color and when $K = 1000$ is with red color. The achieved results are not so sensitive to the value of the parameter A. The role of the parameter A is to guarantee that the value of the temperature is positive. The other algorithm parameters are shown on Table 1.

Regarding the lattice with 1415 nodes (Fig. 3) we observe that the achieved energy is very similar with all values of the coefficient K. This lattice is too small and the possibilities for variation of the solutions are small. Therefore the impact of the temperature is low.

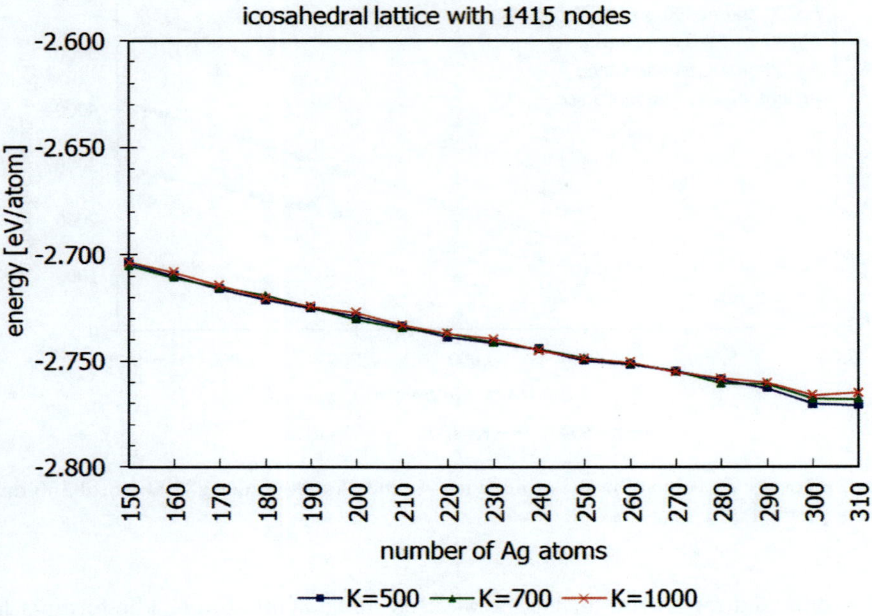

Fig. 3. Influence of the temperature for lattice with 1415 nodes

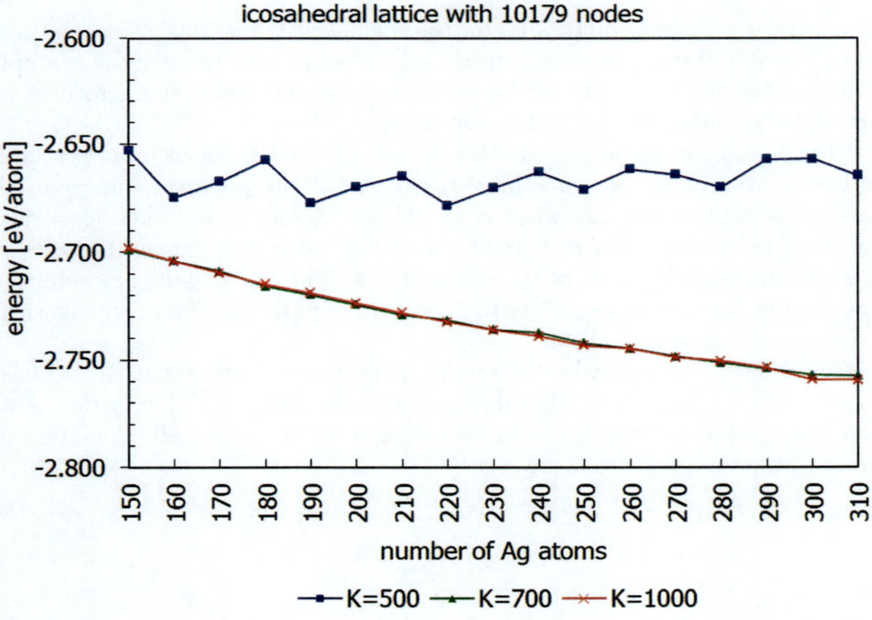

Fig. 4. Influence of the temperature for lattice with 10179 nodes and up to 310 atoms

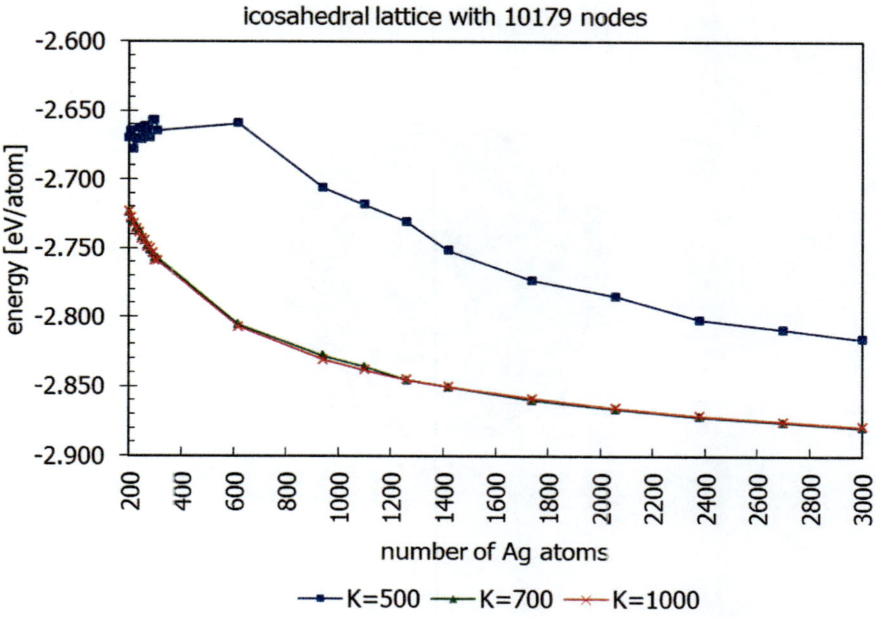

Fig. 5. Influence of the temperature for lattice with 10179 nodes and up to 3000 atoms

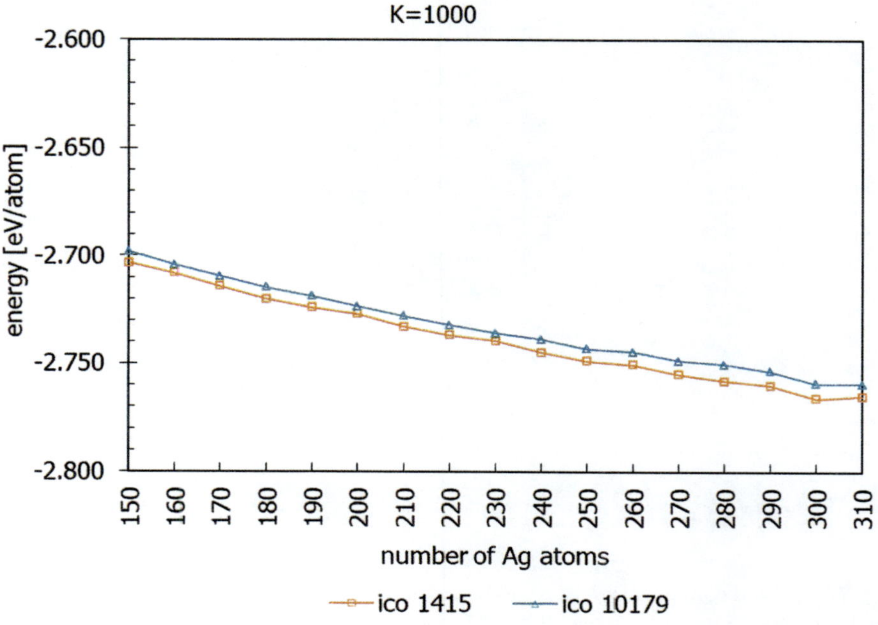

Fig. 6. Comparison between the two lattices

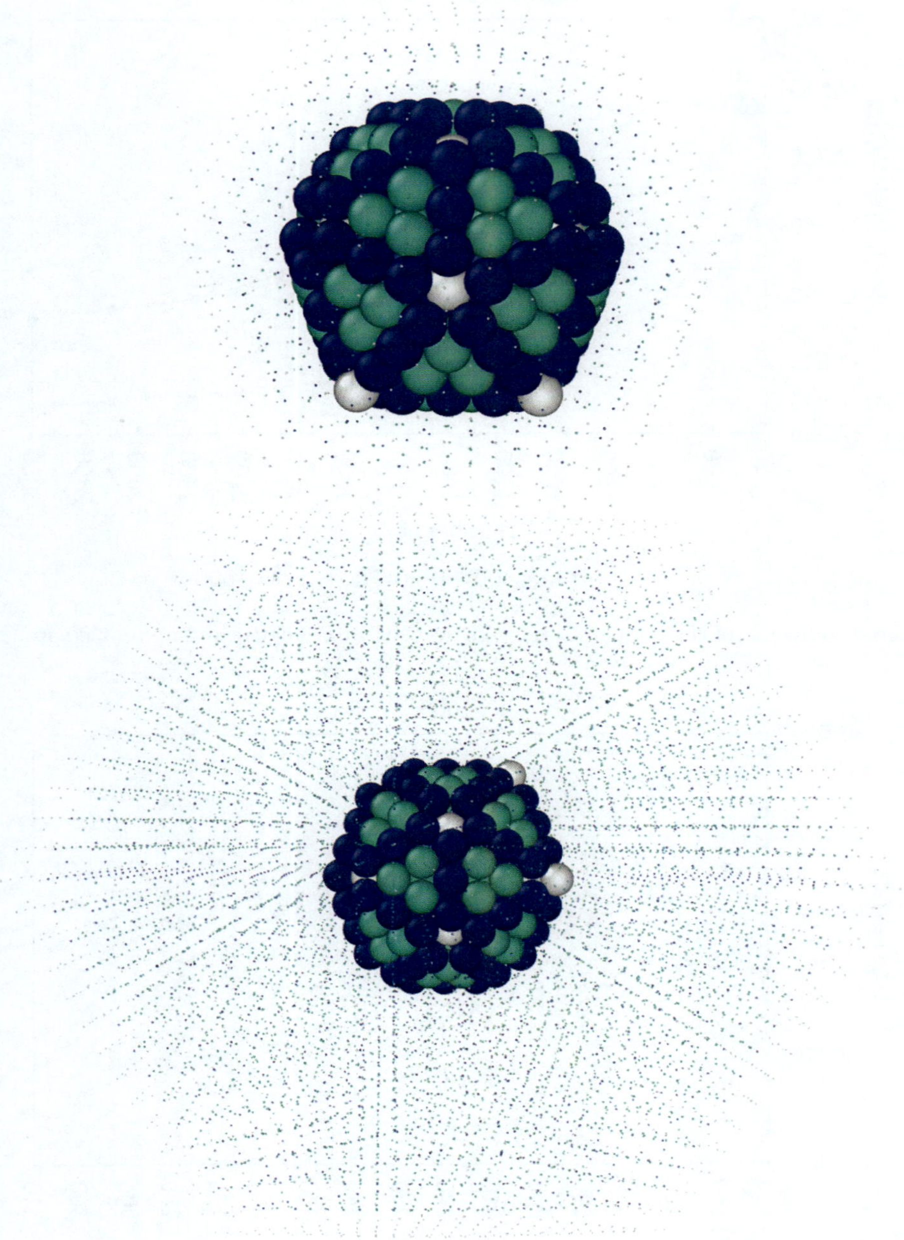

Fig. 7. Optimal configuration of Ag300 obtained on the 1415-node lattice (top) and on the 10179-node lattice (bottom). Green represents fcc atoms, blue – hcp atoms, white – fivefold axes. Empty nodes are represented by dots.

On Figs. 4 and 5 are shown the results when the used lattice has 10179 nodes and the number of atoms is up to 310 and up to 3000 respectively. We observe that when $K < 700$, the achieved energy is much higher than in the case of $K \geq 700$. When $K \geq 700$ the results are similar. Thus we can conclude that the value of the coefficient K needs to be greater than or equal to 700, to guarantee good performance of the algorithm. When the initial temperature is too high the algorithm needs more iterations to achieve good solutions. Thus the optimal value of K to find the best solution with minimal number of iterations is $K = 700$.

On Fig. 6 is a comparison between the two lattices (smaller and larger one), using the same number of atoms. With orange color is the energy achieved with the 1415-node lattice and with blue color is the energy achieved with the 10179-node lattice. We observe that with the smaller lattice we achieve solutions with less energy.

Figure 7 shows an example of an obtained optimal configuration on both lattices.

5 Concluding Remarks

In this paper we apply SA algorithm to find the optimal nanostructure of metals. The algorithm performance is affected by algorithm parameters. The most important between them is the initial temperature. Our idea is the initial temperature to be related with the input data. We represent the initial temperature as a linear function of the energy of the initial solution. We tested our idea on silver atoms. In our future work we plan to prepare tests with atoms of other metals and to find relationship between the atoms of different metals and parameters of temperature representation.

Acknowledgments. The co-authors from Tver University are supported by the Russian Foundation for Basic Research project No. 20-37-70007 and No. 18-03-00132; by the Ministry of Science and Higher Education of the Russian Federation in the framework of the State Program in the Field of the Research Activity (project no. 0817-2020-0007) and the co-authors from IICT are supported by National Scientific Program "Information and Communication Technologies for a Single Digital Market in Science, Education and Security (ICTinSES)", Ministry of Education and Science – Bulgaria; Grant No BG05M2OP001-1.001-0003, financed by the Science and Education for Smart Growth Operational Program and co-financed by the European Union through the European structural and Investment funds.

References

1. Carr, J.M., Mazauric, D., Cazals, F., Wales, D.J.: Energy landscapes and persistent minima. J. Chem. Phys. **144**(5), 054109 (2016)
2. Cleri, F., Rosato, V.: Tight-binding potentials for transition metals and alloys. Phys. Rev. B. **48**(1), 22–33 (1993)
3. Doye, J.P.K.: Physical perspectives on the global optimization of atomic clusters. In: Global Optimization, vol. 85, pp. 103–139. Springer US (2006)

4. Huang, W.Q., Lai, X.J., Xu, R.C.: Structural optimization of silver clusters from Ag141 to Ag310 using a modified dynamic lattice searching method with constructed core. Chem. Phys. Lett. **507**(1), 199–202 (2011)
5. Husic, B.E., Schebarchov, D., Wales, D.J.: Impurity effects on solid-solid transitions in atomic clusters. NANO **8**, 18326–18340 (2016)
6. Iravani, S., Korbekandi, H., Mirmohammadi, S.V., Zolfaghari, B.: Synthesis of silver nanoparticles: chemical, physical and biological methods. Res. Pharm. Sci. **9**(6), 385–406 (2014)
7. Harrison, W.A.: Tight-binding methods. Surf. Sci. **299**(300), 298–310 (1994)
8. Jäger, M., Schäfer, R., Johnston, R.L.: First principles global optimization of metal clusters and nanoalloys. Adv. Phys. X. **3**(1), 1516514 (2018)
9. Kirkpatrick, S., Gellat, C.D., Vecchi, P.M.: Optimization by simulated annealing. Science **220**, 671–680 (1983)
10. Kovacs, G., Kozlov, S., Neyman, K.: Versatile optimization of chemical ordering in bimetallic nanoparticles. J. Phys. Chem. C **121**, 10803–10808 (2017)
11. Kozlov, S.M., Kovacs, G., Ferrando, R., Neyman, K.M.: How to determine accurate chemical ordering in several nanometer large bimetallic crystallites from electronic structure calculations. Chem. Sci. **6**, 3868–3880 (2015)
12. Li, X.J., Fu, J., Qin, Y., Hao, S.Z., Zhao, J.J.: Gupta potentials for five HCP rare earth metals. Comput. Mater. Sci. **112**, 75–79 (2016)
13. Metropolis, N., Rosenbluth, A., Rosenbluth, M., Teller, A., Teller, E.: Equation of state calculations by fast computing machines. J. Chem. Phys. **21**(6), 1087–1092 (1953)
14. Myasnichenko, V., Kirilov, L., Mikhov, R., Fidanova, S., Sdobnyakov, N.: Simulated annealing method for metal nanoparticle structures optimization. In: Georgiev, K., Todorov, M., Georgiev, I. (eds.) Advanced Computing in Industrial Mathematics. Studies in Computational Intelligence, vol. 793, pp. 277–288. Springer (2019)
15. Myasnichenko, V., Sdobnyakov, N., Kirilov, L., Mikhov, R., Fidanova, S.: Monte Carlo approach for modeling and optimization of one-dimensional bimetallic nanostructures. In: Nikolov, G., Kolkovska, N., Georgiev, K. (eds.) Numerical Methods and applications. Lecture Notes in Computer Science, vol. 11189, pp. 133–141. Springer (2019)
16. Myshlavtsev, A.V., Stishenko, P.V.: Modification of the Metropolis algorithm for modeling metallic nanoparticles. Omsk Sci. Newspap. **1**(107), 21–25 (2012). (in Russian)
17. Rene, V.V.: Applied Simulated Annealing. Springer, Berlin (1993)
18. Saleh, H.A., Dare, P.: Effective heuristics for the GPS survey network of malta: simulated annealing and tabu search techniques. J. Heuristics **7**(6), 533–549 (2001)
19. Schön, J.: Nanomaterials – what energy landscapes can tell us. Proc. Appl. Ceram. **9**(3), 157–168 (2015)
20. Vega, L., Aleksandrov, H.A., Neyman, K.: Using density functional calculations to elucidate atomic ordering of Pd-Rh nanoparticles at sizes relevant for catalytic applications. Chin. J. Catal. **40**, 1749–1757 (2019)
21. Wales, D.J., Miller, M.A., Walsh, T.R.: Archetypal energy landscapes. Nature **394**(6695), 758–760 (1998)
22. Wales, D.: Energy landscapes and structure prediction using basin-hopping. In: Oganov, A. (ed.) Modern Methods of Crystal Structure Prediction, pp. 29–54. WILEY-VCH Verlag & Co. KGaA, Weinheim (2011)
23. Wu, X., Sun, Y.: Stable structures and potential energy surface of the metallic clusters: Ni, Cu, Ag, Au, Pd, and Pt. J. Nanopart. Res. **19**, 201 (2017)

An Intuitionistic Fuzzy Zero Suffix Method for Solving the Transportation Problem

Velichka Traneva(✉) and Stoyan Tranev

"Prof. Asen Zlatarov" University, "Prof. Yakimov" Blvd, 8000 Burgas, Bulgaria
veleka13@gmail.com, tranev@abv.bg

Abstract. The transportation problem (TP) is a special class of linear programming problem. Its main objective is to determine the amount of a shipment from source to destination to maintain the supply and demand requirements at the lowest total transportation cost. The TP was originally developed by Hitchcock in 1941. There are different approaches for solving the TP with crisp data, but in many situations the cost coefficients, the supply and demand quantities of the TP may be uncertain. To overcome this Zadeh introduce fuzzy set concepts to deal with an imprecision and a vagueness. The values of membership and non-membership cannot handle such uncertainty involved in real-life problems. Thus Atanassov in 1983 first proposed the concept of intuitionistic fuzzy sets (IFSs).

In this paper, a new type of the TP is formulated, in which the transportation costs, supply and demand quantities are intuitionistic fuzzy pairs (IFPs), depending on the diesel prices, road condition and other market factors. Additional constraints are formulated to the problem: upper limits to the transportation costs for delivery. The main contribution of the paper is that it proposes for the first time the Zero suffix method for finding an optimal solution of the intuitionistic fuzzy TP (IFTP), interpreted by the IFSs and index matrix (IM) concepts, proposed by Atanassov. The example in the paper illustrates the efficiency of the algorithm. An advantage of the proposed algorithm is that it can be applied to problems with imprecise parameters and can be extended in order to obtain the optimal solution for other types of multidimensional transportation problems in fuzzy environment.

1 Introduction

The basic TP originally developed by Hitchcock in 1941 [19]. Dantzig, in 1951, used simplex method to the TP [12]. The first overall, finished method for solving TP ("method of potentials") Kantorovich developed in 1949 [25]. Charnes and Cooper, in 1954, developed the Stepping stone method which provides an alternative way of determining the simplex method information [11].

In classical TP it is assumed that the transportation costs are exactly known. In real-life transportation problems, decision makers may face with many uncertainties on the cost of transportation due to changing weather, road conditions or other economic conditions. The costs are not stable since this imprecision may follow from the lack of exact information or data and uncertainty in judgment. Many TPs have been arisen in unclear circumstances. In 1965, Zadeh introduced the fuzzy set (FS) theory [49].

I. Dimov and S. Fidanova (Eds.): HPC 2019, SCI 902, pp. 73–87, 2021.
https://doi.org/10.1007/978-3-030-55347-0_7

As there is a hesitation in the parameters of TP many authors have solved this problem under intuitionistic fuzzy environment. In 1983, Atanassov proposed the IFSs [4], which is more reliable than the FSs proposed by Zadeh. The major advantage of IFS over FS is that IFS separates the degree of membership (belongingness) and the degree of non membership (non belongingness) of an element in the set.

There are different approaches for solving the TP with fuzzy and intuitionistic fuzzy data. Chanas et al., in 1984, presented a fuzzy linear programming model for solving TPs with crisp cost coefficients and fuzzy supply, and demand values [10]. Gen et al. have been given a genetic algorithm for solving a bicriteria solid TP with fuzzy numbers (FNs) [15]. Jimenez and Verdegay, in 1999, researched fuzzy Solid TP with trapezoidal FNs and proposed a genetic approach for finding a fuzzy solution [24]. Pandian and Natarajan, in 2010, proposed a new algorithm for finding an optimal solution for FTP where all the parameters are trapezoidal FNs (zero point method) [34]. Kaur and Kumar, in 2012, introduced fuzzy least cost method, fuzzy north west corner rule and fuzzy Vogel approximation method to solve FTP [28].

All the previous methods of finding the optimal solution of the TP requires to find initial basic feasible solution. This obtained solution is improved towards the optimal solution by checking the optimality criteria using the modified distribution method of Dantzig [12], Stepping stone method of Charnes and Cooper [11], the modified Stepping stone method of Shih [40], the simplex algorithm of Arsham and Kahn [2], and the dual matrix approach of Ji and Chu [23]. But Sudhakar et al., in 2010, suggested an approach called the Zero suffix method that gives the optimal solution for TPs without finding the initial basic feasible solution initially [41]. A comparative study on the TPs in fuzzy environment [32] has been made and the conclusion has been given that the Zero point method is better than both Vogel's Approximation method, and the modified distribution method. The Zero suffix method uses less iterations than Zero point method, whereas allocation and minimum optimum solution are the same [33]. In [22,41] have been applied Zero suffix method to solve FTP after the its converting into the crisp problem. The method proposed in [37], in 2018, is fuzzified version of Zero suffix method and applicable to FTPs. A modified Zero suffix method for finding an optimal solution for TPs have been defined in [39] and have been researched there that the proposed method is free from the problem of degeneracy, requires least number of iterations to reach optimality than the Zero suffix method for finding optimal solution of TPs. The total transportation cost obtained by the proposed method is better than or equal to that of the Zero suffix method. Gani and Abbas, in 2014 [14], and Kathirvel, and Balamurugun, in 2012 [26,27], proposed a method for solving TP in which all the parameters except transportation costs are represented by trapezoidal intuitionistic FNs (IFNs). Patil and Chandgude, in 2012, performed "Fuzzy Hungarian approach" for TP with trapezoidal FNs [35]. Jahihussain and Jayaraman, in 2013, presented a Zero suffix method for obtaining an optimal solution for FTPs with triangular and trapezoidal FNs [20,21]. Shanmugasundari and Ganesan, in 2013, proposed a fuzzy version of modified distribution algorithm and approximation method of Vogel to solve FTP with FNs [38]. Aggarwal and Gupta, in 2013, proposed algorithm for solving intuitionistic fuzzy TP (IFTP) with trapezoidal IFNs via ranking method [16]. Antony et al., in 2014, used Vogel's approximation method for solving TP with triangular IFNs [1]. Computationally simple method called "PSK method" for finding an optimal solution to IFTPs presented by Kumar and Hussain in 2015 [29]. A new method has been proposed in [13] to solve fully FTPs using the Hungarian and MODI algorithm in 2017.

The methods described above cannot be applied for solving IFTP, where the costs, the demand and supply are IFNs. In [17] has been given a detailed explanation of the shortcomings of these methods:

- Ranking procedures for ordering the triangular and trapezoidal FNs are proposed in some of methods for solving IFTPs. The proposed ranking functions cannot be used for ordering all IFNs.
- Fuzzy methods cannot be used for solving TPs where IFNs are used to represent the imprecise parameters of the TP under consideration.
- Some of the methods proposed a simple approach for solving fully IFTP in which costs, supply and demand all are triangular IFN. The intuitionistic fuzzy optimal value of the objective function must be non-negative, whereas some of the respective IFNs in these methods are negative.

In the papers [46,47], in 2016, we proposed for the first time the method of the potentials for finding an optimal solution of the TP, interpreted by the IFSs and IMs [5] concepts. An advantage of the algorithm is that it can be generalized and can be applied to both the TP with crisp parameters and with intuitionistic fuzzy ones.

In this paper, a new type of TP is formulated, in which the transportation costs, supply and demand quantities are IFPs, depending on the diesel prices [36], road condition and other market factors. Additional constraints are formulated to the problem: upper limits to the transportation costs for delivery. The main contribution of the paper is that it proposes for the first time the intuitionistic fuzzy modified Zero suffix method for finding an optimal solution of the IFTP, interpreted by the IFSs and IMs concepts, proposed by Atanassov. The rest of this paper is structured as follows: Sect. 2 describes the related concepts of the IMs and IFPs. In Sect. 3, we propose an algorithm for IFTP, based on the Zero suffix method, by using the concepts of IMs and IFSs. The effectiveness of the proposed method is demonstrated by an example in Sect. 4. Section 5 offers the conclusion and outlines some aspects for future research.

2 Basic Definitions

This section provides some remarks on intuitionistic fuzzy pairs from [6,8,9,42,44] and on index matrix tools from [7,48].

2.1. Short Notes on Intuitionistic Fuzzy Logic

The **IFP** is an object of the form $\langle a,b\rangle = \langle \mu(p), \nu(p)\rangle$, where $a,b \in [0,1]$ and $a+b \leq 1$, that is used as an evaluation of a proposition p [8,9]. $\mu(p)$ and $\nu(p)$ respectively determine the "truth degree" (degree of membership) and "falsity degree" (degree of non-membership).

Let us have two IFPs $x = \langle a,b\rangle$ and $y = \langle c,d\rangle$. We recall some basic operations

$$\neg x = \langle b,a\rangle; \qquad x \wedge_1 y = \langle \min(a,c), \max(b,d)\rangle;$$
$$x \vee_1 y = \langle \max(a,c)), \min(b,d)\rangle; \qquad x \wedge_2 y = x+y = \langle a+c-a.c, b.d\rangle;$$
$$x \vee_2 y = x.y = \langle a.c, b+d-b.d\rangle; \qquad \alpha.x = \langle 1-(1-a)^{\alpha}, b^{\alpha}\rangle (\alpha \in R);$$
$$x-y = \langle \max(0,a-c), \min(1,b+d,1-a+c)\rangle$$

and relations with IFPs

$$\begin{array}{llll} x \geq y & \text{iff } a \geq c \text{ and } b \leq d; & x \leq y & \text{iff } a \leq c \text{ and } b \geq d; \\ x \geq_{\Box} y & \text{iff } a \geq c; & x \leq_{\Box} y & \text{iff } a \leq c; \\ x \geq_{\Diamond} y & \text{iff } b \leq d; & x \leq_{\Diamond} y & \text{iff } b \geq d; \\ x = y & \text{iff } a = c \text{ and } b = d. & & \end{array} \quad (1)$$

When comparing two IFPs $x = \langle a,b \rangle$ and $y = \langle c,d \rangle$ following (1), we use the degrees of membership and non-membership of IFPs separately, giving priority to one of components, usually the membership one.

Let us use a relation for comparison of two IFPs x and y to handle both components simultaneously using the measure:
$R_{\langle a,b \rangle} = 0.5(1+1-a-b)0.5(abs|1-a|+abs|b|+abs|1-a-b|)$ [42] as follows:

$$x \geq_R y \text{ iff } R_{\langle a,b \rangle} \leq R_{\langle c,d \rangle}. \quad (2)$$

$l_{IFS}(M,x) = 0.5(abs|1-a|+abs|b|+abs|1-a-b|)$ is the distance x from ideal positive intuitionistic fuzzy pair $M(1,0)$ [42].

The IFP x is an **"intuitionistic fuzzy false pair" (IFFP)** if and only if $a \leq b$, while x is a **"false pair" (FP)** iff $a = 0, b = 1$.

Let a set E be fixed. An **"intuitionistic fuzzy set" (IFS)** A in E is an object of the following form (see [6]): $A = \{\langle x, \mu_A(x), \nu_A(x) \rangle | x \in E\}$, where functions $\mu_A : E \to [0,1]$ and $\nu_A : E \to [0,1]$ define the degrees of membership and the non-membership of the $x \in E$, respectively, and for every $x \in E$: $0 \leq \mu_A(x) + \nu_A(x) \leq 1$.

2.2. Definition, Operations and Relations over Intuitionistic Fuzzy Index Matrices

Let $\mathscr{I}$ be a fixed set. By two-dimensional intuitionistic fuzzy index matrix (2D-IFIM) with index sets K and L ($K, L \subset \mathscr{I}$), we denote the object:

$$[K, L, \{\langle \mu_{k_i,l_j}, \nu_{k_i,l_j} \rangle\}]$$

$$\equiv \begin{array}{c|ccccc} & l_1 & \dots & l_j & \dots & l_n \\ \hline k_1 & \langle \mu_{k_1,l_1}, \nu_{k_1,l_1} \rangle & \dots & \langle \mu_{k_1,l_j}, \nu_{k_1,l_j} \rangle & \dots & \langle \mu_{k_1,l_n}, \nu_{k_1,l_n} \rangle \\ \vdots & \vdots & \ddots & \vdots & \ddots & \vdots \\ k_m & \langle \mu_{k_m,l_1}, \nu_{k_m,l_1} \rangle & \dots & \langle \mu_{k_m,l_j}, \nu_{k_m,l_j} \rangle & \dots & \langle \mu_{k_m,l_n}, \nu_{k_m,l_n} \rangle \end{array},$$

where for every $1 \leq i \leq m, 1 \leq j \leq n$: $0 \leq \mu_{k_i,l_j}, \nu_{k_i,l_j}, \mu_{k_i,l_j} + \nu_{k_i,l_j} \leq 1$.

Following [7], we recall some operations over two IMs $A = [K, L, \{\langle \mu_{k_i,l_j}, \nu_{k_i,l_j} \rangle\}]$ and $B = [P, Q, \{\langle \rho_{p_r,q_s}, \sigma_{p_r,q_s} \rangle\}]$.

Negation: $\neg A = [K, L, \{\langle \nu_{k_i,l_j}, \mu_{k_i,l_j} \rangle\}]$.

Addition-$(\circ,*)$**:** $A \oplus_{(\circ,*)} B = [K \cup P, L \cup Q, \{\langle \phi_{t_u,v_w}, \psi_{t_u,v_w} \rangle\}]$, where

$$\langle \phi_{t_u,v_w}, \psi_{t_u,v_w} \rangle = \begin{cases} \langle \mu_{k_i,l_j}, \nu_{k_i,l_j} \rangle, & \text{if } t_u = k_i \in K \text{ and } v_w = l_j \in L - Q \\ & \text{or } t_u = k_i \in K - P \text{ and } v_w = l_j \in L; \\ \langle \rho_{p_r,q_s}, \sigma_{p_r,q_s} \rangle, & \text{if } t_u = p_r \in P \text{ and } v_w = q_s \in Q - L \\ & \text{or } t_u = p_r \in P - K \\ & \text{and } v_w = q_s \in Q; \\ \langle \circ(\mu_{k_i,l_j}, \rho_{p_r,q_s}), & \text{if } t_u = k_i = p_r \in K \cap P \\ *(\nu_{k_i,l_j}, \sigma_{p_r,q_s}) \rangle, & \text{and } v_w = l_j = q_s \in L \cap Q; \\ \langle 0, 1 \rangle, & \text{otherwise.} \end{cases}$$

where $\langle \circ, * \rangle \in \{\langle \max, \min \rangle, \langle \min, \max \rangle, \langle \text{ average,average} \rangle\}$.

Termwise subtraction-(max,min): $A -_{(\max,\min)} B = A \oplus_{(\max,\min)} \neg B$.
Termwise multiplication-$(\min,\max)$:
$A \otimes_{(\min,\max)} B = [K \cap P, L \cap Q, \{\langle \phi_{t_u,v_w}, \psi_{t_u,v_w} \rangle\}]$, where
$\langle \phi_{t_u,v_w}, \psi_{t_u,v_w} \rangle = \langle \min(\mu_{k_i,l_j}, \rho_{p_r,q_s}), \max(\nu_{k_i,l_j}, \sigma_{p_r,q_s}) \rangle$.
Reduction: We use symbol "$\perp$" for lack of some component in the separate definitions. The operations $(k, \perp)$-reduction of a given IM A is defined by:

$$A_{(k,\perp)} = [K - \{k\}, L, \{c_{t_u,v_w}\}],$$

where $c_{t_u,v_w} = a_{k_i,l_j}$ for $t_u = k_i \in K - \{k\}$ and $v_w = l_j \in L$.
Projection: Let $M \subseteq K$ and $N \subseteq L$. Then, $pr_{M,N}A = [M, N, \{b_{k_i,l_j}\}]$, where for each $k_i \in M$ and each $l_j \in N$, $b_{k_i,l_j} = a_{k_i,l_j}$.
Substitution: Let IM $A = [K, L, \{a_{k,l}\}]$ be given. Local substitution over the IM is defined for the couples of indices (p,k) and/or (q,l), respectively, by
$\left[\frac{p}{k}; \perp\right] A = \left[(K - \{k\}) \cup \{p\}, L, \{a_{k,l}\}\right], \left[\perp; \frac{q}{l}\right] A = \left[K, (L - \{l\}) \cup \{q\}, \{a_{k,l}\}\right]$.
Index type operations:
$AGIndex_{\{(\min/\max)/(\min_\square/\max_\square)/(\min_\diamond/\max_\diamond)(\min_R/\max_R)\}(\not\perp)}(A) = \langle k_i, l_j \rangle$, which finds the index of the minimum/ maximum element of A with no empty value in accordance with the relations (1) or (2).
$AGIndex_{\{(\min/\max)/(\min_\square/\max_\square)/(\min_\diamond/\max_\diamond)(\min_R/\max_R)\}(\not\perp)(\notin F)}(A) = \langle k_i, l_j \rangle$, which finds the index of the minimum/ maximum element between the elements of A, whose indexes $\notin F$, with no empty value in accordance with the relations (1) or (2).
$Index_{\{(\min/\max)/(\min_\square/\max_\square)/(\min_\diamond/\max_\diamond)(\min_R/\max_R)\}(\not\perp),k_i}(A)$
$= \{\langle k_i, l_{v_1} \rangle, \ldots, \langle k_i, l_{v_x} \rangle, \ldots, \langle k_i, l_{v_V} \rangle\}$,
where $\langle k_i, l_{v_x} \rangle$ (for $1 \le i \le m, 1 \le v \le n, 1 \le x \le V$) are the indices of the minimum/ maximum IFFP of k_i-th row of A with no empty value in accordance with the relations (1) or (2).
$Index_{(\not\perp)}(A) = \{\langle k_1, l_{v_1} \rangle, \ldots, \langle k_i, l_{v_i} \rangle, \ldots, \langle k_m, l_{v_m} \rangle\}$, where $\langle k_i, l_{v_i} \rangle$ (for $1 \le i \le m$) are the indices of the element of A, whose cell is full.

Aggregation operations
Let us have two IFPs $x = \langle a, b \rangle$ and $y = \langle c, d \rangle$. We use the following three operations $\#_q, (q \le i \le 3)$ for scaling aggregation evaluations [45]:
$x\#_1 y = \langle min(a,c), max(b,d) \rangle$; $x\#_2 y = \langle average(a,c), average(b,d) \rangle$;
$x\#_3 y = \langle max(a,c), min(b,d) \rangle$.

From the theorem in [45] we can see that: $x\#_1 y \le x\#_2 y \le x\#_3 y$.

Let $k_0 \notin K$ be a fixed index. Following [7,45], the definition of the aggregation operation by the dimension K is:

$$\alpha_{K,\#_q}(A, k_0) = \begin{array}{c|ccc} & l_1 & \ldots & l_n \\ \hline k_0 & \overset{m}{\underset{i=1}{\#_q}} \langle \mu_{k_i,l_1}, \nu_{k_i,l_1} \rangle & \ldots & \overset{m}{\underset{i=1}{\#_q}} \langle \mu_{k_i,l_n}, \nu_{k_i,l_n} \rangle \end{array}, \text{ where } 1 \le q \le 3.$$

Aggregate global internal operation: $AGIO_{\oplus_{(\max,\min)}}(A)$.
This operation finds the "$\oplus_{(\max,\min)}$"-operation of all the matrix elements.
Internal subtraction of IMs' components [43,44,48]:

$$IO_{-(\max,\min)}(\langle k_i, l_j, A\rangle, \langle p_r, q_s, B\rangle) = [K, L, \{\langle \gamma_{t_u,v_w}, \delta_{t_u,v_w}\rangle\}]$$

$$\langle \gamma_{t_u,v_w}, \delta_{t_u,v_w}\rangle = \begin{cases} \langle \mu_{t_u,v_w}, \nu_{t_u,v_w}\rangle, & \text{if } t_u \neq k_i \in K, \\ & v_w \neq l_j \in L; \\ \langle \max(0, \mu_{k_i,l_j} - \rho_{p_r,q_s}), & \text{if } t_u = k_i \in K, \\ \min(1, \nu_{k_i,l_j} + \sigma_{p_r,q_s}, 1 - \mu_{k_i,l_j} + \rho_{p_r,q_s})\rangle & v_w = l_j \in L \end{cases}$$

where $k_i \in K,\ l_j \in L;\ p_r \in P,\ q_s \in Q$.
The non-strict relation "inclusion about value" is defined by:

$$A \subseteq_v B \text{ iff } (K = P)\ \&\ (L = Q)\ \&\ (\forall k \in K)(\forall l \in L)(a_{k,l} \leq b_{k,l}).$$

3 An Intuitionistic Fuzzy Approach to the Zero Suffix Algorithm of the Transportation Problem

Let us formulate following IFTP: Let a product be produced at m sources (producers) $\{k_1, \ldots, k_i, \ldots, k_m\}$ in quantities $c_{k_i,R}$ (for $1 \leq i \leq m$). Let n-consumers (destinations) $\{l_1, \ldots, l_j, \ldots, l_n\}$ demand this product in amounts of c_{Q,l_j} (for $1 \leq j \leq n$). Let c_{k_i,l_j} be the intuitionistic fuzzy cost for transporting one unit quantity of the product from the k_i-th source to the l_j-th destination; x_{k_i,l_j} - the number of units of the product, transported from k_i-th source to l_j-th destination and c_{pl,l_j} (for $1 \leq j \leq n$) are intuitionistic fuzzy upper limits to the transportation costs of delivery a particular product from the k_i-th source to the l_j-th destination.

The all parameters of the problem are IFPs. The purpose of the problem is to find the optimal transportation of the product from various sources to various destinations so that the intuitionistic fuzzy transportation cost is minimum according to (1) or (2).

The mathematical model of the above problem is as follows:

$$\begin{aligned} &\text{An objective function: minimize } \sum_{i=1}^{m} \sum_{j=1}^{n} c_{k_i,l_j} x_{k_i,l_j} \\ &\text{Subject to: } \sum_{j=1}^{n} x_{k_i,l_j} = c_{k_i,R}, \qquad i = 1, 2, \ldots, m \\ &\sum_{i=1}^{m} x_{k_i,l_j} = c_{Q,l_j}, \qquad j = 1, 2, \ldots, n \end{aligned} \tag{3}$$

The additional constraint to the problem (3) is: c_{pl,l_j}, for $1 \leq j \leq n$ – an intuitionistic fuzzy upper limit to the corresponding transportation cost of delivery a particular product from the k_i-th source to the l_j-th destination.

Note: The operations "addition" and "multiplication", used in the problem (3) are those for IFPs, defined in Sect. 2.

Let us construct the following IM C, in accordance with the problem (3),

$$C[K, L]$$

$$
= \begin{array}{c|cccccc}
 & l_1 & \dots & l_n & R & pu \\
\hline
k_1 & \langle \mu_{k_1,l_1}, \nu_{k_1,l_1} \rangle & \dots & \langle \mu_{k_1,l_n}, \nu_{k_1,l_n} \rangle & \langle \mu_{k_1,R}, \nu_{k_1,R} \rangle & \langle \mu_{k_1,pu}, \nu_{k_1,pu} \rangle \\
\vdots & \vdots & \ddots & \vdots & \vdots & \\
k_m & \langle \mu_{k_m,l_1}, \nu_{k_m,l_1} \rangle & \dots & \langle \mu_{k_m,l_n}, \nu_{k_m,l_n} \rangle & \langle \mu_{k_m,R}, \nu_{k_m,R} \rangle & \langle \mu_{k_m,pu}, \nu_{k_m,pu} \rangle \\
Q & \langle \mu_{Q,l_1}, \nu_{Q,l_1} \rangle & \dots & \langle \mu_{Q,l_n}, \nu_{Q,l_n} \rangle & \langle \mu_{Q,R}, \nu_{Q,R} \rangle & \langle \mu_{Q,pu}, \nu_{Q,pu} \rangle \\
pl & \langle \mu_{pl,l_1}, \nu_{pl,l_1} \rangle & \dots & \langle \mu_{pl,l_n}, \nu_{pl,l_n} \rangle & \langle \mu_{pl,R}, \nu_{pl,R} \rangle & \langle \mu_{pl,pu}, \nu_{pl,pu} \rangle \\
pu_1 & \langle \mu_{pu_1,l_1}, \nu_{pu_1,l_1} \rangle & \dots & \langle \mu_{pu_1,l_n}, \nu_{pu_1,l_n} \rangle & \langle \mu_{pu_1,R}, \nu_{pu_1,R} \rangle & \langle \mu_{pu_1,pu}, \nu_{pu_1,pu} \rangle
\end{array},
$$

where $K=\{k_1,k_2,\dots,k_m,Q,pl,pu_1\}$, $L=\{l_1,l_2,\dots,l_n,R,pu\}$ and for $1\le i\le m$, $1\le j\le n$, $\{c_{k_i,l_j}, c_{k_i,R}, c_{k_i,pu}, c_{pl,l_j}, c_{pl,R}, c_{pl,pu}, c_{Q,l_j}, c_{Q,R}, c_{Q,pu}, c_{pu_1,l_j}, c_{pu_1,R}, c_{pu_1,pu}\}$ are IFPs.

The expert approach described in detail in the [6] may be used to determine the transportation costs in the form of IFPs. Each expert is asked to evaluate at least a part of the alternatives in terms of their performance with respect to each predefined criterion. Due to several uncertainties (variation in rates of diesel, traffic etc.), the expert is not certain about the transportation costs. Also due to some uncontrollable factors (such as weather), the expert hesitates in prediction of transportation cost from different resources to different destinations. According to the past experience of the expert the transportation costs are estimated as intuitionistic fuzzy numbers after a thorough discussion, interpreted in the intuitionistic fuzzy framework: these numbers express a "positive" and a "negative" evaluations, respectively. With each expert a pair of values is associated, which express the expert's reliability (confidence in her/his evaluation with respect to each criterion). Distinct reliability values are associated with distinct criteria.

Let we denote by $|K|=m+3$ the number of elements of the set K; then $|L|=n+2$. We also define

$$
X[K*,L*] = \begin{array}{c|ccccc}
 & l_1 & \dots & l_j & \dots & l_n \\
\hline
k_1 & x_{k_1,l_1} & \cdots & x_{k_1,l_j} & \cdots & x_{k_1,l_n} \\
\vdots & \vdots & \ddots & \vdots & \ddots & \vdots \\
k_m & x_{k_m,l_1} & \dots & x_{k_m,l_j} & \dots & x_{k_m,l_n}
\end{array},
$$

$K*=\{k_1,k_2,\dots,k_m\}$, $L*=\{l_1,l_2,\dots,l_n\}$, and for $1\le i\le m$, $1\le j\le n$: $x_{k_i,l_j}=\langle \rho_{k_i,l_j}, \sigma_{k_i,l_j}\rangle$.

Let us we define the following auxiliary index matrices:

1) $S=[K,L,\{s_{k_i,l_j}\}]$, such that $S=C$ i.e. $(s_{k_i,l_j}=c_{k_i,l_j}\ \forall k_i\in K, \forall l_j\in L)$;

2) $U[K*,L*,\{u_{k_i,l_j}\}]$ and for $1\le i\le m$, $1\le j\le n$: $u_{k_i,l_j}=\begin{cases} 1, & \text{if } c_{k_i,l_j}<c_{pl,l_j} \\ \perp, & \text{otherwise} \end{cases}$;

3) $Suf=[K*,L*,\{suf_{k_i,l_j}\}]$.

In the begining of the algorithm $suf_{k_i,l_j}=x_{k_i,l_j}=\langle\perp,\perp\rangle$ and $u_{k_i,l_j}=\perp$ $(\forall k_i\in K*, \forall l_j\in L*)$.

We will propose for the first time a new approach for the algorithm for finding the optimal solution of the TP with intuitionistic fuzzy data using the modified Zero suffix method [20,30,39], interpreted with the tools of IMs and IFPs. A part of Microsoft

Visual Studio.NET 2010 C project's program code is used.

Step 1. Construct the IFIM C for the given IFTP and then, convert it into a balanced one ($\sum_{i=1}^{m} c_{k_i,R} = \sum_{j=1}^{n} c_{Q,l_j}$), if it is not.

For this purpose, the following operations are executed:

– We define $2D$-IMs as follows:

$C_1[Q, L/\{R, pu\}] = pr_{Q,L/\{R,pu\}}C$ and $C_2[K/\{Q, pl, pu_1\}, R] = pr_{K/\{Q,pu_1\},R}C$ and let $\{k_{m+1}, l_{n+1}\} \notin K \cup L$.

– If $\alpha_{K,\#_q}(C_1, l_{n+1}) \supset_v \left[\frac{Q}{R}; \perp\right] (\alpha_{L,\#_q}(C_2, l_{n+1}))'$ (i.e. $\sum_{i=1}^{m} c_{k_i,R} > \sum_{j=1}^{n} c_{Q,l_j}$), then *introduce dummy column l_{n+1} having all its costs as $\langle 0, 1\rangle$ and execute operations for finding the demand at this dummy destination:* $c_{Q,l_{n+1}} = \sum_{i=1}^{m} c_{k_i,R} - \sum_{j=1}^{n} c_{Q,l_j}$;

{Let us define $2D$-IMs C_3, C_4, C_5 such that

$C_3 = \alpha_{K,\#_q}(C_1, l_{n+1}) -_{(\max,\min))} \alpha_{L,\#_q}(\left[\frac{Q}{R}; \perp\right] (C_2, l_{n+1}))'$;

$C_4[K/\{Q, pl, pu_1\}, \{l_{n+1}\}, \{\langle 0, 1\rangle\}]$; $C_5 = [K, \{l_{n+1}\}, \{c_{k_i,l_{n+1}}\}] = C_3 \oplus_{(\max,\min))} C_4$;

The new matrix of costs is obtained by carrying out the operation "matrix addition": $C := C \oplus_{(\max,\min))} C_5$, go to *Step 2.* }

– If $\left[\perp; \frac{R}{Q}\right] \alpha_{K,\#_q}(C_1, k_{m+1}))' \subset_v \alpha_{L,\#_q}(C_2, k_{m+1}))'$ (i.e. $\sum_{i=1}^{m} c_{k_i,R} < \sum_{j=1}^{n} c_{Q,l_j}$), then *introduce dummy row k_{m+1} having all its costs as $\langle 0, 1\rangle$ and execute operations for finding the demand at this dummy destination:* $c_{k_{m+1},R} = \sum_{i=1}^{m} c_{k_i,R} - \sum_{j=1}^{n} c_{Q,l_j}$.

{Let us define $2D$-IMs C_3, C_4, C_5 such that

$C_3 = \alpha_{K,\#_q}(C_2, k_{n+1}) -_{(\max,\min))} \left[\perp; \frac{R}{Q}\right] \alpha_{L,\#_q}(C_1, k_{m+1}))'$;

$C_4[\{k_{m+1}\}, L/\{Q, pu\}, \{\langle 0, 1\rangle\}]$; $C_5 = [k_{m+1}, L, \{c_{k_{m+1},l_j}\}] = C_3 \oplus_{(\max,\min))} C_4$;

$C := C \oplus_{(\max,\min))} C_5$, go to *Step 2.* }

Step 2. Verification of compliance with the requirement to limit transport costs;

for (int $i = 0; i < m; i++$)

for (int $j = 0; j < n; j++$)

{If $\left(\left[\frac{k_i}{pl}; \perp\right] pr_{pl,l_j}C\right) \subset_v pr_{k_i,l_j}C$, then $u_{k_i,l_j} = 1$. }

$EG = Index_{(\perp)}(U) = \{\langle k_{i_1}, l_{j_1}\rangle, \langle k_{i_2}, l_{j_2}\rangle, \ldots, \langle\langle k_{i_\phi}, l_{j_\phi}\rangle\}$; for each $\langle k_i, l_j\rangle \in EG$, let us the element s_{k_i,l_j} of S is equal to $\langle 1, 0\rangle$ [31]; Go to *Step 3.*

Step 3. For each row of the matrix S, the smallest element is found in accordance with the relations (1) or (2) and is recorded to the right of the row, in the column pu:

for (int $i = 0; i < m; i++$)

for (int $j = 0; j < n; j++$)

$\{AGIndex_{\{(\min)/(\min_\square)/(\min_\diamond)/(\min_R)\}}\left(pr_{k_i,L/\{R,pu\}}S\right) = \langle k_i, l_{v_j}\rangle$;

If $pr_{k_i,l_{v_j}}S \subseteq_v \left(\left[\frac{k_i}{pl}; \perp\right] pr_{pl,l_{v_j}}S\right)$, then

$S_6[k_i, l_{v_j}] = pr_{k_i,l_{v_j}}S; S_7 = \left[\perp; \frac{pu}{l_{v_j}}\right] S_6; S := S \oplus_{(\max,\min)} S_7.\}$

From each element of the matrix S, subtract the smallest element in the same row:

for (int $i = 0; i < m; i++$)

for (int $j=0; j<n; j++$)
$\{IO_{-(\max,\min)}(\langle k_i,l_j,S\rangle,\langle k_i,pu,pr_{K/\{Q,pl,pu_1\}}S\rangle)\}$;
Go to *Step 4*.
Step 4. For each column of the matrix S, the smallest element is found in accordance with the relations (1) or (2). It is recorded at the bottom of the column, in line pu_1:
for (int $j=0; j<n; j++$)
$\{AGIndex_{\{(\min)/(\min_\square)/(\min_\diamond)/(\min_R)\}}\left(pr_{K/\{Q,pl,pu_1\},l_j}S\right)=\langle k_{w_i},l_j\rangle$;
Let us create two 2D-IMs S_6 and S_7 : $S_6[k_{w_i},l_j]=pr_{k_{w_i},l_j}S; S_7=\left[\frac{pu_1}{k_{w_i}};\perp\right]S_6$;
$S:=S\oplus_{(\max,\min)}S_7.\}$
for (int $j=0; j<n; j++$)
for (int $i=0; i<m; i++$)
$\{IO_{-(\max,\min)}(\langle k_i,l_j,S\rangle,\langle pu_1,l_j,pr_{pu_1,L/\{R,pu\}}S\rangle)\}$;
Go to *Step 5*.
Step 5. Obtain the suffix value suf_{k_i,l_j} of the minimum element s_{k_i,l_j} according to the relations (1) or (2) in corresponding row or column by average of the smallest and the next smallest elements of the corresponding row in accordance with the relations (1) or (2), and the smallest and the next smallest elements in accordance with the relations (1) or (2) of the corresponding column.
for (int $i=0; i<m; i++$)
for (int $j=0; j<n; j++$)
If $s_{k_i,l_j}=\langle 0,1\rangle$ (or $\langle k_i,l_j\rangle\in Index_{\{(\min)/(\min_\square)/(\min_\diamond)/(\min_R)\}}(S)$ then
$\{AGIndex_{\{(\min)/(\min_\square)/(\min_\diamond)/(\min_R)}\left(pr_{k_i,L/\{R,pu\}}S\right)=\langle k_i,l_{v_j}\rangle$;
$AGIndex_{\{(\min)/(\min_\square)/(\min_\diamond)/(\min_R)}\left(pr_{k_i,L/\{R,pu,v_j\}}S\right)=\langle k_i,l_{w_j}\rangle$;
$AGIndex_{\{(\min)/(\min_\square)/(\min_\diamond)/(\min_R)}\left(pr_{K/\{Q,pl,pu_1\},l_j}S\right)=\langle k_{v_e},l_j\rangle$;
$AGIndex_{\{(\min)/(\min_\square)/(\min_\diamond)/(\min_R)}\left(pr_{K/\{Q,pl,pu_1,v_e\},l_j}S\right)=\langle k_{w_g},l_j\rangle$.
Define IMs: $S_r (8\le r\le 16)$ as follows: $S_8[k_i,l_{v_j}]=pr_{k_i,l_{v_j}}S; S_9[k_i,l_{w_j}]=pr_{k_i,l_{w_j}}S$;
$S_{10}=\left[\perp;\frac{pu}{l_{v_j}}\right]S_8\oplus_{(average,average)}\left[\perp;\frac{pu}{l_{w_j}}\right]S_9; S:=S\oplus_{(\max,\min)}S_{10}$;
$S_{11}[k_{v_e},l_j]=pr_{k_{v_e},l_j}S; S_{12}[k_{w_g},l_j]=pr_{k_{w_g},l_j}S$;
$S_{13}=\left[\frac{pu_1}{k_{v_i}};\perp\right]S_{11}\oplus_{(average,average)}\left[\frac{pu_1}{k_{w_g}};\perp\right]S_{12}; S:=S\oplus_{(\max,\min)}S_{13}$;
$S_{14}[k_i,pu]=pr_{k_i,pu}S; S_{15}[pu_1,l_j]=pr_{pu_1,l_j}S$;
$S_{16}=\left[\perp;\frac{l_j}{pu}\right]S_{14}\oplus_{(average,average)}\left[\frac{pu_1}{k_i};\perp\right]S_{15}; Suf=Suf\oplus_{(\max,\min)}S_{16}.\}$
Go to *Step 6*.
Step 6. Search the greatest suffix value of the IM Suf in accordance with the relations (1) or (2) by $AGIndex_{\{(\max)/(\max_\square)/(\max_\diamond)/(\max_R)\},(\not\perp)}Suf=\langle k_e,l_g\rangle$;
Go to *Step 7*.
Step 7. Then assign the minimum of demand and supply to the corresponding s_{k_e,l_g} cell and delete the row/column having supply/demand are exhausted and find the reduced IM S.
Define IMs $S_{17}[k_e,R]=pr_{k_e,R}S$ and $S_{18}[Q,l_g]=pr_{Q,l_g}S$;

If $S_{17} \subseteq_v \left[\frac{k_e}{l_g}; \frac{R}{Q}\right] (S_{18})'$ (i.e. $\min(s_{k_e,R}, s_{Q,l_g}) = s_{k_e,R}$), then
$\{X := X \oplus_{(\max,\min)} \left[\bot; \frac{l_g}{R}\right] S_{17}$;
Ignore the k_e-th row from matrix S to obtain a new matrix with dimensions $(m+2) \times (n+2)$ (operation "reduction" $S_{(k_e,\bot)}$ executes) by the operations:
let us create IM S_{19} as follows: $S_{19}[Q, l_g] = S_{18} -_{(\max,\min)} \left[\frac{Q}{R}; \frac{l_g}{k_e}\right] (S_{17})'$;
Then $S := S \oplus_{(\max,\min)} S_{19}$; }
If $S_{17} \supseteq_v \left[\frac{k_i}{l_j}; \frac{R}{Q}\right] (S_{18})'$ (i.e. $\min(s_{k_i,R}, s_{Q,l_j}) = s_{Q,l_j}$), then {the IM X changes with:
$X := X \oplus_{(\max,\min)} \left[\frac{k_e}{Q}; \bot\right] S_{18}$.
Ignore the l_g-th column from matrix S to obtain a new matrix with dimensions $(m+3) \times (n+1)$ (operation reduction $S_{(\bot,l_g)}$ executes) by the operations:
let us construct IM S_3 as follows: $S_{19}[k_e, R] = S_{17} -_{(\max,\min)} \left[\frac{k_e}{l_g}; \frac{R}{Q}\right] (S_{18})'$;
$S := S \oplus_{(\max,\min)} S_{19}$; }
Go to *Step 8*.
Step 8. Repeat Steps 3–7 until $|S| = 6$ (all the demands are satisfied and all the supplies are exhausted), i.e. S is reduced to the form

$$S[K^r, L^r] = \begin{array}{c|cc} & R & pu \\ \hline Q & \langle \mu_{Q,R}, \nu_{Q,R} \rangle & \langle \mu_{Q,pu}, \nu_{Q,pu} \rangle \\ & & \\ pl & \langle \mu_{pl,R}, \nu_{pl,R} \rangle & \langle \mu_{pl,pu}, \nu_{pl,pu} \rangle \\ pu_1 & \langle \mu_{pu_1,R}, \nu_{pu_1,R} \rangle & \langle \mu_{pu_1,pu}, \nu_{pu_1,pu} \rangle \end{array};$$

Go to *Step 9*.
Step 9. $D = Index_{\not\perp} X = \{\langle k_{i*_1}, l_{j*_1} \rangle, \dots, \langle k_{i*_f}, l_{j*_f} \rangle, \dots, \langle k_{i*_\varphi}, l_{j*_\varphi} \rangle\}$. If the intuitionistic fuzzy feasible solution is degenerated (it contains less than $m+n-1$ (the total number of producers and consumers decreased by 1) occupied cells in the X i.e. $|D| < m+n-1$) [3] then complement the basic cells x_{k_i,l_j} with one to which the minimum transport cost corresponds. Let us the recorded delivery of this cell is $\langle 0, 1 \rangle$. The operations are:
If $|D| < m+n-1$, then
$\{AGIndex_{\{(\min/\max)/(\min_\square/\max_\square)/(\min_\diamond/\max_\diamond)(\min_R/\max_R)\}(\not\perp)(\notin D)}(C) = \langle k_\alpha, l_\beta \rangle$;
$x_{k_{al},l_\beta} = \langle 0, 1 \rangle\}$.
Go to *Step 10*.
Step 10. for (int $i = 0; i < m; i++$)
for (int $j = 0; j < n; j++$)
If $x_{k_i,l_j} \neq \langle \bot, \bot \rangle$ and $\langle k_i, l_j \rangle \in EG$ then the problem has not solution [3] and the algorithm stop else
{all the demands and supplies are exhausted and the algorithm stop. The optimal basic solution $X_{opt}[K*, L*, \{x_{k_i,l_j}\}]$ is obtained.
for (int $i = 0; i < m; i++$)
for (int $j = 0; j < n; j++$)
If $x_{k_i,l_j} = \langle \bot, \bot \rangle$ then $x_{k_i,l_j} = \langle 0, 1 \rangle$.
The optimal intuitionistic fuzzy transportation cost is:
$AGIO_{\oplus_{(\max,\min)})}\left(C_{(\{Q,pl,pu_1\},\{R,pu\})} \otimes_{(\min,\max)} X_{opt}\right)$.

4 An Example of the IFTP

Let us formulate following IFTP: Let a product be produced at producers $\{k_1, k_2, k_3\}$ in quantities $c_{k_i,R}$ (for $1 \leq i \leq 3$). Let consumers ($\{l_1, l_2, l_3, l_4\}$) demand this product in an amount of c_{Q,l_j} (for $1 \leq j \leq 4$) and c_{pl,l_j} (for $1 \leq j \leq 4$) are intuitionistic fuzzy upper limits to the transportation costs of delivery a particular product from the k_i-th source to the l_j-th destination. Let the cost c_{k_i,l_j} for transporting one unit quantity of the product from the k_i-th source to the l_j-th destination is an IFP and is defined as an element of IFIM C, where

$$C[K,L] = \left\{ \begin{array}{c|ccccc} & l_1 & l_2 & l_3 & l_4 & R \\ \hline k_1 & \langle 0.5,0.2\rangle & \langle 0.7,0.1\rangle & \langle 0.3,0.1\rangle & \langle 0.8,0.1\rangle & \langle 0.5,0.2\rangle \\ k_2 & \langle 0.5,0.3\rangle & \langle 0.4,0.1\rangle & \langle 0.5,0.1\rangle & \langle 0.3,0.2\rangle & \langle 0.7,0.1\rangle \\ k_3 & \langle 0.4,0.2\rangle & \langle 0.3,0.2\rangle & \langle 0.6,0.1\rangle & \langle 0.7,0.2\rangle & \langle 0.4,0.5\rangle \\ Q & \langle 0.4,0.2\rangle & \langle 0.5,0.3\rangle & \langle 0.6,0.2\rangle & \langle 0.06,0.02\rangle & \\ pl & \langle 0.55,0.3\rangle & \langle 0.6,0.4\rangle & \langle 0.75,0.2\rangle & \langle 0.65,0.3\rangle & \end{array} \right.$$

Let x_{k_i,l_j} – the number of units of the product, transported from k_i-th source to l_j-th destination (for $1 \leq i \leq 3$ and $1 \leq j \leq 4$) and is an element of IFIM X with initial elements $\langle \perp, \perp \rangle$.

Solution of the problem:

Step 1. The problem is balanced.

Step 2. Verification of compliance with the requirement to limit transport costs.

for (int $i = 0; i < m; i++$)

for (int $j = 0; j < n; j++$)

{If $\left(\left[\frac{k_i}{pl}; \perp \right] pr_{pl,l_j} C \right) \subset_v pr_{k_i,l_j} C$, then $u_{k_i,l_j} = 1$}.

The IM C is transformed in:

$$C[K,L] = \left\{ \begin{array}{c|cccccc} & l_1 & l_2 & l_3 & l_4 & R & pu \\ \hline k_1 & \langle 0.6,0.2\rangle & \langle 1,0\rangle & \langle 0.3,0.1\rangle & \langle 1,0\rangle & \langle 0.5,0.2\rangle & \langle \perp,\perp\rangle \\ k_2 & \langle 0.5,0.3\rangle & \langle 0.4,0.1\rangle & \langle 0.5,0.1\rangle & \langle 0.3,0.2\rangle & \langle 0.7,0.1\rangle & \langle \perp,\perp\rangle \\ k_3 & \langle 0.4,0.2\rangle & \langle 0.3,0.2\rangle & \langle 0.6,0.1\rangle & \langle 0.7,0.2\rangle & \langle 0.4,0.5\rangle & \langle \perp,\perp\rangle \\ Q & \langle 0.4,0.2\rangle & \langle 0.5,0.3\rangle & \langle 0.6,0.2\rangle & \langle 0.06,0.02\rangle & \langle \perp,\perp\rangle & \langle \perp,\perp\rangle \\ pl & \langle 0.7,0.3\rangle & \langle 0.6,0.4\rangle & \langle 0.75,0.2\rangle & \langle 0.65,0.3\rangle & \langle \perp,\perp\rangle & \langle \perp,\perp\rangle \\ pu_1 & \langle \perp,\perp\rangle & \langle \perp,\perp\rangle & \langle \perp,\perp\rangle & \langle \perp,\perp\rangle & \langle \perp,\perp\rangle & \langle \perp,\perp\rangle \end{array} \right.$$

Let us define IM $S = [K, L, \{s_{k_i,l_j}\}]$ such that $S = C$.

Step 3. In each row of the S, the smallest element is found in accordance with the relation (2):

$$\langle a,b \rangle \leq_R \langle c,d \rangle \text{ iff } R_{\langle a,b \rangle} \geq R_{\langle c,d \rangle}$$

and it is subtracted from all elements in the appropriate row go to *Step 4*.

$$S[K,L] = \left\{ \begin{array}{c|cccccc} & l_1 & l_2 & l_3 & l_4 & R & pu \\ \hline k_1 & \langle 0.3,0.3\rangle & \langle 0.7,0.1\rangle & \langle 0,0.2\rangle & \langle 0.7,0.1\rangle & \langle 0.5,0.2\rangle & \langle 0.3,0.1\rangle \\ k_2 & \langle 0.2,0.5\rangle & \langle 0.1,0.3\rangle & \langle 0.2,0.3\rangle & \langle 0,0.4\rangle & \langle 0.7,0.1\rangle & \langle 0.3,0.2\rangle \\ k_3 & \langle 0.1,0.4\rangle & \langle 0,0.4\rangle & \langle 0.3,0.3\rangle & \langle 0.4,0.4\rangle & \langle 0.4,0.5\rangle & \langle 0.3,0.2\rangle \\ Q & \langle 0.4,0.2\rangle & \langle 0.5,0.3\rangle & \langle 0.6,0.2\rangle & \langle 0.06,0.02\rangle & \langle \perp,\perp\rangle & \langle \perp,\perp\rangle \\ pl & \langle 0.55,0.3\rangle & \langle 0.6,0.4\rangle & \langle 0.75,0.2\rangle & \langle 0.65,0.3\rangle & \langle \perp,\perp\rangle & \langle \perp,\perp\rangle \\ pu_1 & \langle \perp,\perp\rangle & \langle \perp,\perp\rangle & \langle \perp,\perp\rangle & \langle \perp,\perp\rangle & \langle \perp,\perp\rangle & \langle \perp,\perp\rangle \end{array} \right.$$

Step 4. For each column of the matrix S, the smallest element is found in accordance with the relation (2):

$$\langle a,b\rangle \leq_R \langle c,d\rangle \text{ iff } R_{\langle a,b\rangle} \geq R_{\langle c,d\rangle}$$

and it is subtracted from all elements in the corresponding column and go to *Step 5.*

$$S[K,L] = \left\{ \begin{array}{c|cccccc} & l_1 & l_2 & l_3 & l_4 & R & pu \\ \hline k_1 & \langle 0.2,0.7\rangle & \langle 0.7,0.3\rangle & \langle 0,0.4\rangle & \langle 0.7,0.3\rangle & \langle 0.5,0.2\rangle & \langle 0.5,0.2\rangle \\ k_2 & \langle 0.1,0.9\rangle & \langle 0.1,0.7\rangle & \langle 0.2,0.5\rangle & \langle 0,0.8\rangle & \langle 0.7,0.1\rangle & \langle 0.3,0.2\rangle \\ k_3 & \langle 0,0.8\rangle & \langle 0,0.8\rangle & \langle 0.3,0.5\rangle & \langle 0.4,0.6\rangle & \langle 0.4,0.5\rangle & \langle 0.3,0.2\rangle \\ Q & \langle 0.4,0.2\rangle & \langle 0.5,0.3\rangle & \langle 0.6,0.2\rangle & \langle 0.06,0.02\rangle & \langle \perp,\perp\rangle & \langle \perp,\perp\rangle \\ pl & \langle 0.55,0.3\rangle & \langle 0.6,0.4\rangle & \langle 0.75,0.2\rangle & \langle 0.65,0.3\rangle & \langle \perp,\perp\rangle & \langle \perp,\perp\rangle \\ pu_1 & \langle 0.1,0.4\rangle & \langle 0,0.4\rangle & \langle 0,0.2\rangle & \langle 0,0.4\rangle & \langle \perp,\perp\rangle & \langle \perp,\perp\rangle \end{array} \right.$$

After execution of Steps 5.-10. we obtain the optimal solution IM X_{opt} as follows:

$$X_{opt}[K*,L*] = \left\{ \begin{array}{c|cccc} & l_1 & l_2 & l_3 & l_4 \\ \hline k_1 & \langle 0,1\rangle & \langle 0,1\rangle & \langle 0.5,0.2\rangle & \langle 0,1\rangle \\ k_2 & \langle 0.4,0.2\rangle & \langle 0.17,0.62\rangle & \langle 0.1,0.4\rangle & \langle 0.03,0.02\rangle \\ k_3 & \langle 0,1\rangle & \langle 0.4,0.5\rangle & \langle 0,1\rangle & \langle 0,1\rangle \end{array} \right. .$$

The total intuitionistic fuzzy cost is:
$AGIO_{\oplus_{(\max,\min)})}\left(C_{(\{Q,pl,pu_1\},\{R,pu\})} \otimes_{(\min,\max)} X_{opt}\right) = \langle 0.4,0.2\rangle.$

The intuitionistic fuzzy optimum solution is not degenerated (it contains $m+n-1$ (the total number of producers and consumers decreased by 1) occupied cells in the X. The ranking function R, defined in (2), we can use to rank alternatives of decision-making process. For this optimal solution $R_{\langle 0.4;0.2\rangle} = 0.42$. The degree of membership (acceptance) of this optimal solution is equal to 0.4 and the its degree of non-membership (non-acceptance) is equal to 0.2.

The example illustrates the efficiency of the algorithm.

5 Conclusion

TPs have wide applications in logistics for reducing the cost. In real world applications, the parameters in TP may not be known precisely due to some uncontrollable factors. In this paper for the first time modified Zero suffix method is used to determine the optimal solution for an IFTP with additional restrictions by using apparatus of IMs and IFPs. Its main advantages are that can be applied to problems with imprecise parameters and can be extended in order to obtain the optimal solution for other types of multidimensional TPs. In future the proposed method will be implemented for various types FTPs. Research will continue in the direction of defining intuitionist fuzzy meta-euristic algorithms [18] to solve the TP.

Acknowledgements. The work is supported by the "Asen Zlatarov" University project under Ref. No. NIX-423/2019 "Innovative methods for extracting knowledge management".

References

1. Antony, R., Savarimuthu, S., Pathinathan, T.: Method for solving the transportation problem using triangular intuitionistic fuzzy number. Int. J. Comput. Algorithm **03**, 590–605 (2014)
2. Arsham, H., Khan, A.: A simplex type algorithm for general transportation problems-an alternative to stepping stone. J. Oper. Res. Soc. **40**(6), 581–590 (1989)
3. Atanassov, B.: Quantitative Methods in Business Management. Publishing House MedIna, Varna (1994). (in Bulgarian)
4. Atanassov K.T.: Intuitionistic Fuzzy Sets. VII ITKR Session, Sofia, 20–23 June 1983 (Deposed in Centr. Sci.-Techn. Library of the Bulg. Acad. of Sci., 1697/84) (in Bulgarian). Reprinted: Int. J. Bioautomation **20(S1)**, S1–S6 (2016)
5. Atanassov, K.: Generalized index matrices. Comptes rendus de l'Academie Bulgare des Sciences **40**(11), 15–18 (1987)
6. Atanassov, K.: On Intuitionistic Fuzzy Sets Theory. STUDFUZZ, vol. 283. Springer, Heidelberg (2012). https://doi.org/10.1007/978-3-642-29127-2
7. Atanassov, K.: Index Matrices: Towards an Augmented Matrix Calculus. Studies in Computational Intelligence, vol. 573. Springer, Cham (2014). https://doi.org/10.1007/978-3-319-10945-9
8. Atanassov, K.: Intuitionistic Fuzzy Logics. Studies in Fuzziness and Soft Computing, vol. 351. Springer (2017). https://doi.org/10.1007/978-3-319-48953-7
9. Atanassov, K., Szmidt, E., Kacprzyk, J.: On intuitionistic fuzzy pairs. Notes Intuitionistic Fuzzy Sets **19**(3), 1–13 (2013)
10. Chanas, S., Kolodziejckzy, W., Machaj, A.: A fuzzy approach to the transportation problem. Fuzzy Sets Syst. **13**, 211–221 (1984)
11. Charnes, A., Cooper, W.: The stepping-stone method for explaining linear programming calculation in transportation problem. Manag. Sci. **1**, 49–69 (1954)
12. Dantzig, G.: Application of the simplex method to a transportation problem. In: Chapter XXIII, Activity Analysis of Production and Allocation, vol. 13, pp. 359-373. Wiley, New York. Cowles Commision Monograph (1951)
13. Dhanasekar, S., Hariharan, S., Sekar, P.: Fuzzy Hungarian MODI Algorithm to solve fully fuzzy transportation problems. Int. J. Fuzzy Syst. **19**(5), 1479–1491 (2017)
14. Gani, A., Abbas, S.: A new average method for solving intuitionistic fuzzy transportation problem. Int. J. Pure Appl. Math. **93**(4), 491–499 (2014)
15. Gen, M., Ida, K., Li, Y., Kubota, E.: Solving bicriteria solid transportation problem with fuzzy numbers by a genetic algorithm. Comput. Ind. Eng. **29**(1), 537–541 (1995)
16. Gupta, G., Kumar, A., Sharma, M.: A note on "a new method for solving fuzzy linear programming problems based on the fuzzy linear complementary problem (FLCP)". Int. J. Fuzzy Syst. **17**, 1–5 (2016)
17. Ebrahimnejad, A., Verdegay, J.: A new approach for solving fully intuitionistic fuzzy transportation problems. Fuzzy Optim. Decis. Making **17**(4), 447–474 (2018)
18. Fidanova, S., Paprzycki, M., Roeva, O.: Hybrid GA-ACO algorithm for a model parameters identification problem. In: Proceedings of the Federated Conference on Computer Science and Information Systems (FedCSIS), pp. 413–420 (2014) https://doi.org/10.15439/2014F373
19. Hitchcock, F.: The distribution of a product from several sources to numerous localities. J. Math. Phys. **20**, 224–230 (1941)
20. Jahirhussain, R., Jayaraman, P.: Fuzzy optimal transportation problem by improved zero suffix method via robust rank techniques. Int. J. Fuzzy Math. Syst. (IJFMS) **3**, 303–311 (2013)

21. Jahihussain, R., Jayaraman, P.: A new method for obtaining an optinal solution for fuzzy transportation problems. Int. J. Math. Archive **4**(11), 256–263 (2013)
22. Jayaraman, P., Jahirhussain, R.: Fuzzy optimal transportation problem by improved zero suffix method via Robust Ranking technique. Int. J. Fuzzy Math. Syst. **3**(4), 303–311 (2013)
23. Ji, P., Chu, K.: A dual matrix approach to the transportation problem. Asia Pacific J. Oper. Res. **19**(1), 35–45 (2002)
24. Jimenez, F., Verdegay, J.: Solving fuzzy solid transportation problems by an evolutionary algorithm based parametric approach. Eur. J. Oper. Res. **117**(3), 485–510 (1999)
25. Kantorovich, L., Gavyrin, M.: Application of mathematical methods in the analysis of cargo flows. Coll. of articles Problems of increasing the efficiency of transport. M.: Publishing House AHSSSR, pp. 110–138 (1949). (in Russian)
26. Kathirvel, K., Balamurugan, K.: Method for solving fuzzy transportation problem using trapezoidal fuzzy numbers. Int. J. Eng. Res. Appl. **2**(5), 2154–2158 (2012)
27. Kathirvel, K., Balamurugan, K.: Method for solving unbalanced transportation problems using trapezoidal fuzzy numbers. Int. J. Eng. Res. Appl. **3**(4), 2591–2596 (2013)
28. Kaur, A., Kumar, A.: A new approach for solving fuzzy transportation problems using generalized trapezoidal fuzzy numbers. Appl. Soft Comput. **12**(3), 1201–1213 (2012)
29. Kumar, P., Hussain, R.: A method for solving unbalanced intuitionistic fuzzy transportation problems. Notes Intuitionistic Fuzzy Sets **21**(3), 54–65 (2015)
30. Kumari, N., Kumar, R.: Zero suffix method for the transshipment problem with mixed constraints. Int. J. Math. Archive **9**(3), 242–251 (2018)
31. Lalova, N., Ilieva, L., Borisova, S., Lukov, L., Mirianov, V.: A Guide to Mathematical Programming. Science and Art Publishing House, Sofia (1980). (in Bulgarian)
32. Mohideen, S., Kumar, P.: A comparative study on transportation problem in fuzzy environment. Int. J. Math. Res. **2**(1), 151–158 (2010)
33. Ngastiti, P., Surarso, B., Sutimin: zero point and zero suffix methods with robust ranking for solving fully fuzzy transportation problems. J. Phys.: Conf. Ser. **1022**, 1–10 (2018)
34. Pandian, P., Natarajan, G.: A new algorithm for finding a fuzzy optimal solution for fuzzy transportation problems. Appl. Math. Sci. **4**, 79–90 (2010)
35. Patil, A., Chandgude, S.: Fuzzy Hungarian approach for transportation model. Int. J. Mech. Ind. Eng. **2**(1), 77–80 (2012)
36. Petkova, St., Tasheva, Y., Petkov, P.: Classification and possibilities of management of primary energy sources. Oxid. Commun. Book **33**(2), 462–469 (2010)
37. Purushothkumar, M., Ananthanarayanan, M., Dhanasekar, S.: Fuzzy zero suffix algorithm to solve fully fuzzy transportation problems. Int. J. Pure Appl. Math. **119**(9), 79–88 (2018)
38. Shanmugasundari, M., Ganesan, K.: A novel approach for the fuzzy optimal solution of fuzzy transportation problem. Int. J. Eng. Res. Appl. **3**(1), 1416–1424 (2013)
39. Sharma, S., Shanker, R.: A modified zero suffix method for finding an optimal solution for transportation problems. Eur. J. Sci. Res. **104**(4), 673–676 (2013)
40. Shih, W.: Modified stepping stone method as a teaching aid for capacitated transportation problems. Decis. Sci. **18**, 662–676 (1987)
41. Sudhagar, V., Navaneethakumar, V.: Solving the multiobjective two stage fuzzy transportation problem by zero suffix method. J. Math. Res. **2**(4), 135–140 (2010)
42. Szmidt, E., Kacprzyk, J.: Amount of information and its reliability in the ranking of Atanassov's intuitionistic fuzzy alternatives. In: Rakus-Andersson, E., Yager, R., Ichalkaranje, N., Jain, L.C. (eds.) Recent Advances in Decision Making. SCI, vol. 222, pp. 7–19. Springer, Heidelberg (2009)
43. Traneva, V.: Internal operations over 3-dimensional extended index matrices. Proc. Jangjeon Math. Soc. **18**(4), 547–569 (2015)

44. Traneva, V., Tranev, S., Atanassova, V.: An intuitionistic fuzzy approach to the Hungarian algorithm. In: Nikolov, G., et al. (eds.) NMA 2018. LNCS, vol. 11189, pp. 1–9. Springer (2019). https://doi.org/10.1007/978-3-030-10692-8_19
45. Traneva, V., Tranev, S., Stoenchev, M., Atanassov, K.: Scaled aggregation operations over two- and three-dimensional index matrices. Soft. Comput. **22**(15), 5115–5120 (2018). https://doi.org/10.1007/s00500-018-3315-6
46. Traneva, V., Marinov, P., Atanassov, K.: Index matrix interpretations of a new transportation-type problem. Comptes rendus de l'Academie Bulgare des Sciences **69**(10), 1275–1283 (2016)
47. Traneva, V., Marinov, P., Atanassov, K.: Transportation-type problems and their index matrix interpretations. Adv. Stud. Contemp. Math. **26**(4), 587–594 (2016)
48. Traneva, V., Tranev, S.: Index Matrices as a Tool for Managerial Decision Making. Publishing House of the Union of Scientists, Bulgaria (2017). (in Bulgarian)
49. Zadeh, L.: Fuzzy sets. Inf. Control **8**(3), 338–353 (1965)

InterCriteria Analysis Implementation for Exploration of the Performance of Various Docking Scoring Functions

Ivanka Tsakovska[1], Petko Alov[1], Nikolay Ikonomov[2], Vassia Atanassova[1], Peter Vassilev[1], Olympia Roeva[1], Dessislava Jereva[1], Krassimir Atanassov[1], Ilza Pajeva[1], and Tania Pencheva[1](✉)

[1] Institute of Biophysics and Biomedical Engineering, Bulgarian Academy of Sciences, Acad. G. Bonchev Bl. 105, 1113 Sofia, Bulgaria
{itsakovska,olympia,dessislava.jereva,pajeva, tania.pencheva}@biomed.bas.bg, petko@biophys.bas.bg, vassia.atanassova@gmail.com, peter.vassilev@gmail.com, krat@bas.bg

[2] Institute of Mathematics and Informatics, Bulgarian Academy of Sciences, Acad. G. Bonchev Bl. 8, 1113 Sofia, Bulgaria
nikonomov@math.bas.bg

Abstract. The present study describes an implementation of InterCriteria Analysis (ICrA) in the field of the computer-aided drug design and computational toxicology. ICrA strives to go beyond the nature of the criteria involved in a process of evaluation of multiple objects against multiple criteria, and, thus to discover some dependencies between the criteria themselves. The approach is based on the apparatus of the index matrices and the intuitionistic fuzzy sets. In this study new software capabilities, implemented in order to apply ICrA to *in silico* drug design, are presented. As a case study, ICrA is implemented to explore the performance of various scoring functions in docking. Docking, which is the most commonly used structure-based drug design method, has been applied to predict the binding mode and to provide a measure for the ligand binding affinity to the protein. In particular, ligands of the peroxisome proliferator-activated nuclear receptor gamma (PPARγ), involved in the control of a number of physiologically significant processes, have been investigated towards prediction of their binding to the protein. A dataset of 160 tyrosine-based PPARγ agonists with experimentally determined binding affinities has been used in this investigation. Docking combined with the *in-house* developed pharmacophore model as a filter has been applied. Various scoring functions and docking protocols have been tested in the molecular modelling platform MOE (v. 2019.01). ICrA has been applied to assess dependencies among the scoring functions. The analysis has demonstrated high positive consonance for two of the scoring functions – London dG and Alpha HB. None of the functions could be distinguished as a good predictor of the experimental binding affinity.

1 Introduction

In silico methods to direct the drug design process depend on the available structural information and are generally classified as ligand- and structure-based. The latter rely

I. Dimov and S. Fidanova (Eds.): HPC 2019, SCI 902, pp. 88–98, 2021.
https://doi.org/10.1007/978-3-030-55347-0_8

on the data for the biomacromolecule (receptor) structure provided by X-ray crystallography and/or NMR (Nuclear Magnetic Resonance) and the core method applied here is docking. Docking is a modelling technique that positions the small molecule (ligand) structure in different orientations and conformations within the binding site of the receptor to calculate optimal binding mode and binding energy. The method is based on the principle of complementarity in the ligand-receptor interaction and applies a scoring function to approximate the change in the free energy of their binding [16]. There are many docking algorithms and the implemented scoring functions play a key role for the algorithm performance. Despite the large number of comparative studies on various scoring functions, the selection of a suitable scoring function for a particular dataset of ligands that bind to a particular receptor is not straightforward. To face this question, InterCriteria analysis (ICrA), recently developed as a new multicriteria decision making approach, is a suitable tool [7]. So far ICrA has been successfully applied in the fields of economics [10], e-learning [23], algorithms performance [25,26], ecology [19,20], medicine [30], etc. In this study the capacity of ICrA as a supportive tool in the selection of a reliable scoring function was explored. To this aim a case docking study was performed on ligands of the peroxisome proliferator-activated nuclear receptor gamma (PPARγ) – one of the important nuclear hormone receptors involved in the control of a number of physiologically significant processes. A number of docking runs were performed using different docking protocols and different scoring functions implemented in Molecular Operating Environment (MOE) software [24]. The results were subjected to ICrA and the dependencies between the scoring functions were analysed.

2 InterCriteria Analysis

2.1 Theoretical Outline

ICrA has been developed in order to discover certain relations existing between comparable criteria involved in multicriteria problems [7]. The method is founded upon the apparatus of the index matrices [5], and intuitionistic fuzzy sets [3,4].

Detailed discussions regarding ICrA exist in a number of papers devoted to different areas of application [2,12,27,29,30].

Further, we provide a brief outline of the theoretical background of ICrA. Each pair of criteria is assigned an Intuitionistic Fuzzy Pair (IFP) [8] as an estimate of their "agreement" and "disagreement", with respect to the different objects they measure. To illuminate the idea behind the approach we provide a concise description of the main steps.

Let O denote the set of all objects $O_1, O_2, \ldots, O_n$ ($n \geq 3$) being evaluated, and $C(O)$ be the set of values computed[1] by a given criteria C for the objects, i.e.,

$$O \stackrel{\text{def}}{=} \{O_1, O_2, O_3, \ldots, O_n\},$$

$$C(O) \stackrel{\text{def}}{=} \{x_1, x_2, x_3, \ldots, x_n\}.$$

[1] On the most abstract level these computed values may be of any type. The only restriction posed on them is that they should be in some sense partitionable in at least two groups - comparable and non-comparable. In practice, the current implementation works only with real numerical values or ordered pairs of real numerical values (in particular IFPs).

where, for brevity, we have put $x_i = C(O_i)$. Then the following set can be defined:

$$C^*(O) \stackrel{\text{def}}{=} \{\langle x_i, x_j\rangle | i \neq j \,\&\, \langle x_i, x_j\rangle \in C(O) \times C(O)\}.$$

In order to find the "agreement" of two criteria, a vector of all internal relations corresponding to a given criterion must be constructed.[2]

Any element $u \in C^*(O)$ must fulfil exactly one of four relations $R, \overline{R}, \tilde{R}$ and $\tilde{I}$. That is, for a fixed criterion C and any ordered pair $\langle x, y\rangle \in C^*(O)$ the following must hold true:

$$\langle x, y\rangle \in R \Leftrightarrow \langle y, x\rangle \in \overline{R}, \tag{1}$$
$$\langle x, y\rangle \in \tilde{R} \Leftrightarrow \langle y, x\rangle \in \tilde{R}, \tag{2}$$
$$\langle y, x\rangle \in \tilde{I} \Leftrightarrow \langle x, y\rangle \notin (R \cup \overline{R} \cup \tilde{R}), \tag{3}$$
$$R \cup \overline{R} \cup \tilde{R} \cup \tilde{I} = C^*(O). \tag{4}$$

In informal statement, the four relations should cover all possible combinations of pairs, and two of them should be "opposites". These may be thought as an analogue of preference relations – "better", "worse" (the opposing ones), "indifferent" and "incomparable".

In the current version of the algorithm the relations considered are $R \equiv >$, $\overline{R} \equiv <$ and $\tilde{R} \equiv =$ and $\tilde{I}$ – incomparable $\not\sim$. These partial orderings are defined as follows:

$$(u_1, u_2) > (v_1, v_2) \Leftrightarrow \begin{cases} u_1 > v_1 \\ u_2 < v_2 \end{cases} \text{or} \begin{cases} u_1 \geq v_1 \\ u_2 < v_2 \end{cases} \text{or} \begin{cases} u_1 > v_1 \\ u_2 \leq v_2 \end{cases},$$

$$(u_1, u_2) < (v_1, v_2) \Leftrightarrow \begin{cases} u_1 < v_1 \\ u_2 > v_2 \end{cases} \text{or} \begin{cases} u_1 \leq v_1 \\ u_2 > v_2 \end{cases} \text{or} \begin{cases} u_1 < v_1 \\ u_2 \geq v_2 \end{cases},$$

$$(u_1, u_2) = (v_1, v_2) \Leftrightarrow \begin{cases} u_1 = v_1 \\ u_2 = v_2 \end{cases},$$

$$(u_1, u_2) \not\sim (v_1, v_2) \Leftrightarrow \begin{cases} (u_1, u_2) \neq (v_1, v_2) \\ (u_1, u_2) \not> (v_1, v_2) \\ (u_1, u_2) \not< (v_1, v_2) \end{cases}.$$

From the discussion above, it is evident that only a subset of $C(O) \times C(O)$ is sufficient for the effective calculation of the vector of internal comparisons (due to Eqs. (1)–(4)). Therefore, we shall consider lexicographically ordered pairs $\langle x, y\rangle$. Denote for brevity: $C_{i,j} = \langle x_i, x_j\rangle$. Then for a criterion C, the vector with $n(n-1)/2$ elements is obtained:

$$V(C) = \{C_{1,2}, C_{1,3}, \ldots, C_{1,n}, C_{2,3}, C_{2,4}, \ldots, C_{2,n}, C_{3,4}, \ldots, C_{3,n}, \ldots, C_{n-1,n}\}.$$

[2] In what follows we provide the description for ordered pairs of real numerical values as they include the real numerical values as a particular case. In future work, tolerance thresholds and other ways to provide more user control are planned to be implemented.

Considering instead of $V(C)$, its simplified version $\hat{V}(C)$, with k-th component $(1 \leq k \leq n(n-1)/2)$ is given by:

$$\hat{V}_k(C) = \begin{cases} 1, \text{ iff } V_k(C) \in R, \\ -1, \text{ iff } V_k(C) \in \overline{R}, \\ 2, \text{ iff } V_k(C) \in \tilde{R}, \\ 0, \text{ otherwise.} \end{cases}$$

When comparing two criteria C and C', the degree of "agreement" $\mu_{C,C'}$ is usually determined as the number of matching non-zero components of the respective vectors (divided by the length of the vector for normalization purposes). The degree of "disagreement" $\nu_{C,C'}$ is the number of odd-valued components of opposing signs in the two vectors.

It is obvious (from the way of calculation) that for $\mu_{C,C'}$ and $\nu_{C,C'}$ we have $\mu_{C,C'} = \mu_{C',C}$ and $\nu_{C,C'} = \nu_{C',C}$. As an immediate consequence of that fact we have that $\langle \mu_{C,C'}, \nu_{C,C'} \rangle$ is an IFP. For further details, regarding the different algorithms implemented in the software, we refer the interested reader to [28].

In essence, ICrA may be viewed as a transformation of the index matrix of criteria and objects:

$$\begin{array}{c|cccc} & O_1 & O_2 & \dots & O_m \\ \hline C_1 & C_1(O_1) & C_1(O_2) & \dots & C_1(O_m) \\ C_2 & C_2(O_1) & C_2(O_2) & \dots & C_2(O_m) \\ \vdots & \vdots & \vdots & \ddots & \vdots \\ C_n & C_n(O_1) & C_n(O_2) & \dots & C_n(O_m) \end{array} \tag{5}$$

into the symmetric index matrix of criteria:

$$\begin{array}{c|cccc} & C_1 & C_2 & \dots & C_n \\ \hline C_1 & \langle 1,0 \rangle & \langle \mu_{C_1,C_2}, \nu_{C_1,C_2} \rangle & \dots & \langle \mu_{C_1,C_n}, \nu_{C_1,C_n} \rangle \\ C_2 & \langle \mu_{C_2,C_1}, \nu_{C_2,C_1} \rangle & \langle 1,0 \rangle & \dots & \langle \mu_{C_2,C_n}, \nu_{C_2,C_n} \rangle \\ \vdots & \vdots & \vdots & \ddots & \vdots \\ C_n & \langle \mu_{C_n,C_1}, \nu_{C_n,C_1} \rangle & \langle \mu_{C_n,C_2}, \nu_{C_n,C_2} \rangle & \dots & \langle 1,0 \rangle \end{array}, \tag{6}$$

Informally speaking the closer the values $\langle \mu_{C_i,C_j}, \nu_{C_i,C_j} \rangle$ is to $\langle 1,0 \rangle$, the more similar the behaviour of the C_i and C_j. For the more formal considerations we refer the interested reader to [6,9,18].

The concepts that will be used further are "consonance" and "dissonance". Let $0 \leq \beta \leq \alpha \leq 1$ be given, such that $\alpha + \beta \leq 1$. Following [6], we say that criteria C_i and C_j are in:

- (α, β)-positive consonance, if $\mu_{C_i,C_j} > \alpha$ and $\nu_{C_i,C_j} < \beta$;
- (α, β)-negative consonance, if $\mu_{C_i,C_j} < \beta$ and $\nu_{C_i,C_j} > \alpha$;
- (α, β)-dissonance, otherwise.

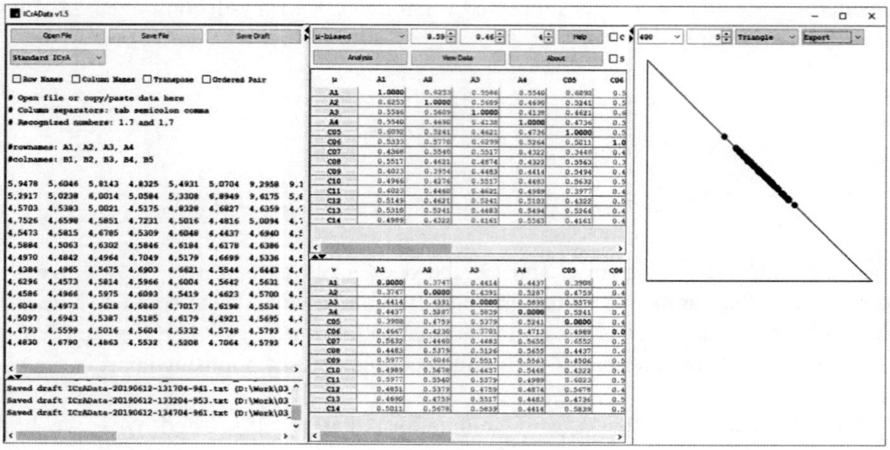

Fig. 1. A view of the user interface

2.2 Software Description

Here we provide a quick overview of the software implementing the ICrA approach. It is called ICrAData and is freely available for use from [17]. It is written in the Java programming language [21], and requires the installation of Java Virtual Machine from [22]. This makes it usable under Linux and Windows environment. The application is started by "ICrAData.jar". The user interface is shown on Fig. 1 – the left panel displays the input data, the central panel – the result of ICrA in a coloured table view, and the rightmost panel shows the graphical interpretation of the results.

Among the recent additions to the software the most notable are the following:

In order to easily load data from other software products, the capability to load CSV (comma separated values) files with headers (row and column) which are taken as names for objects and criteria, was added to the software. This allows loading of tables from MS Excel/LibreOffice Calc.

For better visualization of the results, table cell colours were added, according to the following rules, depending on the user defined α and β thresholds:

- The results are displayed in dark-green colour in case of positive consonance ($\mu_{C_i,C_j} > \alpha$ and $\nu_{C_i,C_j} < \beta$);
- The results are displayed in light-blue colour in case of negative consonance ($\mu_{C_i,C_j} < \beta$ and $\nu_{C_i,C_j} > \alpha$);
- Otherwise, in case of dissonance – violet colour.

The default values used by the software ICrAData are $\alpha = 0.75$ and $\beta = 0.25$, respectively.

An example of the coloured feature (for user defined α and β) augmenting the visualization of the output results (degree of "agreement" μ_{C_i,C_j} and degree of "disagreement" ν_{C_i,C_j}) is presented on Fig. 2.

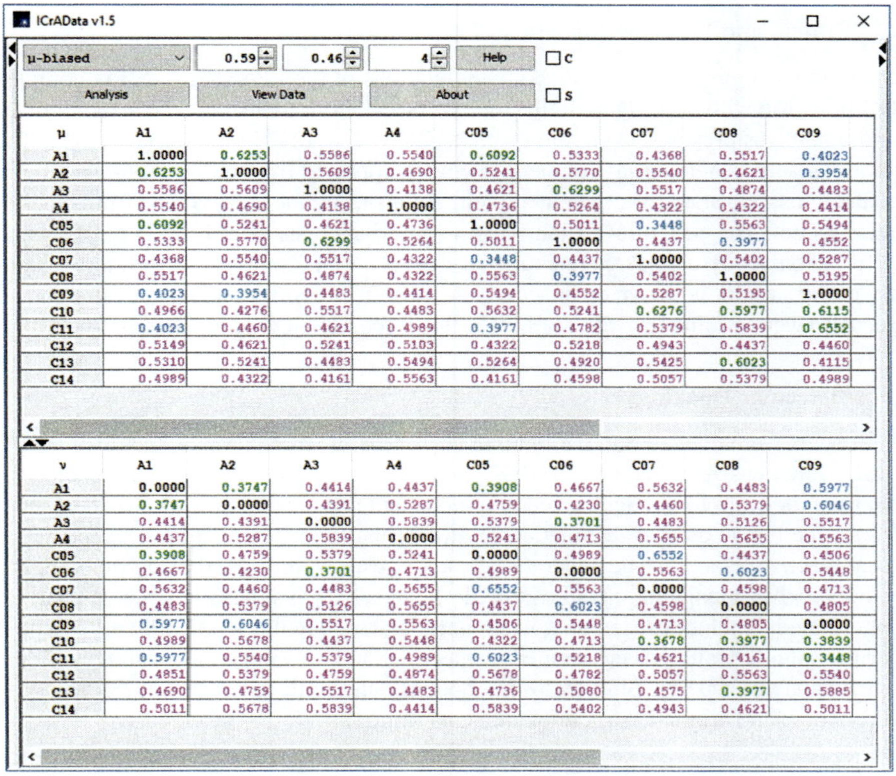

μ	A1	A2	A3	A4	C05	C06	C07	C08	C09
A1	1.0000	0.6253	0.5586	0.5540	0.6092	0.5333	0.4368	0.5517	0.4023
A2	0.6253	1.0000	0.5609	0.4690	0.5241	0.5770	0.5540	0.4621	0.3954
A3	0.5586	0.5609	1.0000	0.4138	0.4621	0.6299	0.5517	0.4874	0.4483
A4	0.5540	0.4690	0.4138	1.0000	0.4736	0.5264	0.4322	0.4322	0.4414
C05	0.6092	0.5241	0.4621	0.4736	1.0000	0.5011	0.3448	0.5563	0.5494
C06	0.5333	0.5770	0.6299	0.5264	0.5011	1.0000	0.4437	0.3977	0.4552
C07	0.4368	0.5540	0.5517	0.4322	0.3448	0.4437	1.0000	0.5402	0.5287
C08	0.5517	0.4621	0.4874	0.4322	0.5563	0.3977	0.5402	1.0000	0.5195
C09	0.4023	0.3954	0.4483	0.4414	0.5494	0.4552	0.5287	0.5195	1.0000
C10	0.4966	0.4276	0.5517	0.4483	0.5632	0.5241	0.6276	0.5977	0.6115
C11	0.4023	0.4460	0.4621	0.4989	0.3977	0.4782	0.5379	0.5839	0.6552
C12	0.5149	0.4621	0.5241	0.5103	0.4322	0.5218	0.4943	0.4437	0.4460
C13	0.5310	0.5241	0.4483	0.5494	0.5264	0.4920	0.5425	0.6023	0.4115
C14	0.4989	0.4322	0.4161	0.5563	0.4161	0.4598	0.5057	0.5379	0.4989

ν	A1	A2	A3	A4	C05	C06	C07	C08	C09
A1	0.0000	0.3747	0.4414	0.4437	0.3908	0.4667	0.5632	0.4483	0.5977
A2	0.3747	0.0000	0.4391	0.5287	0.4759	0.4230	0.4460	0.5379	0.6046
A3	0.4414	0.4391	0.0000	0.5839	0.5379	0.3701	0.4483	0.5126	0.5517
A4	0.4437	0.5287	0.5839	0.0000	0.5241	0.4713	0.5655	0.5655	0.5563
C05	0.3908	0.4759	0.5379	0.5241	0.0000	0.4989	0.6552	0.4437	0.4506
C06	0.4667	0.4230	0.3701	0.4713	0.4989	0.0000	0.5563	0.6023	0.5448
C07	0.5632	0.4460	0.4483	0.5655	0.6552	0.5563	0.0000	0.4598	0.4713
C08	0.4483	0.5379	0.5126	0.5655	0.4437	0.6023	0.4598	0.0000	0.4805
C09	0.5977	0.6046	0.5517	0.5563	0.4506	0.5448	0.4713	0.4805	0.0000
C10	0.4989	0.5678	0.4437	0.5448	0.4322	0.4713	0.3678	0.3977	0.3839
C11	0.5977	0.5540	0.5379	0.4989	0.6023	0.5218	0.4621	0.4161	0.3448
C12	0.4851	0.5379	0.4759	0.4874	0.5678	0.4782	0.5057	0.5563	0.5540
C13	0.4690	0.4759	0.5517	0.4483	0.4736	0.5080	0.4575	0.3977	0.5885
C14	0.5011	0.5678	0.5839	0.4414	0.5839	0.5402	0.4943	0.4621	0.5011

Fig. 2. Example of the colouring feature (a closer view of the middle panel from Fig. 1

ICrAData saves a draft automatically each 15 min and on program exit in order to prevent accidental loss or overwriting of data.

The features outlined above allow for better automation in working process with program and additional improvements in that regard are also planned in the future.

3 Case Study

3.1 Datasets Preparation

Two datasets were used in this study:

(i) 160 tyrosine-based PPARγ agonists with experimentally determined binding affinities (K_i) collected from the literature [13–15];
(ii) 21 PPARγ full agonists extracted from the Protein Data Bank (PDB, www.rcsb.org) [11]. The preparation, optimisation, and docking of the chemical structures were performed in MOE v. 2019.01.

3.2 Protein and Ligands' Structures Preparation

As a template complex for the docking simulations in the PPARγ, the X-ray complex PDB ID 1FM6 with the drug Rosiglitazone was used. Prior to docking the receptor binding domain was prepared using the Protonate3D tool in MOE as presented in [1]. The tool assigns hydrogens to structures following the optimal free energy proton geometry and the ionisation states of titratable protein groups using the Generalized Born electrostatics model. The structures of the tyrosine-based agonists were built in MOE, while the full agonists' structures were extracted from the PDB complexes. The stereochemistry of the structures was fixed in accordance with the reported stereochemistry and all structures were minimised with the MMFF94s force field including electrostatics.

3.3 Molecular Docking

The ligands were docked into the binding site of the prepared protein and all available scoring functions implemented in MOE were explored. The scoring functions have a dual role: they aim to predict the binding mode and at the same time to provide a measure for the ligand affinity (estimation of affinity and/or differentiating between active and non-active ligands). The generated docking poses were filtered using the *in-house* developed pharmacophore model of PPARγ ligands [31]. The most restrictive pharmacophore model includes seven pharmacophoric features. In this study only the essential pharmacophoric features were applied, namely: F1 and F2 – polar atoms and functional groups capable of performing hydrogen bonding and ionic interactions; F3 and F5 – hydrophobic and aromatic structural elements (Fig. 3).

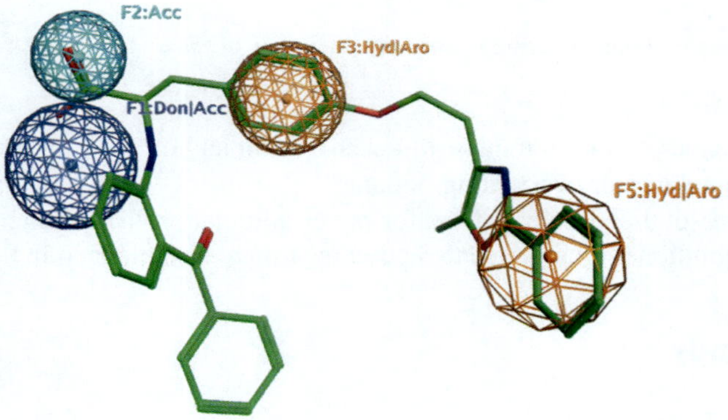

Fig. 3. Four-features pharmacophore model of PPARγ full agonists

The investigated dataset of 160 tyrosine-based PPARγ agonists was subjected to docking applying the following protocols:

(i) *Virtual screening*, where the structure of the receptor is kept rigid during docking;
(ii) *Induced fit* that allows the protein side chains to move during docking adopting more adequate geometries toward the ligand structure.

3.4 Intercriteria Analysis Implementation

ICrA was applied to evaluate the ability of the scoring functions to predict binding affinity of the studied ligands toward PPARγ. Further, the ligands' binding mode prediction was investigated. As expected, the *Induced fit* protocol required much more computational time than the *Virtual screening* protocol, however no significant improvement in the predictions was achieved. Thus, all further investigations were performed applying *Virtual screening* protocol. Generated poses were filtered using the *in-house* developed pharmacophore model of PPARγ full agonists (Fig. 3) that guaranteed further precision in the selection of the final poses. This two-step virtual screening procedure was applied on all 160 tyrosine-based PPARγ agonists and the performance of the scoring functions was evaluated toward prediction of their binding affinity. In the analysis lower scores indicated more favourable poses for all scoring functions. Docking was performed applying all five scoring functions available in MOE: Affinity dG, Alpha HB, ASE, GBVI/WSA dG, London dG. The top ten poses of each ligand were saved. All obtained results were collected and structured in an appropriate manner for further implementation of ICrA approach. Two scoring functions, namely GBVI/WSA dG and Affinity dG, were excluded from further analyses because of unrealistic (negative values close to zero and positive values) binding affinities predicted. The performance of ASE, Alpha HB and London dG scoring functions was further subjected to ICrA for different scenarios, as follows:

(i) using the score (calculated interaction energy) of the top ranked docking pose of each ligand;
(ii) using the mean of the scores of the top five docking poses of each ligand;
(iii) using the mean of the scores of the top ten docking poses of each ligand.

No significant consonance between the calculated binding affinities using any of the investigated scoring functions and experimental binding affinities was found, no matter the top one or the mean of the top five/top ten poses were explored. Further the relations between the scoring functions themselves were investigated using ICrA. For any of the scenarios the results revealed positive consonance between London dG and Alpha HB scoring functions, as shown in Table 1 and Fig. 4.

Table 1. ICrA implementation based on the mean of the top ten ranked binding poses (for the meaning of the colours refer to Fig. 4)

μ_{C_i,C_j}	Alpha HB	ASE	London DG	pK$_i$
Alpha HB	1.000	0.716	0.876	0.411
ASE	0.716	1.000	0.714	0.335
London DG	0.876	0.714	1.000	0.401
pK$_i$	0.411	0.335	0.401	1.000

The results suggest that there is a positive consonance between the predicted binding affinity values calculated by London dG and Alpha HB scoring functions. Thus for

A

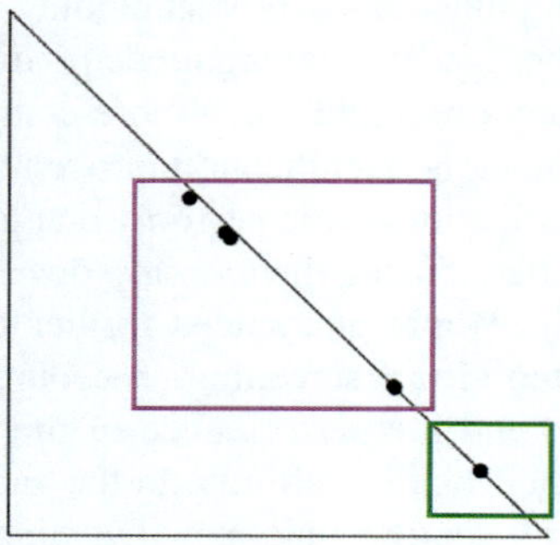

B

Meaning	Interval of μ
strong positive consonance	(0.95 - 1.00]
positive consonance	(0.85 - 0.95]
weak positive consonance	(0.75 - 0.85]
weak dissonance	(0.67 - 0.75]
dissonance	(0.57 - 0.67]
strong dissonance	(0.43 - 0.57]
dissonance	(0.33 - 0.43]
weak dissonance	(0.25 - 0.33]
weak negative consonance	(0.15 - 0.25]
negative consonance	(0.05 - 0.15]
strong negative consonance	(0.00 - 0.05]

Fig. 4. A. Geometrical interpretation of the results presented in Table 1 by the intuitionistic fuzzy triangle; B. ICrA scale according to the μ_{C_i,C_j} values [6]

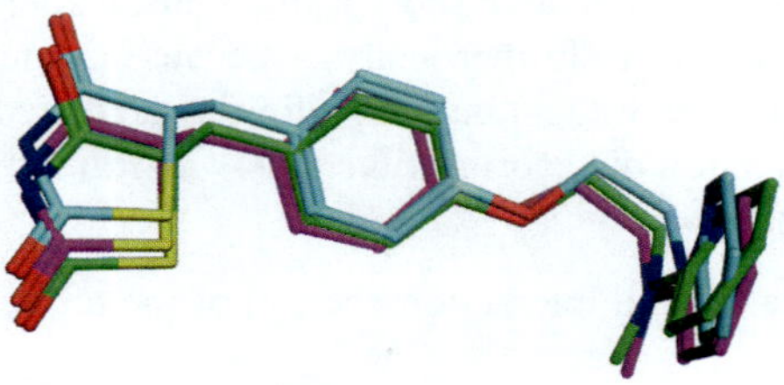

Fig. 5. Binding poses of Rosiglitasone resulting from the prediction using Alpha HB (green), London dG (light blue) scoring functions and PDB structure (magenta, template complex PDB ID 1FM6)

the particular class of tyrosine-based PPARγ agonists both functions produce similar results. To investigate further the ability of London dG and Alpha HB to predict the ligands' binding mode the performance of both functions was investigated on crystal complexes of 21 PPARγ full agonists extracted from PDB and compared to the other scoring functions in MOE. For this purpose, the ligands were extracted from their binding site and re-docked, applying the *Virtual screening* protocol and PPARγ pharmacophore as a filter. The top ten poses of each ligand were analysed and the one best fitted to the ligand structure from the PDB crystal complex was selected (Fig. 5).

The comparison of the RMSD (root-mean-square deviation) values demonstrated the best performance of Alpha HB in the prediction of the binding mode, followed by London dG. It is known that Alpha HB is a linear combination of two terms, the first one measuring the geometrical fit of the ligand to the binding site, and the second one reflecting hydrogen bonding effects in the ligand-receptor interactions. Therefore, the best performance of this scoring function on the investigated PDB complexes is in agreement with the experimental knowledge about the role of the hydrogen bonding in the stabilization of helix 12 of the ligand-binding domain of PPARγ.

4 Conclusion

The paper presents an implementation of InterCriteria Analysis as a decision making approach in the field of the *in silico* drug design and computational toxicology. To illustrate its potential, a case study on docking of ligands in the PPARγ binding domain was performed and the applicability of ICrA was investigated toward evaluation of various scoring functions performance. In the performed analyses, two scoring functions (London dG and Alpha HB) demonstrated positive consonance, while none of the functions was outlined as a good predictor of the binding affinity. Further analyses outlined Alpha HB as a good predictor of the experimental ligands' binding mode in the PPARγ binding site. For the purposes of *in silico* drug design studies the ICrAData software was upgraded to cover some new features of the ICrA as well as to propose user-friendly view for subsequent analysis.

Acknowledgements. Funding from the National Science Fund of Bulgaria, under grant DN 17/6 "A New Approach, Based on an Intercriteria Data Analysis, to Support Decision Making in *in silico* Studies of Complex Biomolecular Systems", is gratefully acknowledged.

References

1. Al Sharif, M., Tsakovska, I., Pajeva, I.: The application of molecular modelling in the safety assessment of chemicals: a case study on ligand-dependent PPARγ dysregulation. Toxicology **392**, 140–154 (2017)
2. Angelova, N., Atanassov, K., Riecan, B.: InterCriteria analysis of the intuitionistic fuzzy implication properties. Notes IFSs **21**(5), 20–23 (2015)
3. Atanassov, K.: Intuitionistic fuzzy sets, VII ITKR Session, Sofia, 20–23 June 1983, Reprinted: Int. J. Bioautom. **20**(S1), S1–S6 (2016)
4. Atanassov, K.: On Intuitionistic Fuzzy Sets Theory. Springer, Cham (2012)
5. Atanassov, K.: Index Matrices: Towards An Augmented Matrix Calculus. Springer, Berlin (2014)
6. Atanassov, K., Atanassova, V., Gluhchev, G.: InterCriteria analysis: ideas and problems. Notes IFSs **21**(1), 81–88 (2015)
7. Atanassov, K., Mavrov, D., Atanassova, V.: Intercriteria decision making: a new approach for multicriteria decision making, based on index matrices and intuitionistic fuzzy sets. Issues IFSs GNs **11**, 1–8 (2014)
8. Atanassov, K., Szmidt, E., Kacprzyk, J.: On intuitionistic fuzzy pairs. Notes IFSs **19**(3), 1–13 (2013)
9. Atanassova, V.: Interpretation in the intuitionistic fuzzy triangle of the results, obtained by the InterCriteria analysis. In: Proceedings of 9th Conference of the European Society for Fuzzy Logic and Technology (EUSFLAT), pp. 1369–1374 (2015)
10. Atanassova, V., Doukovska, L., Atanassov, K., Mavrov, D.: InterCriteria decision making approach to EU member states competitiveness analysis. In: Proceedings of the International Symposium on Business Modeling and Software Design – BMSD 2014, pp. 289–294 (2014)
11. Berman, H.M., Westbrook, J., Feng, Z., Gilliland, G., Bhat, T.N., Weissig, H., Shindyalov, I.N., Bourne, P.E.: The protein data bank. Nucl. Acids Res. **28**, 235–242 (2000)
12. Bureva, V., Sotirova, E., Sotirov, S., Mavrov, D.: Application of the InterCriteria decision making method to Bulgarian universities ranking. Notes IFSs **21**(2), 111–117 (2015)

13. Cobb, J.E., Blanchard, S.G., Boswell, E.G., et al.: N-(2-Benzoylphenyl)-L-tyrosine PPARγ agonists. 3. Structure-activity relationship and optimization of the N-aryl substituent. J. Med. Chem. **41**, 5055–5069 (1998)
14. Collins, J.L., Blanchard, S.G., Boswell, E.G., et al.: N-(2-Benzoylphenyl)-L-tyrosine PPARγ agonists. 2. Structure-activity relationship and optimization of the phenyl alkyl ether moiety. J. Med. Chem. **41**, 5037–5054 (1998)
15. Henke, B.R., Blanchard, S.G., Brackeen, M.F., et al.: N-(2-Benzoylphenyl)-L-tyrosine PPARγ agonists. 1. Discovery of a novel series of potent antihyperglycemic and antihyperlipidemic agents. J. Med. Chem. **41**(25), 5020–5036 (1998)
16. Höltje, H.D., Sippl, W., Rognan, D., Folkers D.: Molecular Modeling: Basic Principles and Applications. 3rd Revised and Expanded edition, Wiley-VCH Verlag GmbH & Co. KGaA, Weinheim (2008)
17. InterCriteria.net – ICrAData software. http://intercriteria.net/software/. Accessed 31 May 2019
18. Ikonomov, N., Vassilev, P., Roeva, O.: ICrAData - Software for InterCriteria analysis. Int. J. Bioautom. **22**(1), 1–10 (2018)
19. Ilkova, T., Petrov, M.: Application of InterCriteria analysis to the Mesta river pollution modelling. Notes IFSs **21**(2), 118–125 (2015)
20. Ilkova, T., Petrov, M.: InterCriteria analysis for evaluation of the pollution of the Struma river in the Bulgarian section. Notes IFSs **22**(3), 120–130 (2016)
21. Java platform SE 8, API documentation. https://docs.oracle.com/javase/8/docs/api/. Accessed 31 May 2019
22. Java virtual machine. https://java.com/. Accessed 31 May 2019
23. Krawczak, M., Bureva, V., Sotirova, E., Szmidt, E.: Application of the InterCriteria decision making method to universities ranking. Adv. Intell. Syst. Comput. **401**, 365–372 (2016)
24. Molecular operating environment. Chemical computing group. http://www.chemcomp.com. Accessed 31 May 2019
25. Pencheva, T., Angelova, M.: InterCriteria analysis of simple genetic algorithms performance. Stud. Comput. Intell. **681**, 147–159 (2017)
26. Pencheva, T., Angelova, M., Vassilev, P., Roeva, O.: InterCriteria analysis approach to parameter identification of a fermentation process model. Adv. Intell. Syst. Comput. **401**, 385–397 (2016)
27. Roeva, O., Fidanova, S., Luque, G., Paprzycki, M.: Intercriteria analysis of ACO performance for workforce planning problem. Stud. Comput. Intell. **795**, 47–67 (2019)
28. Roeva, O., Vassilev, P., Ikonomov, N., Angelova, M., Su, J., Pencheva, T.: On different algorithms for InterCriteria relations calculation. Stud. Comput. Intell. **757**, 143–160 (2019)
29. Sotirov, S., Sotirova, E., Melin, P., Castillo, O., Atanassov, K.: Modular neural network preprocessing procedure with intuitionistic fuzzy InterCriteria analysis method. In: Proceedings of Flexible Query Answering Systems, pp. 175–186 (2016)
30. Todinova, S., Mavrov, D., Krumova, S., Marinov, P., Atanassova, V., Atanassov, K., Taneva, S.G.: Blood plasma thermograms dataset analysis by means of InterCriteria and correlation analyses for the case of colorectal cancer. Int. J. Bioautom. **20**(1), 115–124 (2016)
31. Tsakovska, I., Al Sharif, M., Alov, P., Diukendjieva, A., Fioravanzo, E., Cronin, M., Pajeva, I.: Molecular modelling study of the PPARγ receptor in relation to the mode of action/adverse outcome pathway framework for liver steatosis. Int. J. Mol. Sci. **15**, 7651–7666 (2014)

Improvement of Traffic in Urban Environment Through Signal Timing Optimization

Yordanka Boneva(✉) and Vladimir Ivanov

Institute of Information and Communication Technologies, Bulgarian Academy of Sciences, Acad. G.Bonchev bl 2, Sofia, Bulgaria
boneva.yordanka@gmail.com, ivanov.vladi@gmail.com

Abstract. A mean to control and optimize traffic in urban environment is adjusting signal timings of traffic lights. This paper presents a study of selected network in the city of Sofia, Bulgaria which network was modeled in the software environment AIMSUN, then optimized in the software environment TRANSYT and exported back to AIMSUN for validation of results. The experiment consisted in optimization of the signal timings – green splits and offsets that lead to improvement of eleven selected indicators. As the improvement of traffic have many aspects it is worth mentioning that this paper discuses only indicators of traffic such as queues, speed, travel time etc., but not the aspects of fuel consumption and environmental pollution. The later will be subject of future research as they are important issues in urban settings.

1 Introduction

Adjustment of traffic lights signaling at junctions in cities is a mean to control traffic flow. The parameters, which define the operation of traffic lights, are green light, traffic lights cycle (total time of green, red and amber lights of a junction) and offset of time between the cycles of neighbor junctions [1]. The optimization of the green signal duration affects the flow capacity of junctions and thereby the formation of queues that influence air pollution. Therefore, a solution needs to be found for junctions that do not function at an optimal level, so that there are fewer queues and less environmental pollution. The vehicle traffic management can also be improved through information for the location of the moving objects via the usage of navigation systems [2,3,10]. The information systems for the optimization of vehicle traffic can also work in cloud network [4].

According to an observation and a study with micro-simulation model made in AIMSUN (Advanced Interactive Microscopic Simulator for Urban and Non-urban Networks) [5] a network that does not function at its optimum was identified. The network comprises of 4 connected signalized intersections along the Shipchenski Prohod Blvd. In the city of Sofia, Bulgaria. The objective of the experiment described in this paper is to optimize the car flow in the selected network with regard to several indicators such as queues, speed, total travel time, delays, stops etc. As the selected network is in a densely populated area, infrastructure changes are not considered as an option. The goal of the researchers was to achieve an improvement in the network performance with the current state of the infrastructure only by adjusting the signal timings trough software simulation.

I. Dimov and S. Fidanova (Eds.): HPC 2019, SCI 902, pp. 99–107, 2021.
https://doi.org/10.1007/978-3-030-55347-0_9

The experiment uses AIMSUN for micro-simulation model of the selected network of junctions and TRANSYT for optimization of the fixed-time signal plan. The result was an improvement in the network performance and change in signal timings that was consequently validated in AIMSUN environment.

2 Study Workflow

Firstly, the target network was selected. The network consists of 4 connected signalized intersections along the Shipchenski Prohod Blvd., Sofia, Bulgaria (Fig. 2). It was

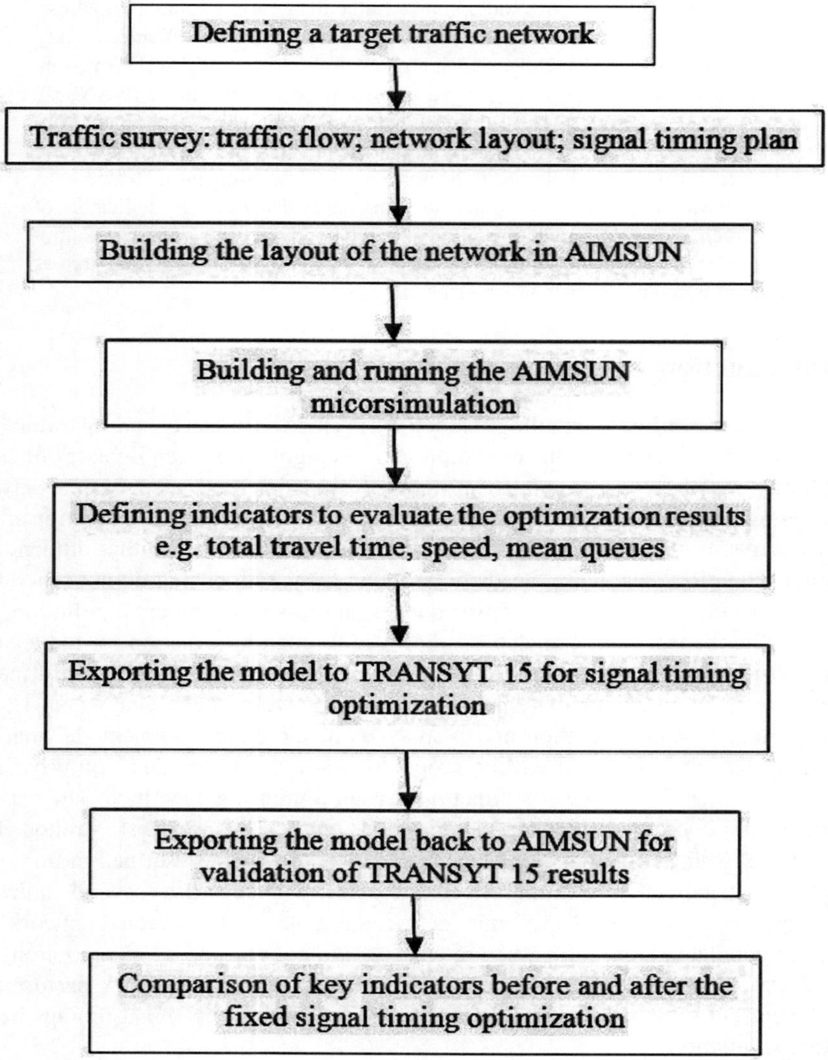

Fig. 1. Study workflow

selected because it is connecting some neighborhoods of the city with the city center and therefore there is a constant moderate to intense traffic. Moreover, there are many point of interest in the area like offices, shopping malls and other shopping opportunities, schools and a medical center. Secondly, after a location for the study was chosen the researchers proceeded to the next step – a traffic survey that consisted of traffic flow and signal timing data collection, as well as collecting data about the network layout to ensure valid model of the road lanes and turns at a later stage. The network layout was imported from Internet and a microsimulation was built and ran in AIMSUN environment. Indicators of traffic were selected that would later serve to measure the optimization improvements. These indicators include total travel time, speed, mean queues etc. The model was exported in TRANSYT for signal timing optimization where the mean queue was the selected indicator for optimization apart from the Performance Index (PI) which objective function uses delay and stops. After the optimization process had finished the model with the optimized signal timings was exported back in AIMSUN for validation of the results. The final result of the study consists in a comparison of chosen indicators before and after the optimization in TRANSYT.

3 Research Methodology

Software products were used for the modeling and optimization of signal timings of 4 intersections. The software environment that was used for the modeling of the network is called AIMSUN. With the help of AIMSUN the network layout was modeled and the traffic demand was filled in according to a field survey, and a simulation was performed. As it was supposed that the network does not work at it optimum the model was exported to another software product – TRANSYT that handles the optimization of signal timings. Eleven indicators of traffic were selected according to which the improvement in the network performance was measured.

The traffic demand was entered in AIMSUN based on a field survey that included direct observation of the traffic at the selected intersections, manual counting of traffic flow, signal timing detection and network layout identification. The studied area as mentioned included 4 signalized intersections along the Shipchenski Prohod Blvd. with many points of interest situated in the area. During the field survey because of the manual counting a researcher of the team considered the idea about collection of data about traffic indicators and adaptive management of traffic lights in real time [6].

A method in case of limited input flow data is to use the capacity of a graph for each arc (also called edge or line). The capacity defines the maximum traffic that can pass through a given arc. In the case of urban traffic management, the capacity is defined as: vehicle/time unit [7]. This method might be used in future study of the examined network as an alternate to the collected traffic flow data.

A significant limitation of the field study is that it could not identify if the signal timings are already synchronized and each intersection with its characteristics was studied separately. But later when the experiment was performed the synchronization between the intersections was accomplished through TRANSYT and because of this

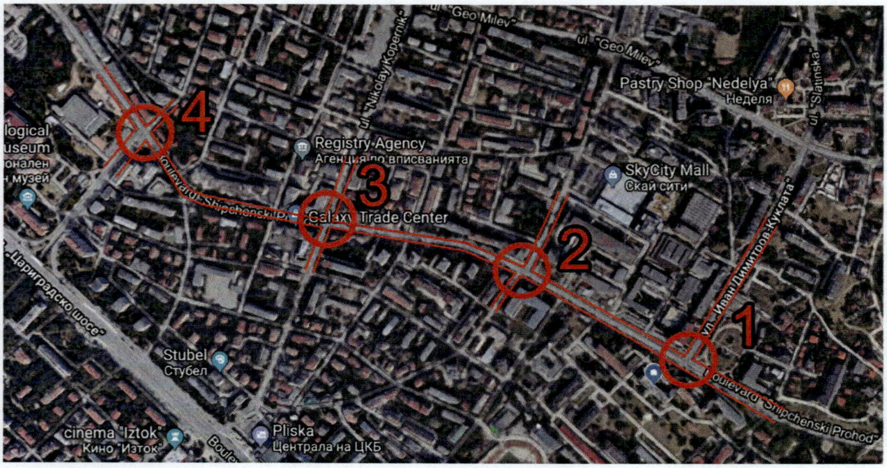

Fig. 2. The network of four intersections along Shipchenski prohod Blvd

synchronization the traffic was significantly improved. An important issue was that there is a legal prescription about the traffic light cycle duration of which was obeyed in the presented experiment. Although, it should be mentioned that the cited study points out that the formula for the cycle duration in the legislation is not applicable in all cases of saturation of the intersections [8]. A previous study with the same network had been performed but it relayed on manual change in signal timings. [9]. The difference and significance in the current paper is the use of TRANSYT which lead to much better performance of the network and also optimized the offsets of traffic light which was unthinkable to lead to good results with only manual changes.

TRANSYT environment works by optimizing delays and stops in its objective function. On the basis of these indicators it calculates the so called Performance Index (PI) [10]. “A traffic model of the network calculates a Performance Index (P.I.) in monetary terms, which, in its simplest form, is a weighted sum of all vehicle delay and stops. A number of available optimizing routines systematically alter signal offsets and/or allocation of green times to search for the timings which reduce the P.I. to a minimum value”. “The TRANSYT signal optimizers attempt to minimize the P.I. If it were possible to reduce the delay and number of stops to zero, then this would mean that every vehicle which approached a traffic signal would find the signal at green and would thus proceed to its destination without any delay or stops and hence with minimum journey time. The TRANSYT optimizers are therefore seeking to produce multi-directional ‘green waves’. Clearly, it is usually not possible to eliminate all delay or stops within signal network, but the minimum total value will be sought automatically by the TRANSYT program.” [11].

4 Shipchenski Prohod Blvd. Case Study

The selected network consists of four signalized intersections along the Shipchenski prohod Blvd., Sofia, Bulgaria. The simulated network is 1.5 km long. It is an urban area that is densely populated. Part of traffic is generated by locals as well as non-local population who visit mainly different points of interest in the area as offices where they work, shopping centers and other shopping opportunities, schools, a medical center in the neighborhood. Large traffic is generated by people working in the area and by people heading for the city center or in the opposite direction. Allowed network speed is 50 km/h. This is a small network with 4 signaling crossroads in close proximity to one another. For this reason, the acceleration and deceleration of cars often occur in the simulated network and this leads to an increase in emissions from cars. A further problem is the constant increase of the car ownership, which leads to an increase in the intensity of the movement and hence to the variable mode of movement, frequent stops and accumulation of cars at intersections. Eleven indicators were selected to monitor improvement in traffic before and after the optimization with TRANSYT. The indicators are as follows: 1. Delay time 2. Density 3. Flow 4. Harmonic speed 5. Mean queue 6. Number of stops 7. Speed 8. Stop time 9. Total number of stops 10. Total travel time 11. Travel time.

After the optimization with TRANSYT an additional adjustment was necessary at intersection No. 1 – the duration of the green light was reduced with 4 s as yellow with of one of the signals overlapped with the green light of another signal.

5 Results and Discussion

There was an improvement in all investigated indicators after the optimization with TRANSYT (Fig. 6). The signal timings were change through the optimization with the software and exported back to AIMSUN. In particular there was a change in green splits and offsets of signal timings which change lead to improvement of the performance of the selected network. In Fig. 3, Fig. 4 and Fig. 5 views from AISMUN and TRANSYT regarding signal timings are presented. The figures represent only one of the intersections (Intersection No. 2 in Fig. 2) but similar data is available for each intersections. As it is obvious from Fig. 3, Fig. 4 and Fig. 5 there is a slight difference of several seconds in green splits and significant difference in offsets due to the optimization process. After the optimization Signal 1 and Signal 3 from Fig. 5 were moved to the end of the cycle whereas Signal 2 and Signal 4 were moved in the beginning compared to Fig. 3. The reason is better synchronization with the other intersections.

Figure 6 presents the indicators, their units of measurement, the values and standard deviation of the indicators before and after the optimization as well as the difference in value and percentage before and after the optimization with TRANSYT. Following the field study some changes were made to the traffic lights on all four intersections – but they seem to affect the pedestrians more than car traffic. Still there might be slight changes to traffic lights that may affect the performance of the network in a positive manner. These possible changes are not taken into account in this study which may present a limitation of the study.

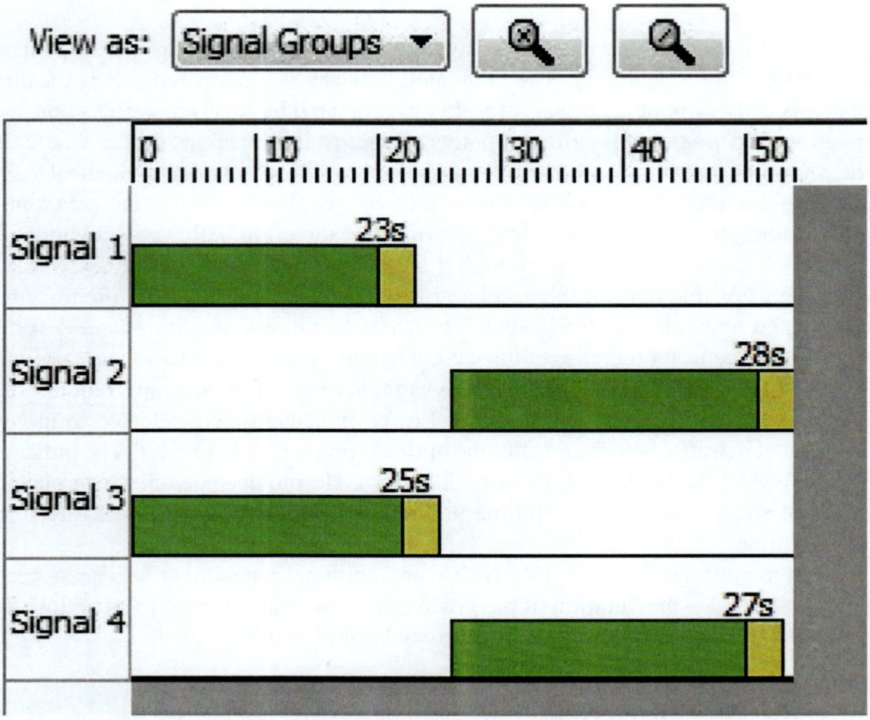

Fig. 3. Signal timings before the optimization – AIMSUN view

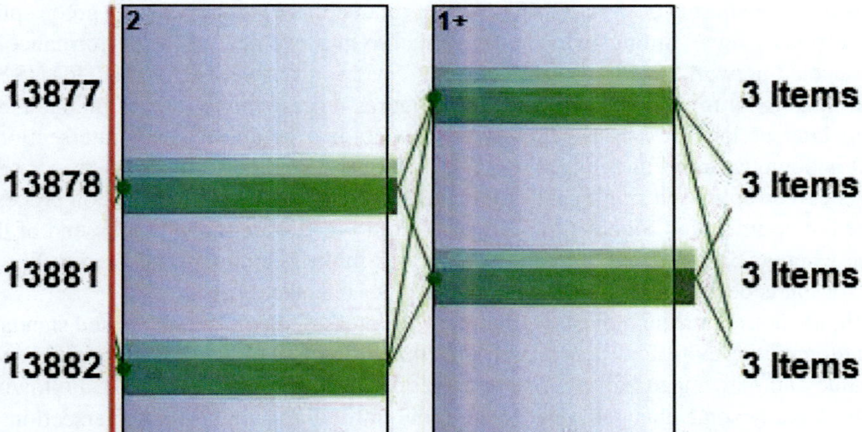

Fig. 4. Signal timings after the optimization – TRANSYT view

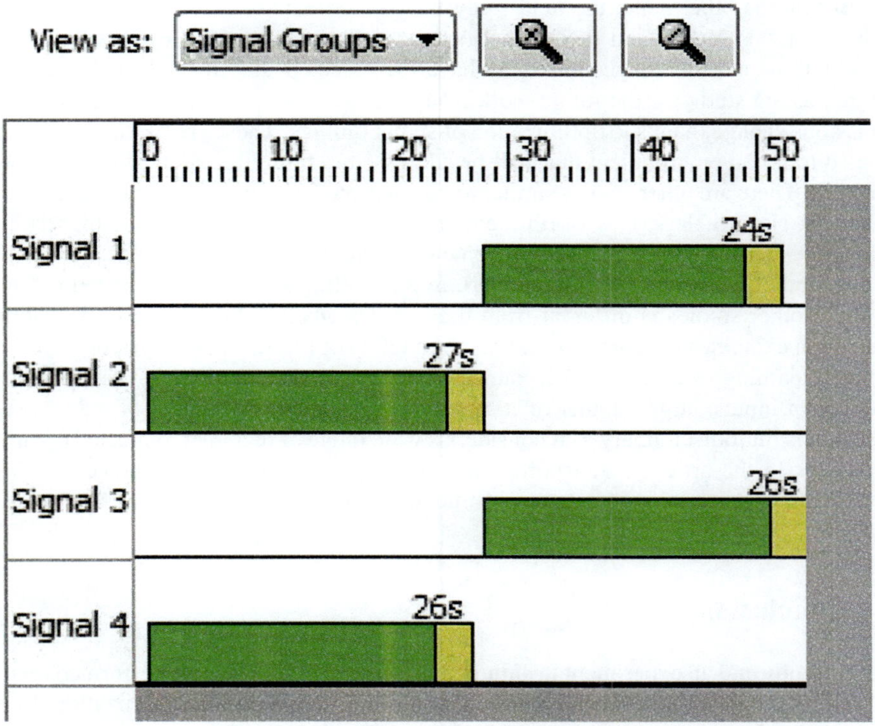

Fig. 5. Signal timings after the optimization – AIMSUN view

Indicators		Before optimization with TRANSYT		After optimization with TRANSYT			
	Units	**Value**	**Standard Deviation**	**Value**	**Standard Deviation**	**Difference**	**Difference in %**
Delay Time - Car	sec/km	51.16	40.3	36.33	35.88	-14.83	-28.99%
Density - Car	veh/km	7.69	N/A	6.53	N/A	-1.16	-15.08%
Flow - Car	veh/h	4095	N/A	4126	N/A	31	0.76%
Harmonic Speed - Car	km/h	30.57	10.61	34.97	10.66	4.4	14.39%
Mean Queue - Car	veh	29.85	N/A	19.08	N/A	-10.77	-36.08%
Number of Stops - Car	#/veh/km	0.16	N/A	0.12	N/A	-0.04	-25.00%
Speed - Car	km/h	34.25	11.56	38.21	10.37	3.96	11.56%
Stop Time - Car	sec/km	38.42	35.17	26.34	32.84	-12.08	-31.44%
Total Number of Stops - Car		8322.87	N/A	6367.76	N/A	-1955.11	-23.49%
Total Travel Time - Car	h	94.76	N/A	81.2	N/A	-13.56	-14.31%
Travel Time - Car	sec/km	117.76	40.53	102.96	35.96	-14.8	-12.57%

Fig. 6. Indicators before and after optimization

But on the other hand, the field study was performed about four year prior to this paper. For this reason and due to the tendency of increase in the ownership of cars the car flow and the congestions may have increased. Beside the selected indicators an interesting study for the future work would be how fuel consumption and pollution emissions change thanks to optimization of signal timings. These are two key indicators related to urban environment that will be addressed in future research of the presented network. There are other studies similar to the presented here but they use other software products. Nevertheless, it is worth mentioning that other researchers in Bulgaria also perform experiments related to improvement of traffic conditions. Likewise, the object of the studies is a transportation network in Sofia, Bulgaria. But the selected network in these other studies is different from the one described in this paper. [12]. Another study that examines the complex nature of modeling of transportation systems is related to deep learning of complex interconnected processes. The complexity includes large number of inputs, large number of state parameters, and large number of outputs. The research is multidisciplinary as it considers a wide range of real interconnected complex processes in the areas of transportation and the environment. These processes often operate under uncertainty due to environmental disturbances, vague or incomplete data [13].

6 Conclusion

A field study and an experiment in simulation software products were performed related to traffic optimization in urban environment through signal timing optimization. In the experiment the software products AIMSUN and TRANSYT were used respectively for modeling and for optimization of signal timings. The improvement was measured by eleven traffic indicators such as queues, speed, travel time etc. Two cases were compared – before optimization and after optimization of signal timings with TRANSYT. The results show that after the optimization the traffic indicators have improved significantly with some of the indicators reaching an improvement of over 30% compared to the case before optimization. An objective of this study was the improvement in indicators of traffic but other important issues are the change in fuel consumption and environmental pollution. These issues will be examined in future work.

Acknowledgements. This work has been partly supported by project KP-06-27/9, 17.12.2018 of the Bulgarian National Science fund: "Contemporary digital methods and tools for exploring and modeling transport flows"

References

1. Stoilov, T., Stoilova, K., Stoilova, V.: Bi-level formalization of urban area traffic lights control. In: Margenov, S., et al. (eds.) Innovative Approaches and Solutions in Advanced Intelligent Systems. Studies in Computational Intelligence, vol. 648, pp. 303-318. SJR:0.19. Springer (2016). https://doi.org/10.1007/978-3-319-32207-0-20. ISBN 978-3-319-32206-3, ISSN 1860-949X

2. Garvanov, I., Kabakchiev, C., Behar, V., Garvanova, M.: Target detection using a GPS Forward-Scattering Radar. In: 2015 International Conference Engineering and Telecommunications, Moscow, Dolgoprudny, Russia, pp. 29–33 (2015)
3. Garvanov, I., Kabakchiev, C., Shishkov, B., Garvanova, M.: Towards context-aware vehicle navigation in urban environments: modeling challenges. In: Shishkov, B. (ed.) Business Modeling and Software Design BMSD. Lecture Notes in Business Information Processing, vol. 319, pp. 382–389. Springer, Cham (2018)
4. Pavlova, D.: Exploration of a distance learning toolkit through integration the capabilities of public and private clouds in a heterogeneous environment. In: 13th International Technology, Education and Development Conference, INTED2019 Proceedings, Valencia, Spain, 11th–13th March 2019, pp. 8061–8066. https://doi.org/10.21125/inted.2019., ISBN 978-84-09-08619-1, ISSN 2340-1079
5. Aimsun Microscopic 8 Macrosopic Modelling Manual, Copyright 2005–2013 TSS Transport Simulation Systems, October 2013
6. Ivanov, V., Stoilova, K.: Traffic lights control using measured characteristics of urban traffic in real time. In: Scientific Proceedings of XIV International Congress MACHINES. TECHNOLGIES. MATERIALS. 2017 - Summer Session, Year I, vol. VI, pp. 435–438 (2017). Print ISSN 2535-0021. (in Bulgarian)
7. Pavlova, K., Vatchova, B., Paunova, E.: Quantitative evaluation of throughput capabilities in transportation graph under limited output data conditions. J. Bul. Sci. (107), 45–52 (2018). ISSN 1314-1031. (in Bulgarian)
8. Todorova, M.: Improvement of regulations aimed to improve traffic management in cities. Sci. J. Mech. Transp. Commun. **14**(3/1), I-16– I-22, paper 1308 (2016). ISSN 2367-6620 (online), ISSN 1312-3823 (print). https://mtc-aj.com/library/1308.pdf. (in Bulgarian)
9. Boneva, Y.: Optimization of car traffic on traffic light signalized intersections through the simulation environment AIMSUN. Sci. J. Mech. Transp. Commun. **16**(2), I-1–I-9, paper 1663 (2018). ISSN 1312-3823 (print), ISSN 2367-6620 (online). (in Bulgarian)
10. Madjarov, A.N. One algorithm for the joint use of the ICI and GPS information. In: Benkovski, G., Mitropoliya, D. (eds.) Collection of Reports at Scientific Session 9–10 December 1997 "85th Anniversary of the Participation of Bulgarian Aviation in the Balkan War", BBU, vol. 2, pp. 260–270 (1998). (in Bulgarian)
11. Transyt-Aimsun Link (Version 2), User Guide, Issue B by James C Binning, Copyright Copyright TRL Limited 2010 (2013)
12. Trendafilov, Z.: Analyses on the methods for estimating the phases of the traffic lights. Acad. J. Mech. Transp. Commun. **15**(3), I-21–I-27, Article 1446 (2017). ISSN 2367-6620 (online), ISSN 1312-3823 (print). (in Bulgarian)
13. Vatchova, B., Pavlova, K., Paunova, E., Stoilova, K.: Deep learning of complex interconnected processes for bi-level optimization problem under uncertainty. "Industry 4.0", vol. III, no. 1. Scientific Technical Union of Mechanical Engineering, Bulgaria (2018). ISSN (Print) 2543-8582

Teaching Supercomputers

Stefka Fidanova[1(✉)] and Velislava Stoykova[2]

[1] Institute of Information and Communication Technologies, Bulgarian Academy of Sciences, Acad. G. Bonchev 25A, 1113 Sofia, Bulgaria
stefka.fidanova@gmail.com, stefka@parallel.bas.bg

[2] Institute for Bulgarian Language, Bulgarian Academy of Sciences, 52, Shipchensky prohod street, bl. 17, 1113 Sofia, Bulgaria
vstoykova@yahoo.com

Abstract. The paper presents results of semantic search analysis of the key educational concepts and their conceptual relations presented in the course "Supercomputers I and II" taught at the School of Media, Arts and Technology at the Solent Southampton University during the 2015–2019 academic years. The study uses Big Data Analytics approaches to extract the domain key concepts and to outline their pedagogical interconnections which reflect the structure of course lectures, so to teach 'supercomputers' as an university subject with its specificity and to allow students to acquire that knowledge comprehensively by having the insight of its key concepts as well as of its practical applications. The study uses the Sketch Engine software's statistical approaches, so to extract the key terms (related to the key concepts in the field) from the course texts lectures and to compare their semantic content with that produced by the students (feedback) in the related courses taught at some other universities in UK. It, also, outlines the importance of the 'activity-based learning approach' used for supporting the teaching process, so to improve the acquired knowledge by the use of Solent Online Learning system allowing interactive and distant web-based learning.

Keywords: Supercomputers · High performance computing · Teaching approaches · Educational data processing

1 Introduction

Modern academic teaching uses extensively complex approaches combining traditional teaching methods with that inspired from the IT-based perspective, so to give the students the best insight about the knowledge they need to acquire and to master. The so-called 'activity-based approach to teaching and learning' [7] uses extensively the technologies from knowledge management, so to maximize the students knowledge outcome.

However, to optimize the teaching process and to make it more adequate to the students needs and more competitive among universities, a comparison of

I. Dimov and S. Fidanova (Eds.): HPC 2019, SCI 902, pp. 108–117, 2021.
https://doi.org/10.1007/978-3-030-55347-0_10

different courses in the same domain with respect to their semantic content (key domain concept hierarchy) coverage is needed. Also, the domain terminology and the related semantic relations among the domain terms are important part of that unification which may facilitate the university ranking according to the semantic content of subjects and courses they offer.

2 Related Approach

We are going to present the results of the study related to semantic search analysis of the key educational concepts and their conceptual relations presented in the course "Supercomputers I and II" taught at the School of Media, Arts and Technology at the Solent Southampton University (SSU) during the 2015–2019 academic years. For that study, we have basically used and compared the (teaching) electronic corpus containing the course texts lectures (on 'Supercomputers') and the students electronic corpus containing written students works related to the same academic domain ('Supercomputers') taught at some other universities in UK.

The study approach we use is based on the extensive application of the Distributional Semantic Model frameworks which regard the use of statistical similarity (or distance) as relevant for semantic similarity (or distance) [2]. Thus, we evaluate statistically-significant words (key concepts) as semantically-relevant and we extract key words which function as typical key domain concepts showing additional information about their variability and functionality. The techniques used adopt metrics for extraction of different types of word semantic relations by estimation of word similarity measure [1].

Also, we have used the Sketch Engine software's statistical approaches for storing, sampling and searching, so to extract the key words (terms) (related to the key concepts in the field of *supercomputers*). We present several search techniques and discuss the received results of using statistical functions for concordances and collocations generation, thesauri generation and key words (terms) extraction.

3 Sketch Engine

The SE software [5] allows approaches to extract semantic properties of words and most of them are with multilingual application. Generating keywords is a widely used technique to extract terms of particular studied domain. Also, semantic relations can be extracted by generation of related word contexts through word concordances which define context in quantitative terms and a further work is needed to be done to extract semantic relations by searching for co-occurrences and collocations of related keyword.

Co-occurrences and collocations are words which are most probably to be found with a related keyword. They assign the semantic relations between the keyword and its particular collocated word which might be of a similarity or of a distance. The statistical approaches used by SE to search for co-occurrence and collocated words are based on defining the probability of their co-occurrences and collocations. We use techniques of $T - score$, $MI - score$ and $MI^3 - score$ for corpora processing and searching. For all, the following terms are used: N - corpus size, f_A - number of occurrences of keyword in the whole corpus (the size of concordance), f_B - number of occurrences of collocated keyword in the whole corpus, f_{AB} – number of occurrences of collocate in the concordance (number of co-occurrences). The related formulas for defining $T - score$, $MI - score$ and $MI^3 - score$ are as follows:

$$T - Score = \frac{f_{AB} - \frac{f_A f_B}{N}}{\sqrt{f_{AB}}} \quad (1)$$

$$MI - Score = log_2 \frac{f_{AB} N}{f_A f_B} \quad (2)$$

$$MI^3 - Score = log_2 \frac{f_{AB}^3 N}{f_A f_B} \quad (3)$$

The $T - score$, $MI - score$ and $MI^3 - score$ are applicable for processing educational corpora as well.

Collocations have been regarded as statistically similar words which can be extracted by using techniques for estimation the strength of association between co-occurring words. Recent developments improved that techniques with respect to various application areas.

Further, we shall present and analyze the search results for extracting the key *supercomputers* domain concepts using the SE software and shall compare received results with respect to their semantic relations which are important for teaching.

4 Educational SuperComputer Source (SCS) and British Academic Written English (BAWE) Corpora

Educational electronic text corpora were primarily designed to study language of a certain domain by mainly analyzing related concepts (terminology). Later, the analysis of linguistic data was focused on exploration of specific vocabulary – Academic Word Lists (AWL) which are typical for that domain [10], and a more general term 'academic literacy' [3] was accepted introducing the standards for optimal AWL acquisition and their correct usage. The standards were tested and established by using a specific software [6] for corpora processing.

The most representative electronic educational corpora for British English are: (i) the British Academic Spoken English (BASE) Corpus – containing academic texts transcripts of course lectures (1,186,280 words) and (ii) the British Academic Written English (BAWE) Corpus – containing students written texts (6,506,995 words). Both corpora are uploaded into SE and are freely available for search and retrieval. However, for our research, we have compiled the Super-Computer Source (SCS) Corpus.

4.1 SuperComputer Source (SCS) Corpus

In the context of electronic text corpora studies, we have compiled the SuperComputer Source (SCS) Corpus consisting of academic texts transcripts of lectures of the course "Supercomputers I and II" taught at the School of Media, Arts and Technology at the Solent Southampton University (SSU) during the 2015–2019 academic years (app. 2,000 words). The compiling methodology did not used annotation allowing statistical search and retrieval approaches for deriving and comparing the domain key concepts. The SCS Corpus was uploaded into SE, so to use its statistical approaches for storing, sampling and searching in order to extract the key terms (related to the key concepts in the field of *supercomputers*). The SCS Corpus academic texts content can be analyzed for studying the structure, semantic content and delivery stile of academic lectures.

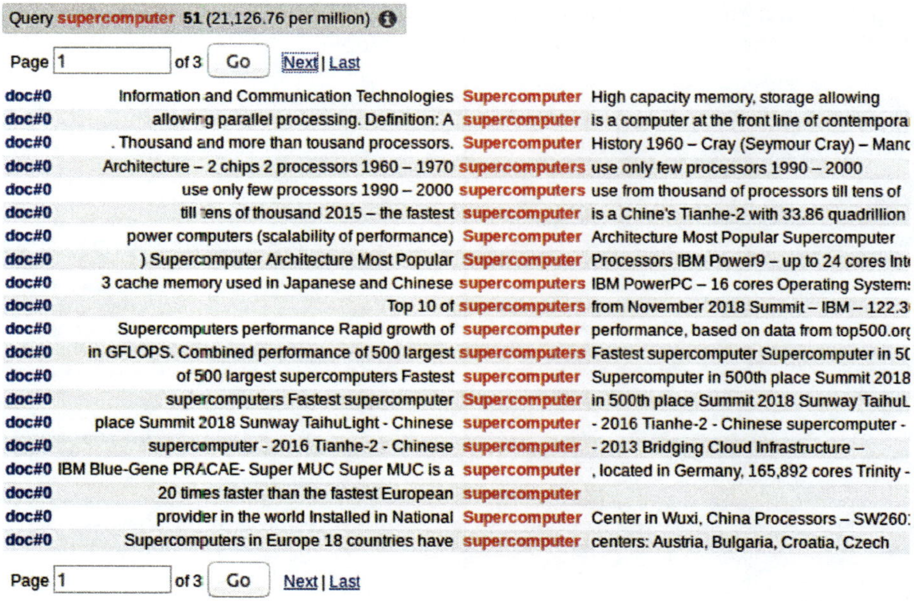
Query supercomputer 51 (21,126.76 per million)

Page 1 of 3 Go Next | Last

doc#0	Information and Communication Technologies	Supercomputer	High capacity memory, storage allowing
doc#0	allowing parallel processing. Definition: A	supercomputer	is a computer at the front line of contempora
doc#0	. Thousand and more than tousand processors.	Supercomputer	History 1960 – Cray (Seymour Cray) – Manc
doc#0	Architecture – 2 chips 2 processors 1960 – 1970	supercomputers	use only few processors 1990 – 2000
doc#0	use only few processors 1990 – 2000	supercomputers	use from thousand of processors till tens of
doc#0	till tens of thousand 2015 – the fastest	supercomputer	is a Chine's Tianhe-2 with 33.86 quadrillion
doc#0	power computers (scalability of performance)	Supercomputer	Architecture Most Popular Supercomputer
doc#0	) Supercomputer Architecture Most Popular	Supercomputer	Processors IBM Power9 – up to 24 cores Int
doc#0	3 cache memory used in Japanese and Chinese	supercomputers	IBM PowerPC – 16 cores Operating System
doc#0	Top 10 of	supercomputers	from November 2018 Summit – IBM – 122.3
doc#0	Supercomputers performance Rapid growth of	supercomputer	performance, based on data from top500.or
doc#0	in GFLOPS. Combined performance of 500 largest	supercomputers	Fastest supercomputer Supercomputer in 5(
doc#0	of 500 largest supercomputers Fastest	supercomputer	Supercomputer in 500th place Summit 2018
doc#0	supercomputers Fastest supercomputer	Supercomputer	in 500th place Summit 2018 Sunway TaihuL
doc#0	place Summit 2018 Sunway TaihuLight - Chinese	supercomputer	- 2016 Tianhe-2 - Chinese supercomputer -
doc#0	supercomputer - 2016 Tianhe-2 - Chinese	supercomputer	- 2013 Bridging Cloud Infrastructure –
doc#0	IBM Blue-Gene PRACAE- Super MUC Super MUC is a	supercomputer	, located in Germany, 165,892 cores Trinity -
doc#0	20 times faster than the fastest European	supercomputer	
doc#0	provider in the world Installed in National	Supercomputer	Center in Wuxi, China Processors – SW260
doc#0	Supercomputers in Europe 18 countries have	supercomputer	centers: Austria, Bulgaria, Croatia, Czech

Page 1 of 3 Go Next | Last

Fig. 1. Concordance search results for the key word *supercomputer(s)* from the SCS Corpus.

4.2 British Academic Written English (BAWE) Corpus

The British Academic Written English Corpus (BAWE) was collected as part of the project *An Investigation of Genres of Assessed Writing in British Higher Education* funded by the Economic and Social Research Council (ESRC) (2004–2007, project number RES-000-23-0800). The corpus contains records (written texts) of proficient university-level student writings at the turn of the 21st century. It contains just under 3000 good-standard student assignments (6,506,995 words).

Holdings are fairly evenly distributed across four broad disciplinary areas: (i) Arts and Humanities, (ii) Social Sciences, (iii) Life Sciences and (iv) Physical Sciences and across four levels of study (undergraduate and taught masters level). Thirty main disciplines are represented. The corpus is available free of charge to researchers. The BAWE corpus contains 2761 pieces of proficient assessed student writing, ranging in length from about 500 words to about 5000 words. The assignments have been annotated using a system devised in accordance with the Text Encoding Initiative guidelines with related DTD definitions and mark-up conventions.

The BAWE Corpus educational content can be analyzed with respect to: studying the meaning and use of concepts and multi-word units, studying the frequency and range of academic lexis, etc. It can also be searched on-line via the SE open site https://ca.sketchengine.co.uk/open/.

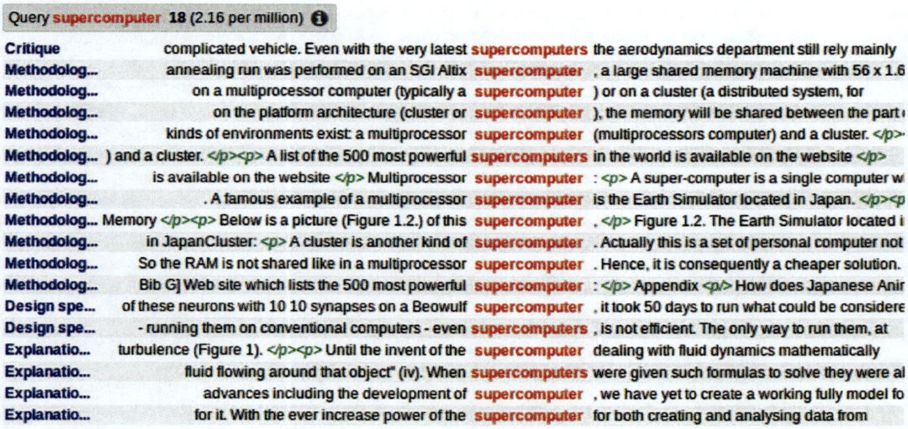

Fig. 2. Concordance search results for the key word *supercomputer(s)* from the BAWE Corpus.

5 Semantic Search Results and Analysis

The SCS Corpus and the BAWE Corpus are uploaded into the SE software allowing key word (key concepts) semantic search according to several criteria, so to outline the semantic, structural and functional variability of their electronic textual content. For the entire study, we have employed the SE options to generate concordances, collocations, distributional thesauri and keywords generation by performing statistical search. We have searched both corpora in order to generate all that types of results for the key word (key concept) *supercomputer(s)*, so to outline the differences of its semantic relations from both corpora.

The concordances present all occurrences of a given key word within the corpus and outline its statistical distribution within all corpus texts. Figure 1 shows the concordance search results for the keyword *supercomputer(s)* from SCS Corpus and outlines that the key word (key concept) *supercomputer(s)* has 51 occurrences within the whole corpus which appear together with their related statistical contexts.

Figure 2 presents the concordance search results for keyword *super computer(s)* from the BAWE Corpus and provides additional information about the texts included in the corpus (genre specification) relaying on its annotation scheme. The corpus texts which include the key word (key concept) *supercomputer(s)* are from the genres: *Critique*, *Methodology recount*, *Design specification* and *Explanation*.

Collocation candidates

		Cooccurrence count	Candidate count	T-score	MI	MI3	log likelihood
P | N	located	7	7	2.589	5.564	11.179	54.988
P | N	processors	5	5	2.188	5.564	10.208	39.068
P | N	have	4	5	1.947	5.242	9.242	26.208
P | N	performance	5	7	2.169	5.079	9.723	30.770
P | N	processing	4	6	1.936	4.979	8.979	23.613
P | N	in	19	38	4.174	4.564	13.060	102.428
P | N	a	11	27	3.144	4.269	11.188	51.413
P | N	is	10	27	2.981	4.131	10.775	44.198
P | N	Supercomputer	4	11	1.883	4.105	8.105	17.028
P | N	.	16	47	3.751	4.010	12.010	69.604
P | N	Germany	4	12	1.873	3.979	7.979	16.212
P | N	2013	3	9	1.622	3.979	7.149	12.103
P | N	)	4	14	1.852	3.757	7.757	14.816
P | N	(	4	14	1.852	3.757	7.757	14.816
P | N	are	4	14	1.852	3.757	7.757	14.816
P | N	data	3	11	1.597	3.690	6.860	10.750
P | N	on	3	11	1.597	3.690	6.860	10.750
P | N	2018	3	12	1.585	3.564	6.734	10.186
P | N	the	11	51	2.991	3.351	10.270	35.544
P | N	cores	3	14	1.561	3.342	6.512	9.214

Fig. 3. Collocation search results for the key word *supercomputer(s)* from the SCS Corpus.

Collocations are words which are most probably to be found with a related key word and their statistical significance and ranking are evaluated by computing the probability of their co-occurrence in the text using $MI - score$ criterion. Some previous work [4] and [8] studied the collocations in science writing as well as the collocations representing key concepts (as terms being multi-word expressions) outlining their semantic relations of equivalence or hierarchy.

Collocation candidates

	Cooccurrence count	Candidate count	T-score	MI	MI3	log likelihood
P | N multiprocessor	4	5	1.999	18.499	22.499	100.325
P | N cluster	3	148	1.731	13.196	16.366	49.475
P | N on	5	38,754	2.198	5.900	10.544	32.562
P | N :	3	47,060	1.673	4.883	8.053	15.011
P | N a	8	125,736	2.732	4.881	10.881	42.680
P | N)	5	91,843	2.147	4.656	9.299	24.100
P | N is	6	110,721	2.351	4.649	9.819	29.262
P | N (	4	90,538	1.902	4.354	8.354	17.417
P | N of	7	270,136	2.425	3.585	9.199	24.680
P | N the	9	429,343	2.690	3.279	9.619	29.388
P | N ,	7	391,643	2.326	3.049	8.663	19.814
P | N .	5	313,580	1.933	2.884	7.528	12.529

Fig. 4. Collocation search results for the key word *supercomputer(s)* from the BAWE Corpus.

Figure 3 shows collocation search results for the key word *supercomputer(s)* from the SCS Corpus. It outlines the semantic relations between the keyword (key concept) *supercomputer(s)* and its collocated words (related concepts) *processor*, *performance*, *processing* and *cores* showing their semantic and functional hierarchical relations extracted from the academic lectures corpus texts.

Figure 4 shows collocation search results for the key word *supercomputer(s)* from the BAWE Corpus. It presents the ranking list of collocation candidates for the key word *supercomputer(s)* from the BAWE Corpus generated according to $MI - score$ criterion. The results outline the semantic relations between the keyword (key concept) *supercomputer(s)* and its collocated words (related concepts) *multiprocessor* and *cluster* showing their semantic hierarchical relations extracted from the students writing corpus texts. The received results show a difference with related results obtained from SCS Corpus regarding the semantically related concepts – which are similar but less in number and do not contain the concepts relating to the functionality characteristics of the studied key concept *supercomputer(s)* (as *processing* and *performance* from SCS Corpus) outlining mostly its structural characteristics (hardware architecture configuration) as *multiprocessor* and *cluster*.

If two key words (two key concepts) share the common collocations, thus they share a distributional thesauri semantic relation [9] and can be regarded as belonging to each one's thesauri. Such distributional thesauri relations can be generated using SE (by estimation of common key words collocations). Thesauri generation search shows more complex semantic relations between a key word (key concept) and the related generated words (related concepts).

Figure 5 presents thesauri generation results for the keyword (key concept) *supercomputer(s)* from the SCS Corpus. It outlines the concept *simulation* as semantically related to the studied key concept *supercomputer(s)* (by having common collocations with it). The related concept *simulation* reveals another characteristic of the key concept – its application and usage, i.e. *supercomputer(s)* are used for *simulation*.

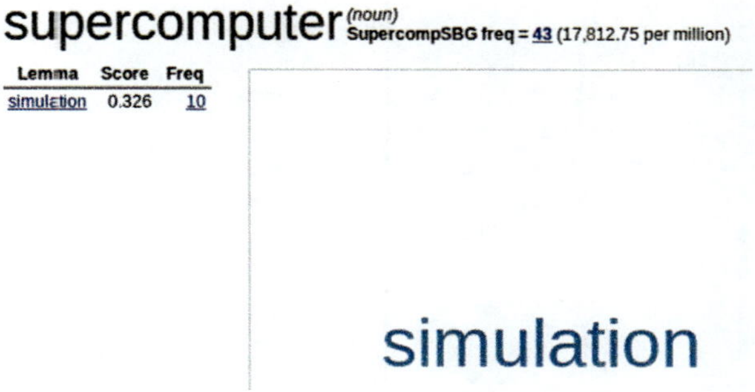

Fig. 5. Thesauri generation results for the keyword *supercomputer(s)* from the SCS Corpus.

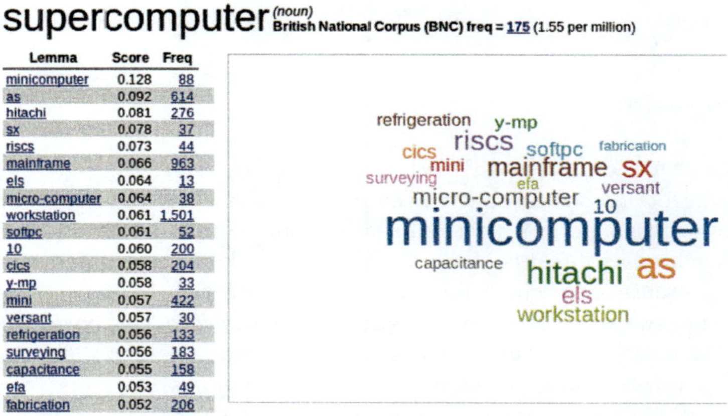

Fig. 6. Thesauri generation results for the keyword *supercomputer(s)* from the BNC.

The presented results for thesauri generation (Fig. 6) received from British National Corpus (BNC) (which is not the educational but general-purpose corpus) do not outline such conceptual semantic relation. They imply the difference between *knowledge* and *information*.

As for the linguistic structure of domain concepts (represented by terms), it is important to study not only single-word terms but also multi-word terms since they represent domain concepts as well. For that, we have used the *keyness sore* criterion which evaluate statistical distance instead of statistical similarity.

Thus, we have processed the SCS Corpus using the *keyness sore* option of the SE, so to extract the basic concepts (terms) related to the key concept *supercomputer(s)*. The part of received results is presented at Fig. 7 and contains single-word and multi-word terms. They include terms related to the key concept *supercomputer(s)* like: *parallelism*, *cache memory*, *parallel processing*, *parameter space*, *aerodynamic design*, *drug design*, etc.

Single-word	Score	F	RefF	Multi-word	Score	F	RefF
supercomputer	9,913.48	43	18,113	drug design	1,966.08	5	13
PFlop	4,138.95	10	25	cache memory	1,204.02	3	8
Cray	3,396.48	11	7,771	aerodynamic design	1,193.62	3	10
Supercomputers	3,292.12	8	158	small molecule	970.92	3	67
Supercomputer	3,142.85	8	1,245	primary memory	829.50	2	0
TaihuLight	2,900.75	7	0	perspective parallelism	829.50	2	0
Pflop	2,896.67	7	32	pollution level prediction	829.50	2	0
parallelism	2,513.55	9	10,993	chinese supercomputer	829.50	2	0
Sunway	2,286.51	6	1,988	level prediction	826.20	2	1
Xeon	1,772.09	6	9,163	pollution level	806.12	2	7
MUC	1,625.04	4	461	parameter space	806.12	2	7
PRACAE	1,243.75	3	0	parallel processing	759.62	2	22

Fig. 7. Key words (term) extraction results for the keyword *supercomputer*(*s*) from the SCS Corpus.

6 Conclusion

The search results of SCS and BAWE Corpora for the *supercomputer*(*s*) related domain concepts and their specific semantic relations outline some conceptual differences between the two electronic educational academic resources. The SCS Corpus (teaching lectures) contains well structured knowledge (hierarchical conceptual representation) which includes concepts relating to *supercomputer*(*s*) structure (hardware architecture and configuration: *processor*, *core*, etc.), functionality (*processing*, *parallel processing*, *performance*, etc.) and usage and applications (*simulation*, *aerodynamic design*, *drug design*, etc).

The BAWE Corpus search results reflecting the content (students written texts) present the domain key concept *supercomputer*(*s*) mainly with relational concepts outlining the structural characteristics (hardware architecture and configuration: *multiprocessor*, *cluster*, etc). The presented results are promising that the study approach can be used to evaluate and compare the semantic content of educational course materials by comparing the semantic relations of the key concepts included.

The teaching approaches used in the course "Supercomputers I and II" at SSU, the content and structure model of teaching materials (the combination of lectures and Computer Labs exercises) can be considered as specific methodological approach to teach *supercomputer*(*s*), and might be used as a general framework for other IT courses which require up-to-date teaching.

Acknowledgments. The authors are supported by Ministry of Education and Science – Bulgaria, Grant No BG05M2OP001-1.001-0003, financed by the Science and Education for Smart Growth Operational Program and co-financed by the European Union through the European structural and Investment funds and by the National Scientific Fund of Bulgaria under grant DFNI DN12/5 "Efficient Stochastic Methods and Algorithms for Large-Scale Problems".

References

1. Baroni, M., Evert, S.: Statistical methods for corpus exploitation. In: Lüdeling, K., Kytö, M. (eds.) Corpus Linguistics: An International Handbook, vol. 2, pp. 777–803. Mouton De Gruyter, Berlin/New York (2008)
2. Baroni, M., Lenci, A.: Distributional memory: a general framework for corpus-based semantics. Comput. Linguist. **36**(4), 673–721 (2010)
3. Blue, G.: Developing Academic Literacy. Peter Lang (2010)
4. Gledhill, Ch.: Collocations in Science Writing. Gunter Narr Verlag (2000)
5. Killgarriff, A., et al.: The sketch engine: ten years on. Lexicography **1**, 17–36 (2014)
6. Nation, P., Heatley, A.: Range: A Program for the Analysis of Vocabulary in Texts (2002). http://vuw.ac.nz/lals/staff/paul-nation/nation.aspx
7. Penev, K., Rees, J.: Development of a 'sticky' virtual community on the Internet environment. In: 9th European Conference on Knowledge Management, pp. 645–652. Academic Publishing Limited, UK (2008)
8. Stoykova, V.: Extracting academic subjects semantic relations using collocations. EAI Endorsed Trans. Energy Web Inf. Technol. **4, 17**(14) (2017a). https://doi.org/10.4108/eai.4-10-2017.153161
9. Stoykova, V.: Discovering distributional Thesauri semantic relations. In: 2017 Proceedings of the 3rd International Workshop on Knowledge Discovery on the Web, CEUR-WS, Cagliari, Italy (2017b). http://ceur-ws.org/Vol-1959/paper-11.pdf
10. Thompson, P.: Changing the bases for academic word lists. In: Thompson, P., Diani, G. (eds.) English for Academic Purposes: Approaches and Implications, pp. 317–342. Cambridge Scholars, Newcastle-upon-Tyne (2015)

Timeline Event Analysis of Social Network Communications Activity: The Case of Ján Kuciak

Kristina G. Kapanova[1] and Velislava Stoykova[2](✉)

[1] School of Computer Science and Statistics, Trinity College Dublin, Dublin, Ireland
kkapanova@gmail.com, kapanova@parallel.bas.bg
[2] Institute for Bulgarian Language, Bulgarian Academy of Sciences, 52, Shipchensky Prohod str., bl. 17, 1113 Sofia, Bulgaria
vstoykova@yahoo.com

Abstract. The paper presents a Complex Network Analysis of Twitter communication activities invoking the political protests caused by the murder of Slovakian journalist Ján Kuciak in February 2018 and followed by a governmental resignation. It offers an overview of existing approaches to analyze the social networks communication strategies and the related specific linguistic markers (hashtags) used during the network communications sessions in similar political protests and outlines that which are typical for the protests in Slovakia. The approach used is the application of Artificial Neural Networks and Big Data Analysis techniques employing *complex network analysis* by utilizing word vector representations within the *neural embedding model*. We estimate the statistical similarity of most popular hashtags used during the protests communications network activities and evaluate their ability to add a meaning and to mobilize discussing participants to act as protesting activists. Finally, we outline how the specific hashtags (slogans) function as clear and mobilizing political messages.

Keywords: Complex Network Analysis · Neural embedding model · Hashtags · Social media

1 Introduction

Recent years improved social media platforms as a specific way of multi-modal information exchange challenging the language in its societal and communicative functions by extending them significantly and unlocking the language's latent power. There are various reasons for that starting from the fact, that using social media is a fast and often free of charge way to obtain, give and exchange various types of information (irrespectively of users localization) and going beyond by allowing everyone to express freely beliefs, ideas, likes, etc.

Thus, social media became a complex phenomenon, itself, allowing people both to behave as a personality and to act as a part of a chosen community, to

I. Dimov and S. Fidanova (Eds.): HPC 2019, SCI 902, pp. 118–131, 2021.
https://doi.org/10.1007/978-3-030-55347-0_11

share and express freely opinions and at the same time, to act as belonging to an organized virtual community without being engaged with certain political views and believes (even anonymous or using a fake personality). With respect to that, the power of social networks for organizing and motivating people is manifested by means of natural language use (in a new socio-digital context) empowered by a digitally mediated transmission.

That functionalities of social media combined with their large-scale networking information processing architecture have been used both for probing new ideas and for studying the language changes connected to human communications and resulting actions. Moreover, using Complex Network Analysis and Big Data Analytics approaches, it is possible to obtain a clear and detailed picture of networking activities connected to places, time, language used, etc., so to study the driving power and on-line timeline development of various social phenomena.

The integration of linguistic and social functions of hashtags necessitates their study as a community building and community mobilizing marker (message code). Our study adopts the toolbox from *complex network analysis* and *neural embedding model*, so to study the virtual community language conventions by means of tracking hashtags virality and density within the social network interactions.

Further, we are going to present the results of studying the role of language communications (mainly, the use of hashtags) in social media networks activities invoking massive civil protests caused by the murder of Slovakian investigative journalist Ján Kuciak in February 2018 and consequently, resulting in a governmental resignation.

2 Digital Media Platforms as Tools for Communicating by Presenting Ideas and Discussing

2.1 Creating Value and Meaning by Sharing, Linking and Liking

Various research have been done to study how digital social media were used to present and discuss political ideas or consumers opinions and how they resulted in a real motivated community actions. Fundamental difference with traditional 'collective' actions was outlined by analyzing the creative power of their networking mechanism which transforms social media into 'connective' community actions generator [1].

The term Digital Networked Action (DNA) [1] was introduced to define that specific amalgamation which functions as an independent socio-digital phenomenon by retrieving various personal opinions, adding value and creating meaning through sharing, linking and liking [8] and transforming the digital community into networked social organism by spreading information (news, ideas, etc.) [14]. The process of creating meaning, itself, is based on linguistic and conceptual blending empowered by the media networking mechanism which transform the social movement into self-organized protest.

The social networking has been used during Arab Spring (Arab Uprising) and Occupy movements and was studied [17] outlining its primary role in mediation, organization and maintaining the protests. Some researchers who studied the use of Twitter during Arab Spring pointed out that 'revolutions were tweeted' [12]. Thus, by studying the paths that information follows in networks and its related linguistic 'expression', it is possible to capture timeline and differences about various types of social interactions, even at a very fine granularity.

2.2 Hashtags and Their Contextual Functionalities

Hashtags are central feature in our Complex Network Analysis because of their multi-functionality. They originally identify channels and topics, or they mark a message for a particular group. Thus, hashtags are analyzed for identifying and propagating messages in the network that can impact wider media public and may influence beyond the specific network.

Popular hashtags have transformed into media friendly monikers for appearing to (momentarily) capture the zeitgeist of the on-line world. So, hashtag has proven itself to be extraordinarily high in its capacity for 'cultural generativity', e.g. coordination of emergency relief (Fukushima, earthquake), political activity (Arab Spring, Occupy), celebrities, etc. Searching for messages via popular hashtags enables the content of posts to be readily discovered and followed (as hashtag searches can be saved). Hashtags, also, have been studied in the formation of ad hoc publics [3].

Particularly, in Twitter hashtags are playing role to organize and categorize information rending its topicality [4] and facilitating participation in conversations. Hashtags are studied for information retrieval as well: amplifying the significance of a collection of messages and render them more readily visible and findable improving contextualization, content filtering and exploratory serendipity. Hashtags allow users to create communities of people interested in the same topic [20] by making it easier for them to find and share information related to it [11].

From a network standpoint, social media researchers examine how users communicate information on social networks through hashtags, so to mobilize individuals to action. Measuring and mapping the propagation of hashtags in complex networks is important challenge to researchers hoping to have advantage from the wealth of data currently available.

Through *hashtag network propagation* research we see the emergence of networked social movements and a paradigm shift from 'collective' to 'connective' action. Hashtags afford new ways of citizens engagement through personalized information sharing. Using hashtags for sampling or for analyzing the diffusion of one particular hashtag, it is possible to uncover the dynamics of a social movement [12,17].

Hashtags influence the process of network information diffusion triggering transformation of linguistic expressions by relating the notions and ideas discussed by virtual communities. Thus, using a shared set of hashtags, a distributed community was able to locate, self-organize and collectively contribute to the information stream about a networked social movement [1,2,14]. The use of hashtags undoubtedly facilitates the information flow to the targeted audience.

The linguistic analysis of hashtags uncover how community ideas relate to each other, revealing their timeline dynamics and change. It shows how the ideas are transformed into 'attention mobilizer' for networked social movement, which may increase the virality of movement messages transforming it into protest.

Hence, we are going to examine how different characteristics of hashtags may explain their structural virality in the hashtag co-occurrence network, and how the role may manifest itself differently during a social movement.

3 Data Structure

The very large amount of linguistic data available from sources such as Twitter, Facebook, YouTube, various blogs and discussion forums make it possible to examine relations between demographic information, language use, and social interaction. Artificial Neural Networks (ANN) have been applied to study the linguistic data, so to discover (visualize) unanticipated patterns, e.g. pattern recognition in larger networks.

For our study, we have extracted data from Twitter: the case of Ján Kuciak, and we have used an event-based approach to on-line data collection. We choose Twitter as it is a popular microblogging platform for users in Slovakia (Fig. 1).

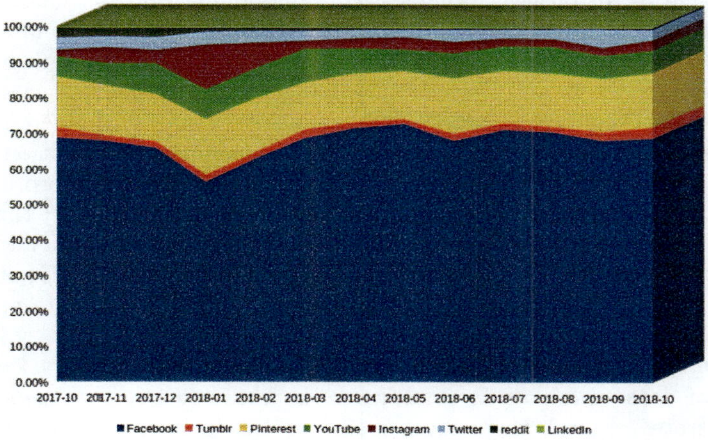

Fig. 1. Social Media Statistics of Slovakia (Slovak Republic) Oct 2017–Oct 2018.

A Twitter posting is called a tweet and a tweet is restricted to 280 characters in length. Length constraint makes characters 'expensive', hence tweets present an informal, sometimes ungrammatical, language as well as introducing many abbreviations. Twitter allows the usage of two meta characters: @ marking a user name (e.g. @*BarackObama*), and # marking a hashtag: a sequence of non-whitespace characters preceded by the hash character (e.g. #*healthCareReform*).

Thus, we collected data, Twitter messages, containing hashtag $\#AllforJan$ and the temporal observation span for our dataset: 28 February 2018–28 July 2018. The data collection comprises a total of 4611 tweets and 595 unique users. The data extraction method was custom crawler collecting: date of post, hashtags, id of tweet, number of likes, links, location, replies, retweets, time of post, timezone, the tweet body, user id, and username.

4 Methodology

The linguistic and cultural influence accomplished by the propagation and adoption of hashtags from various on-line communities has been examined in [6] and a dependency between the hashtag distribution and the frequency ranking of the adopted hashtags was found. Romero [18] reviewed the model of information spread on-line, summarizing its dependence on the manner of hashtag propagation. The results of our previous research [9,10] also show the dependency between the use of hashtags and the communication patterns of certain networks virtual communities.

4.1 Using Hashtags both as Indexical Tools and as Linguistic Tags

The dual nature of hashtags enable their functioning both as indexical tools and as carriers of linguistic information in ongoing networks communication activities. The later accentuate the importance of topics and meaning in social media communities as well as their linguistic and cognitive boundaries through time. Following the interpretation that language is 'the place where our sense of ourselves, our subjectivity is constructed', we examine the evolution of social networks language during the process of communication and interpretation of media texts, carried forward through hashtags. Considering the fact that during human interactions, people tend to align their language across all linguistic levels, the developed network provides a way to understand tagging behavior of network users, their ability and knowledge of available resource manipulation, and the ability to build collective meaning through shared tags.

The analytical framework of our study, combined with the social contextualization of data enable us to investigate the characteristics of users communication strategies by distinguishing their linguistic behaviors (evaluating *likes* and *retweets*, so to associate and connect with others who are similar to them in their interests). The underlying semantics of hashtags provides important insights into user connectivity patterns. $Co - occurrence\ networks$ and *neural embedding model* analysis underly emblematic linguistic characteristics, shaped by prior texts and intertextual references.

4.2 Neural Embedding Model Analysis

In the context of fan media studies, we have utilized a methodology by [7] deriving a hashtag $co - occurrence\ network$ to study the interaction between the users' cultural generativity of hashtags and their semantic content.

A network represents the components of a system – called nodes or vertices, and the direct interaction between them, known as links or edges. There are two main network parameters. The number of nodes represent the aggregation of elements in the system. The set of links represents the total number of interactions between the nodes. In this particular case, the hashtag co-occurrence network is an undirected graph $G = (V, E)$ with V being the number of nodes and E – the number of links developed from a set of V hashtags parsed from the collected data. Each node $v \in V$ represents a hashtag from set V. The edges define the different semantic associations (adjacency relations) of the hashtags formed in the post's tagging space, i.e. $e \equiv (v_i, v_j) \in E$ being the association between the hashtags v_i and v_j. The network is weighted and the edge value depicts the frequency of two hashtags co-occurring in a post.

From a parametric point of view, we focus on several properties of the network. One of the key properties of each node is its degree – representing the number of connections to other nodes. The degree centrality measure is often very effective in determining the influence of a node to the structure of the system. In the case of an undirected network, the computation of the indicator is straightforward since if node A is connected to node B, then B is by definition connected to A. Let the degree of the ith node in the network be denoted by k, with $k_1 = 2, k_2 = 3, k_3 = 4, k_4 = 5$. Contingent upon our undirected network the total number of links L is expressed as the sum of the node degrees, such as:

$$L = \frac{1}{2}\sum_{i=1}^{E} k_i \tag{1}$$

where the factor $\frac{1}{2}$ represents a correction for the fact that we count each link twice due to its lack of directionality.

We also examine the network's clustering coefficient to discern the structure of the network and establish the main hashtags and their semantic relationship. Having a network of n nodes, the c_{ij} denoting the links between nodes i and j ($c_{ij} = cji \geq 0$). With s_{ij} we depict the association strength of nodes i and j, given by

$$s_{ij} = \frac{2mc_{ij}}{c_i c_j} \tag{2}$$

with c_i being the total number of links of node i, and m the absolute number of links in the network, giving

$$c_i = \sum_{j \neq i} c_{ij} \tag{3}$$

and

$$m = \frac{1}{2}\sum_{i} c_i \tag{4}$$

For the mapping purposes, one has to establish for each node i a vector $x_i \in R^p$, showing the location of node i in a 2-dimensional map. In terms of clustering, for each node i, one needs to find a positive integer x_i which serves as indicator of the cluster that i belongs to. The approach used to mapping and clustering is based on [19] where we need to minimize

$$V(x_1, \ldots, x_n) = \sum_{i<j} s_{ij} d_{ij}^2 - \sum_{i<j} d_{ij} \tag{5}$$

where d_{ij} with respect to $x_1, \ldots, x_n$ depicts the distance between nodes i and j provided by

$$d_{ij} = |x_i - x_j| = \sqrt{\sum_{k=1}^{p} (x_{ij} - x_{jk})^2} \tag{6}$$

where mapping is concerned and by

$$d_{ij} = \begin{cases} 0 & x_i = x_j \\ \frac{1}{\gamma} & x_i \neq x_j \end{cases} \tag{7}$$

for the clustering. The parameter γ in this case indicates the resolution where $\gamma > 0$. The larger γ illustrates the bigger number of clusters that are obtained.

We have moreover implemented models for learning vector representations of words, i.e. word embeddings, as outlined by [16] and computed by a neural embedding model. In our case [9] the hashtags constitute a large corpus, described as a vector space of several hundred dimensions, with each unique hashtag being represented by a corresponding vector in the space.

Vectors in turn reflect the structure of each hashtag in terms of context, morphology, and global corpus statistics. As such, hashtags that share common semantic context are allocated closer in proximity.

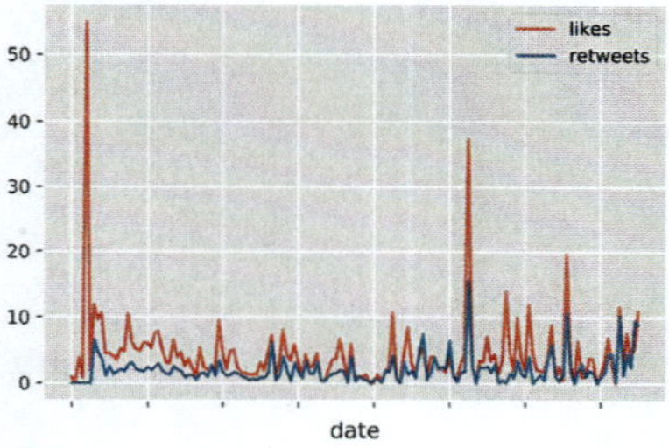

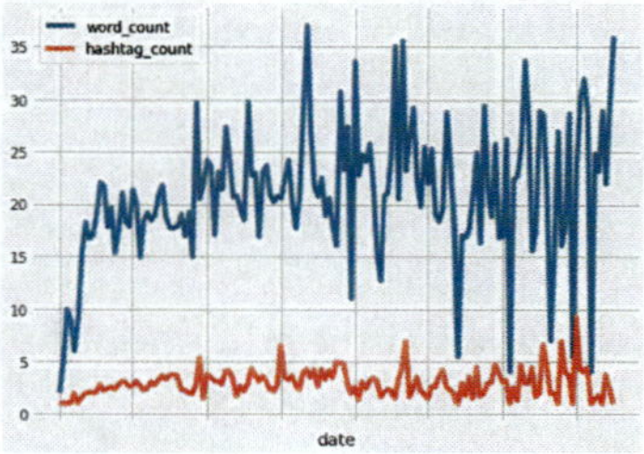

Fig. 2. *Likes* and *retweets* for the studied period (left) and Average Word Count (18.85)/Hashtag Count (2.6) (right).

5 Complex Network Analysis Results

In further analysis, we shall study the formation of virtual network community as relating to the process of network information diffusion and transformation of its linguistic expressions, hashtags. We, also, regard *likes* and *retweets* as an expression of that process and we evaluate them for the studied timeline span. Figure 2 (left) shows that results. The related graphic points out two peaks of *likes* – at the beginning of the event.

Figure 2 (right) presents the Average Word Count (AWC) related to the Hashtag Count (HC) showing that hashtags can be regarded as a summarization of network users discussions and vise versa expressing its topicality.

However, the more precise measurement which can be interpreted semantically is a generation of hashtags co-occurrence network. It gives a detailed picture of the most frequent and semantically interconnected hashtags of the communication network activities.

Fig. 3. Visualization of word cloud for hashtag *#allforjan*.

5.1 The 'allforjan' Co-occurrence Network

Thus, we generate a hashtag co-occurrence network consisting of related to *#allforjan* hashtags whose word cloud (cluster of semantically connected hashtags) is visualized at Fig. 3. In fact, the resulting co-occurrence network is topically related as well. Since hashtags are created in a freely established manner, we can expect a simultaneous emergence of multiple different tags, each embodying the distinct information which relate to the topic *#allforjan*.

However, not all hashtags are equally accepted by the virtual community users. While some propagate the network and progress, others are restricted to only few messages. This implies the realization of power law distribution of the hashtags – there are dominant hashtags that emerge and are the most frequently adopted, with a rapid decrease in frequency and formation of a long tail for

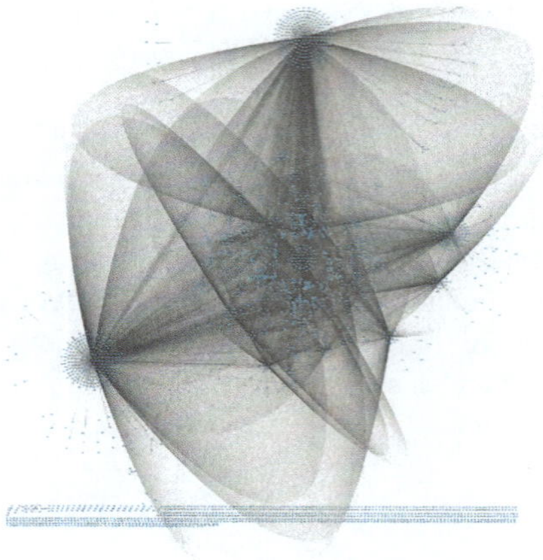

Fig. 4. Visualization of the whole network for hashtag *#allforjan*.

the rest of the hashtags. To test the supposition we compute the Heavy-Tailed Distribution, confirming the presence of the Zipf's Law.

Consequently, we outline in the co-occurrence network several related to *#allforjan* hashtags, exemplifying a rich hashtags vocabulary which interrelate topically as being produced by clustering. The most frequent among all are: *#Slovensko* (*Slovakia*), *#zaslusneslovensko* (For a Decent Slovakia), *#protest*, *#Bratislava*, *#PressFreedom*, *#AllForMartina*, *#corruption*, *#predcasne volby* (early elections), etc.

Also, the analyzed hashtag *#allforjan* appears in several versions without being distracted. Similarly, its semantically related hashtag *#zaslusneslovensko* (For a Decent Slovakia) also appears in some other graphical versions without being decomposed or distracted, and a further subsetting of the data leads to the conclusion in favor of topically centralized and well focused network communications.

In fact, the whole network remains larger and comprises more hashtags (Fig. 4) representing a semantic correlation to *#allforjan* with direct correspondence to the presence of topical relations, or the tendency of people to associate with others and participate in virtual community discussions activities.

5.2 Word Embeddings

Hashtags can be analyzed in a closer perspective by studying their nearest neighbors words/hashtags as well. Figure 5 (left) presents a visualization of generated nearest neighbors of the word/hashtag *#allforjan*.

The results show that generated nearest neighbors hashtags are: *#hnevamsa* (I am Angry), *#zaslusneslovensko* (For a Decent Slovakia), *#ichkrvou* (Their Blood), *#niejenamtojedno*, etc. Also, some other words which are not hashtags remains – *facebook*, *instagram*, *https*, *status*, etc. Both initial and resulting hashtags are topically interrelated.

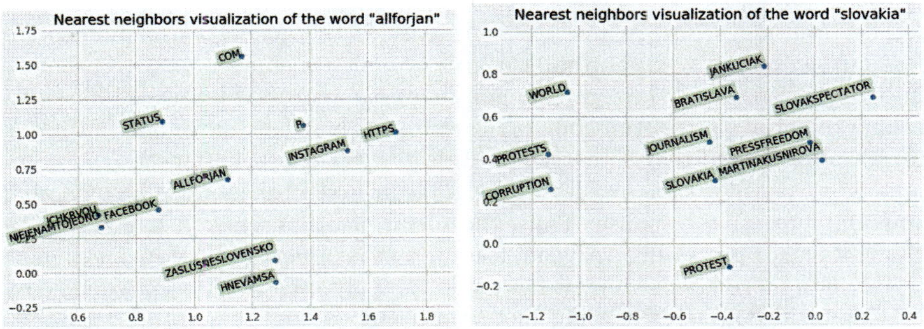

Fig. 5. Nearest neighbors visualization of the word *#allforjan* (left) and Nearest neighbors visualization of the word *#Slovakia* (right).

Figure 5 (right) presents a visualization of generated nearest neighbors of the word/hashtag *#Slovakia*. The related results show that the received nearest neighbors hashtags are: *#Slovakia*, *#Bratislava*, *#PressFreedom*, *#protest*, *#journalism*, *#MartinaKushnirova*, *#corruption*, *#protests*, *#world*, etc., which are not randomly distributed but are topically interrelated with the initial hashtag.

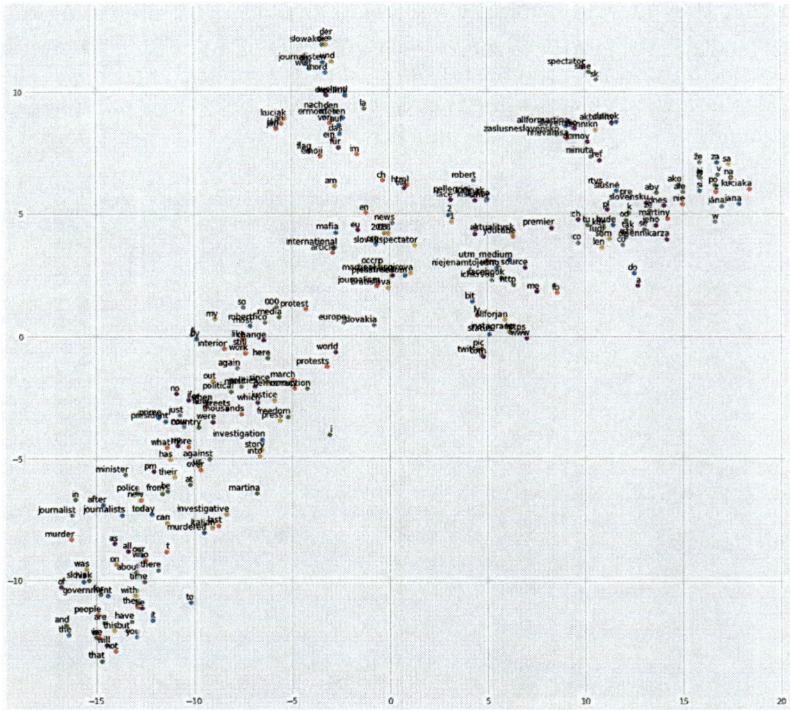

Fig. 6. t–Distributed Stochastic Neighbor Embedding (t–SNE).

The resulting hashtags are strictly related to the generation of immediate interactive community users responses to a specific timeline context to promote multiple conversations through common linguistic representations. As hashtags become essential part of the communication process, their meaning transforms into a rhetorical strategy into organized protest whose social factors and communal ideas construction are playing a significant role in the formation of virtual community reactive response. Thus, the social media discussion has become a driving force for protesting activism, leading to the opportunity of social change.

Taking a more closer view, we are able to analyze the network structure by studying it in details estimating more word embeddings. Figure 6 presents t–Distributed Stochastic Neighbor Embedding (t–SNE). The results present some more relatively frequent words (the majority of which are not hashtags) topically related to the studied hashtags among which are: justice, journalist, journalists, investigation, government, people, against, premier, minister, mafia, youtube, twitter, facebook, etc., which also are topically interrelated, i. e. frequently used in the virtual community discussions.

5.3 Translocality

Among received word embeddings results we found the names of other social media platforms like youtube, facebook, instagram and some https localities suggesting that they are also topically interrelated. Thus, it is interesting to study the underlying translocality of social networking sites for *#allforjan* hashtag. A density visualization of the hashtags network is presented at Fig. 7. The blue regions depict lower density, while the red ones describe regions of higher density (i.e. frequency of usage) and interconnectivity.

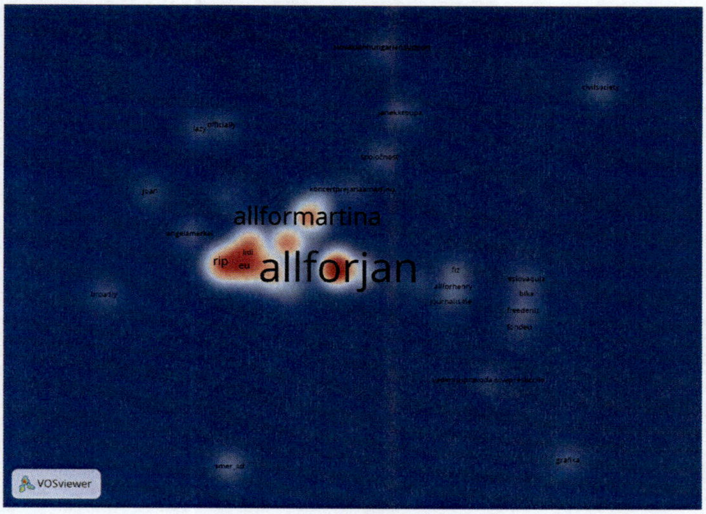

Fig. 7. Term map of the hashtag network. Colors indicate the density of the hashtags.

As presented at Fig. 7, two close to each other (centrally located) clusters represent hashtags topically relating to the virtual community users discussions – *#allforjan* and *#allformartina*.

5.4 Linguistic Analysis

A simple linguistic analysis, instead of the stochastic one (which reveal the combinatorial functionality of words) can reveal typological and semantic word properties. Among the received results, we can outline three general types of words:

(i) expressions (single or as part of larger tags (hashtags)) which are the main hashtags.
(ii) named entities (names of countries, towns, organizations, persons, etc. (also single or as part of larger tags (hashtags) which interrelate topically) like: Slovakia, Bratislava, Nitra, Ján Kuciak, Martina Kusnirova, Europe, EU, etc.
(iii) terms, mainly political (single or as part of larger tags (hashtags) which interrelate topically) like: justice, government, press freedom, etc.

It is important to note, that the above types of words can be, additionally, combined with certain prepositions or prefixes forming more complex and semantically meaningful expressions (hashtags slogans) like:

All (all) – AllAgainst – allAgainstMafia – AllFor (allfor) – AllForJan (AllForMartina), AllForJanHead, AllForJanDead – For – justiceForJan – WorldPressFreedom, etc.

However, the most interesting for the linguistic analysis are hashtags which presents slogans and function as 'attention mobilizing messages' and 'acting codes' which are formally produced in the same way. Some data extracted examples are: *#zaSlusneSlovensko* (For a Decent Slovakia) – *#chcemeverit* (We want to Believe) – *#nestaci* (Not Enough), etc.

They can be regarded as specific ambigrams as 'AllFor' construction may refer to 'news, facts about' as well as to 'support, identify with' which refer to the fact that news headlines and slogans have very much in common, however, the language of printed media (news) only affects but does not has such a mobilizing power as slogans. Thus, hashtags have semantic structure of linguistic units having also the properties like synonymy, ambiguity, variability, etc. and behaving very much like linguistic units functioning by unlocking the language potential power to mobilize for action.

6 Conclusion

The presented results of complex network analysis utilize neural embedding model to study the hashtags behavior in virtual community communications activity connected to the Case of Ján Kuciak revealing important and emblematic social contextualization related to the way the language data were used.

The approach allows us to investigate specific linguistic characteristics of virtual community users communications. It shows that in multiple instances of hashtags usage, virtual community users are able to exchange and discuss important event related timeline specific information on shared pool of community discussions activities, and as a result to organize themselves to act as protesting activists.

The timeline development of the event presents the following stages: communication – organization – reactive response. All that stages are mediated by the virtual communication activities subsumming: opinion self-expression (identifying by attachment to a relates hashtag) – discussion – action (protest). In fact, the main hashtags usage and evolution (as being the mobilizing slogans) turn out the virtual community discussion into organized civil protest.

The linguistic expression of hashtags understood as mobilizing messages and slogans can reveal the mechanism of networking communicative power, giving much insights to the 'logic of connective actions' network users activity within the respective timeline span which transforms the virtual community communications into organized civil protest movement.

Acknowledgements. This work is partially supported by the Ireland Research Center Programme (Grant $13/RC/2106$), by HELIOS $H2020$ grant No 825585, and by DFNI-DN-12/5 project.

References

1. Bennett, W.L., Segerberg, A.: The logic of connective action: digital media and the personalization of contentious politics. Inf. Commun. Soc. **15**(5), 739–768 (2012)
2. Bruns, A., Burgess, J.E.: The use of Twitter hashtags in the formation of at hoc publics. In: Proceedings of the 6th European Consortium for Political Research (ECPR) General Conference, Reykjavik, Iceland (2011)
3. Bruns, A., Burgess, J.E.: Twitter hashtags from at hoc to calculated publics. In: Rambukkana, N. (ed.) Hashtag Publics: The Power and Politics of Discursive Networks, pp. 13–28. Peter Lang, New York (2015)
4. Chang, H.C.: A new perspective on Twitter hashtag use: diffusion of innovation theory. Proc. Am. Soc. Inf. Sci. Technol. **47**(1), 1–4 (2010)
5. Clark, E. et al.: Creative writing with a machine in the loop: case studies on slogans and stories. In: IUI 2018 23rd International Conference on Intelligent User Interfaces, pp. 329–340 (2018)
6. Cunha, E., Magno, G., Comarela, G., Almeida, V., Goncalves, M.A., Benevenuto, F.: Analyzing the dynamic evolution of hashtags on Twitter: a language-based approach. In: Proceedings of the Workshop on Languages in Social Media, pp. 58–65. Association for Computational Linguistics (2011)
7. Ince, J., Rojas, F., Davis, C.A.: The social media response to black lives matter: how Twitter users interact with black lives matter through hashtag use. Ethnic Racial Stud. **40**(11), 1814–1830 (2017)
8. Jenkins, H., Ford, S., Green, J.: Spreadable Media: Creating Value and Meaning in a Networked Culture. New York University Press, New York (2013)

9. Kapanova, K.G., Fidanova, S.: Generalized nets: a new approach to model a hashtag linguistic network on Twitter. In: Advanced Computing in Industrial Mathematics. Studies in Computational Intelligence, vol. 793, pp. 211–221. Springer (2018)
10. Kapanova, K.G., Stoykova, V.: Social network analysis of 'Clexa' community interaction patterns. In: Ntalianis, K., Vachtsevanos, G., Borne, P., Croitoru, A. (eds.) Applied Physics, System Science and Computers III. APSAC 2018. Lecture Notes in Electrical Engineering, vol. 574, pp. 257–264. Springer (2019)
11. Kricfalusi, E.: The Twitter Hash Tag: What Is It and How Do You Use It? (2009). https://techforluddites.com/hashtag/
12. Lotan, G., et al.: The revolutions were tweeted: information flows during the 2011 Tunisian and Egyptian revolutions. Int. J. Commun. **5**(5), 1375–1405 (2011)
13. Margan, D., Martincic-Ipsic, S., Mestrovic, A.: Preliminary report on the structure of Croatian linguistic co-occurrence networks. arXiv preprint arXiv:1405.4433 (2014)
14. Meraz, S., Papacharissi, Z.: Networked gatekeeping and networked framing on #Egypt. Int. J. Press/Politics **18**(2), 138–166 (2013)
15. Meraz, S.: Hashtags Wars and Networking Framing. In: Telleria, A.S. (ed.) Between the Public and Private in Mobile Communication, pp. 303–323. London, Routledge (2017)
16. Mikolov, T., Chen, K., Corrado, G., Dean, J.: Efficient estimation of word representations in vector space. arXiv preprint arXiv:1301.3781 (2013)
17. Papacharissi, Z., de Fatima Oliveira, M.: Affective news and networked publics: the rhythms and news storytelling on #Egypt. J. Commun. **62**(2), 266–282 (2012)
18. Romero, D.M., Meeder, B., Kleinberg, J.: Differences in the mechanics of information diffusion across topics: idioms, political hashtags, and complex contagion on Twitter. In: Proceedings of the 20th International Conference on World Wide Web, pp. 695–704. ACM (2011)
19. Waltman, L., van Eck, N.J., Noyons, E.C.: A unified approach to mapping and clustering of bibliometric networks. J. Inform. **4**(4), 629–635 (2010)
20. Zappavigna, M.: Ambient affiliation: a linguistic perspective on Twitter. New Media Soc. **13**(5), 788–806 (2011)

Can the Artificial Neural Network Be Applied to Estimate the Atmospheric Contaminant Transport?

A. Wawrzynczak[1,2](✉) and M. Berendt-Marchel[1]

[1] Institute of Computer Sciences, Siedlce University, Siedlce, Poland
{anna.wawrzynczak-szaban,monika.berendt-marchel}@uph.edu.pl
[2] National Centre for Nuclear Research, Swierk-Otwock, Poland

Abstract. Release of hazardous materials in chemical industries is a significant threat to surrounding areas, especially in the urban terrain. To face this thread, there are proposed frameworks capable of estimating the contamination source parameters like localization and release rate based on the substance concentrations recorded by the sensors network. In such systems, the parameters of the given dispersion model are sampled to fit the model output to the registrations. Application of such an approach in the urbanized terrain requires the use of the complicated and computationally expensive dispersion models taking into account the complicated wind field. As a result, the localization systems require to much time to be applied in the real toxin threat situation, when the reaction time is the most crucial.

We examine the possibility of training an artificial neural network (ANN) so that it could effectively simulate the atmospheric toxin transport. The use of a fast neural network in place of costly computational dispersion models in systems localizing the source of contamination might significantly improve their efficiency (speed). In this paper, we present results of training the ANN with the use of dataset covering the contamination source term parameters and point output concentrations generated by The Quick Urban & Industrial Complex (QUIC) Dispersion Modeling System. We test various ANN structures, i.e., numbers of ANN layers, neurons, and activation functions to achieve the ANN capable of estimating the contaminant concentration. The performed tests confirm that trained ANN has the potential to replace the dispersion model in the contaminant source localization systems. However, its efficiency requires the carefully generated training data.

1 Motivation

The transport and storage of the toxic materials entails a constant risk of releasing dangerous materials into the atmosphere. Of course, all precautions are taken to minimize all risks to people and the environment. However, it is impossible

I. Dimov and S. Fidanova (Eds.): HPC 2019, SCI 902, pp. 132–142, 2021.
https://doi.org/10.1007/978-3-030-55347-0_12

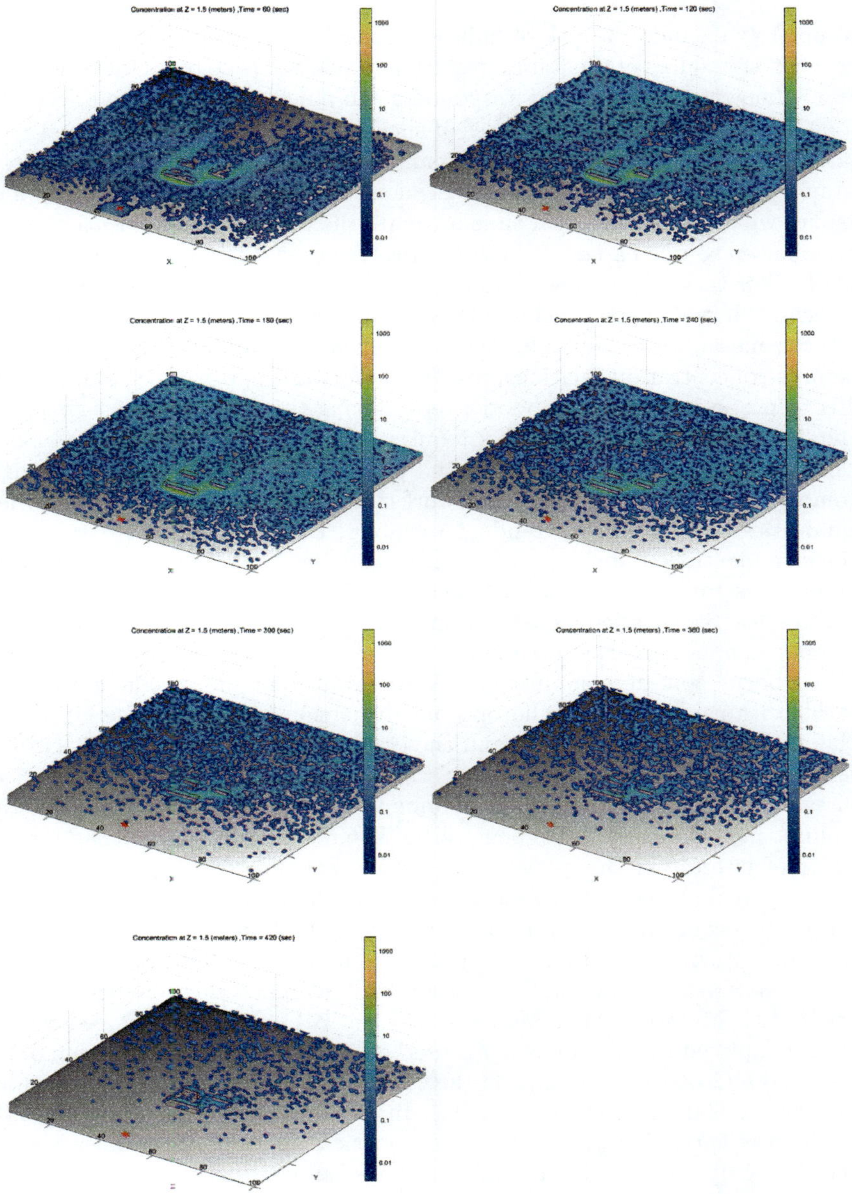

Fig. 1. The concentration [g/m^3] of the dense gas chlorine during seven minutes after the release of 10 Mg of gas. The source location is marked by red cross.

to completely exclude the risk of failure of storage or transport systems. In the cases when the source of the failure resulting in releasing the toxin into the atmosphere is known, the emergency responders can quickly undertake all necessary actions to minimize the consequences of such release. The more problematic are situations when the sensors report the non-zero concentration of the dangerous substance, which source is not known. In such cases, it would be valuable to have a system, which could, in a real-time estimate the most probable location of the contamination source based on the concentration data reported from the sensors network. This task is quite tricky and can be done by using backward or forward approaches, in which source characteristics are inferred from concentration or deposition measurements at different locations and time intervals by establishing source-concentration relationships. In the forward approach, the underlying goal is to propose a dispersion model reproducing the contamination registered by the sensor network. This, in turn, requires defining the set of employed model parameters. Such inverse problem has no unique analytical solution, but might be analyzed with probabilistic frameworks, as the Bayesian approach, where all quantities are modeled as random variables. Bayesian approach transforms the inverse mentioned above problem into searching for a posterior distribution based on the sampling of an ensemble of simulations using a priori knowledge and observed data. Stochastic reconstruction of the contamination source consists of two important mechanisms. One is the dispersion model suitable for modeling of the airborne contaminant in considered terrain and the second is the sampling algorithm able to find the optimal dispersion model parameters basing on the model output comparison and the contaminant registrations. Regarding the efficiency of the applied parameter scanning algorithm, each reconstruction requires multiple runs of the dispersion model. In open terrain, the simple Gaussian plume models (e.g. [1]) can be applied. The reconstruction in urban terrain, which is of interest in this paper, requires more advanced models taking into account the turbulence of the wind field around the buildings.

The first reconstructions in urban scales using building models was reported in [10] and [4]. In [10], authors used an adjoint representation of the source-receptor relationship and applied a Bayesian inference methodology in conjunction with Markov Chain Monte Carlo (MCMC) sampling procedures. In [4] authors applied the methodology presented in [9] to the reconstruction of the flow around an isolated building and the flow during IOP3 and IOP9 of the Joint Urban 2003 Oklahoma City experiment. In this reconstruction, the FEM3MP [3] model was applied to predict the atmospheric dispersion of the released substance. Recently, in [11] the approximate bayesian computation algorithm was applied to localize the source of contamination in the highly urbanized terrain of the center of London utilizing the real field experiment data from DAPPLE experiment [16]. As the forward dispersion model, the Quick Urban Industrial Complex (QUIC) Dispersion Modeling System was applied [15]. The successful estimation of the release source required over 10000 runs of the dispersion model. Even though the QUIC model is able to simulate the airborne contaminant transport in the city relatively quickly, a single simulation over the $800\,\mathrm{m} \times 800\,\mathrm{m}$

domain takes as minimum 2 min. Thus, reconstruction on a single computer requires over 330 h. Computation time can be shortened by using a distributed system, but it still will be challenging to apply it in the real-time working localization system, when the answer time is crucial. The required simulation time will increase with the enlargement of the considered terrain e.g. for the whole city. From the other side, using the simple Gaussian plume model will make the localization system not effective in the city domain.

In this situation, the artificial neural networks (ANN) might be helpful. If we could train the ANN to be efficient in the simulation of the airborne contaminant transport, we could use it as the forward model in the system localizing the contamination source in real time. Of course, the ANN has to be trained on the fixed city topology using the real wind conditions. This process requires lots of simulations being the training data-sets for the ANN. The process of training the ANN is computationally expensive but, ones trained the ANN would be a high-speed tool for estimation the point-concentrations for a given contamination source.

In this paper we present the attempt to train the ANN to work as a dispersion model in the urbanized area. The paper is constructed as follows: Sect. 2 introduce the ANN background assumptions, the details of generation the training data-set are given in the Sect. 2.1. The topology of the specified network is given in Sect. 2.2. The obtained results are presented and discussed in Sect. 3.

2 Artificial Neural Network

Artificial neural networks (ANN) are computational models that consist of interconnected elements called neurons. They are modeled on the construction of natural neurons and synapses connecting them [2]. ANN are capable of learning from training samples without knowing any laws or equations. There are several types of neural networks. In this paper, we used one of the simplest and widely used ANNs, the feed-forward neural network e.g. [7]. This network has a one-way structure, i.e., the signal flows only in one direction from input nodes to output nodes. Feedforward neural networks are used, among others in recognition of image and speech, in the classification. They were also successfully used to predict the transport of pollutant in open areas e.g. [8,12].

The ANN contaminant dispersion model is considered as a system that receives information from n distinct sets of inputs $X_i (i = 1, \ldots, n)$, namely contaminant source parameters and sensor location, and produces a specific output, in our case the concentration of the chlorine. No prior knowledge about the relationship between input and output variables is assumed. The input variables should be independent from each other, and each one is represented by its own input neuron $i = 1, \ldots, n$. Each neuron calculates a linear combination of the weighted inputs ω_{ij}, including a bias term b_i, from the links feeding into it and the corresponding summed value $C_j = \sum_i \omega_{ij} X_i + b_i$ is transformed using a function f, either linear or non-linear for example log-sigmoid or hyperbolic tangent. The bias term is included in order to allow the activation functions to be offset from zero and it can be set randomly or to the desired value. The output obtained is then passed as a new input $\tilde{X}_j = f(C_j)$ to other nodes in the following layer, usually named hidden layer. Though one is allowed to use several neurons in this hidden layer, it is generally advantageous to somehow minimize the number of hidden neurons, in order to improve the generalization capabilities of the model and also to avoid over-fitting. Having such a framework of input variables and sets of functions, the ANN has to be trained in order to obtain the best estimate for each weight ω. The weight values are determined by an optimization procedure, the so-called learning algorithm [6]. The result of the neural network is compared with the target in order to calculate a predefined value of the error function. The error is sent over the network, and the algorithm adjusts the weights of each connection, respectively, to reduce the value of the error function. This repeated process corresponds to the number of training iterations that causes the network result to coincide with a state where the error between the output and the target is very small.

To train the ANN to work as a dispersion model we require the suitable training data-set. The next section presents details of the considered domain in which were performed the dense gas simulations used to train the ANN.

2.1 Generation of the Training Data

Training ANN requires a large reliable set of data. In this case, it should be measurements of the contaminant being a result of various release rates under different meteorological conditions. It is not possible to obtain such a set of data from real gas releases in urban areas. We decided to train ANN using the data from the simulations of the chlorine release in the urbanized terrain under steady meteorological conditions. To generate the data-set utilized to train, test, and validate the ANN we have used the Quick Urban Industrial Complex (QUIC)

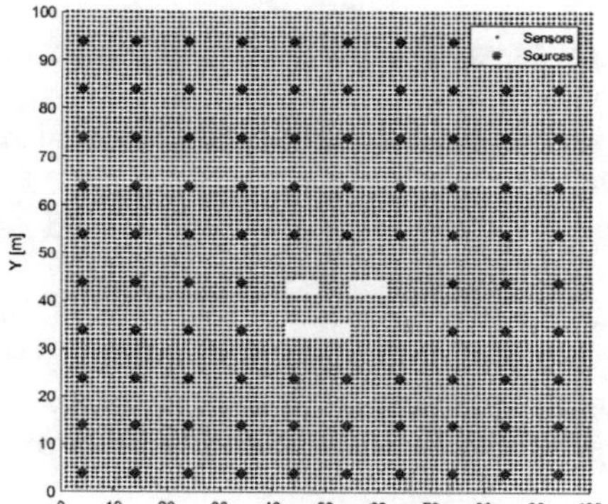

Fig. 2. The domain assumed during the preparation of the testing data-set. The considered contamination source locations are marked in red, the locations of the sensors are marked in blue. The white rectangles represents the buildings.

Dispersion Modeling System. QUIC is intended for applications where the dispersion of air pollutants released near buildings must be computed very quickly [15]. The QUIC system comprises of a wind model QUIC-URB, a dispersion model QUIC-PLUME, and a graphical user interface. The modeling strategy adopted in QUIC-URB was originally developed by Rockle [13] and uses a 3D mass-consistent wind model to combine properly resolved time-averaged wind fields around buildings [14]. The mass-consistent technique is based on a 3D complex terrain diagnostic wind model. The basic methodology involves first generating an initial wind field that includes various empirical parameterizations to account for the physics of flow around buildings. Next, this velocity field is forced to be divergence free, subject to the weak constraint that the variance of the difference between the initial velocity field and mass consistent final velocity field is minimized. The QUIC-PLUME is a Lagrangian particle model which describes gas dispersion by simulating the release of particles and moving them with an instantaneous wind composed of mean and turbulent components. A dense gas cloud model has been incorporated in order to evaluate the effects of heavier-than-air chemical industrial gas dispersion. The code has been tested for both idealized and real-world cases (e.g., [5,15]).

To test the possibility of applying the ANN to simulate the airborne contaminant dispersion in the urbanized terrain we have prepared the domain of size $100\,\mathrm{m} \times 100\,\mathrm{m} \times 20\,\mathrm{m}$ in which we have placed three buildings: one of size $12.2\,\mathrm{m} \times 2.4\,\mathrm{m} \times 2.6\,\mathrm{m}$, and two of size $6.1\,\mathrm{m} \times 2.4\,\mathrm{m} \times 2.6\,\mathrm{m}$. In this domain, we have set the simulations of a dense gas chlorine release at an industrial facility and registered its concentration for seven minutes. The sample simulation results are presented in Fig. 1. To prepare a reliable data-set for training the ANN we have placed the grid of sensors every 1 m inside the domain at the height of 1, 2 and 3 m. Inside the domain, we have considered the various locations of the release source every 10 m and release mass from 1 Mg up to 10 Mg. The domain and locations of sensors and considered sources are shown in Fig. 3. The performed simulations resulted in about 30 000 000 data that have been used for training the ANN.

2.2 The Selected ANN Topology

During ANN setup phase, numerous tests were performed with different combinations of the hidden layers, neuron number, learning rates, and activation functions. Figure 3 illustrates the selected topology. In the input layer, we introduce seven neurons representing the coordinates of the contamination source, release rate and the coordinates of the registration point (sensor) within the domain (Fig. 2) in which the tracer concentration will be given in the output layer. It appeared that the ANN performed best when the four hidden layers were introduced, with 49, 34, 20 and 4 neurons in subsequent layers. The Levenberg-Marquardt learning method was used, which minimizes an error function in "damped" procedures, i.e., select steps proportional to the gradient of the error function. The learning rate was set to 0.001. We have tested various activation functions, and the best results were achieved using the hyperbolic tangent *tanh*

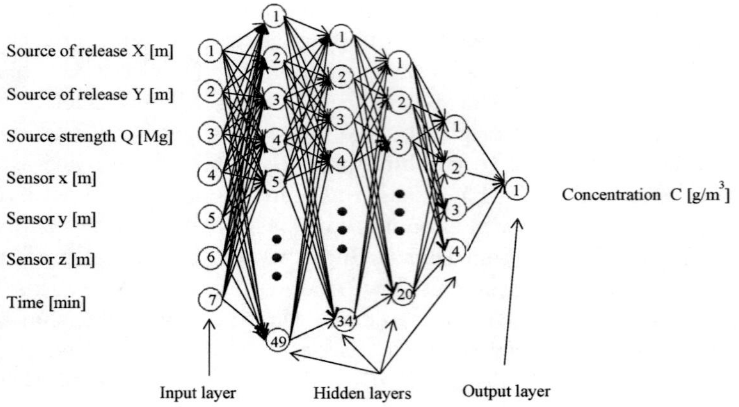

Fig. 3. The topology of the six-layered feed-forward applied ANN with listed input and output neurons.

function in all hidden layers, and linear function in the output layer. The crucial for ANN better performance occurred scaling of all input parameters to be in the interval $(0, 1)$. Scaling allowed escaping from the problem of different scales and model instability. Additionally, the output concentrations were logarithmized. The input data set described in Sect. 2 was divided to training data set - 70%, validation and testing data-sets 15% each. The objective function describing the mean squared error between the real concentration and network output was set to reach $1e{-}04$ value.

3 Results and Summary

The ANN was set under the Deep Learning Toolbox of the Matlab software. The results of the ANN training present Fig. 4. The quality of the trained ANN is quite good. The R-value for training, test and validation data equals 0.95 which indicates a strong relationship between the outputs and targets. The regression lines show that ANN slightly underestimates the higher concentrations. More detailed analysis of the ANN performance is shown in Fig. 5. Figure 5 presents the comparison of the simulated by QUIC dispersion model one-minute concentrations of the chlorine versus the concentrations estimated by the trained ANN for the same release scenario. The scatter plots display that the best agreement is achieved at the beginning of the release, with time the ANN starts to underpredicts the concentrations. The correlation coefficient for the first minutes equals 0.96 and decreases for subsequent minutes reaching value 0.83 for the seventh minute. Moreover, it is visible that ANN has a tendency to underpredicts the smaller concentrations, while for greater concentrations level of agreement increases. A possible reason is that there is a larger variety of scenarios leading to smaller concentrations and fewer favorable to increased concentrations.

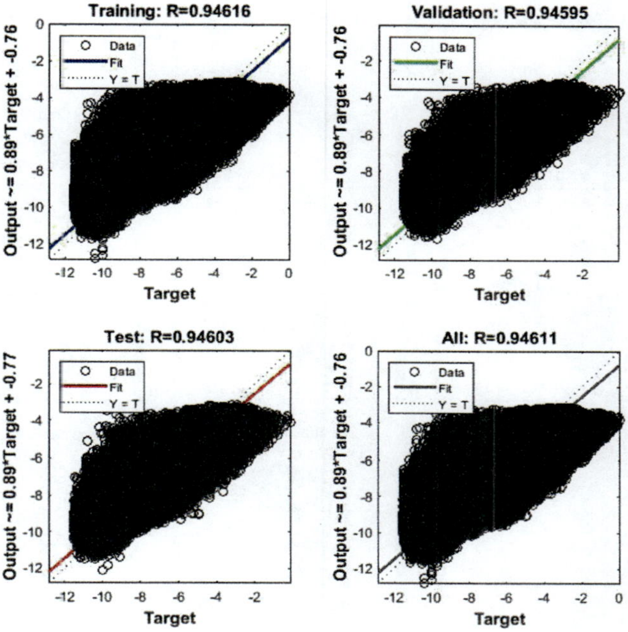

Fig. 4. The scatter plots representing results of training, testing, and validation process of the ANN. The dashed line represents the ideal fit.

Nevertheless, does the ANN learned the physics standing behind the dense gas dispersion over the urbanized area? Figures 6 and 7 presents the contour plots for the chlorine dispersion during seven minutes after two release scenarios. One can see that the simulated by the ANN dispersion of the chlorine is in the agreement with the QUIC simulation presented in Fig. 1. In the initial phase, the gas is stopped in front of the buildings and then spreads to the sides, flowing from both sides of the buildings, of which less gas passes through the narrow passages between the buildings. The gas then travels in the wind direction set to 180° (along the $\overrightarrow{Y}$ axis). As a result, the concentration of chlorine slowly decreases. Comparison of Fig. 6 where the released mass equals 1 Mg and Fig. 7 for released mass equal to 10 Mg confirms the correct estimation of chlorine concentrations by ANN. The concentrations predicted by ANN for release of 10 Mg of chlorine are larger than in the case of 1 Mg release. The regions of higher concentrations stay longer in the vicinity of buildings. Simulated by the ANN transport of chlorine in the vicinity of buildings agrees with the physics of the dense gas dispersion.

Presented results confirm that trained ANN can properly simulate the transport of dense gas around the buildings. Even though we do not obtain the one-to-one agreement between the QUIC and ANN model concentrations, the trajectory of gas particles and gradient of concentrations predicted by ANN agree

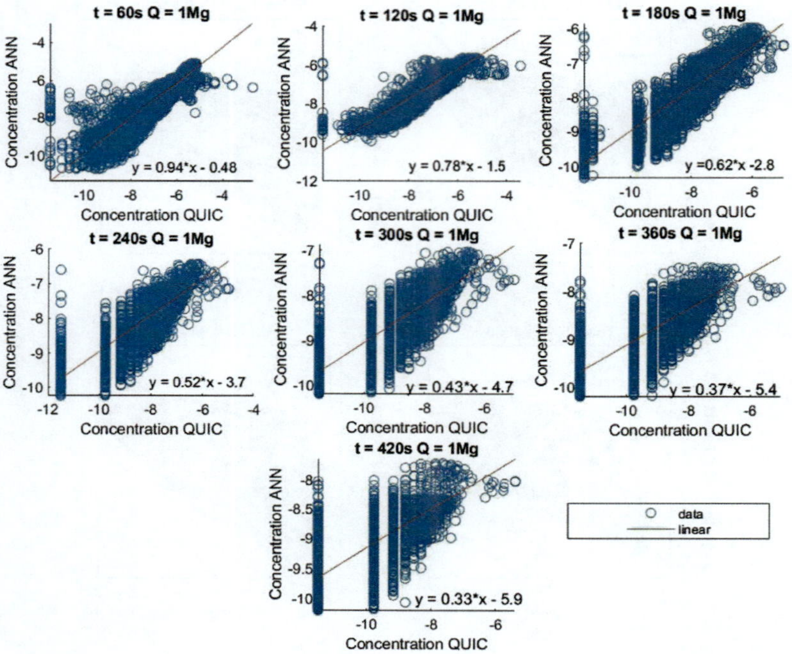

Fig. 5. The regression lines for the simulated by the QUIC model and the trained ANN concentrations during consecutive seven minutes after the release of 1 Mg chlorine.

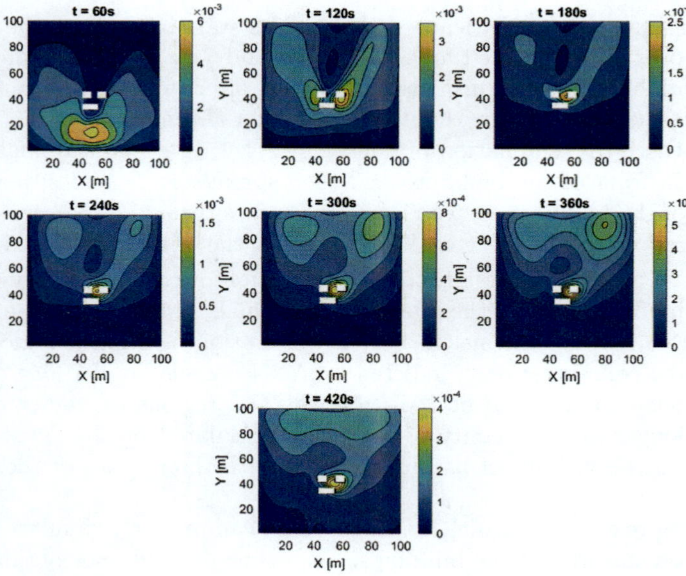

Fig. 6. The dispersion of the dense gas simulated by the ANN during consecutive seven minutes after the release of 1 Mg chlorine

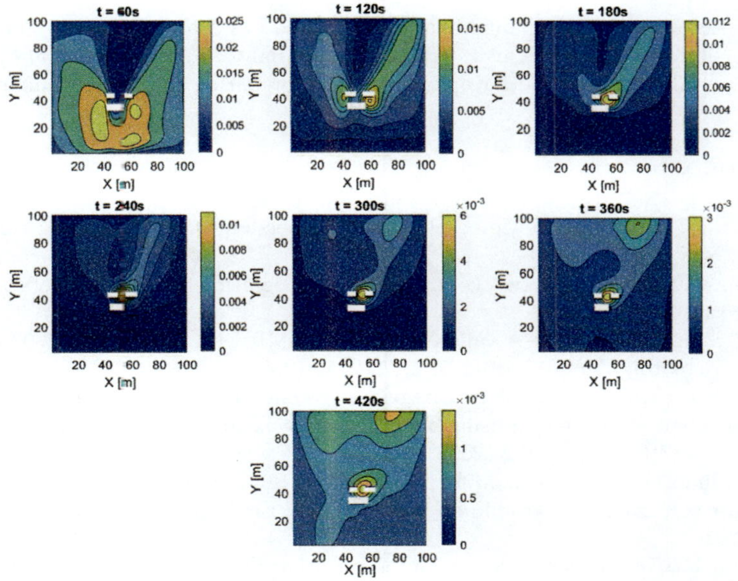

Fig. 7. The dispersion of the dense gas simulated by the ANN during consecutive seven minutes after the release of 10 Mg chlorine

with the expectations. Obtained results suggest that the trained ANN can be successfully used in the contaminant source localization system as the forward dispersion model. In such systems, the contaminant source is estimated based on minimizing the distance measure between the real concentrations from the sensors network and concentrations expected from the forward dispersion model. Taking that into account, more important is that ANN should correctly estimate the concentration gradients than its exact values. Additionally, the application of ANN in such localization system will enable its operation in real time. The time required by the presented in this paper ANN to estimate gas concentrations in a single point was 2.41 s, while the QUIC model it is estimated as 60 s, this gives us 20 times speed up.

Of course, presented in this paper study requires further development. To apply ANN as a forward model in the localization system dedicated to working at given city/terrain, the ANN must be trained on a dedicated data-set. This set should be created by taking into account the specific terrain topology and most common meteorological conditions. In such cases different setup of ANN might be more effective. Moreover, the presented analysis showed, that included release scenarios should immensely widen to generate the concentrations of given substances in a large range of values. The process of generating such data and training ANN will be very time consuming, but well train ANN applied in emergency system localizing the source of abrupt contamination makes possible its work in a real-time.

Acknowledgements. We thank Michael Brown and Los Alamos National Laboratory for the possibility to use the Quick Urban & Industrial Complex Dispersion Modeling System. Special thanks to Piotr Kopka for help in preparing the QUIC simulation.

References

1. Arya, S.P.: Air Pollution Meteorology and Dispersion, vol. 310. Oxford University Press, New York (1999)
2. Bishop, C.M.: Neural Networks for Pattern Recognition. Oxford University Press, Oxford (1995)
3. Chan, S.T., Leach, M.J.: A validation of FEM3MP with joint urban 2003 data. J. Appl. Meteorol. Climatol. **46**(12), 2127–2146 (2007)
4. Chow, F.K., Kosovia, B., Chan, S.: Source inversion for contaminant plume dispersion in urban environments using building-resolving simulations. J. Appl. Meteorol. Climatol. **47**(6), 1553–1572 (2008)
5. Gowardhan, A.A., Brown, M.J., Pardyjak, E.R.: Evaluation of a fast response pressure solver for flow around an isolated cube. Environ. Fluid Mech. **10**(3), 311–328 (2010)
6. Haykin, S.: Neural Networks: A Comprehensive Foundation. Prentice Hall PTR (1994)
7. Hecht-Nielsen, R.: Theory of the backpropagation neural network. In: Neural Networks for Perception, pp. 65–93. Academic Press (1992). ISBN 9780127412528
8. Hossain, K.: Predictive ability of improved neural network models to simulate pollutant dispersion. Int. J. Atmos. Sci. **2014**, 141923 (2014). https://doi.org/10.1155/2014/141923
9. Johannesson, G., Hanley, B., Nitao, J.: Dynamic Bayesian models via Monte Carlo-an introduction with examples (No. UCRL-TR-207173). Lawrence Livermore National Laboratory, Livermore, CA, USA (2004)
10. Keats, A., Yee, E., Lien, F.S.: Bayesian inference for source determination with applications to a complex urban environment. Atmos. Environ. **41**(3), 465–479 (2007)
11. Kopka, P., Wawrzynczak, A.: Framework for stochastic identification of atmospheric contamination source in an urban area. Atmos. Environ. **195**, 63–77 (2018)
12. Ma, D., Zhang, Z.: Contaminant dispersion prediction and source estimation with integrated Gaussian-machine learning network model for point source emission in atmosphere. J. Hazard. Mater. **311**, 237–245 (2016)
13. Röckle, R.: Bestimmung der Strömungsverhältnisse im Bereich komplexer Bebauungsstrukturen. na (1990)
14. Sherman, C.A.: A mass-consistent model for wind fields over complex terrain. J. Appl. Meteorol. **17**(3), 312–319 (1978)
15. Williams, M.D., Brown, M.J., Singh, B., Boswell, D.: QUIC-PLUME theory guide. Los Alamos National Laboratory, vol. 43 (2004)
16. Wood, C.R., Arnold, S.J., Balogun, A.A., Barlow, J.F., Belcher, S.E., Britter, R.E., Neophytou, M.K.: Dispersion experiments in central London: the 2007 DAPPLE project. Bull. Am. Meteorol. Soc. **90**(7), 955–970 (2009)

Advanced HPC Monte Carlo and Quasi-Monte Carlo Applications

Optimization of the Direction Numbers of the Sobol Sequences

Emanouil Atanassov, Sofiya Ivanovska, and Aneta Karaivanova(✉)

Institute of Information and Communication Technologies,
Bulgarian Academy of Sciences, Sofia, Bulgaria
{emanouil,anet}@parallel.bas.bg

Abstract. The quasi-Monte Carlo methods use specially crafted sequences instead of the usual pseudo-random sequences used in Monte Carlo methods, in order to improve on the rate of convergence.

Most of the time these sequences are of low-discrepancy, with the Sobol sequences being one of the most widely used in practice, especially in Financial Mathematics. The definition of the Sobol sequences offers substantial freedom in choosing the so-called direction numbers, so that the basic properties of the sequences are ensured, while additional equi-distribution properties are achieved through some optimization procedure. In a particular computation when the number of dimensions is relatively high compared with the number of terms of the sequence, certain artefacts in the distribution of the sequences become difficult to avoid. In this work we describe our approach on how to define additional equi-distribution criteria and the optimization procedure that we used in order to obtain new sets of direction numbers usable in such settings. Numerical results demonstrating the performance of the sequences using these direction numbers are also discussed.

1 Introduction

The quasi-Monte Carlo methods are built upon the use of specially designed sequences instead of the usual pseudo-random numbers in the Monte Carlo methods. The majority of the Monte Carlo methods can be considered to be solving the problem of numerical integration of a suitable function in the s-dimensional unit cube E^s. Following Sobol' [12] we call s the constructive dimension of the algorithm. Each deterministic sequences in E^s gives rise to a corresponding quasi-Monte Carlo method for numerical integration. The most popular measure of the irregularity of distribution of a sequence is its discrepancy [9], which is related to the integration error through the Koksma-Hlawka inequality [3], when the sub-integral function is of bounded variation in the sense of Hardy and Krause. Many families of sequences that attain the presumed best possible rate of convergence of their discrepancy $\mathscr{O}\left(N \log^s N\right)$ are known, for example the sequences of Sobol', Halton, Faure. Even when other measures of irregularity of

I. Dimov and S. Fidanova (Eds.): HPC 2019, SCI 902, pp. 145–154, 2021.
https://doi.org/10.1007/978-3-030-55347-0_13

distribution are studied, these sequences or some modified versions of them are obvious candidates to achieve lower numbers.

One important factor to consider in the use of low-discrepancy sequences is the addition of some randomisation procedure. Even the simplest randomization procedures solve some practical and theoretical problems of the original sequences: they avoid with probability 1 singularities that may be present in the sub-integral function, e.g., at the bounds of E^s; provide for statistical a-posteriory error estimation similar to the one available for usual Monte Carlo methods. The scrambling procedure introduced by Owen [10] also offers better convergence in some settings.

In this paper we concentrate on the Sobol sequences, with two types of scrambling - the Matousek scrambling and the Owen scrambling [1]. Our aim is to optimise the free parameters that are available within their definition in order to obtain better distribution and smaller integration errors.

There has been significant body of research on the problem of numerical integration when the sub-integral function has some additional smoothness properties, especially when weighted function spaces are considered [5]. In a practical setting one can obtain some information about the relative impact of the different variables and their interactions on the value of the subintegral function by analysing the so-called Sobol' sensitivity indices. Since such information can be obtained numerically, we are motivated to optimise the low-discrepancy sequences in order to achieve the minimum possible integration error when the sub-integral function is from a suitably parametrized weighted function space. In the next section we introduce the optimisation problem that we consider and our approach to solving it numerically.

In the following section we describe our approach for parallel implementation of the algorithm on CPU and GPU computational devices. Then we discuss the efficiency of our implementation and the numerical results that have been achieved.

2 Optimization Problem

2.1 The Scrambled Sobol' Sequences

The Sobol' sequences have been defined by Sobol' in [13]. The purpose of the next definition (equivalent to the original with the addition of either Matousek or Owen scrambling) is to clarify what are the parameters that we consider for optimisation and the particular scrambling procedures that we use in our tests.

Definition 1. Sobol sequences are multi-dimensional sequences, obtained by expanding the index N as

$$\sum_{i=0}^{n} b_i 2^i$$

and then computing

$$\sum_{j=0}^{n}\left(\oplus_{i=0}^{n} b_i m_{ij}^{(k)}\right) 2^{-j-1}.$$

For each dimension there is a primitive polynomial

$$p_k(x) = x^r + a_1^{(k)} x^{r-1} \cdots + a_{r-1}^{(k)} x + 1$$

over the corresponding finite field $GF(2)$, so that the numbers $m_{ij}^{(k)}$ are either 0 or 1 and satisfy the following properties:

- $m_{ij}^{(k)} = 0$ for $j > i$
- $m_{ij}^{(k)} = 1$ for $j = i$
- $m_{ij}^{(k)} = \oplus$ for $j < i$

The numbers $m_{ij}^{(k)}$ can be set freely for $i < r, j < i$, where r is the degree of the corresponding primitive polynomial. The scrambled Sobol' sequence when using Matousek scrambling is obtained by sampling one random number for each dimension and then performing bitwise **xor** operation. The scrambled Sobol' sequence when using Owen scrambling is obtained by sampling a tree of random bits, where each level m (starting from 0) has 2^m bits indexed with the numbers from 0 to $2^m - 1$. In order to decide whether to invert the bit in position j, we use the random bit from the level m whose position is determined by the previous bits of the sequence before level j. For level 0 there is only one possible random bit. The procedure stops when we have reached the desired numerical accuracy (52 bits for double precision).

The Sobol' low-discrepancy sequences are hugely popular in Mathematical Finance.

The direction numbers $m_{ij}^{(k)} \in \{0, 1\}$ determine the quality of the sequence. For an s-dimensional sequence, generated by polynomials of degrees $g_1, \ldots, g_s$, there are

$$\sum_{i=1}^{s} \frac{g_i (g_i - 1)}{2}$$

degrees of freedom. The problem is to optimise over these binary values, aiming at lower integration error.

$$m_{k,j} := 2a_{1,j} m_{k-1,j} \oplus 2^2 a_{2,j} m_{k-2,j} \oplus \cdots \oplus 2^{s_j-1} a_{s_j-1,j} m_{k-s_j+1,j} \\ \oplus 2^{s_j} m_{k-s_j,j} \oplus m_{k-s_j,j} \quad (1)$$

2.2 Optimisation Problem

The idea is to define a measure of the quality of distribution of the sequence, which is oriented towards the function space from which the sub-integral function is taken, and optimise over the binary free variables. We use the system of Walsh functions [4] which is an extension of the Rademacher's incomplete system of orthogonal functions of the type

$$t- > r_m(t) = \text{sign}\left(\sin 2^{m+1}\pi t\right)$$

and is a complete orthogonal system. The Walsh system can be constructed as a product of Rademacher functions and obviously extends to multivariate functions. Each non-negative number n encodes a Walsh function ψ_n, obtained as product of those Rademacher's functions, which correspond to the indices of the non-zero bits in the expansion of n.

The absolute value of the integration error of the Sobol' sequences when computing an integral of a multivariate Walsh function, corresponding to the numbers $(m_1, \ldots, m_s)$ is either 0 or 1, for a fixed $N = 2^k$, and can be easily computed from the direction numbers. See, for example, [11].

If we associate a system of weights for to the Walsh system, we can consider the following measure that depends on the direction numbers of the sequence σ:

$$R(\sigma) = \sum_{(m_1,\ldots,m_s)\in Z'_{\geq 0}} w(m_1, \ldots, m_s)\, r(\sigma, m_1, \ldots, m_s),$$

It is natural to consider weights of the type

$$w(m_1, \ldots, m_s) = \prod_{i=1}^{s} 2^{-k_i\alpha}.$$

where $2^{k_i} \leq \max(m_i, 1) < 2^{k_i+1}$.

Similar measures involving the so-called trigonometric sums, have been considered for Korobov nets, while the dyadic diaphony is well known measure frequently used for digital nets, [2].

It has been shown that low-discrepancy sequences are successful in problems where the low-dimensional interactions of variables are more prominent. A suitable choice of weights that assigns higher weights to such interactions will produce a measure that is more relevant for solving such problems.

By fixing a sampling algorithm for the Walsh functions we can consider to have a stochastic optimisation problem, or alternatively, a noisy optimisation problem. Since all Walsh functions encoded as a number between 2^k and 2^{k+1} have the same norm in our setting, it is natural to first sample the number k appropriately and then select each of these numbers with equal probability. Thus for a particular selection of weights the measure can be evaluated by using an appropriate distribution for the sampling of the number k and setting the weights for the computed values in order to minimize the variance. The selection of weights in the measure would depend on our view of the function space and the relative importance of the different dimensions. If the total number of points is 2^n then it is not advantageous to consider functions for which $k_1 + \ldots k_s > n$, since their weights would be too small. In our particular implementation we fixed the maximum number of dimensions of the Walsh functions sampled to be 5 and we sampled the number k uniformly, but still setting the weight to be 2^{-k} for each used dimension. Even though we did not optimise for interactions for more than 5 dimensions, we still obtained good results for integrals that include significant such interactions. Our objective function has high-dimensional binary inputs and real-valued output. In some previous research we used Genetic Algorithms via GA-lib, which seemed to yield acceptable results. However, for this work we implemented the Binary Particle Swarm Optimisation Algorithm as described in [8], with some modifications. The velocity formula was as usual given by:

$$\hat{v}_i^{(j)} = \omega v_i^{(j)} + c_1 r_1 \left(\bar{g}^{(j)} - g_i^{(j)}\right) + c_2 r_2 \left(\bar{g}_i^{(j)} - g_i^{(j)}\right)$$

whereas the new positions are sampled from the set 0, 1 with

$$P\left(\hat{g}_i^{(j)} = 1\right) = \frac{1}{1 + \exp\left(-\hat{v}_i^j\right)},$$

but only for the values corresponding to sufficiently high dimensions, while the smaller dimensions become fixed after certain number of steps. For example, after the first 500 steps we would fix one degree of freedom for each 5 steps. The fixed value is taken from the current best position. The idea is to achieve a natural setup of the dimensions of the Sobol' points, where the first dimensions have better distribution than the later dimensions. We obtained our best results with the constants c_1 and c_2 and the inertia w set to 1, while using a velocity limit of 6. Setting c_1 and c_2 to 2 and the maximum velocity to 4 also achieves good results.

The objective function is computed by sampling a fixed number of Walsh functions (in practice reaching 100 000 such functions). When a supposedly better solution is found, it is compared with the current best using even more functions (20 times more in our implementation). In this way the current best solution has been computed using at least 2×10^6 functions and since the solution does not change at every step, even more. We tested population sizes between 100 to 10000. It is not necessary to go for larger population sizes but a size between 100 and 500 produced acceptable results.

3 Numerical Results

3.1 Optimization Progress

In our experience we noticed a tradeoff between speed of computing and number of steps necessary when varying the population sizes. In the following figure one can see how some particular computations proceeded for varying population sizes.

Although we can not be sure that a much lower minimum of the objective function can not be attained, we obtained good practical results with the developed algorithm, as shown in the next section.

3.2 Numerical Integration Comparison

We considered different test functions, which are usually tested when using low-discrepancy sequences. Here we only show some of these results, the other being qualitatively similar. Since we observed substantially better accuracy when using full Owen scrambling, for any kind of direction numbers that we compared, we consider the results with full Owen scrambling included to be the most relevant. We compute integrals of the type:

$$\int_{E^s} F_i(x)\, dx$$

where the sub-integral function is:

- $F_1(x) = \frac{4}{(s-1)} \sum_{i=1}^{s-1} x_i x_{i+1}$
- $F_2(x) = \frac{8}{s-2} \sum_{i=1}^{s-2} x_i x_{i+1} x_{i+2}$
- $F_3(x) = \prod_{i=1}^{s} (4 \min(x_i, 1 - x_i))$
- $F_4(x) = \left(1 + \frac{1}{s}\right)^s \left(\prod_{i=1}^{s} x_i\right)^{\frac{1}{s}}$

The exact values of these integrals are always 1. In order to obtain the error we run at least 1000 tests with different seeds for the scrambling, which resulted in stable error estimates (Tables 1 and 2).

In the next table one can compare results obtained with randomly selected direction numbers (no optimisation), optimised direction numbers with our algorithm and highly optimised (fixed) direction numbers, used in financial applications, available from Broda, Inc., [15] and Joe and Kuo, [7], which improves upon [6]. We note that by randomly selected bits for the direction numbers are mean only those that correspond to the degrees of freedom. Most of the bits are still obtained using the polynomials as described above. If all the bits are randomly selected, the results would be significantly worse. We can see how for the Integral 3 the random selection of the bits corresponding to the available degrees of freedom gives better results than the widely used tables of direction numbers. We suppose that the reason for this is that they are optimised for smaller correlations between pairs of dimensions, which is more relevant for Integral 1 and to smaller extent to Integral 2.

Table 1. Mean squared errors in dimension 64, 2^{16} number of points

Method	Integral 1	Integral 2	Integral 3
Optimised	3.4×10^{-8}	3.3×10^{-7}	1.7
Random	7.2×10^{-7}	1.7×10^{-6}	2.7
Broda	8.9×10^{-8}	2.1×10^{-6}	22.2
Kuo	1.2×10^{-7}	4.4×10^{-6}	16.8

Table 2. Mean squared errors for Integral 4, number of points 2^{16}, dimension 16

Method	Matousek	Owen
Optimised	1.1×10^{-5}	6.4×10^{-6}
Kuo	1.3×10^{-5}	8.9×10^{-6}
Broda	1.5×10^{-5}	7.5×10^{-6}

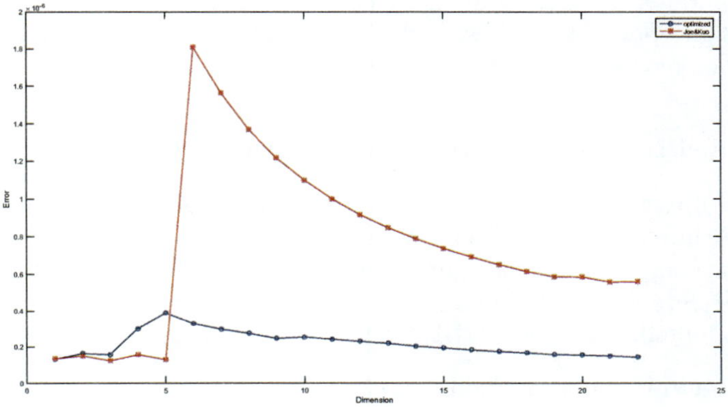

Fig. 1. Comparison of precision for Integral 2, 2^{16} points, dimension between 1 and 24.

In the next figure we show comparison between the integration errors for Integral 1, obtained with our optimised numbers compared with the numbers from J. and Kuo and Broda, when varying the dimensions (Fig. 1).

We can see that via the optimisation one can avoid certain "bumps" which happens for some dimensions in the fixed tables.

In our testing without using the Owen scrambling all implementations have much smaller precision, which was to be expected. In the next table we compare results when using the simpler Matousek scrambling and the full Owen scrambling. With the simpler scrambling algorithm our optimisation still provides better results

4 Parallel Implementation of the Optimization Algorithm

4.1 Parallel Implementation on CPU Using OpenMP

We only consider in this work the problem of parallel implementation of the whole optimisation procedure, since our goal is to produce good sets of direction numbers. The actual computation of integrals when the direction numbers are fixed is a different problem which will be addressed in following works.

The most important phases of the optimisation which are repeated until satisfactory convergence is achieved are as follows:

- Managing the initialisation and evolution of the population
- Generating random Walsh functions
- Evaluating values of the sampled functions for each of the particles

On a regular server using multiple CPU cores we achieved natural parallelisation using the OpenMP standard by adding directives like *"#pragma omp parallel for"* where applicable. The resulting implementation is practically usable, with good parallel efficiency, but relatively slow for the desired numbers of functions, steps and population sizes, with the evaluations of the objective functions being the most expensive part.

4.2 Parallel Implementation Using NVIDIA CUDA

The type of operations in our algorithm are suitable for parallisation on GPUs. Some initial implementation was developed using OpenCL, but our estimation was that the complexity of the code would be very similar between OpenCL and CUDA, therefore we decided to switch to CUDA in order to be able to optimise further if required. Optimised CUDA kernels were created for

- generation of random functions
- expansion of bits from the population into direction numbers.
- evaluation of integration errors

We observed that the actual computation of objective function values became extremely fast and the computation time becomes determined by auxiliary steps with theoretically lower complexity. The optimised code produces reproducible results due to the way we use the *xoroshift/* family of pseudorandom generators on both the CPU and the GPU, which is useful because any particular computation can be repeated to obtain some additional information about its progress.

4.3 Results Analysis

We tested two target systems:

- System A - HP Proliant SL390s Gen8, CPU 2x Intel Xeon E5-2650, 8 NVIDIA Tesla M2090 GPUs.
- System B - Fujitsu Primergy RX 2540 M4, NVIDIA Tesla V100 32GB, 128 GB RAM, CPU 2x Intel Xeon Gold 5118 2.30 GHz 24 cores, 2x800 GB SSD

On the older System A we used only one of the GPUs, even though it would not be hard to split the computation across multiple GPUs. We show computational times for optimizing with a population of 1000 points, 500 steps, dimension 256. The total degrees of freedom in this case are 10648 (Table 3).

Table 3. Computational time for 256 dimension and up to 2^{10} points

Parallelisation	System A	System B
OpenMP	1834 s	542 s
CUDA	758 s	**17 s**

We also obtained the distribution of the time spent in various parts of the code for the CUDA implementation on the modern system:

- expand population time 0.93 s
- evolution of the population (CPU) 6 s
- random function generation 0.43
- evaluation of objective function on GPU - 3.25 s
- memory transfers CPU/GPU - 1.24, 2.18,
- processing of objective function results - 2.18

It is obvious that the evolution of the population should also be moved to the GPU, but we are currently evaluating some possible improvements to the whole optimisation procedure for which the CPU implementation is more flexible.

5 Conclusions

The chosen optimisation approach is viable and produces better results than widely used tables of direction numbers. The use of GPUs speeds up the convergence of the optimisation and produces usable direction numbers in high dimensions within minutes. There are ample possibilities to tailor the direction numbers for specific function spaces, especially taking into account different weightings of the different dimensions or considering only some kinds of interactions between dimensions. From the numerical results that we have seen we are confident that sufficiently general purpose tables can also be obtained, since it seems that we can find direction numbers that are good not only for one final dimension and number of points, but also for any of the lower dimensions and smaller number of points.

Acknowledgements. This work has been accomplished with the financial support of the MES by the Grant No. D01-221/03.12.2018 for NCHDC - part of the Bulgarian National Roadmap on RIs.

References

1. Atanassov, E.I.: A new efficient algorithm for generating the scrambled Sobol' sequence. LNCS, vol. 2542, pp. 83–90. Springer, Heidelberg (2003)
2. Atanassov, E.I.: On the dyadic diaphony of the Sobol' sequences. LNCS, vol. 2179, pp. 133–140 (2001)

3. Caflisch, R.: Monte Carlo and quasi-Monte Carlo methods. Acta Numerica **7**, 1–49 (1998)
4. Dick, J., Pillichshammer, F.: Multivariate integration in weighted Hilbert spaces based on Walsh functions and weighted Sobolev spaces. J. Complex. **21**(2), 149–195 (2005). https://doi.org/10.1016/j.jco.2004.07.003
5. Dick, J., Sloan, I.H., Wang, X., Woźniakowski, H.: Good lattice rules in weighted Korobov spaces with general weights. Numerische Mathematik **103**(1), 63–97 (2006)
6. Joe, S., Kuo, F.Y.: Remark on algorithm 659: implementing Sobol's quasirandom sequence generator. ACM Trans. Math. Softw. **29**, 49–57 (2003)
7. Joe, S., Kuo, F.Y.: Constructing Sobol sequences with better two-dimensional projections. SIAM J. Sci. Comput. **30**, 2635–2654 (2008)
8. Lee, S., Soak, S., Oh, S., Pedrycz, W., Jeon, M.: Modified binary particle swarm optimization. Prog. Nat. Sci. **18**(9), 1161–1166 (2008). https://doi.org/10.1016/j.pnsc.2008.03.018. ISSN 1002-0071
9. Niederreiter, H.: Random Number Generation and Quasi-Monte Carlo Methods. Society for Industrial and Applied Mathematics, Philadelphia (1992)
10. Owen, A.B.: Scrambling Sobol' and Niederreiter-Xing points. J. Complex. **14**(4), 466–489 (1998)
11. Schmid, W.C.: Projections of digital nets and sequences. Math. Comput. Simul. **55**(1–3), 239–247 (2001)
12. Sobol', I.M.: On the distribution of points in a cube and the approximate evaluation of integrals. Comput. Math. Math. Phys **7**, 86–112 (1967)
13. Sobol, I.M.: Uniformly distributed sequences with an additional uniform property. USSR Comput. Math. Math. Phys. **16**, 236–242 (1976)
14. Sobol', I., Asotsky, D., Kreinin, A., Kucherenko, S.: Construction and comparison of high-dimensional Sobol generators. Wilmott J. **56**, 64–79 (2011)
15. Sobolseq Sobol sequence generator (2019). www.broda.co.uk/software.html

Advanced Quasi-Monte Carlo Algorithms for Multidimensional Integrals in Air Pollution Modelling

Venelin Todorov[1,2(✉)], Ivan Dimov[1], Tzvetan Ostromsky[1], and Zahari Zlatev[3]

[1] Institute of Information and Communication Technologies, Department of Parallel Algorithms, Bulgarian Academy of Sciences, 1113 Sofia, Bulgaria
{venelin,ceco}@parallel.bas.bg, vtodorov@math.bas.bg, ivdimov@bas.bg
[2] Institute of Mathematics and Informatics, Department of Information Modeling, Bulgarian Academy of Sciences, 1113 Sofia, Bulgaria
[3] National Centre for Environment and Energy, University of Århus, Frederiksborgvej 399, P. O. Box 358, 4000 Roskilde, Denmark
zz@dmu.dk

Abstract. Sensitivity analysis is a powerful tool for studying and improving the reliability of mathematical models. Air pollution and meteorological models are in front places among the examples of mathematical models with a lot of natural uncertainties in their input data sets and parameters. In this work some results of the global sensitivity study of the Unified Danish Eulerian Model (UNI-DEM) have been presented. One of the most attractive features of UNI-DEM is its advanced chemical scheme – the Condensed CBM IV, which consider a large number of chemical species and numerous reactions between them, of which the ozone is one of the most important pollutants for its central role in many practical applications of the results. A comprehensive experimental study of quasi-Monte Carlo algorithms based on lattice sequences with different generating vectors for multidimensional numerical integration has been done. The Fibonacci based lattice rule is compared with other types of lattice rules. The performance of a lattice rule depends on the choice of the generator vectors. When the integrand is sufficiently regular the lattice rules outperform not only the standard Monte Carlo methods, but also other types of methods using low discrepancy sequences.

1 Introduction

A systematic approach for sensitivity analysis studies in the area of air pollution modelling has been presented. The Unified Danish Eulerian Model (UNI-DEM) [30,31] is used in this particular study. The reason to choose this model as a case study here is its sophisticated chemical scheme, where all relevant chemical processes in the atmosphere are accurately represented.

The main goal of the present work is to develop and investigate numerical algorithms for multiple numerical integration and software tool providing

I. Dimov and S. Fidanova (Eds.): HPC 2019, SCI 902, pp. 155–167, 2021.
https://doi.org/10.1007/978-3-030-55347-0_14

sensitivity analysis (SA) that means evaluating Sobol' sensitivity indices (SIs) [5,23,24]. While the classical deterministic grid methods are efficient for low dimensional integrands [4], they become computationally intensive and even impracticable for high dimensions s because the number of required integrand evaluations grows exponentially. In contrast, the convergence rate of the plain Monte Carlo (MC) integration methods [4] does not depend on the number of dimensions s. That is why the Monte Carlo method is a power tool in sensitivity analysis of large-scale systems [8,17].

Several efficient Quazi-Monte Carlo algorithms – Halton, Fibonacci based lattice rule and two specific types of lattice rules have been used in our sensitivity studies of the model output results for some air pollutants with respect to the emission levels and some chemical reactions rates. A comparison between the Halton sequence and the different lattice rules has been given. The Halton sequence is completely described in [9,10]. The Halton sequence used here is not scrambled, a comparison with the scrambled Halton sequence will be an object of a future study.

2 An Overview of UNI-DEM and Its Sensitivity Analysis Version

UNI-DEM is a powerful large-scale air pollution model for calculation the concentrations of a large number of pollutants and other chemical species in the air. The calculations are done in a large spatial domain, which covers completely the European region and the Mediterranean, for certain time period (meteorological data must be available for it) [5]. In this particular study we use them for two of the most dangerous pollutants: the ozone (O_3) and the ammonia (NH_3). Other accumulative functions related to sone specific applications, maximal values, etc. are also calculated, exported and used in various application areas (environmental protection, agriculture, health care, etc.).

UNI-DEM is mathematically represented by the following system of partial differential equations (PDE), in which the unknown concentrations c_s of a large number of chemical species (pollutants and other chemically active components) take part. The main physical and chemical processes (advection, diffusion, chemical reactions, emissions and deposition) are represented in that system [6,7]:

$$\frac{\partial c_s}{\partial t} = -\frac{\partial(uc_s)}{\partial x} - \frac{\partial(vc_s)}{\partial y} - \frac{\partial(wc_s)}{\partial z} \tag{1}$$

$$+ \frac{\partial}{\partial x}\left(K_x \frac{\partial c_s}{\partial x}\right) + \frac{\partial}{\partial y}\left(K_y \frac{\partial c_s}{\partial y}\right) + \frac{\partial}{\partial z}\left(K_z \frac{\partial c_s}{\partial z}\right) \tag{2}$$

$$+ E_s + Q_s(c_1, c_2, \ldots, c_q) - (k_{1s} + k_{2s})c_s, \quad s = 1, 2, \ldots, q. \tag{3}$$

where c_s are the concentrations of the chemical species; u, v, w are the wind components along the coordinate axes; K_x, K_y, K_z – the diffusion coefficients; E_s – the emissions; k_{1s}, k_{2s} – dry/wet deposition coefficients; $Q_s(c_1, c_2, \ldots c_q)$ –

non-linear functions describing the chemical reactions between species under consideration. The above PDE system is non-linear and stiff. Both non-linearity and stiffness are introduced mainly by the chemical scheme: the condensed CBM-IV (Carbon Bond Mechanism) [31]. It is quite detailed and accurate, but computationally expensive as well.

For the purpose of efficient numerical treatment, the system (1–3) is split according to the major physical and chemical processes and the following 3 submodels are formed [5–7]: ***Advection-diffusion***, ***Chemistry*** & ***deposition*** and ***Vertical transport (vertical wind and convection)***.

Two additional levels of parallelism are introduced in SA-DEM: top-level(MPI) and bottom-level (OpenMP). They allow to use efficiently the computational power of the contemporary cluster supercomputers with multicore nodes.

3 Quasi-Monte Carlo Methods for Numerical Integration Based on Lattice Sequences

Lattice rules are based on deterministic sequences rather than random sequences which are a feature of the Monte-Carlo methods. When the integrand is sufficiently regular the lattice rules using special types of sequences with a low discrepancy generally outperform the basic Monte Carlo methods. The monographs of Sloan and Kachoyan [26], Niederreiter [19], Hua and Wang [12] and Sloan and Joe [27] provide comprehensive expositions on the theory of integration lattices. Applications of the methods for multidimensional numerical integration are studied in Paskov [22], Wang and Hickernell [29], Bakhvalov [2], F. Y. Kuo and D. Nuyens [16]. Consider the quadrature formula

$$I_N(f) = \frac{1}{N}\sum_{i=0}^{N-1} f(x_i), \tag{4}$$

where $P_N = \{x_0, x_1, \dots, x_{N-1}\}, x_i \in [0,1)^s$ are the integration nodes of the formula. The choice of the nodes is essential, because it determines the discrepancy of the sequence and the accuracy of the quadrature. The integration nodes, of the lattice rules proposed by Korobov [14,15], are defined by the following formula:

$$x_k = \left(\left\{\frac{kz_1}{N}\right\}, \left\{\frac{kz_2}{N}\right\}, \dots, \left\{\frac{kz_s}{N}\right\}\right), \; k = 1, 2, \dots, N, \tag{5}$$

where N is the number of the nodes, z is an s-dimensional generating vector of the lattice set and $\{a\} = a - [a]$ is the fractional part of a. The lattice rules with nodes (5) and generators z are called "rank 1" rules. The class $E_s^\alpha(c)$ is defined by Koborov.

Definition 1 [26,27]**.** *We say that $f(x)$ belongs to the class of functions $E_s^\alpha(c)$ for $\alpha > 1$ and $c > 0$, if f is a periodic function with period 1 for every of its* *components $x_i, i = 1, 2 \ldots, s$, defined over the unit cube $[0,1]^s$ and its Fourier coefficients satisfy the following inequalities:*

$$|a(m)| < \frac{c}{(\overline{m}_1 \ldots \overline{m}_s)^\alpha}, \tag{6}$$

where

$$\overline{m} = \begin{cases} |m|, & |m| \neq 0, \\ 0, & m = 0, \end{cases}$$

and the constant c does not depend on $m_1, \ldots, m_s$.

Theorem 1 ([1]). *There exists an optimal choice of the generating vector z, for which the error of integration satisfies*

$$\left| \frac{1}{N} \sum_{k=0}^{N-1} f\left(\left\{\frac{k}{N} z\right\}\right) - \int\limits_{[0,1)^s} f(u)du \right| \leq cd(s,\alpha) \frac{(\log N)^{\beta(s,\alpha)}}{N^\alpha}, \tag{7}$$

for the function $f \in E_s^\alpha(c)$, where $\alpha > 1$ and $d(s,\alpha), \beta(s,\alpha)$ do not depend on N. Moreover, if N is a prime number, then $\beta(s,\alpha) = \alpha(s-1)$.

The generating vector z, for which inequality (7) is satisfied, is an optimal generating vector in the sense of Korobov. The point set P_N is a set of good integration points and in this case the numerical integration method is called Good Lattice Point method (GLP). While the theoretical result establish the existence of optimal generating vectors the difficulty of the construction of GLPs is in the construction of the optimal vectors. The problem is especially difficult for multiple integrals in high dimensions. The discrepancy and the "worst case" error are two important characteristics for the quality of the lattice sequences.

Definition 2 [21]. *Consider the point set $X = \{x_i \mid i = 1, 2, \ldots N\}$ in $[0,1)^s$ and $N > 1$. Denote by $x_i = (x_i^{(1)}, x_i^{(2)}, \ldots, x_i^{(s)})$ and $J(v) = [0, v_1) \times [0, v_2) \times \ldots \times [0, v_s)$. Then the discrepancy of the set is defined as*

$$D(N) := \sup_{0 \leq v_j \leq 1} \left| \frac{\#\{x_i \in J(v)\}}{N} - \prod_{j=1}^{s} v_j \right|.$$

Definition 3. *For $f \in E_\alpha^s(c)$ the worst case error is defined as [29]*

$$P_\alpha(z, N) = \sum_{z.a \equiv 0 \ (mod N), a \neq 0} \frac{c}{(\overline{m}_1 \ldots \overline{m}_s)^\alpha}.$$

The quantity $P_\alpha(N, z)$ and the discrepancy are similar measures of the quality of the lattice point set. In fact the proofs of existence of lattice point sets with low discrepancy and low worst case error are related. The proofs involve well-chosen averages over the space of the feasible lattice points. The following theorem is proved by Korbov and Bakhvalov (1959) [1,13].

Theorem 2. *If N is a prime number, then there is an optimal choice of the generating vector z, so that*

$$D(N) = O(N^{-1} \log^s N),$$
$$P_\alpha(z, N) = \mathcal{O}(N^{-\alpha} \log^{\alpha s} N).$$

Niederreiter (1978) showed that [18]:

Theorem 3. *If N is a composite number, then there exist lattice rules for which:*

$$P_\alpha(z, N) = \mathcal{O}(N^{-\alpha}(\log N)^{\alpha(s-1)+1}(\frac{N}{\phi(N)})), s \geq 2,$$
$$P_\alpha(z, N) = \mathcal{O}(N^{-\alpha}(\log N)^{\alpha}(\frac{N}{\phi(N)} + \frac{\tau(N)}{log(N)})), s = 2,$$
$$P_\alpha(z, N) = \mathcal{O}(N^{-\alpha}(\log N)^{\alpha}(s-1)(1 + \frac{\tau(N)}{log^{s-1}(N)})), s \geq 3,$$

where $\phi(N)$ is the Euler totient function and $\tau(N)$ is the number of positive divisors of N. When N is a prime number there is a generating vector z, so that

$$P_\alpha(z, N) = \mathcal{O}(N^{-\alpha} \log^{\alpha(s-1)}(N)).$$

3.1 Lattice Sequence Based on Fibonacci Generating Vector

The efficient lattice rules require an optimal generating vector z, which ensures that the integration nodes are evenly distributed over the unit cube. Now we present an explicit construction of a generating vector with good properties. In 1981 Hua and Wang [12] generalized the Fibonacci numbers for any dimension

$$F_{l+s}^{(s)} = F_l^{(s)} + F_{l+1}^{(s)} + ... + F_{l+s-1}^{(s)}, l = 0, 1, \ldots$$

with initial conditions:

$$F_0^{(s)} = F_1^{(s)} = \ldots = F_{s-2}^{(s)} = 0, F_{s-1}^{(s)} = 1.$$

The generating vector is constructed based on the generalized Fibonacci numbers of the corresponding dimension.

$$z = \left(1, F_{l+1}^{(s)}, ..., F_{l+s-1}^{(s)}\right), \quad n_l = F_l^{(s)}, \tag{8}$$

where $F_j^{(s)}$ is the generalized Fibonacci number of dimension s. The discrepancy of the lattice set with a generating vector (8) is asymptotically estimated in a theorem by Hua and Wang given in [29]. We refer to the lattice method with a generating vector (8) as Fibonacci lattice rule (FIBO). The advantages of the FIBO method are the linear complexity of the algorithm and fast computation of the generating vector.

3.2 Lattice Sequences Based on Optimal Generating Vectors

First, we consider standard lattice rule in base two [16], which is suitable only for periodic functions F with a period one, in all variables. For a given number of lattice points N, the performance of the routine is affected by the choice of the generator vector z. Optimal generating vectors of the standard lattice rules (LAT) are constructed by Dirk Nuyens [20,32] and their computation involves fast component by component operations. The method is improved by generating the points from a lattice sequence in base 2 in gray coded radical inverse ordering. While the construction of the generating vector is a modification of the radical inverse ordering its computation is more expensive. The special choice of the generating vector is better than the vector from generalized Fibonacci numbers for higher dimensions, as can be seen from the experiments in the next section.

The modified lattice sequence (LATM) applies the transformation function

$$\varphi(T) = 3T^2 - 2T^3$$

to a nonperiodic integrand to make it suitable for applying a lattice rule. The effect of the generating vector on the performance of the lattice rule is studied in [20,32]. The routine LATM uses a special type of uniformly distributed integration nodes. Given a generating vector z of integers, the k-th point of the sequence is given by $x_k := \varphi(k)\ z \, mod\, n$ for $k = 0, 1, 2, \ldots$, where φ is the gray coded radical inverse function in the base of the lattice sequence. The radical inverse of an integer $k = (k_{m-1}k_{m-2}\ldots k_0)_b$, which has m digits in base b is obtained by mirroring the digits at the fractional part, i.e. $\varphi(k) = (0k_0k_1\ldots k_{m-1})_b$. The algorithm for reversing the digits of an integer number is computationally efficient.

The novelty of the proposed approaches is that the modified lattice rule is applied for the first time to sensitivity studies of the particular air pollution model. This is also the first time a standard lattice rule with this generating vector is applied to the problem under consideration.

4 Sensitivity Studies with Respect to Emission Levels

Results for the sensitivity of UNI-DEM output (in particular, the ammonia mean monthly concentrations) with respect to the anthropogenic emissions input data variation are shown and discussed in this section. The anthropogenic emissions input consists of 4 different components

$\mathbf{E} = (\mathbf{E}^{\mathbf{A}}, \mathbf{E}^{\mathbf{N}}, \mathbf{E}^{\mathbf{S}}, \mathbf{E}^{\mathbf{C}}):$
$\mathbf{E}^{\mathbf{A}}$ – ammonia (NH_3);
$\mathbf{E}^{\mathbf{N}}$ – nitrogen oxides $(NO + NO_2)$;
$\mathbf{E}^{\mathbf{S}}$ – sulphur dioxide (SO_2);
$\mathbf{E}^{\mathbf{C}}$ – anthropogenic hydrocarbons.

The domain under consideration is the 4-dimensional hypercube $[0.5, 1]^4$). Polynomials of second degree have been used as an approximation tool [7]. The

Table 1. Relative error for the evaluation of $f_0 \approx 0.048$.

# of samples n	LAT	Halton	FIBO	LATM
	Relative error	Relative error	Relative error	Relative error
2^{10}	2.80e−04	3.15e−005	2.09e−004	8.46e−004
2^{12}	7.73e−005	1.14e−004	4.32e−005	1.79e−004
2^{14}	1.37e−005	1.27e−005	2.25e−005	2.62e−006
2^{16}	6.66e−006	8.20e−006	8.70e−006	4.14e−007
2^{18}	1.30e−006	2.40e−006	1.79e−006	1.17e−006
2^{20}	3.52e−007	1.03e−006	4.21e−007	1.15e−006

input data have been generated by the improved SA-DEM version, developed for the purpose of our sensitivity studies (see the previous section).

Results of the relative error estimation for the quantities f_0, the total variance $\mathbf{D}$, first-order (S_i) and total (S_i^{tot}) sensitivity indices are given in Tables 1, 2, 3, respectively. f_0 is presented by a 4-dimensional integral, while the rest of the above quantities are presented by 8-dimensional integrals, following the ideas of *correlated sampling* technique to compute sensitivity measures in a reliable way (see [11,28]). The four different stochastic approaches used for numerical integration are presented in separate columns of the tables.

A study of the computational efficiency of the stochastic algorithms under consideration for evaluating sensitivity measures presented by multidimensional integrals (total variance) or a ratio of multidimensional integrals (Sobol global sensitivity indices) have been made. The results show that the computational efficiency of the algorithms depends on integrand dimension and magnitude of estimated quantity. One can notice this fact from results in Table 3 too. The order of relative error is different for different quantities of interest (see column *Reference value*) for the same sample size. From Tables 1, 2 and 3 we can conclude that lattice sequence LAT gives the most reliable relative errors for sufficiently large number of samples.

Table 3 is similar to Table 4, with the only difference – the increased number of samples $n = 2^{20}$ (instead of $n = 2^{16}$ in Table 7). In general, this increases the accuracy of the estimated quantities by 1 order - see for example the relative errors for S_1, S_2, S_3 and S_4.

For lower dimensions it can be seen that the modified lattice sequence LATM gives better results for lower dimensions, while Halton sequence and LATM produces results of same order for the 8-dimensional integrals - see Table 3.

Most influential emissions about ammonia output concentrations are ammonia emissions themselves (about 89% for Milan). The second most influential emissions about ammonia output are sulphur dioxide emissions (about 11%) [7].

Table 2. Relative error for the evaluation of the total variance $\mathbf{D} \approx 0.0002$.

# of samples n	LAT	Halton	FIBO	LATM
	Relative error	Relative error	Relative error	Relative error
2^{10}	6.86e−03	1.40e−02	1.63e−01	1.54e−02
2^{12}	3.75e−03	7.81e−03	2.39e−02	3.67e−03
2^{14}	2.64e−04	1.77e−03	2.90e−03	1.49e−03
2^{16}	1.28e−04	5.96e−04	2.65e−04	1.61e−03
2^{18}	2.18e−05	1.48e−04	3.01e−04	1.48e−03
2^{20}	2.73e−05	4.77e−05	1.19e−04	1.46e−03

Table 3. Relative error for estimation of sensitivity indices of input parameters using various Monte Carlo and quasi-Monte Carlo approaches ($n \approx 65536$).

Est. quantity	Ref. value	LAT	Halton	FIBO	LATM
S_1	9e−01	4.47e−05	2.95e−04	3.62e−04	7.27e−04
S_2	2e−04	5.28e−03	3.49e−02	1.74e−01	2.76e−02
S_3	1e−01	2.61e−04	2.30e−03	3.22e−03	4.24e−03
S_4	4e−05	7.44e−02	1.21e−01	4.87e−01	1.65e−02
S_1^{tot}	9e−01	3.51e−05	2.97e−04	4.61e−04	5.14e−04
S_2^{tot}	2e−04	3.31e−02	3.24e−02	3.45e−01	2.21e−01
S_3^{tot}	1e−01	2.62e−04	2.25e−03	1.96e−03	6.41e−03
S_4^{tot}	5e−05	2.83e−02	1.20e−01	5.06e−01	1.60e−01

Table 4. Relative error for estimation of sensitivity indices of input parameters using various Monte Carlo and quasi-Monte Carlo approaches ($n \approx 1048576$).

Est. quantity	Ref. value	LAT	Halton	FIBO	LATM
S_1	9e−01	2.95e−06	1.83e−05	5.29e−08	2.78e−03
S_2	2e−04	3.78e−05	1.83e−03	3.17e−03	2.32e−02
S_3	1e−01	3.38e−05	1.58e−04	6.88e−05	4.28e−03
S_4	4e−05	4.04e−03	5.72e−04	1.88e−01	3.46e−02
S_1^{tot}	9e−01	4.53e−06	2.01e−05	2.14e−05	5.20e−04
S_2^{tot}	2e−04	3.09e−03	2.78e−03	4.56e−03	2.19e−01
S_3^{tot}	1e−01	2.82e−05	1.41e−04	4.69e−05	6.41e−03
S_4^{tot}	5e−05	1.12e−02	4.77e−03	6.08e−02	1.66e−01

5 Sensitivity Studies with Respect to Chemical Reactions Rates

Another part of our research was to study the sensitivity of the ozone concentration values in the air over Genova with respect to the rate variation

Table 5. Relative error for the evaluation of $f_0 \approx 0.27$.

# of samples n	LAT	Halton	FIBO	LATM
	Relative error	Relative error	Relative error	Relative error
2^{10}	1.35e−04	1.60e−04	2.08e−03	7.12e−03
2^{12}	1.56e−05	5.55e−05	1.40e−04	1.80e−03
2^{14}	9.23e−06	2.70e−05	3.98e−04	4.04e−05
2^{16}	2.99e−06	1.60e−06	2.61e−04	9.91e−06
2^{18}	9.71e−08	1.02e−06	7.29e−06	7.24e−06
2^{20}	4.56e−07	5.56e−07	4.57e−07	7.04e−06

Table 6. Relative error for the evaluation of the total variance $\mathbf{D} \approx 0.0025$.

# of samples n	LAT	Halton	FIBO	LATM
	Relative error	Relative error	Relative error	Relative error
2^{10}	8.58e−03	4.86e−02	6.73e+00	3.11e−02
2^{12}	5.67e−03	1.25e−03	5.27e−01	8.76e−02
2^{14}	9.13e−05	1.65e−03	1.02e−01	7.54e−04
2^{16}	2.03e−06	4.34e−04	1.97e−03	9.13e−04
2^{18}	1.85e−05	3.79e−04	4.53e−03	2.22e−03
2^{20}	5.92e−06	3.34e−05	9.33e−03	2.22e−03

of some chemical reactions of the condensed CBM-IV scheme ([30]), namely: # 1, 3, 7, 22 (time-dependent) and # 27, 28 (time independent). The simplified chemical equations of those reactions are:

[**#1**] $NO_2 + h\nu \Longrightarrow NO + O$; [**#22**] $HO_2 + NO \Longrightarrow OH + NO_2$;
[**#3**] $O_3 + NO \Longrightarrow NO_2$; [**#27**] $HO_2 + HO_2 \Longrightarrow H_2O_2$;
[**#7**] $NO_2 + O_3 \Longrightarrow NO_3$; [**#28**] $OH + CO \Longrightarrow HO_2$.

The domain under consideration is the 6-dimensional hypercube $[0.6, 1.4]^6$). Polynomials of second degree have been used for approximation again (see [6]).

Homma and Saltelli discuss in [11] which is the better estimation of

$$f_0^2 = \left(\int_{U^d} f(\mathrm{x})\mathrm{dx}\right)^2 \tag{9}$$

in the expression for total variance and Sobol global sensitivity measures. The first formula is

$$f_0^2 \approx \frac{1}{n}\sum_{i=1}^{n} f(\mathrm{x}_{i,1}, \ldots, \mathrm{x}_{i,d})\ f(\mathrm{x}'_{i,1}, \ldots, \mathrm{x}'_{i,d}) \tag{10}$$

Table 7. Relative error for estimation of sensitivity indices of input parameters using various Monte Carlo and quasi-Monte Carlo approaches ($n \approx 65536$).

Est. quantity	Ref. value	LAT	Halton	FIBO	LATM
S_1	4e−01	1.57e−03	2.87e−03	3.82e−02	1.50e−02
S_2	3e−01	2.70e−04	3.76e−03	1.03e−02	2.14e−02
S_3	5e−02	9.73e−05	7.27e−03	5.48e−01	8.28e−02
S_4	3e−01	1.07e−03	2.19e−03	1.07e−02	6.81e−03
S_5	4e−07	6.06e+00	3.68e+01	3.40e+03	2.07e+03
S_6	2e−02	5.95e−03	1.30e−02	1.32e+00	1.19e−02
S_1^{tot}	4e−01	5.29e−04	2.79e−03	7.92e−02	1.07e−02
S_2^{tot}	3e−01	1.14e−03	3.26e−03	3.06e−02	2.28e−02
S_3^{tot}	5e−02	1.66e−03	6.43e−03	1.31e+00	4.92e−02
S_4^{tot}	3e−01	1.01e−03	2.11e−03	3.84e−01	1.93e−02
S_5^{tot}	2e−04	2.79e−01	1.38e−02	8.85e+01	6.78e+00
S_6^{tot}	2e−02	5.40e−03	1.04e−02	2.15e+00	7.63e−02
S_{12}	6e−03	4.52e−02	7.92e−03	3.21e+00	2.21e−01
S_{14}	5e−03	9.01e−02	9.12e−03	8.64e+00	1.31e+00
S_{15}	8e−06	9.35e+02	9.36e+02	9.19e+02	9.62e+02
S_{24}	3e−03	4.67e−02	1.83e−02	1.37e+01	5.63e−01
S_{45}	1e−05	2.97e−02	9.08e−01	4.25e+01	3.87e+01

and the second one is

$$f_0^2 \approx \left\{ \frac{1}{n} \sum_{i=1}^{n} f(\mathrm{x}_{i,1}, \dots, \mathrm{x}_{i,d}) \right\}^2 \tag{11}$$

where x and x′ are two independent sample vectors. In case of estimating sensitivity indices of a fixed order, formula (10) is better (as recommended in [11]), here we use it too.

The relative error estimation for the quantities f_0, the total variance $\mathbf{D}$ and some sensitivity indices are given in Tables 5, 6 and 7 respectively. The four different stochastic approaches used for numerical integration are presented in separate columns of the tables. The quantity f_0 is presented by 6-dimensional integral, while the rest are presented by 12-dimensional integrals, following the ideas of *correlated sampling*.

Table 7 is similar to Table 8, with the only difference – the increased number of samples $n = 2^{20}$ (instead of $n = 2^{16}$ in Table 7). In general, this increases the accuracy of the estimated quantities. Exceptions are S_5 and S_{15}, which have extremely small reference values. None of the 4 methods estimates S_{15} reliably, which has extremely small reference values. Note that S_5 is estimated reliably only with the lattice algorithm LAT and Halton sequence for number of samples $n = 2^{20}$. This natural "size effect" does not destroy the accuracy of the

Table 8. Relative error for estimation of sensitivity indices of input parameters using various Monte Carlo and quasi-Monte Carlo approaches ($n \approx 1048576$).

Est. quantity	Ref. value	LAT	Halton	FIBO	LATM
S_1	4e−01	5.30e−05	1.65e−04	9.21e−03	1.49e−02
S_2	3e−01	8.06e−05	2.01e−04	1.47e−02	2.11e−02
S_3	5e−02	6.97e−04	4.90e−04	6.50e−01	8.25e−02
S_4	3e−01	1.63e−06	1.06e−04	1.53e−01	5.68e−03
S_5	4e−07	3.64e−01	9.92e−01	2.68e+03	2.08e+03
S_6	2e−02	2.27e−04	3.01e−04	1.13e+00	1.55e−02
S_1^{tot}	4e−01	3.29e−05	1.65e−04	9.69e−03	1.08e−02
S_2^{tot}	3e−01	1.40e−04	1.81e−04	3.01e−02	2.19e−02
S_3^{tot}	5e−02	1.41e−04	4.63e−04	1.37e+00	4.96e−02
S_4^{tot}	3e−01	3.63e−05	9.42e−05	3.67e−01	1.91e−02
S_5^{tot}	2e−04	2.75e−04	2.29e−03	3.90e+01	1.45e+01
S_6^{tot}	2e−02	5.52e−04	4.44e−04	1.76e+00	7.68e−02
S_{12}	6e−03	7.66e−04	8.54e−04	8.40e−02	2.14e−01
S_{14}	5e−03	1.16e−03	1.34e−04	1.85e−01	1.30e+00
S_{15}	8e−06	9.34e+02	9.34e+02	9.46e+02	9.62e+02
S_{24}	3e−03	2.16e−02	1.27e−04	1.41e+01	6.63e+00
S_{45}	1e−05	2.65e−01	6.20e−02	2.60e+01	3.85e+01

corresponding total sensitivity indices (which are much larger, so the influence of S_5 and S_{15} is negligible).

From these tables we can see that LAT gives better results than Halton and the difference is 1–2 orders. The LATM produce better results for 4 and 6-dimensional integrals in comparison with 8 and 12-dimensional integrals. In one case the results for LATM are better than those for the Halton sequence - see the value for S_6 in Table 7. The lattice sequence LAT gives better results than Halton sequence by 1 order and even by 2 order - see the values of S_3 in Table 7, and in very few cases Halton is slightly better - see the values of S_5^{tot} and S_{14} in Table 7. Overall the standard lattice rule gives the best results in terms of relative error and computational efficiency.

6 Conclusion

The computational efficiency (in terms of relative error and computational time) of several stochastic algorithms for multidimensional numerical integration has been studied to analyze the sensitivity of UNI-DEM model output to variation of input emissions of the anthropogenic pollutants and of rates of several chemical reactions. The algorithms have been successfully applied to compute global Sobol sensitivity measures corresponding to the influence of several input parameters on the concentrations of important air pollutants. The study has been done for the areas of several European cities with different geographical locations. The novelty is that this is the first the two lattice sequences - standard with a special choice of the generating vector and modified lattice rule are applied to a problem in air pollution modelling and a comparison with the low discrepancy of Halton has been made. Overall the lattice sequence gives the best results in terms of accuracy. The numerical tests show that the stochastic algorithms under consideration are efficient for the multidimensional integrals under consideration and especially for computing small by value sensitivity indices. Clearly, the progress in the area of air pollution modeling, is closely connected with the progress in reliable algorithms for multidimensional integration. In the future work we plan to test the lattice rule with different optimal generating vectors which are supposed to give even lower relative errors.

Acknowledgements. Venelin Todorov is supported by the Bulgarian National Science Fund under Young Scientists Project KP-06-M32/2 - 17.12.2019 "Advanced Stochastic and Deterministic Approaches for Large-Scale Problems of Computational Mathematics" and by the National Scientific Program "Information and Communication Technologies for a Single Digital Market in Science, Education and Security (ICT in SES)", Task 1.2.5., financed by the Ministry of Education and Science in Bulgaria. The work is also supported by the Bulgarian National Science Fund under Project DN 12/5-2017 "Efficient Stochastic Methods and Algorithms for Large-Scale Problems".

References

1. Bahvalov, N.: On the approximate computation of multiple integrals. In: Vestnik Moscow State University, Ser. Mat., Mech., vol. 4, pp. 3–18 (1959)
2. Bakhvalov, N.: J. Complex. **31**(4), 502–516 (2015)
3. Bratley, P., Fox, B.: Algorithm 659: implementing Sobol's quasirandom sequence generator. ACM Trans. Math. Softw. **14**(1), 88–100 (1988)
4. Dimov, I.T., Atanassov, E.: Exact error estimates and optimal randomized algorithms for integration. LNCS, vol. 4310, pp. 131–139 (2007)
5. Dimov, I., Georgieva, R.: Monte Carlo algorithms for evaluating Sobol' sensitivity indices. Math. Comput. Simul. **81**(3), 506–514 (2010). https://doi.org/10.1016/j.matcom.2009.09.005. ISSN 0378-4754
6. Dimov, I.T., Georgieva, R., Ostromsky, Tz., Zlatev, Z.: Variance-based sensitivity analysis of the unified Danish Eulerian Model according to variations of chemical rates. In: Dimov, I., Faragó, I., Vulkov, L. (eds.) Proceedings of the NAA 2012. LNCS, vol. 8236, pp. 247–254. Springer (2013)
7. Dimov, I.T., Georgieva, R., Ostromsky, Tz., Zlatev, Z.: Sensitivity studies of pollutant concentrations calculated by UNI-DEM with respect to the input emissions. Cent. Eur. J. Math. **11**(8), 1531–1545 (2013). Numerical Methods for Large Scale Scientific Computing
8. Fidanova, S.: Convergence proof for a Monte Carlo method for combinatorial optimization problems. In: International Conference on Computational Science, pp. 523–530. Springer, Heidelberg (2004)
9. Halton, J.: On the efficiency of certain quasi-random sequences of points in evaluating multi-dimensional integrals. Numerische Mathematik **2**, 84–90 (1960)

10. Halton, J., Smith, G.B.: Algorithm 247: radical-inverse quasi-random point sequence. Commun. ACM **7**, 701–702 (1964)
11. Homma, T., Saltelli, A.: Importance measures in global sensitivity analysis of non-linear models. Reliab. Eng. Syst. Saf. **52**, 1–17 (1996)
12. Hua, L.K., Wang, Y.: Applications of Number Theory to Numerical Analysis. Springer, Heidelberg (1981)
13. Korobov, N.M.: Dokl. Akad. Nauk SSSR **124**, 1207–1210 (1959)
14. Korobov, N.M.: Soviet Math. Dokl. **1**, 696–700 (1960)
15. Korobov, N.M.: Number-theoretical methods in approximate analysis. Fizmatgiz, Moscow (1963)
16. Kuo, F.Y., Nuyens, D.: Found. Comput. Math. **16**(6), 1631–1696 (2016)
17. Myasnichenko, V., Sdobnyakov, N., Kirilov, L., Mikhov, R., Fidanova, S.: Structural instability of gold and bimetallic nanowires using Monte Carlo simulation. In: Recent Advances in Computational Optimization, pp. 133–145. Springer, Cham (2020)
18. Niederreiter, H.: Monatsh. Math **86**, 203–219 (1978)
19. Niederreiter, H., Talay, D.: Monte Carlo and Quasi-Monte Carlo Methods. Springer, Heidelberg (2002)
20. Nuyens, D.: The magic point shop of QMC point generators and generating vectors. https://people.cs.kuleuven.be/~dirk.nuyens/qmc-generators/
21. Owen, A.: Monte Carlo and Quasi-Monte Carlo methods in Scientific Computing. Lecture Notes in Statistics, vol. 106, pp. 299–317 (1995)
22. Paskov, S.H.: Computing high dimensional integrals with applications to finance. Technical report CUCS-023-94, Columbia University (1994)
23. Poryazov, S.: A suitable unit of sensitivity in telecommunications. In: TELECOM 2011, Sofia, 13–14 October 2011, pp. 165–172 (2011). ISSN 1314-2690
24. Poryazov, S., Saranova, E., Ganchev, I.: Conceptual and analytical models for predicting the quality of service of overall telecommunication systems. In: Autonomous Control for a Reliable Internet of Services, pp. 151–181. Springer, Cham (2018)
25. Sharygin, I.F.: Zh. Vychisl. Mat. i Mat. Fiz. **3**, 370–376 (1963)
26. Sloan, I.H., Kachoyan, P.J.: SIAM J. Numer. Anal. **24**, 116–128 (1987)
27. Sloan, I.H., Joe, S.: Lattice Methods for Multiple Integration. Oxford University Press, Oxford (1994)
28. Sobol, I.M., Tarantola, S., Gatelli, D., Kucherenko, S., Mauntz, W.: Estimating the approximation error when fixing unessential factors in global sensitivity analysis. Reliab. Eng. Syst. Saf. **92**, 957–960 (2007)
29. Wang, Y., Hickernell, F.J.: An historical overview of lattice point sets. In: Monte Carlo and Quasi-Monte Carlo Methods 2000, Proceedings of a Conference held at Hong Kong Baptist University, China (2000)
30. Zlatev, Z.: Computer Treatment of Large Air Pollution Models. KLUWER Academic Publishers, Dordrecht (1995)
31. Zlatev, Z., Dimov, I.T.: Computational and Numerical Challenges in Environmental Modelling. Elsevier, Amsterdam (2006)
32. Fast component-by-component constructions. https://people.cs.kuleuven.be/~dirk.nuyens/fast-cbc/

Treatment of Large Scientific and Engineering Problems Challenges and Their Solutions

Numerical Modeling of Extreme Wind Profiles Measured with SODAR in a Coastal Area

Damyan Barantiev[1(✉)], Ekaterina Batchvarova[2], Hristina Kirova[3], and Orlin Gueorguiev[4]

[1] Climate, Atmosphere and Water Research Institute - Bulgarian Academy of Sciences (CAWRI - BAS), 66, Blvd Tzarigradsko chaussee, 1784 Sofia, Bulgaria
barantiev@gmail.com
[2] CAWRI - BAS, Sofia, Bulgaria
[3] National Institute of Meteorology and Hydrology (NIMH), Sofia, Bulgaria
[4] NIMH, Sofia, Bulgaria

Abstract. The extreme wind events could be dangerous and can cause significant infrastructure damages, economic losses and even loss of human life. Therefore, the results of research in this field are of great importance for decision-making purposes related to better management and reduction of risks. Effective and accurate systems for observations and forecasting of hazardous meteorological events are society requirements for the prevention of socio-economic costs. The mesoscale numerical simulations in coastal areas are challenging research task due to the theoretical description intricacy of the Atmospheric Boundary Layer (ABL) processes. The variability of atmospheric stratification and meteorological fields in horizontal and vertical direction in complex orography impose the necessity of continuous improvement of the parametrizations in numerical simulations on one side and high-resolution data for evaluation, on the other side. In this study we use sodar measurements (MFAS SCINTEC) at the Black sea coast in Bulgaria to evaluate the performance of the Weather Research and Forecasting (WRF) mesometeorological model to represent the evolution of vertical wind field structure during extreme wind conditions. The acoustic sounding with high vertical (10 m) and temporal (10 min) resolution was performed at Ahtopol for a long-term period: August 2008-October 2016. A reference extreme wind speed profile was obtained applying criteria for "rare" events of the Intergovernmental Panel on Climate Change (IPCC). The WRF simulations were run on the Supercomputer System "AVITOHOL" at the Institute of Information and Communication Technologies – Bulgarian Academy of Sciences (IICT – BAS). The model was used with fine horizontal grid of 1 km with the local ABL scheme of Mellor-Yamada-Janjic (MYJ) for a period with extreme wind events. The numerical simulations were evaluated through Pearson's correlation coefficients between sodar measurements and model simulations of wind profiles. The results are important for the appropriate choice of WRF setup to deal with extreme wind synoptic situations on Bulgarian coastal site.

I. Dimov and S. Fidanova (Eds.): HPC 2019, SCI 902, pp. 171–183, 2021.
https://doi.org/10.1007/978-3-030-55347-0_15

1 Introduction

The extreme phenomena and the thresholds for the values of corresponding parameters, by definition, vary from place to place for natural reasons (different climates), because the extreme value of given meteorological parameter at one place, can be within the normal range at other place. Wind storms (extremely strong winds) in nature are associated with hazardous meteorological conditions which could cause property damages and human life loss [12]. Hence, studying extremely strong winds is of vital importance both from economical and human point of view. Therefore, to study, understand and to characterize the extreme meteorological events and the climate of extremes in time and space with using specific methodologies for their definition and to identify corresponding thresholds is of primary interest for meteorologists worldwide.

For prevention of high damages and socio-economic costs caused by hazardous meteorological events which happen more often during last years (heavy rains, extreme temperatures, high winds), many governments invest in development of effective systems for observations and forecasting of such events, taking into account the variability of climate, climate change and the need to adapt to them. The results of research in this field are of great importance for adequate action of governments and business to reach better management and reduction of risks.

The ABL structure in coastal regions is studied in many countries as many cities and industrial zones are located at the coasts of oceans, seas, or large lakes. The variety of physical, geographical and climate conditions, as well as the meteorology is large, so there is constant need for improvement of the weather forecasts. Within the ABL the exchange of energy between the surface and the atmosphere is taking place and all processes are influenced by the characteristics of the surface. The abrupt change in surface (such as field-city, sea-land, field-forest, etc.) leads to complexity in atmospheric processes in coastal areas, which is characterized by transformation of the air masses (AM) to the physical characteristics of the new surface [6, 7]. The theoretical description of such type atmospheric processes under complex terrain and local circulations is difficult due to the complex layering of the ABL in coastal areas and therefore the numerical modelling of the meteorological processes is difficult as well. Most ABL parametrizations are derived and tested for flat homogeneous terrain and they are incorporated in mesoscale models as such. Hence, in coastal areas additional validation with high temporal and spatial resolution observations is needed.

The advancement of remote sensing technologies of the atmosphere during the last decade allows to improve the quality of data for model validation. The technological development of different types of ground-based remote sensing instruments for observation of meteorological parameters within the atmospheric boundary layer (ABL) and above it, allow to obtain more accurate spatial qualitative and quantitative assessments of the state of the lower layers of the atmosphere at given moment and to understand the physical processes in it. These instruments provide wide range of data for initialization, assimilation and evaluation of the performance of new generation of weather forecast and climate models of high resolution. The ground-based remote sensing instruments data are also important for climatological analysis and a number of applications that need high spatial and temporal resolution of data.

2 Methodology

Extreme events (wind speed profiles exceeding specific threshold) have been identified in the framework of the project REPLICA (extReme Events and wind ProfiLe In a Coastal Area) using high spatial and temporal resolution data from a modern remote sensing technology for wind and turbulence profiles measurements. Following the criteria of the Intergovernmental Panel on Climate Change (IPCC) for "rare" events [12], a reference extreme wind speed profile has been obtained based on data accumulated during the first eight years of acoustic soundings and defining the 90th percentile of the wind speed statistical distribution for all heights up to 600 m (Fig. 2a). Through a comparison of the reference extreme wind speed profile with all measured profiles during the eight-year period an "extreme winds" data base has been created. The assessment of the performance of mesometeorological Weather Research and Forecasting (WRF) model in reproducing vertical structure of the wind field during meteorological situations with extreme wind speed profiles and at least four hours duration is one of the primary tasks of the REPLICA project. The date 01 April 2012 from the extreme winds data base is studied in this paper.

2.1 Study Area and Equipment

The Experimental polygon (EP) Ahtopol is a synoptic station of the Bulgarian National Meteorological Service located at the south-eastern part of the Bulgarian Black Sea coast at about 2 km southeast of the town of Ahtopol (Fig. 1- middle). Its geographical location falls into the Black Sea coastal Strandzha climate region, which is a part of the Black Sea climatic sub - region of Continental-Mediterranean climatic zone in Bulgaria [23]. According to the climatic characteristics (averaged on the territory of the climatic zones) in the Black Sea coastal Strandzha climate region's average annual temperature is 13.0 °C, the amplitude of annual temperature is 19.6 °C, the average annual wind speed is 3.5 ms^{-1}, the number of days with wind over 14 ms^{-1} are 17, the annual rainfall is 550 lm^{-2}, the annual number of cloudy days are 100, the annual number of sunny days are 80 [27]. Local breeze circulation is typical for the climate in the region and it is well expressed in the warm half of the year, whereas during the cold half lower frequency and smaller temporal and spatial scales of breeze circulations are registered [3].

EP Ahtopol is situated primarily on a flat grassland (Fig. 1 - left, up) at about 400 m inland and 30 m height above sea level. The coast line stretches out from NW to SE (Fig. 1 - right) with a steep about 10 m high cost (Fig. 1 - left, down). Data collection and analysis in the coastal ABL using acoustic sounding starts in Bulgaria in summer 2008 at the EP Ahtopol. These high spatial and temporal resolution measurements allow to perform climatological studies of the coastal ABL in Bulgaria [2–5,20]. The sodar (SCINTEC Flat Array Sodar MFAS), gives information about a number of wind and turbulence parameters and reveals the vertical structure of the coastal ABL up to about 1000 m. The Doppler sounding system provides data of 20-min running average at every 10 min with 10 m vertical resolution starting from 30 m above ground level (AGL) and reaching different height between 400 and 800 m depending on atmospheric conditions.

Fig. 1. Location of EP Ahtopol in Bulgaria (middle) and close look on google map (right) with views of the terrene (left)

The spatialized characteristics of the sodar are: frequency range 1650–2750 Hz, 9 emission/reception angles (0°, ±9.3°, ±15.6°, ±22.1°, ±29°), maximum 100 vertical layers, range between 150–1000 m, accuracy of horizontal wind speed 0.1–0.3 ms^{-1}, range of horizontal wind speed ±50 ms^{-1}, accuracy of vertical velocity 0.03–0.1 ms^{-1}, range of vertical velocity ±10 ms^{-1}, accuracy of wind direction 2–3 ° [24].

2.2 Numerical Modeling Setup

The numerical simulation was performed with the mesoscale non-hydrostatic Advanced Research WRF (ARW), Version 3.8 [25]. The model was initialized with US National Center for Environmental Prediction Final Analyses (FNL) of 1 km spatial and 6 h temporal resolution. The model was run on three 2-way nested domains (centered at 42.084 °N and 27.951°E) with horizontal grid spacing of 25, 5 and 1 km and number of grid points 26 × 21, 36 × 36, 111 × 111, respectively Fig. 2b.

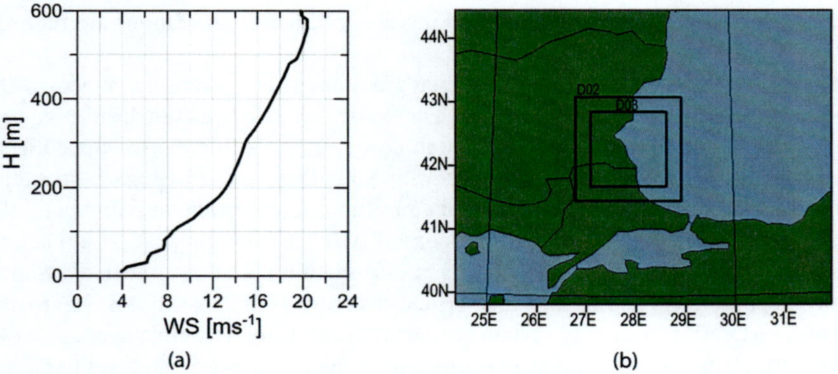

Fig. 2. Study of EP Ahtopol region under REPLICA project: (a) reference extreme wind speed profile; (b) configuration of model domains;

Twenty four categories for land use according USGS (US Geological Survey) were used. Mellor-Yamada-Janjic (MYJ) local scheme which implied 1.5-order (level 2.5), local turbulence closure model of Mellor and Yamada [17] was used as ABL scheme. The choice of the physical processes' parameterizations is listed in Table 1. The model top was set at 50-hPa and 42 terrain-following vertical levels were used, 28 of which were below 1300 m.

Table 1. Physical processes/characteristics and parameterization

microphysics	10 = Morrison 2-moment scheme [19]
lw radiation	1 = RRTM: Rapid Radiative Transfer Model [18]
sw radiation	2 = Goddard [10]
surface layer	2 = Eta similarity [14, 15]
land surface	2 = Noah LSM [26]
ABL	2 = MYJ: Mellor-Yamada-Janjic TKE [15–17]
cumulus convection	5 (D1 and D2) = Grell3D [11]

The model was run on the Supercomputer System "AVITOHOL" at IICT - BAS [1, 21]. The simulations started at 12 UTC on March 31, 2012 with duration of 36 h. The first 10 h were not included in the analysis (spin-up). The output was written at every 10 min to coincide with sodar data. The studied parameters are wind speed (WS), wind direction (WD) and the wind vector, respectively east component (U), north component (V) and vertical velocity (W). The model output profiles of the analyzed parameters were linearly interpolated to the levels at which the sodar performed measurements.

2.3 Approach

Correlation coefficients are used as a criterion for the strength of the linear relationship between two random variables (interpolated model data and the measurements from the sodar) of five wind characteristics' pairs (WS, WD, U, V and W). The level of the linear dependence is calculated with Pearson's correlation coefficient Eq. (1) [22]:

$$r = \frac{\sum_{i=1}^{n}(x_i - \bar{x})(y_i - \bar{y})}{\sqrt{\sum_{i=1}^{n}(x_i - \bar{x})^2 \sum_{i=1}^{n}(y_i - \bar{y})^2}} \quad (1)$$

where $\bar{x}$ and $\bar{y}$ are the arithmetic mean of the two variables. This correlation coefficient may take values from -1 to 1 and present the dependence, in which the individual changes of a variable, are accompanied by an average stable change of the second variable [9]. The negative linear relationship means that the variable x increases while y decreases (or vice versa) and positive that both variables increases or decreases together. For values of the coefficient $|r| = 1$ is assumed that perfect functional relationship exists between the two random variables which could be positive (correlation case) or negative (anticorrelation case) and at r = 0 there is no linear correlation between the variables (uncorrelated case). For interpretation of its values the following is used: if $0 < |r| < 0.3$ the linear relationship is very weak, if $0.3 < |r| < 0.5$ the correlation is moderate, if $0.5 < |r| < 0.7$ is significant, if $0.7 < |r| < 0.9$ is strong and very strong if $0.9 < |r| < 1$. In this paper the spatial and temporal correlation coefficient changes are studied, respectively through, time series comparison at a certain height and individual profiles comparison at a given moment. Equation (1) can be used for calculation of linear relationship of random variables, such as wind speed values and components of its vector. To obtain a correlation coefficient for angular scale data, such as wind direction, the following circular statistic equation is used Eq. (2) [8, 13]:

$$r_c = \frac{\sum_{i=1}^{n} sin(\alpha_i - \bar{\alpha}) sin(\beta_i - \bar{\beta})}{\sqrt{\sum_{i=1}^{n} sin^2(\alpha_i - \bar{\alpha}) \sum_{i=1}^{n} sin^2(\beta_i - \bar{\beta})}} \tag{2}$$

where α_i and β_i are wind directions measured by the sodar and determined through the numerical modeling at a given moment, respectively, $\bar{\alpha}$ and $\bar{\beta}$ are their mean angular values.

3 Results

On 1 April 2012, a well-expressed cold front had passed through Bulgaria from the NW to the SE (Fig. 3). It was cloudy with heavy rainfall, and in the evenings in northeastern Bulgaria the precipitations passed into snow. The frontal system was accompanied by

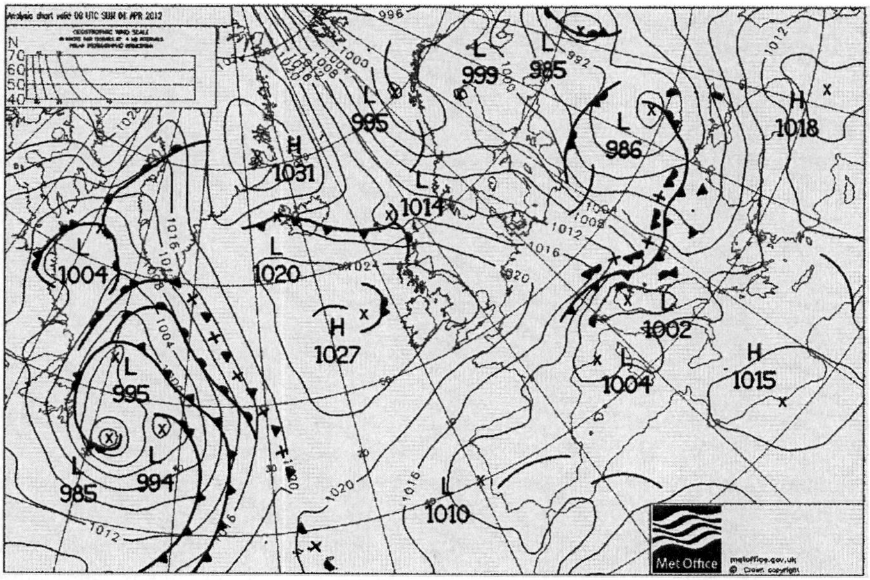

Fig. 3. Surface pressure charts showing pressure and weather fronts over Europe for April 1, 2012 - 00 UTC (UK Met Office)

strong SW wind that changed its direction to the NW, with which colder air mass has settled over the territory - the temperatures at 15 h local time (Eastern European Summer Time - EEST) were from 5 to 8° and in places up to 10° lower than in the morning hours. In Ahtopol on that date the registered temperature amplitude was 18.8 °C, the precipitation amount was 7 lm^{-2} with duration of nearly 9 h, starting at 14:10 EEST.

3.1 Remote Sensing Measurements

Based on the measurements the period started with winds from SW quarter (until 13:30 EEST) in the entire vertical range of the sodar (Fig. 4a), followed by a rapid change in the direction from NE and N (13:40 h - 14:20 EEST) associated with the crossing of the synoptic front, Fig. 3. During almost the entire duration of the precipitation event the sodar vertical range had been decreased to about 200 m and wind flow from NW quarter was established. After the rainfall due to the low wind speed, the wind direction was variable (Fig. 4b). The highest WS of 28–31 ms^{-1} were observed in the layer 300–510 m between 3:40 and 10:10 EEST and maximum WS of 31 ms^{-1} was observed at 510 m early in the morning (4:50 EEST - Fig. 4b). Based on measurements the studied period started with extreme winds (Fig. 4f) and its duration was 12 h without interrup-

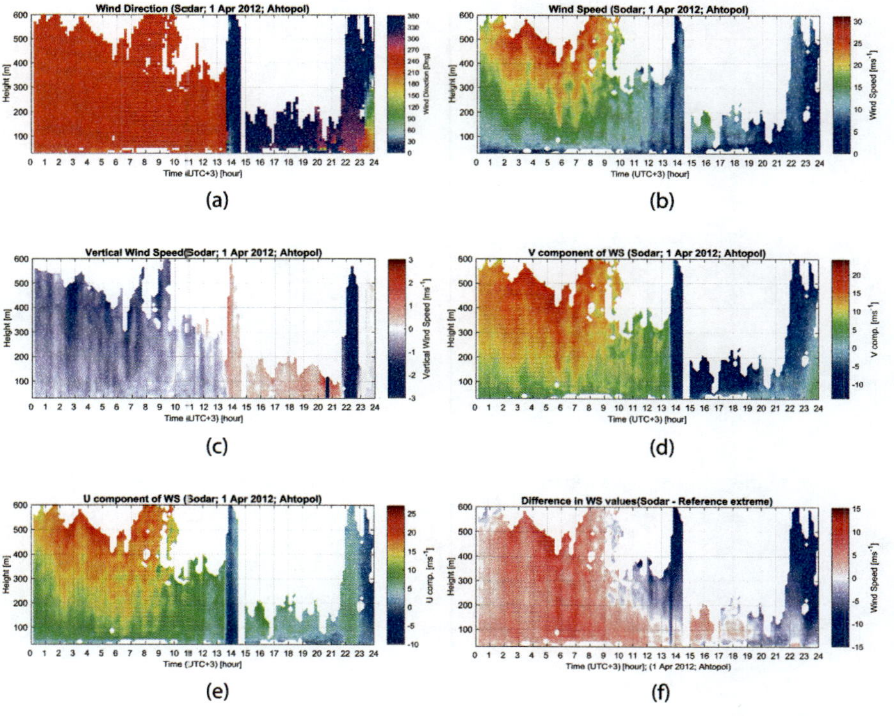

Fig. 4. Sodar measurements on 1 April 2012 in EP Ahtopol: (a) WD; (b) WS; (c) W; (d) V - positive value indicates WD from south to north; (e) U - positive value indicates WD from west to east; (f) difference between measured WS and obtained extreme profile for WS at EP Ahtopol;

tion. Almost the entire range of the measured wind profiles were extreme (until 9:00 EEST) after that the height of extreme wind declined up to 130 m. Almost the entire period with extreme WS was characterized with downdrafts (Fig. 4c).

3.2 Model Results

The period with modelled extreme WS started at 00 EEST with duration of 13 h 40 min (without interruptions). The most of the differences between measured and modelled WD during extreme wind event were up to 20–30° (Fig. 5b). Almost entire precipitation period with NW winds was simulated with difference in WD between 60 and 90 degrees (Fig. 5a and Fig. 5b after 15 EEST). The maximum WS of 28–31 ms^{-1} was modelled in the layer 300–500 m between 4:40 and 10:10 EEST and the value of maximum modelled WS was with 0.8% lower than the measured one. It was modelled at 390 m (8:20 EEST, Fig. 5c). The difference between measured and modelled WS (Fig. 5d) showed that during the period with extreme winds WS was better captured close to the ground (mostly within the range 1–5 ms^{-1} difference) in comparison with the layer between 100 m and 400 m, where the model overestimated WS mostly within

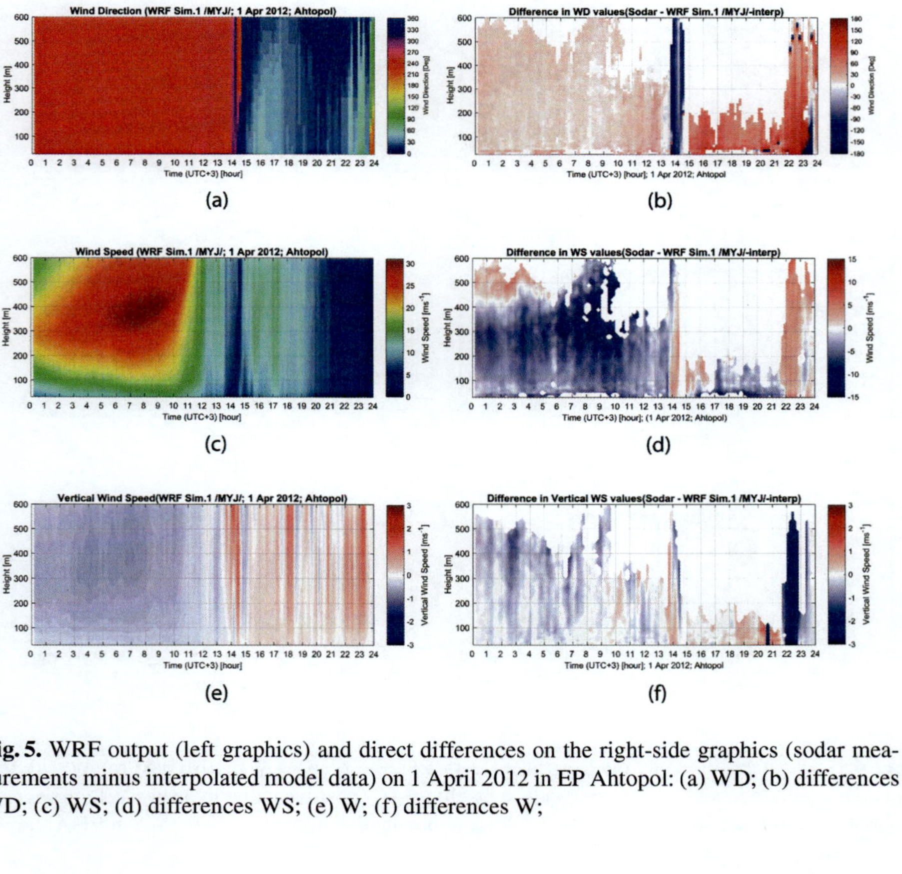

Fig. 5. WRF output (left graphics) and direct differences on the right-side graphics (sodar measurements minus interpolated model data) on 1 April 2012 in EP Ahtopol: (a) WD; (b) differences WD; (c) WS; (d) differences WS; (e) W; (f) differences W;

the range 8–12 ms^{-1}. The used configuration reproduced the observed downdrafts during the studied period with extreme winds and updraft after that (with exception around 20:30 and 22:00–22:40 EEST) (Fig. 5e). In general W was overestimated during the extreme weather event and it was underestimated during the precipitation period by the model on April 1(Fig. 5f).

3.3 Correlation Analysis

The correlation coefficients from Eq. (1) and Eq. (2) were calculated for two data sets (Fig. 6). One was constructed by all time series at each level between 30 to 600 m with step of 10 m and the other one consists of data from all levels at given time.

The first set consisted up to 58 values of the correlation coefficient of up to 144 pair (Fig. 6 - blue lines with multicolored triangles) and it is used to study how well the model represents the vertical structure of the chosen field. Considering the WS (Fig. 6b) there was positive correlation for entire profile, moderate correlation for the level of 40 m, strong for up to 180 m and for the heights in the layer between 440 m and 600 m. For the rest of levels (from 190–430 m) there was very strong correlation between measured and modelled WS. Considering the WD (Fig. 6a) only 510 m level was simulated with weak correlation, two levels (460 m and 500 m) with moderate for the rest of the heights $r_c > 0.51$ as for the most of the levels (30 in total) $0.7 < r_c < 0.9$. Considering the U component (Fig. 6e) r > 0.59 for the entire profile, as for 13 levels (320–430 m and 570) 0.9 < r < 1 and for 35 levels 0.7 < r < 0.9. For the entire profile of V component (Fig. 6d) r > 0.8, as for the 34 levels (60–300 m, 320–350 m and above 560) 0.9 < r < 1. The lowest values of r were obtained for the W (Fig. 6c) - for 38 of the levels were simulated with weak and only for the last 3 levels (580 m, 590 m and 600 m) r > 0.9.

The second dataset consisted of 144 values of the correlation coefficient of up to 58 pairs (Fig. 6 - red lines with multicolored dots) and it is constructed to study how the temporal structure of the chosen field was simulated by the model. Considering the WS (Fig. 6b) there was positive correlation for almost all observational times (except 13:50, 14:00, 14:10, 20:30, 22:30 and 23:40 EEST). There was strong correlation between modeled and observed WS for most observational times (68 in total), very strong for 28 observational times and at16:50 EEST a perfect positive linear relationship is observed but only from 2 pairs measured at 50 and 60 m AGL. Considering the WD (Fig. 6a) a sharp alternation of positive and negative correlations are observed as the number of positive is 67. The circular correlations of 42 observational times (mainly after midday) have met the condition $|r| > 0.5$, 13 of them was simulated with strong correlation, 6 with very strong and one was with perfect positive circular relationship (again at 16:50 EEST). Predominant positive correlations were observed before the rainfall and during the precipitation period the predominant correlations were negative considering the U (Fig. 6e) wind component. There was strong correlation for 68 observational times, very strong for 26 observational times and one perfect negative linear relationship. Considering the V wind component (Fig. 6d) during the extreme wind event (until 12:40 EEST) for 13 times $|r| > 0.7$ and only for 2 of them r > 0.9. Relatively higher values of linear

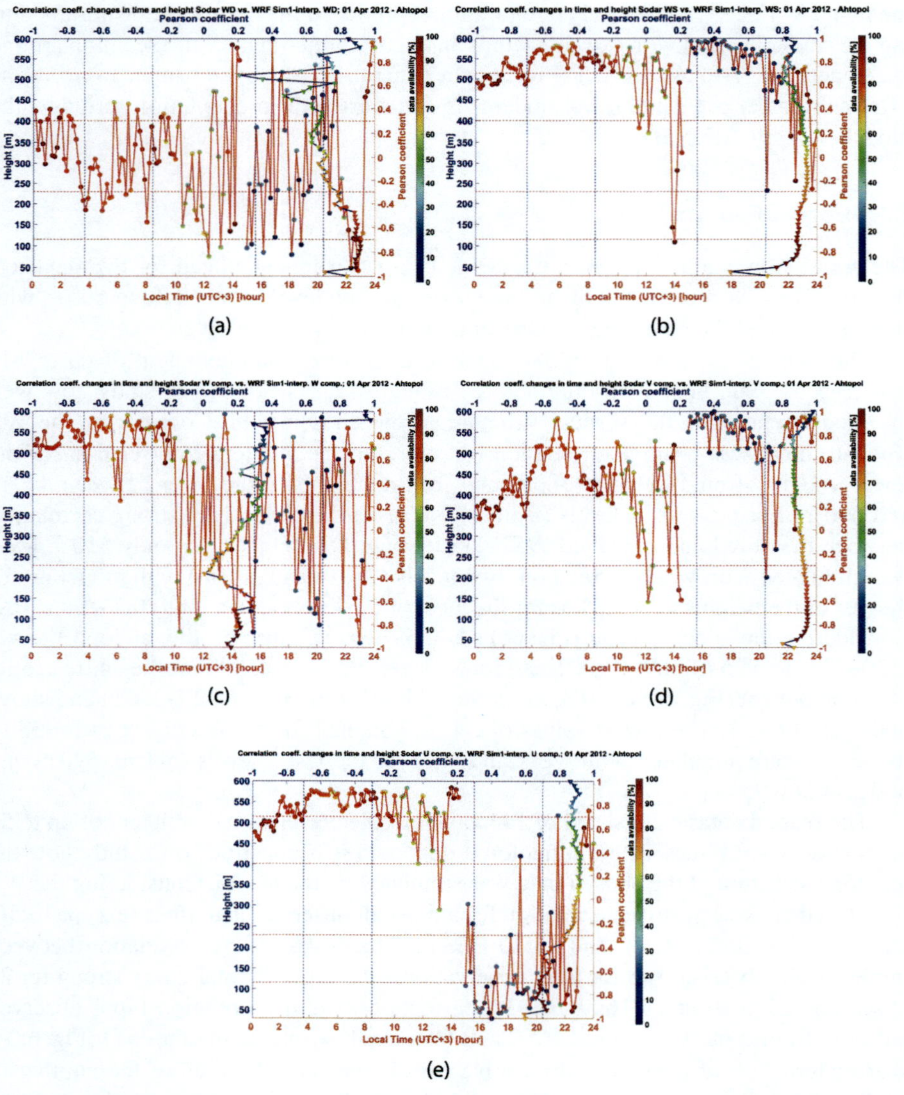

Fig. 6. Spatial (top horizontal axis and blue line with multicolored triangles) and temporal (right vertical axis and red line with multicolored dots) variations of the correlation coefficient. The color of the triangles and the dots corresponds to the data availability in the correlation coefficient calculations (color bar on the right) on 1 April 2012 in EP Ahtopol: (a) WD; (b) WS; (c) W; (d) V; (e) U;

correlation coefficients were observed during the precipitation period. Considering the W (Fig. 6c) during the extreme wind period (until 13:30 EEST) for 8 times $|r| < 0.3$, for 54 times $|r| > 0.7$, for 11 of which $|r| > 0.9$.

4 Conclusion

A case period (01 April 2012) with extreme winds at the southern Bulgarian Black Sea coast was investigated based on sodar data and WRF modeling results with MYJ ABL scheme. The presence of passing cold front system (transition period) on that date and a subsequent precipitation period after it makes that case even more challenging in terms of assessing the operation of the mesometeorological model in a complex area. The simulations with MYJ capture the observed duration of the extreme winds closely as the passage of the frontal system occurs 20 min later in the model data. A positive correlation was found for the entire profiles of modelled and measured WS values. Strong correlation was found for the layers 50 m to 180 m and 440 m to 600 m, and very strong correlations for levels from 190 to 430 m. Strong correlation was found also for WD, U and V for most of the heights, and only for the top 3 layers for W. The temporal structure of the chosen fields was simulated by the model with strong correlation as well. The study identified a WRF model setup for extreme wind case at a Black Sea coastal site for the vertical structure of the wind field. Evaluation of the model results was possible using comprehensive sodar data. Further research will include simulations of other extreme wind speed periods.

Acknowledgments. This study is performed in the framework of REPLICA (extReme Events and wind ProfiLe In a Coastal Area) project, funded by the Bulgaria National Science Fund (Contract № DM 14/1 26-05-2020). The WRF simulations were calculated at the multifunctional high-performance computing complex "AVITOHOL" and would not be possible without innovative data services and tools provided by the Advanced Computing and Data Centre of the Grid Technologies and Applications Department at the at the Institute of Information and Communication Technologies – Bulgarian Academy of Sciences.

References

1. Atanassov, E., Gurov, T., Karaivanova, A., Ivanovska, S., Durchova, M., Dimitrov, D.: On the parallelization approaches for Intel MIC architecture. In: AIP Conference Proceedings: 8th International Conference for Promoting the Application of Mathematics in Technical and Natural Sciences - AMiTaNS 2016, Albena, Bulgaria, 22–27 June 2016, vol. 1773, vol. 070001 (2016)
2. Barantiev, D., Batchvarova, E., Novitsky, M.: Exploration of the coastal boundary layer in Ahtopol through remote acoustic sounding of the atmosphere. In: Section: Physics of Earth, Atmosphere and Space ed., Conference Proceedings: 2nd National Congress on Physical Sciences. Heron Press, Sofia (2013)
3. Barantiev, D., Batchvarova, E., Novitsky, M.: Breeze circulation classification in the coastal zone of the town of Ahtopol based on data from ground based acoustic sounding and ultrasonic anemometer. Bul. J. Meteorol. Hydrol. (BJMH) **22**(5) (2017)
4. Barantiev, D., Novitsky, M., Batchvarova, E.: Meteorological observations of the coastal boundary layer structure at the bulgarian black sea coast. Adv. Sci. Res. (ASR) **6**, 251–259 (2011). https://doi.org/10.5194/asr-6-251-2011
5. Batchvarova, E., Barantiev, D., Novitsky, M.: Costal boundary layer wind profile based on SODAR data – Bulgarian contribution to COST Acton ES0702. In: Conference Proceedings: The 16th International Symposium for the Advancement of Boundary-Layer Remote Sensing – ISARS, Boulder, Colorado, USA (2012)

6. Batchvarova, E., Cai, X., Gryning, S.-E., Steyn, D.: Modelling internal boundary-layer development in a region with a complex coastline. Bound.-Layer Meteorol. **90**(1), 1–20 (1999). https://doi.org/10.1023/A:1001751219627
7. Batchvarova, E., Gryning, S.-E.: Wind climatology, atmospheric turbulence and internal boundary-layer development in Athens during the MEDCAPHOT-TRACE experiment. Atmos. Environ. **32**(12), 2055–2069 (1998). https://doi.org/10.1016/S1352-2310(97)00422-6
8. Berens, P.: CircStat: a MATLAB toolbox for circular statistics. J. Stat. Softw. **31**(10), 21 (2009)
9. Brozek, K., Kogut, J.: Analysis of Pearson's linear correlation coefficient with the use of numerical examples. In: Conference Proceedings: The 10th International Days of Statistics and Economics, Prague, Czech Republic (2016)
10. Chou, M.-D., Suarez, M.J.: An efficient thermal infrared radiation parameterization for use in general circulation models. NASA technical memorandum no. 10460, vol, 3, p. 85 (1994)
11. Grell, G.A., Devenyi, D.: A generalized approach to parameterizing convection combining ensemble and data assimilation techniques. Geophys. Res. Lett. **29**(14) (2002). https://doi.org/10.1029/2002GL015311. Article 1693
12. IPCC: Climate Change 2001: The Scientific Basis. Contribution of Working Group I to the Third Assessment Report of the Intergovernmental Panel on Climate Change. Cambridge University Press, United Kingdom and New York, NY, USA (2011)
13. Jammalamadaka, S.R., Sengupta, A.: Topics in Circular Statistics. Series on Multivariate Analysis, vol. 5. World Scientific (2001)
14. Janjic, Z.I.: The step-mountain eta coordinate model: further developments of the convection, viscous sublayer and turbulence closure schemes. Mon. Weather Rev. (MWR) **122**(5), 927–945 (1994). https://doi.org/10.1175/1520-0493
15. Janjic, Z.I.: The surface layer in the NCEP Eta Model. In: Conference Proceedings: 11th Conference on Numerical Weather Prediction. American Meteorological Society, Boston (1996)
16. Janjic, Z.I.: Nonsingular implementation of the Mellor–Yamada Level 2.5 scheme in the NCEP Meso model, vol. NCEP Office Note #437. National Centers for Environmental Prediction (2001)
17. Mellor, G.L., Yamada, T.: Development of a turbulence closure model for geophysical fluid problems. Rev. Geophys. **20**(4), 851–875 (1982). https://doi.org/10.1029/RG020i004p00851
18. Mlawer, E.J., Taubman, S.J., Brown, P.D., Iacono, M.J., Clough, S.A.: Radiative transfer for inhomogeneous atmosphere: RRTM, a validated correlated-k model for the longwave. J. Geophys. Res.: Atmos. **102**(D14), 16663–16682 (1997). https://doi.org/10.1029/97JD00237
19. Morrison, H., Thompson, G., Tatarskii, V.: Impact of cloud microphysics on the development of trailing stratiform precipitation in a simulated squall line: comparison of one- and two-moment schemes. Mon. Weather Rev. **137**(3), 991–1007 (2009). https://doi.org/10.1175/2008MWR2556.1
20. Novitsky, M., Kulizhnikova, L., Kalinicheva, O., Gaitandjiev, D., Batchvarova, E., Barantiev, D., Krasteva, K.: Characteristics of speed and wind direction in atmospheric boundary layer at southern coast of Bulgaria. Russ. Meteorol. Hydrol. **37**(3), 159–164 (2012). https://doi.org/10.3103/S1068373912030028
21. Radenski, A., Gurov, T., Kaloyanova, K., Kirov, N., Nisheva, M., Stanchev, P., Stoimenova, E.: Big data techniques, systems, applications, and platforms: case studies from academia. In: Conference Proceedings: Federated Conference on Computer Science and Information Systems (FedCSIS). Annals of Computer Science and Information Systems (ACSIS), 11–14 September 2016, Gdansk, Poland, vol. 8 (2016)
22. Radilov, D., Hadzhiev, V., Zhekova, S.: Statistics. Textbook for the Students of the University of Economics - Varna. University Edition - Science and Economics, Varna (2010)

23. Sabev, L., Stanev. S.: Climate regions of Bulgaria and their climate., vol. V. State Publishing House "Science and Art", Sofia (1959)
24. ScintecAG: Scintec Flat Array Sodars - Hardware Manual (SFAS, MFAS, XFAS) including RASS RAE1 and windRASS, version 1.03 ed. Scintec AG, Germany (2011)
25. Skamarock, W.C., Klemp, J.B., Dudhia, J., Gill, D.O., Barker, D.M., Duda, M.G., Huang, X.-Y., Wang, W., Powers, J.G.: A description of the advanced research WRF version 3, vol. NCAR Technical Note, vol. NCAR/TN–475+STR. NCAR Scientific Divisions, Boulder, Colorado, USA (2008)
26. Tewari, M., Chen, F., Wang, W., Dudhia, J., LeMone, M.A., Mitchell, K.E., Ek, M.B., Gayno, G., Wegiel, J., Cuenca, R.H.: Implementation and verification of the united NOAH land surface model in the WRF model. In: Conference Proceedings: AMS 20th Conference on Weather Analysis and Forecasting/16th Conference on Numerical Weather Prediction, Seattle, Washington, USA (2004)
27. Velev, S.: The climate of Bulgaria. Geography and Earth Sciences. The National Publishing House "National education", Sofia (1990)

Sensitivity Studies of an Air Pollution Model by Using Efficient Stochastic Algorithms for Multidimensional Numerical Integration

Tzvetan Ostromsky[1(✉)], Venelin Todorov[1,2], Ivan Dimov[1], and Zahari Zlatev[3]

[1] Institute of Information and Communication Technologies, Department of Parallel Algorithms, Bulgarian Academy of Sciences, 1113 Sofia, Bulgaria
{ceco,venelin}@parallel.bas.bg, vtodorov@math.bas.bg, ivdimov@bas.bg
[2] Institute of Mathematics and Informatics, Department of Information Modeling, Bulgarian Academy of Sciences, 1113 Sofia, Bulgaria
[3] National Centre for Environment and Energy, University of Århus, Frederiksborgvej 399, P.O. Box 358, 4000 Roskilde, Denmark
zz@dmu.dk

Abstract. An important issue when large-scale mathematical models are used to support decision makers is their reliability. Sensitivity analysis of model outputs to variation or natural uncertainties of model inputs is very significant for improving the reliability of these models. A comprehensive experimental study of Monte Carlo algorithm based on adaptive Monte Carlo approach for multidimensional numerical integration has been done. A comparison with Latin Hypercube Sampling and a particular quasi-Monte Carlo lattice rule based on generalized Fibonacci numbers has been presented. Such comparison has been made for the first time and this motivates the present study. The concentration values have been generated by the specialized modification SA-DEM of the Unified Danish Eulerian Model. Its parallel efficiency and scalability will be demonstrated by experiments on some of the most powerful supercomputers in Europe. The algorithms have been successfully applied to compute global Sobol sensitivity measures corresponding to the influence of six chemical reaction rates and four different groups of pollutants on the concentrations of important air pollutants.

1 Introduction

We discuss a systematic approach for sensitivity analysis studies in the area of air pollution modelling. The Unified Danish Eulerian Model (UNI-DEM) [27,28] is used in this particular study. Different parts of the large amount of output data, produced by the model, were used in various practical applications, where the reliability of this data should be properly estimated [6,21]. Another reason to

I. Dimov and S. Fidanova (Eds.): HPC 2019, SCI 902, pp. 184–195, 2021.
https://doi.org/10.1007/978-3-030-55347-0_16

choose this model as a case study here is its sophisticated chemical scheme, where all relevant chemical processes in the atmosphere are accurately represented. We study the sensitivity of concentration variations of some of the most dangerous air pollutants with respect to the anthropogenic emissions levels and with respect to some chemical reactions rates. A special version of UNI-DEM (called SA-DEM) was developed for the purpose of this study. Description of UNI-DEM, SA-DEM and their parallel computer implementations will be given in the next section.

Different efficient stochastic algorithms for multidimensional integration have also been applied on a further stage of these sensitivity studies. Among them are two adaptive Monte Carlo algorithms, described in more detail in Sect. 3. These will be compared with two quasi-Monte Carlo (QMC) algorithms, namely Fibonacci lattice rule and Latin hypercube sampling. Fibonacci lattice rule is completely investigated in [12,22,23,26] and Latin hypercube sampling is described in detail in [15–17].

2 Description and Implementation of UNI-DEM

UNI-DEM is a powerful large-scale air pollution model for calculation of the concentrations of a large number of pollutants and other chemical species in the air along a certain time period. Its results can be used in various application areas (environmental protection, agriculture, health care, etc.). The large computational domain covers completely the European region and the Mediterranean.

UNI-DEM is mathematically represented by the following system of partial differential equations (PDE), in which the unknown concentrations c_s of a large number of chemical species (pollutants and other chemically active components) take part. The main physical and chemical processes (advection, diffusion, chemical reactions, emissions and deposition) are represented in that system.

$$
\begin{aligned}
\frac{\partial c_s}{\partial t} = & -\frac{\partial (uc_s)}{\partial x} - \frac{\partial (vc_s)}{\partial y} - \frac{\partial (wc_s)}{\partial z} \\
& + \frac{\partial}{\partial x}\left(K_x \frac{\partial c_s}{\partial x}\right) + \frac{\partial}{\partial y}\left(K_y \frac{\partial c_s}{\partial y}\right) + \frac{\partial}{\partial z}\left(K_z \frac{\partial c_s}{\partial z}\right) \qquad (1) \\
& + E_s + Q_s(c_1, c_2, \dots c_q) - (k_{1s} + k_{2s})c_s, \quad s = 1, 2, \dots q \,.
\end{aligned}
$$

where c_s are the concentrations of the chemical species; u, v, w are the wind components along the coordinate axes; K_x, K_y, K_z – the diffusion coefficients; E_s – the emissions; k_{1s}, k_{2s} – dry/wet deposition coefficients; $Q_s(c_1, c_2, \dots c_q)$ – non-linear functions describing the chemical reactions between the species under consideration. The above PDE system is non-linear and stiff. Both non-linearity and stiffness are introduced mainly by the chemical scheme: the condensed CBM-IV (Carbon Bond Mechanism) [27,28]. It is quite detailed and accurate, but computationally expensive as well.

For the purpose of efficient numerical treatment, the system (1) is split according to the major physical and chemical processes and the following 3 submodels are formed: ***Advection-diffusion, Chemistry & deposition*** and ***Vertical transport (vertical wind and convection)***.

The following methods are used in the numerical solution of the submodels:

- *Advection-diffusion part:* Finite elements, followed by predictor-corrector schemes with several different correctors.
- *Chemistry-deposition part:* An improved version of the QSSA (Quasi Steady-State Approximation) [10].
- *Vertical transport:* Finite elements, followed by theta-methods.

Spatial and time discretization makes each of the submodels a tough computational task even for the most advanced supercomputer systems. Efficient parallelization has always been a crucial point in the computer implementation of UNI-DEM. The task became much more challenging with development of the sensitivity analysis version of the code, SA-DEM [4,5]. It consists of the following three parts:

- A modification of UNI-DEM with ability to modify certain parameters, subject to SA study. By now we have been interested in some chemical rate constants as well as in the input data for the anthropogenic emissions. A small number of input parameters is reserved for this purpose.
- A driver routine that automatically generates a set of tasks to produce the necessary results for a particular SA study. It allows to perform in parallel a large number of runs with common input data (reusing it), producing at once a whole set of values on a regular mesh (used later for calculating the sensitivity indices).
- An additional program for extracting the necessary mean monthly concentrations and computing the normalised ratios (to be analysed further on).

Significant improvements of the earlier versions of SA-DEM were made by introducing two additional levels of parallelism: top-level(MPI) and bottom-level (OpenMP) [18–20]. They allow us to use efficiently the computational power of the contemporary cluster supercomputers with multicore nodes. Additional improvements in the data management strategy, reducing the number of I/O operations and pipelining most of them with the computationally intensive stages, have also been done.

In Table 1 we show some scalability results from experiments with SA-DEM on one of the largest supercomputers in Europe – IBM MareNostrum III (in BSC, Barcelona, Spain). It consists of 3028 16-core nodes with 32 GB RAM per node. It is seen from Table 1 that the chemical stage (the most expensive in terms of computations) scales pretty well (with almost linear speed-up in the whole range of experiments). Advection stage scales also very well, especially in the first half of the table, with certain "asymptotic" effect in the highly parallel experiments. This is due to the inner boundary strips overlapping between adjacent subdomains with increasing weight when approaching certain partitioning limitations. In general, the code performs quite well and shows relatively high scalability on such a large supercomputing system [18–20].

Table 1. Time (T) and speed-up **Sp** of SA-DEM (MPI only) on the Spanish supercomputer IBM MareNostrum III at BSC, Barcelona

Time and **speed-up** of SA-DEM on IBM MareNostrum III (480 × 480 × 1) grid, 35 chemical species								
# CPU	# nodes	Advection		Chemistry		Total		
		T [s]	**Sp**	T [s]	**Sp**	T [s]	**Sp**	E [%]
10	1	83469	**10**	76988	**10**	169872	**10**	100%
40	3	20875	**40**	17382	**44**	40865	**42**	104%
80	5	10823	**77**	9142	**84**	21971	**77**	97%
160	10	5624	**148**	4625	**166**	12385	**137**	86%
320	20	2889	**288**	2396	**321**	8165	**208**	65%
640	40	1565	**533**	1192	**646**	5117	**332**	52%
960	60	1252	**667**	818	**941**	3925	**433**	45%
1600	100	816	**1023**	495	**1555**	2894	**587**	37%

3 Adaptive Monte Carlo Algorithm

Monte Carlo and quasi-Monte Carlo simulations are the prevailing methods used to solve high-dimensional problems in different areas. Both methods require efficient calculation of multidimensional integrals.

One group of algorithms, widely used for numerical calculation of multidimensional integrals, are the adaptive algorithms [7,13,14]. Most of the adaptive algorithms use a sequence of increasingly finer subdivisions of the original region, chosen to concentrate integrand evaluations on subregions with difficulties. Two main types of subdivision strategies are in common use: local and global subdivision. The main disadvantage of local subdivision strategy is that it needs a local absolute accuracy requirement, which will be met after the achievement of the global accuracy requirement. The main advantage of the local subdivision strategy is that it allows a very simple subregion management. Globally adaptive algorithms usually require more working storage than locally adaptive routines, and accessing the region collection is slower. These algorithms try to minimize the global error as fast as possible, independently of the specified accuracy requirement.

They show advantage when the integrand functions are not smooth, but need a small number of samples to obtain good accuracy. A Monte Carlo integration embedded in a globally adaptive algorithm is able to provide an unbiased estimate of the integral and also probabilistic error bounds for the estimate. In the mean-time it has higher accuracy and faster convergence than the plain Monte Carlo integration as can be seen from the tables below. The only drawback is the higher computational time.

The Adaptive Monte Carlo algorithm that we developed is based on the ideas and results of the importance separation [2,3,7–9], a method that combines the idea of separation of the domain into uniformly small subdomains with the importance sampling approach. The Adaptive method does not use any a priori information about the smoothness of the integrand, but it uses a posteriori information about the variance. The idea of the method consists of the following: the domain of integration G is separated into subdomains with identical volume. The interval [0,1] on every dimension coordinate is partitioned into M subintervals, i.e.

$$G = \sum_j G_j, \ j = 1, M^d.$$

We start with a relatively small number M which is given as input data. For every subdomain the integral I_{G_j} and the variance are evaluated. After that the variance is compared with a preliminary given value ε. If the calculated variance in any region is greater than some constant ε, then this region is divided to new M^d subregions, again by partitioning the segment of the region on every coordinate to M subintervals. The algorithm is described below.

Algorithm

1. **Input data**: *number of points* N*,constant* ε *(estimation for the variance), constant* δ *(stop criterion; estimation for the length of subintervals on every coordinate).*
2. **For** $j = 1, \ M^d$:
   - 2.1. **Calculate** *the approximation of* I_{Ω_j} *and the variance* $\mathbf{D}_{\Omega_j}$ *in subdomain* Ω_j *based on* N *independent realizations of random variable* θ_N;
   - 2.2. **If** $(\mathbf{D}_{\Omega_j} \geq \varepsilon)$ **then**
     - 2.2.1. **Choose** *the axis direction on which the partition will perform,*
     - 2.2.2. **Divide** *the current domain into two* (G_{j_1}, G_{j_2}) *along the chosen direction,*
     - 2.2.3. **If** *the length of obtained subinterval is less than* δ **then go to** *step 2.2.1* **else** $j = j_1$ (G_{j_1} *is the current domain) and* **go to** *step 2.1*;
   - 2.3. **Else if** $(\mathbf{D}_{\Omega_j} < \varepsilon)$ *but an approximation of* $I_{G_{j_2}}$ *has not been calculated yet,* **then** $j = j_2$ (G_{j_2} *is the current domain along the corresponding direction) and* **go to** *step 2.1*;
   - 2.4. **Else if** $(\mathbf{D}_{\Omega_j} < \varepsilon)$ *but there are subdomains along the other axis directions,* **then go to** *step 2.1*;
   - 2.5. **Else** *Accumulation in the approximation* I_N of I.

Two adaptive approaches ADAPT1 (M = 1) and ADAPT2 (M = 2) will be compared with Fibonacci based lattice rule (FIBO) and Latin hypercube sampling (LHS).

First we will briefly discuss the computational complexity of the Crude Monte Carlo algorithm [1,2]. Let ξ be a random point with probability density function $p(x)$. Introducing the random variable

$$\theta = f(\xi)$$

with mathematical expectation equal to the value of the integral I_{G_j}, we have

$$E\theta = \int_{G_j} f(x)p(x)dx.$$

Let $\xi_1, \xi_2, \ldots, \xi_N$ be independent realizations of the random point ξ with probability density function $p(x)$ and $\theta_1 = f(\xi_1), \ldots, \theta_N = f(\xi_N)$. Then an approximate value of I_{G_j} is

$$\hat{\theta}_N = \frac{1}{N}\sum_{i=1}^{N} \theta_i.$$

The computational complexity of the Crude Monte Carlo is linear, because in this simple case we have to choose N random points in the domain and every such choice is at the cost of $\mathcal{O}(1)$ operations. One single evaluation of the function in any of these points is also at the cost of $\mathcal{O}(1)$ operations.

In the Adaptive Monte Carlo algorithm we are doing the same number of operations. For the simple case when $N = 2$ on the first step we have 4 subdomains and

$$\hat{\theta}_N = \frac{1}{N_1}\sum_{i=1}^{N_1} \theta_i + \frac{1}{N_2}\sum_{i=1}^{N_2} \theta_i + \frac{1}{N_3}\sum_{i=1}^{N_3} \theta_i + \frac{1}{N_4}\sum_{i=1}^{N_4} \theta_i,$$

where $N_1 + N_2 + N_3 + N_4 = N$, so we have the same number of operations as the Crude Monte Carlo to estimate an approximate value of I_{G_j}.

In practice when we divide the domain, we choose only $\mathcal{O}(1)$ subdomains where the variance is greater than the parameter ε and this choice is independent of the dimensionality N. One can easily see that on every step adaptiveness is not in all subdomains, but only in $\mathcal{O}(1)$ subdomains. In the beginning we choose $\frac{N}{k_0}$ random points. On the next step, after we divide the domain into 2^N subdomains, we choose only $\mathcal{O}(1)$ subdomains where the variance is greater than the parameter ε and it is important that this choice is independent of the dimensionality N. In these subdomains we choose $\frac{N}{k_1}$ points. To summarize, on the jth step of the adaptive algorithm we choose $\mathcal{O}(1)$ subdomains and in them we choose $\frac{N}{k_j}$ points. We have that $\sum_{j=0}^{i} \frac{1}{k_j} = 1$. So for the computational complexity we obtain

$$\frac{N}{k_0} + \mathcal{O}(1)\frac{N}{k_1} + \cdots + \mathcal{O}(1)\frac{N}{k_i} = N\mathcal{O}(1)\left(\sum_{j=0}^{i} \frac{1}{k_j}\right) = N\mathcal{O}(1) = \mathcal{O}(N).$$

4 Sensitivity Studies with Respect to Emission Levels

The huge output data stream of UNI-DEM contains are the mean monthly concentrations of more than 30 pollutants. We consider 2 of them: *ozone* (O_3)

and *ammonia* (NH_3). In particular, we present some results of a sensitivity study of the mean monthly concentrations of ammonia.

Here we present some results of our research on the sensitivity of UNI-DEM output (in particular, the ammonia mean monthly concentrations) with respect to the anthropogenic emissions input variation. The anthropogenic emissions input consists of 4 different components $\mathbf{E} = (\mathbf{E^A}, \mathbf{E^N}, \mathbf{E^S}, \mathbf{E^C})$ as follows:

$\mathbf{E^A}$ – ammonia (NH_3);
$\mathbf{E^N}$ – nitrogen oxides ($NO + NO_2$);
$\mathbf{E^S}$ – sulphur dioxide (SO_2);
$\mathbf{E^C}$ – anthropogenic hydrocarbons.

The domain is the 4-dimensional hypercube $[0.5, 1]^4$. Polynomials of 2-nd degree have been used as an approximation tool [5]. The input data have been generated by the improved version SA-DEM code, specialized for sensitivity studies (see the previous section).

The results for relative errors for evaluation of the quantities f_0, total variances and first-order and total sensitivity indices using various stochastic approaches for numerical integration are presented in Tables 2, 3, 4, respectively. The quantity f_0 is presented by a 4-dimensional integral whereas the rest of the quantities under consideration are presented by an 8-dimensional integrals following the ideas of *correlated sampling* technique to compute sensitivity measures in a reliable way [11, 25].

Table 2. Relative error for evaluation of $f_0 \approx 0.048$.

# of samples n	**ADAPT1**	**ADAPT2**	**FIBO**	**LHS**
	Relative error	Relative error	Relative error	Relative error
2^{10}	1.88e−03	1.03e−03	2.09e−04	5.37e−04
2^{12}	2.05e−04	5.05e−04	4.32e−05	2.27e−04
2^{14}	1.83e−04	1.38e−05	2.25e−05	6.28e−05
2^{16}	9.89e−05	4.05e−04	8.70e−06	7.74e−05
2^{18}	3.95e−05	3.83e−06	1.79e−06	3.80e−06
2^{20}	4.99e−05	2.93e−05	4.21e−07	7.16e−06

The results in Table 2 show that the algorithms using generalized Fibonacci numbers and LHS simulate the behaviour of the Adaptive Monte Carlo algorithm, but for higher dimensions their efficiency decreases. The particular case study confirms the conclusion that these algorithms are suitable and more efficient for smooth functions with relatively low dimensions. From Tables 2 and 3 we can conclude that all stochastic approaches under consideration give reliable relative errors for a sufficiently large number of samples. This is not the case for some sensitivity indices, which are very small in absolute value (see Table 4), but fortunately, these are of low importance too. The most efficient in terms of computational complexity is the FIBO algorithm, followed very closely by the

Table 3. Relative error for evaluation of the total variance $\mathbf{D} \approx 0.0002$.

# of samples n	**ADAPT1**	**ADAPT2**	**FIBO**	**LHS**
	Relative error	Relative error	Relative error	Relative error
2^{10}	1.56e−03	4.76e−03	1.63e−01	1.74e−02
2^{12}	2.58e−03	4.28e−04	2.39e−02	1.04e−02
2^{14}	6.03e−04	2.79e−04	2.90e−03	1.04e−02
2^{16}	1.83e−04	5.12e−04	2.65e−04	3.65e−04
2^{18}	5.77e−05	1.21e−04	3.01e−04	1.21e−05
2^{20}	3.42e−05	3.28e−05	1.19e−04	5.96e−05

Table 4. Relative error for estimation of sensitivity indices of the input anthropogenic emissions by using various Monte Carlo and quasi-Monte Carlo approaches ($n = 2^{16} = 65536$).

Est. quantity	Ref. value	**ADAPT1**	**ADAPT2**	**FIBO**	**LHS**
S_1	9e−01	7.67e−04	1.22e−03	3.62e−04	9.79e−03
S_2	2e−04	1.47e−03	4.96e−02	1.74e−01	6.60e−01
S_3	1e−01	4.11e−03	1.59e−03	3.22e−03	8.65e−03
S_4	4e−05	1.04e−01	1.69e−01	4.87e−01	6.70e−01
S_1^{tot}	9e−01	4.99e−05	5.36e−05	4.61e−04	4.31e−04
S_2^{tot}	2e−04	5.23e−01	5.00e+00	3.45e−01	2.94e+01
S_3^{tot}	1e−01	1.15e−02	1.28e−02	1.96e−03	1.10e−02
S_4^{tot}	5e−05	1.88e+01	3.43e+01	5.06e+01	2.41e+02

LHS algorithm. The Adaptive algorithm gives results of the same order as LHS and FIBO, and sometimes even outperforms them – see for example the relative errors for S_1^{tot} in Table 4.

5 Sensitivity Studies with Respect to Chemical Reactions Rates

We will also study the sensitivity of the ozone concentration values in the air over Genova with respect to the rate of variation of some chemical reactions of the condensed CBM-IV scheme [27], namely: # $1, 3, 7, 22$ (time-dependent) and # $27, 28$ (time independent). The simplified chemical equations of those reactions are as follows:

[**#1**] $NO_2 + h\nu \Longrightarrow NO + O$;
[**#3**] $O_3 + NO \Longrightarrow NO_2$;
[**#7**] $NO_2 + O_3 \Longrightarrow NO_3$;
[**#22**] $HO_2 + NO \Longrightarrow OH + NO_2$;
[**#27**] $HO_2 + HO_2 \Longrightarrow H_2O_2$;
[**#28**] $OH + CO \Longrightarrow HO_2$.

The domain under consideration is the 6-dimensional hypercube $[0.6, 1.4]^6$). Polynomials of second degree have been used for approximation again (see [4]).

Homma and Saltelli discuss in [11] which is the better estimation of $f_0^2 = \left(\int_{U^d} f(\mathrm{x})\mathrm{dx}\right)^2$ in the expression for total variance and Sobol global sensitivity measures. In case of estimating sensitivity indices of a fixed order, the formula $f_0^2 \approx \frac{1}{n}\sum_{i=1}^{n} f(\mathrm{x}_{i,1},\ldots,\mathrm{x}_{i,d})\, f(\mathrm{x}'_{i,1},\ldots,\mathrm{x}'_{i,d})$, where x and x′ are two independent sample vectors, is better (as recommended in [11,25]).

Table 5. Relative error for evaluation of $f_0 \approx 0.27$.

# of samples n	ADAPT1	ADAPT2	FIBO	LHS
	Relative error	Relative error	Relative error	Relative error
2^{10}	2.74e−04	3.21e−04	2.08e−03	3.73e−04
2^{12}	9.55e−05	4.43e−05	1.40e−04	2.41e−04
2^{14}	1.20e−04	5.64e−05	3.98e−04	7.53e−05
2^{16}	3.49e−05	3.72e−05	2.61e−04	2.02e−04

Table 6. Relative error for evaluation of the total variance $\mathbf{D} \approx 0.0025$.

# of samples n	ADAPT1	ADAPT2	FIBO	LHS
	Relative error	Relative error	Relative error	Relative error
2^{10}	9.67e−04	1.18e−03	6.73e+00	1.91e−02
2^{12}	9.10e−04	2.24e−03	5.27e−01	9.99e−02
2^{14}	1.40e−04	1.86e−04	1.02e−01	1.62e−02
2^{16}	3.01e−05	1.48e−04	1.97e−03	3.56e−05

The relative errors for evaluation of the quantities f_0, total variances, first and second order sensitivity indices by using various stochastic approaches for numerical integration are presented in Tables 5, 6, 7 respectively. Here the quantity f_0 is presented by a 6-dimensional integral, whereas the total variance and the sensitivity indices are presented by 12-dimensional integrals, following the ideas of *correlated sampling.*

QMC lattice rule based on generalized Fibonacci numbers and Latin hypercube sampling produce better results for 6-dimensional integrals in comparison with 12-dimensional integrals. It is clear that by increasing the dimensionality of the integral the Adaptive method produces more accurate results than both FIBO and LHS. The Adaptive Monte Carlo algorithm gives better results in

Table 7. Relative error for estimation of sensitivity indices of several chemical reaction rate parameters by using various Monte Carlo and quasi-Monte Carlo approaches ($n = 2^{16} = 65536$).

Est. quantity	Ref. value	**ADAPT1**	**ADAPT2**	**FIBO**	**LHS**
S_1	4e-01	1.55e−04	3.48e−04	3.82e−02	3.04e−02
S_2	3e-01	4.34e−04	1.58e−04	1.03e−02	7.35e−04
S_3	5e-02	3.42e−04	8.09e−05	5.48e−01	2.33e−02
S_4	3e-01	4.75e−04	9.04e−04	1.07e−02	2.47e−02
S_5	4e-07	1.31e+01	1.07e+01	3.40e+03	9.25e+02
S_6	2e-02	1.08e−03	4.54e−04	1.32e+00	3.81e−02
S_{12}	6e-03	1.30e−02	7.92e−03	3.21e+00	8.99e−02
S_{14}	5e-03	5.30e−03	1.81e−03	8.64e+00	2.74e−01
S_{15}	8e-06	9.34e+02	9.34e+02	9.19e+02	9.21e+02
S_{24}	3e-03	1.26e−03	7.24e−03	1.37e+01	7.10e−01
S_{45}	1e-05	9.93e−02	8.55e−02	4.25e+01	1.05e+01

case of higher dimensional integrals and lower number of samples. For most of the sensitivity indices the Adaptive Monte Carlo algorithm gives more accurate results than FIBO and LHS by at least 2 orders of degree.

6 Conclusion

The numerical tests show that the stochastic algorithms under consideration are efficient for multidimensional numerical integration. The Adaptive MC algorithm outperforms the Fibonacci based lattice rule and Lain hypercube sampling in the case of small number of samples and higher dimensions. In order to achieve a more accurate distribution of inputs influence and a more reliable interpretation of the mathematical model results, sometimes the computation of small in absolute value sensitivity indices is significant. Clearly, the progress in the area of sensitivity analysis of atmospheric pollution modeling is closely connected with the development of efficient stochastic algorithms for multidimensional integration.

Acknowledgements. Venelin Todorov is supported by the Bulgarian National Science Fund under Project KP-06-M32/2 - 17.12.2019 "Advanced Stochastic and Deterministic Approaches for Large-Scale Problems of Computational Mathematics" and by the National Scientific Program "Information and Communication Technologies for a Single Digital Market in Science, Education and Security (ICT in SES)", contract No D01—205/23.11.2018, financed by the Ministry of Education and Science in Bulgaria. This work is partially supported by the Bulgarian National Science Fund under Project DN 12/5-2017 and DN 12/4-2017.

References

1. Dimov, I.: Monte Carlo Methods for Applied Scientists, p. 291p. World Scientific, London/Singapore/New Jersey (2008)
2. Dimov, I.T., Georgieva, R.: Monte Carlo method for numerical integration based on Sobol' sequences. In: Dimov, I., Dimova, S., Kolkovska, N. (eds.) Numerical Methods and Applications. LNCS. Springer, vol. 6046, pp. 50–59 (2011)
3. Dimov, I.T., Georgieva, R.: Multidimensional sensitivity analysis of large-scale mathematical models. In: Iliev, O.P., et al. (eds.) Numerical Solution of Partial Differential Equations: Theory, Algorithms, and Their Applications, vol. 45, pp. 137–156. Springer, New York (2013)
4. Dimov, I.T., Georgieva, R., Ostromsky, T., Zlatev, Z.: Variance-based sensitivity analysis of the unified Danish Eulerian model according to variations of chemical rates. In: Dimov, I., Faragó, I., Vulkov, L. (eds.) Proceedings of NAA 2012. LNCS, vol. 8236, pp. 247–254. Springer (2013)
5. Dimov, I., Georgieva, R., Ostromsky, Tz., Zlatev, Z.: Sensitivity studies of pollutant concentrations calculated by UNI-DEM with respect to the input emissions. In: Central European Journal of Mathematics, "Numerical Methods for Large Scale Scientific Computing", vol. 11, no. 8, pp. 1531–1545 (2013)
6. Fidanova, S.: Convergence proof for a Monte Carlo method for combinatorial optimization problems. In: International Conference on Computational Science, pp. 523–530. Springer, Heidelberg (2004)
7. Dimov, I., Georgieva, R.: Monte Carlo algorithms for evaluating sobol' sensitivity indices. Math. Comput. Simul. **81**(3), 506–514 (2010)
8. Dimov, I., Karaivanova, A.: Error analysis of an adaptive Monte Carlo method for numerical integration. Math. Comput. Simul. **47**, 201–213 (1998)
9. Dimov, I., Karaivanova, A., Georgieva, R., Ivanovska, S.: Parallel importance separation and adaptive Monte Carlo algorithms for multiple integrals. In: 5th International Conference on NMA. Lecture Notes in Computer Science, vol. 2542, pp. 99–107. Springer, Heidelberg (2002/2003)
10. Hesstvedt, E., Hov, Ø., Isaksen, I.A.: Quasi-steady-state approximations in air pollution modeling: comparison of two numerical schemes for oxidant prediction. Int. J. Chem. Kinet. **10**, 971–994 (1978)
11. Homma, T., Saltelli, A.: Importance measures in global sensitivity analysis of nonlinear models. Reliab. Eng. Syst. Saf. **52**, 1–17 (1996)
12. Hua, L.K., Wang, Y.: Applications of Number Theory to Numerical Analysis (1981)
13. Karaivanova, A.: Stochastic numerical methods and simulations (2012)
14. Karaivanova, A., Dimov, I., Ivanovska, S.: A Quasi-Monte Carlo method for integration with improved convergence. In: LNCS, vol. 2179, pp. 158–165 (2001)
15. McKay, M.D., Beckman, R.J., Conover, W.J.: A comparison of three methods for selecting values of input variables in the analysis of output from a computer code. Technometrics **21**(2), 239–245 (1979)
16. Minasny, B., McBratney, B.: A conditioned Latin hypercube method for sampling in the presence of ancillary information. J. Comput. Geosci. Arch. **32**(9), 1378–1388 (2006)
17. Minasny, B., McBratney, B.: Conditioned Latin hypercube sampling for calibrating soil sensor data to soil properties. In: Proximal Soil Sensing, Progress in Soil Science, pp. 111–119 (2010)

18. Ostromsky, Tz., Dimov, I.T., Georgieva, R., Zlatev, Z.: Parallel computation of sensitivity analysis data for the Danish Eulerian model. In: Proceedings of 8th International Conference on LSSC 2011. LNCS, vol. 7116, pp. 301–309. Springer (2012)
19. Ostromsky, Tz., Dimov, I.T., Georgieva, R., Zlatev, Z.: Air pollution modelling, sensitivity analysis and parallel implementation. Int. J. Environ. Pollut. **46**(1/2), 83–96 (2011)
20. Ostromsky, Tz., Dimov, I.T., Marinov, P., Georgieva, R., Zlatev, Z.: Advanced sensitivity analysis of the Danish Eulerian Model in parallel and grid environment. In: Proceedings of 3rd International Conference on AMiTaNS 2011, AIP Conference on Proceedings, vol. 1404, pp. 225–232 (2011)
21. Poryazov, S., Saranova, E., Ganchev, I.: Conceptual and Analytical Models for Predicting the Quality of Service of Overall Telecommunication Systems. Autonomous Control for a Reliable Internet of Services, pp. 151–181. Springer, Cham (2018)
22. Sloan, I.H., Joe, S.: Lattice Methods for Multiple Integration. Oxford University Press, Oxford (1994)
23. Sloan, I.H., Kachoyan, P.J.: Lattice methods for multiple integration: theory, error analysis and examples. S.I.A.M J. Num. Anal. **14**, 116–128 (1987)
24. Sobol, I.M.: Sensitivity estimates for nonlinear mathematical models. Math. Model. Comput. Exp. **1**(4), 407–414 (1993)
25. Sobol, I.M., Tarantola, S., Gatelli, D., Kucherenko, S., Mauntz, W.: Estimating the approximation error when fixing unessential factors in global sensitivity analysis. Reliab. Eng. Syst. Saf. **92**, 957–960 (2007)
26. Wang, Y., Hickernell, F.J.: An historical overview of lattice point sets (2002)
27. Zlatev, Z.: Computer Treatment of Large Air Pollution Models. KLUWER Academic Publishers, Dorsrecht/Boston/London (1995)
28. Zlatev, Z., Dimov, I.T.: Computational and Numerical Challenges in Environmental Modelling. Elsevier, Amsterdam (2006)

A Multistep Scheme to Solve Backward Stochastic Differential Equations for Option Pricing on GPUs

Lorenc Kapllani, Long Teng, and Matthias Ehrhardt(✉)

Applied Mathematics and Numerical Analysis, University of Wuppertal,
Gaussstrasse 20, 42119 Wuppertal, Germany
{kapllani,teng,ehrhardt}@math.uni-wuppertal.de

Abstract. The goal of this work is to parallelize the multistep method for the numerical approximation of the Backward Stochastic Differential Equations (BSDEs) in order to achieve both, a high accuracy and a reduction of the computation time as well. In the multistep scheme the computations at each grid point are independent and this fact motivates us to select massively parallel GPU computing using CUDA. In our investigations we identify performance bottlenecks and apply appropriate optimization techniques for reducing the computation time, using a uniform domain. Finally, a Black-Scholes BSDE example is provided to demonstrate the achieved acceleration on GPUs.

1 Introduction

In this work we parallelize the multistep scheme developed in [16] to approximate numerically the solution of the following (decoupled) *forward backward stochastic differential equation (FBSDE)*:

$$\begin{cases} dX_t &= a(t,X_t)dt + b(t,X_t)dW_t, \quad X_0 = x_0, \\ -dy_t &= f(t,X_t,y_t,z_t)dt - z_t\,dW_t, \\ y_T &= \xi = g(X_t), \end{cases} \tag{1}$$

where $X_t, a \in \mathbb{R}^n$, b is a $n \times d$ matrix, W_t is a d-dimensional Brownian motion, $f(t,X_t,y_t,z_t) : [0,T] \times \mathbb{R}^n \times \mathbb{R}^m \times \mathbb{R}^{m\times d} \to \mathbb{R}^m$ is the driver function and ξ is the terminal condition. We see that the terminal condition y_T depends on the final value of a forward stochastic differential equation (SDE). For $a = 0$ and $b = 1$, namely $X_t = W_t$, one obtains a *backward stochastic differential equation (BSDE)* of the form

$$\begin{cases} -dy_t &= f(t,y_t,z_t)dt - z_t\,dW_t, \\ y_T &= \xi = g(W_T), \end{cases} \tag{2}$$

where $y_t \in \mathbb{R}^m$ and $f(t,y_t,z_t) : [0,T] \times \mathbb{R}^m \times \mathbb{R}^{m\times d} \to \mathbb{R}^m$. In the sequel of this work, we investigate the acceleration of numerical scheme developed in [16] for solving (2). Note that the developed schemes can be applied also for solving (1), where the general Markovian diffusion X_t can be approximated, e.g., by using the Euler-Scheme.

I. Dimov and S. Fidanova (Eds.): HPC 2019, SCI 902, pp. 196–208, 2021.
https://doi.org/10.1007/978-3-030-55347-0_17

The existence and uniqueness of the solution of (2) were proven by Pardoux and Peng [10]. Peng [11] obtained a direct relation between forward-backward stochastic differential equations (FBSDEs) and partial differential equations (PDEs). Based on this relationship, many numerical methods are proposed, e.g. probabilistic based methods in [1,2,6,9,17], tree-based methods in [3,14] etc. El Karoui, Peng and Quenez [5] showed that the solution of a linear BSDE is in fact the pricing and hedging strategy of an option derivative. This was the first claim of application of BSDEs in finance.

In general the solution of BSDEs cannot be established in a closed form. Therefore, a numerical method is mandatory. There are two main classes of numerical methods for approximating the solution of BSDEs. The first class is related with the PDE equivalent based on the Feynman-Kac formula and the second is based on the BSDE. Many methods have been developed, but one of the most interesting (due to the ability to achieve very high accuracy) is developed by Zhao, Zhang and Ju [16]. They used Lagrange interpolating polynomials to approximate the integrals, given the values of integrands at multiple time levels. One of the drawbacks of their method is the computation time. However, the method is highly parallel. Hence, due to simple and intense calculations, the best computing environment is the one offered by GPUs rather than CPUs.

Many acceleration strategies have been developed to solve option pricing problems on the GPU with different mathematical models. However, little work is based on BSDEs. Dai, Peng and Dong [4] solved a linear BSDE on the GPU with the theta-scheme method. They analyzed the effects of the thread number per block to increase the speedup. The parallelized program using CUDA achieved high speedups and showed that the GPU architecture is well suited for solving the BSDEs in parallel. Later in 2011, they developed acceleration strategies for option pricing with non-linear BSDEs using a binomial lattice based method [12]. To increase the speedup, they reduce the global memory access frequency by avoiding the kernel invocation on each time step. Also, due to the load imbalance produced by the binomial grid, they provided load-balanced strategies and showed that the acceleration algorithms exhibit very high speedup over the sequential CPU implementation and therefore suitable for real-time application.

In 2014, Peng, Liu, Yang and Gong [13] considered solving high dimensional BSDEs on GPUs with application in high dimensional American option pricing. A *Least Square Monte-Carlo (LSMC)* method based numerical algorithm was studied, and summarised in four phases. Multiple factors which affect the performance (task allocation, data store/access strategies and the thread synchronisation) were taken into consideration. Results showed much better performance than the CPU version. In 2015, Gobet, Salas, Turkedjiev and Vasquez [7] designed a new algorithm for solving BSDEs based on LSMC. Due to stratification, the algorithm is very efficient especially for large scale simulations. They showed big speedups even in high dimensions.

Next we introduce some preliminary elements which are needed to understand the multistep scheme. We start with the relation of BSDEs and PDEs. Let $(\Omega, \mathscr{F}, \mathbb{P}, \{\mathscr{F}_t\}_{0\leq t\leq T})$ be a complete, filtered probability space. In this space a standard d-dimensional Brownian motion W_t is defined, such that the filtration $\{\mathscr{F}_t\}_{0\leq t\leq T}$ is generated. We define $\|\cdot\|$ as the standard Euclidean norm in the Euclidean space $\mathbb{R}^m$ or $\mathbb{R}^{m\times d}$ and $L^2 = L^2_F(0,T;\mathbb{R}^d)$ the set of all $\{\mathscr{F}_t\}$-adapted and square integrable processes valued in $\mathbb{R}^d$. A pair of processes $(y_t, z_t) : [0,T] \times \Omega \to \mathbb{R}^m \times \mathbb{R}^{m\times d}$ is the solution of BSDE (2) if it is $\{\mathscr{F}_t\}$-adapted, square integrable, and satisfies (2) in the sense of

$$y_t = \xi + \int_t^T f(s, y_s, z_s)\, ds - \int_t^T z_s\, dW_s, \quad t \in [0,T), \tag{3}$$

where $f(t, y_t, z_t) : [0,T] \times \mathbb{R}^m \times \mathbb{R}^{m\times d} \to \mathbb{R}^m$ is $\{\mathscr{F}_t\}$-adapted and the third term on the right-hand side is an Itô-type integral. This solution exist under "reasonable" regularity conditions [10]. Let us consider the following:

$$y_t = u(t, W_t), \quad z_t = \nabla u(t, W_t) \quad \forall t \in [0,T), \tag{4}$$

where ∇u denotes the derivative of $u(t,x)$ with respect to the spatial variable x and $u(t,x)$ is the solution of the following (backward in time) parabolic PDE:

$$\frac{\partial u}{\partial t} + \frac{1}{2}\sum_{i=1}^{d} \frac{\partial^2 u}{\partial x_i^2} + f(t, u, \nabla u) = 0, \tag{5}$$

with the terminal condition $u(T,x) = \phi(x)$. Under "reasonable" conditions, the PDE (5) possess a unique solution $u(t,x)$. Therefore, for $\xi = \phi(W_T)$, the pair (y_t, z_t) is the unique solution of BSDE (3).

Due to conditional expectations that compound the numerical method, the following notations will be used. Let $\mathscr{F}_s^{t,x}$ for $t \le s \le T$ be a σ-field generated by the Brownian motion $\{x + W_r - W_t, t \le r \le s\}$ starting from the time-space point (t,x). We define $E_s^{t,x}[X]$ as the conditional expectation of the random variable X under the filtration $\mathscr{F}_s^{t,x}$, i.e. $E_s^{t,x}[X] = E[X|\mathscr{F}_s^{t,x}]$.

This work is organized as follows. In Sect. 2 we introduce the multistep scheme. Next, in Sect. 3 our algorithmic framework for using GPU is presented. In Sect. 4 we illustrate our findings with the Black-Scholes example.

2 The Multistep Scheme

In this section we briefly present the multistep scheme. This is done in two steps, the first corresponds to the derivation of the stable semi-discrete scheme, as only the time domain is discretized. Furthermore, the space is discretized and the fully stable multistep scheme is achieved. Note that the scheme will be presented for the one-dimensional case (but recall that in principle it can be generalized for the d-dimensional case).

2.1 The Stable Semi-discrete Scheme

Let N be a positive integer and $\Delta t = T/N$ the step size that partitions uniformly the time interval $[0,T]$: $0 = t_0 < t_1 < \cdots < t_{N-1} < t_N = T$, where $t_i = t_0 + i\Delta t$, $i = 0, 1, \ldots, N$. Let k and K_y be two positive integers such that $1 \le k \le K_y \le N$. The BSDE (3) can be expressed as

$$y_{t_n} = y_{t_{n+k}} + \int_{t_n}^{t_{n+k}} f(s, y_s, z_s)\, ds - \int_{t_n}^{t_{n+k}} z_s\, dW_s. \tag{6}$$

In order to approximate y_{t_n} based on the later information $[t_n, t_{n+k}]$, we need to adapt it to the filtration (that is already generated, since we are solving it backwards). Therefore, taking the conditional expectation $E_{t_n}^x[\cdot]$ in (6), we have

$$y_{t_n} = E_{t_n}^x\left[y_{t_{n+k}}\right] + \int_{t_n}^{t_{n+k}} E_{t_n}^x\left[f(s, y_s, z_s)\right] ds \tag{7}$$

where the third term of (6) is disappeared as it is an Itô-type integral. In order to approximate the integral part of (7), Zhao [16] considered the Lagrange interpolating method, since the $E_{t_n}^x\left[f(s, y_s, z_s)\right]$ is a deterministic function of s. Given the values of $\left(t_{n+i}, E_{t_n}^x\left[f(t_{n+i}, y_{t_{n+i}}, z_{t_{n+i}})\right]\right)$ and using Lagrange interpolating polynomial, (7) becomes

$$y_{t_n} = E_{t_n}^x\left[y_{t_{n+k}}\right] + k\Delta t \sum_{i=0}^{K_y} b_{K_y,i}^k E_{t_n}^x\left[f(t_{n+i}, y_{t_{n+i}}, z_{t_{n+i}})\right] + R_y^n, \tag{8}$$

where $b_{K_y,i}^k$ are the coefficients derived from the integration of Lagrange interpolating polynomial [16] and R_y^n is the error due to the former.

Next, we derive a semi-discretized form for the z_t process. Let $\Delta W_s = W_s - W_{t_n}$ for $s \geq t_n$. Then ΔW_s is a standard Brownian motion with mean 0 and standard deviation $\sqrt{s - t_n}$. Let l and K_z be two positive integers such that $1 \leq l \leq K_z \leq N$. Using l instead of k in (6), multiplying both sides by $\Delta W_{t_{n+l}}$, taking the conditional expectation $E_{t_n}^x[\cdot]$ and using the Itô isometry we obtain

$$0 = E_{t_n}^x\left[y_{t_{n+l}} \Delta W_{t_{n+l}}\right] + \int_{t_n}^{t_{n+l}} E_{t_n}^x\left[f(s, y_s, z_s)\Delta W_s\right] ds - \int_{t_n}^{t_{n+l}} E_{t_n}^x\left[z_s\right] ds. \tag{9}$$

Using again the Lagrange interpolation method to approximate the two integrals in (9), we have

$$\begin{aligned} 0 = E_{t_n}^x\left[y_{t_{n+l}} \Delta W_{t_{n+l}}\right] &+ l\Delta t \sum_{i=0}^{K_z} b_{K_z,i}^l E_{t_n}^x\left[f(t_{n+i}, y_{t_{n+i}}, z_{t_{n+i}})\Delta W_{t_{n+i}}\right] + R_{z1}^n \\ &- l\Delta t \sum_{i=0}^{K_z} b_{K_z,i}^l E_{t_n}^x\left[z_{t_{n+i}}\right] - R_{z2}^n, \end{aligned} \tag{10}$$

where $b_{K_z,i}^l$ are the coefficients derived from the integration of the Lagrange interpolating polynomial and (R_{z1}^n, R_{z2}^n) are the errors for the first and second integrals in (9).

Consider (y^n, z^n) as an approximation of (y_t, z_t), the semi-discrete scheme is defined as follows: Given random variables (y^{N-i}, z^{N-i}), $i = 0, 1, \ldots, K-1$ with $K = \max\{K_y, K_z\}$, find the random variables (y^n, z^n), $n = N-K, \ldots, 0$ such that

$$\begin{aligned} y^n &= E_{t_n}^x\left[y^{n+k}\right] + k\Delta t \sum_{i=0}^{K_y} b_{K_y,i}^k E_{t_n}^x\left[f(t_{n+i}, y^{n+i}, z^{n+i})\right] \\ 0 &= E_{t_n}^x\left[z^{n+l}\right] + \sum_{i=1}^{K_z} b_{K_z,i}^l E_{t_n}^x\left[f(t_{n+i}, y^{n+i}, z^{n+i})\Delta W_{t_{n+i}}\right] - \sum_{i=0}^{K_z} b_{K_z,i}^l E_{t_n}^x\left[z^{n+i}\right]. \end{aligned} \tag{11}$$

Table 1. The coefficients $\{b^{K_y}_{K_y,i}\}^{K_y}_{i=0}$ until $K_y = 3$.

K_y	$b^{K_y}_{K_y,i}$			
	$i=0$	$i=1$	$i=2$	$i=3$
1	$\frac{1}{2}$	$\frac{1}{2}$		
2	$\frac{1}{6}$	$\frac{4}{6}$	$\frac{1}{6}$	
3	$\frac{1}{8}$	$\frac{3}{8}$	$\frac{3}{8}$	$\frac{1}{8}$

Table 2. The coefficients $\{b^{1}_{K_z,i}\}^{K_z}_{i=0}$ for $K_z = 1,2,3$

K_z	$b^{1}_{K_z,i}$			
	$i=0$	$i=1$	$i=2$	$i=3$
1	$\frac{1}{2}$	$\frac{1}{2}$		
2	$\frac{5}{12}$	$\frac{8}{12}$	$-\frac{1}{12}$	
3	$\frac{9}{24}$	$\frac{19}{24}$	$-\frac{5}{24}$	$\frac{1}{24}$

Zhao [16] showed that in order to have a stable semi-discrete scheme, the following should hold:

$$\begin{aligned} k &= K_y, \qquad \text{with} \quad K_y = 1,2,\ldots,7 \quad \text{and} \quad K_y = 9 \\ l &= 1, \quad \text{with} \quad K_z = 1,2,3. \end{aligned} \tag{12}$$

The coefficients are presented in the following Table 1 and 2.

2.2 The Stable Fully Discrete Scheme

Let $\mathbb{R}_{\Delta x}$ denote a partition of the real axis, i.e. $\mathbb{R}_{\Delta x} = \{x_i | x_i \in \mathbb{R}, i \in \mathbb{Z}, x_i < x_{i+1}, \lim_{i\to+\infty} x_i = +\infty, \lim_{i\to-\infty} x_i = -\infty\}$. The fully discrete scheme is defined as (cf. [16]): Given random variables (y_i^{N-l}, z_i^{N-l}), $l = 0,1,\ldots,K-1$ with $K = \max\{K_y, K_z\}$, find the random variables (y_i^n, z_i^n), $n = N-K,\ldots,0$ such that

$$\begin{aligned} y_i^n &= \hat{E}^{x_i}_{t_n}\left[\hat{y}^{n+K_y}\right] + K_y\Delta t \sum_{j=1}^{K_y} b^{K_y}_{K_y,j} \hat{E}^{x_i}_{t_n}\left[f(t_{n+j}, \hat{y}^{n+j}, \hat{z}^{n+j})\right] + K_y\Delta t b^{K_y}_{K_y,0} f(t_n, y_i^n, z_i^n) \\ 0 &= \hat{E}^{x_i}_{t_n}\left[\hat{z}^{n+1}\right] + \sum_{j=1}^{K_z} b^{1}_{K_z,j} \hat{E}^{x_i}_{t_n}\left[f(t_{n+j}, \hat{y}^{n+j}, \hat{z}^{n+j})\Delta W_{t_{n+j}}\right] \\ &\quad - \sum_{j=1}^{K_z} b^{1}_{K_z,j} \hat{E}^{x_i}_{t_n}\left[\hat{z}^{n+j}\right] - b^{1}_{K_z,0} z_i^n, \end{aligned} \tag{13}$$

where (y_i^n, z_i^n) denotes the approximation of $\big(y(t_n,x_i), z(t_n,x_i)\big)$, $\hat{E}^{x_i}_{t_n}[\cdot]$ is the approximation of $E^{x_i}_{t_n}[\cdot]$ and $(\hat{y}^{n+j}, \hat{z}^{n+j})$ are the interpolating values from (y^{n+j}, z^{n+j}) at the space point $x_i + W_{t_{n+j}} - W_{t_n}$. In order to approximate the conditional expectations, the Gauss-Hermite quadrature rule is used, due to the high accuracy that can be achieved only with a few points. Therefore, the conditional expectation can be expressed as

$$\hat{E}_{t_n}^{x_i}\left[\hat{y}^{n+k}\right] = \frac{1}{\sqrt{\pi}} \sum_{j=1}^{L} \omega_j \hat{y}^{n+k}\left(x_i + \sqrt{2k\Delta t}\, a_j\right), \tag{14}$$

where (ω_j, a_j), for $j = 1, \ldots, L$ are the weights and roots of the Hermite polynomial of degree L (see [16]). In the same way, one can express the other conditional expectations in (13).

The error of the method exhibits different behaviours due to the time-space discretization. However, the maximal order of convergence is 3 for both processes. For technical details, we refer to [16]. Due to the high order of convergence, the numerical method can achieve very high accuracy. It can be observed from (13) that the calculations on each point are independent. Therefore parallelization techniques can be easily adapted. In the next Sect. 3, we discuss the algorithmic framework.

3 The Algorithmic Framework

3.1 The Algorithm

According to Sect. 2, the whole process for solving (2) is divided into 3 steps.

1. **Construct the time-space discrete domain.**
   We divide the time period $[0, T]$ into N time steps using $\Delta t = T/N$ and get $N+1$ time layers. Moreover, in order to balance the errors in time and space directions, we adjust the space step size Δx and the time step size Δt such that they satisfy the equality $(\Delta x)^r = (\Delta t)^{q+1}$, where $q = \min(K_y + 1, K_z)$ and r denotes the global error from the interpolation method used to generate the non-grid points when calculating the conditional expectations.
2. **Calculate K initial solutions with $K = \max\{K_y, K_z\}$.**
   Since only the terminal value is given, one needs to generate the other $K-1$ values. This can be done by running a 1-step scheme for $[t_{N-K+1}, t_{N-1}]$ with a higher number of time points such that the $K-1$ produced initial values will have neglectable error.
3. **Calculate the numerical solution (y_0, z_0) backward using equation (13).**
   Note that the calculation for the y_t process is done implicitly by Picard iteration.

3.2 Preliminary Considerations

In the numerical experiments, we have considered the following points:

- The space domain needs to be truncated. Since the space domain represent the Brownian motion discretization, in our test we use $[-16, 16]$.
- When generating the non-grid points for the calculation of conditional expectations, some will be outside of the domain. For such points, the value on the boundaries is considered, as the desired solution will not be affected.

- Due to uniformity of the grid, one does not need to consider $2K$ (K for y_t and K for z_t) interpolations for each new calculation, but only 2. This is due to the following: Suppose we are at time layer t_{n-K}. To calculate y_t and z_t values on this time layer, one needs the calculation of conditional expectations for K time layers. The cubic spline interpolation is used to find the non-grid values, and the necessary linear systems are solved. For instance, the coefficients for y_t process are $A_y \in \mathbb{R}^{K \times M}$. All the spline coefficients are stored. When we are at time layer t_{n-K-1}, only the spline interpolation corresponding to the previous calculated values is considered. Then, the columns of matrix A_y are shifted $+1$ to the right in order to delete the last column and enter the current calculated coefficients in the first column. The new A_y is used for the current step. The same procedure is followed until t_0. This reduces the amount of work for the algorithm.
- There is a very important benefit from the uniformity of the grid. When we need to find the position of the non-grid point, a naïve search algorithm is to loop over the grid points. In the worst case, a O(M) work is needed. However, this can be done in O(1), i.e. the for loop is removed. Recall that each new point is generated as $X_j = x_i + \sqrt{2\Delta t k} a_j$. This means that taking $\text{int}((X_j - x_{min})/\Delta x)$ gives the left boundary of the grid interval that X_j belongs to. This reduces the total computation time substantially, as it will be demonstrated in the numerical experiments.

3.3 The Parallel Implementation

In this Section we present the naive parallelization of the multistep scheme. Nevertheless, we have kept into attention the optimal CUDA execution model, i.e. creating arrays such that the access will be aligned and coalesced, reducing the redundant access to global memory, using registers when needed etc.

The first and second steps of the algorithm are implemented in the host. The third step is fully implemented in the device. Recall from (13) that the following steps are needed to calculate the approximated values on each time layer backward:

- **Generation of non-grid points** $X_j = x_i + \sqrt{2\Delta t k} a_j$.
  In the uniform domain, the non-grid points need to be generated only once. To do this, a kernel is created where each thread generates L points.
- **Calculation of the values** $\hat{y}$ and $\hat{z}$ **at the non-grid points.**
  This is the most time consuming part of the algorithm, since it involves the solution of two linear systems (see third point in Subsect. 3.2) arising from the spline interpolation.
  We used the BiCGSTAB iterative method since the matrix is tridiagonal. To apply the method, we considered the cuBLAS and cuSPARSE libraries. For the inner product, second norm and addition of vectors, we use the cuBLAS library.
  For the matrix vector multiplication, we use the cuSPARSE library with the compressed sparse row format, due to the structure of the system matrix. Moreover, we created a kernel to calculate the spline coefficients based on the solved systems under the spline interpolation idea.
  Finally, a kernel to apply the last point in Subsect. 3.2 was created to find the values at non-grid points. Note that each thread is assigned to find the values.

- **Calculation of the conditional expectations.**
For the first conditional expectations in the right hand side of (13), we created one kernel, where each thread calculates one value by using (14). Moreover, we merged the calculation of three conditional expectation in one kernel, namely

$$\hat{E}_{t_n}^{x_i}\left[\hat{z}^{n+j}\right],\quad \hat{E}_{t_n}^{x_i}\left[f(t_{n+j},\hat{y}^{n+j},\hat{z}^{n+j})\right],\quad \hat{E}_{t_n}^{x_i}\left[f(t_{n+j},\hat{y}^{n+j},\hat{z}^{n+j})\Delta W_{t_{n+j}}\right],$$

for $j = 1,2,\ldots,K$. This reduces the accessing of data multiple times from the global memory. Note that one thread calculates three values as in (14).
- **Calculation of the z_t values.**
The second equation in (13) is used and each thread calculates one value.
- **Calculation of the y_t values.**
The first equation in (13) is used and each thread calculates one value, using the Picard iterative process.

4 Numerical Results

We implement the parallel algorithm using CUDA C programming. The parallel computing times are compared with the serial ones on a CPU. Furthermore, the speedups are calculated. The CPU is Intel(R) Core(TM) i5-4670 3.40 Ghz with 4 cores. The GPU is a NVIDIA GeForce 1070 Ti with a total 8 GB GDDR5 memory.

In the following we consider an option pricing example, the Black-Scholes model. Consider a security market that contains one bond with price p_t and one stock with price S_t. Therefore, their dynamics are described by:

$$\begin{cases} dp_t &= r_t p_t\,dt, \quad t \geq 0, \\ p_0 &= p, \end{cases} \tag{15}$$

$$\begin{cases} dS_t &= \mu_t S_t dt + \sigma_t S_t\,dW_t, \quad t \geq 0, \\ S_0 &= x, \end{cases} \tag{16}$$

where r_t denotes the interest rate of the bond, p is its current value, μ_t is the expected return on the stock S_t, σ_t is the volatility of the stock, x is its current value and W_t denotes the Brownian motion.

An European Option is a contract that gives the owner the right, but not the obligation, to buy or sell the underlying security at a specific price, known as the strike price K, on the option's expiration date T. A European call option gives the owner the right to purchase the underlying security, while a European put option gives the owner the right to sell the underlying security. Let us take the European call option as an example. The decision of the holder will depend on the stock price at maturity T. If the value of the stock $S_T < K$, then the holder would discard the option; whereas if $S_T > K$, the holder would use the option and make a profit of $S_T - K$. Therefore, the payoff of a call option is $(S_T - K)^+$ and for a put option $(K - S_T)^+$, where $(f)^+ = \max(0, f)$. The option pricing problem of the writer (or seller) is to determine a premium for this contract at present time t_0. Note that the payoff function is an $\{\mathcal{F}_T\}$-measurable random variable.

Suppose that an agent sells the option at price y_t and then invests it in the market. Denote his wealth on each time by y_t. Assume that at each time the agent invests a portion of his wealth in an amount given by π_t into the stock, and the rest $(y_t - \pi_t)$ into the bond. Now the agent has a portfolio based on the stock and the bond. Considering a stock that pays a dividend $\delta(t, S_t)$, the dynamics of the wealth process y_t are described by

$$\begin{aligned} dy_t &= \frac{\pi_t}{S_t} dS_t + \frac{y_t - \pi_t}{p_t} dp_t + \pi_t \delta(t, S_t)\, dt \\ &= \frac{\pi_t}{S_t}(\mu_t S_t\, dt + \sigma_t S_t\, dW_t) + \frac{y_t - \pi_t}{p_t}(r_t p_t\, dt) + \pi_t \delta(t, S_t)\, dt \\ &= \big(r_t y_t + \pi_t(\mu_t - r_t + \delta(t, S_t))\big) dt + \pi_t \sigma_t\, dW_t. \end{aligned} \tag{17}$$

Let $z_t = \pi_t \sigma_t$, then

$$-dy_t = -\Big(r_t y_t + \big(\mu_t - r_t + \delta(t, S_t)\big)\frac{z_t}{\sigma_t}\Big)\, dt + z_t\, dW_t. \tag{18}$$

For a call option, one needs to solve a Forward Backward Stochastic Differential Equation (FBSDE), where the forward part is given from the SDE modelling of the stock price dynamics.

Example 1. Let us consider the Black-Scholes FBSDE

$$\begin{cases} dS_t &= \mu_t S_t dt + \sigma_t S_t\, dW_t, \quad S_0 = x, \quad t \in [0, T] \\ -dy_t &= -\Big(r_t y_t + \big(\mu_t - r_t + \delta(t, S_t)\big)\frac{z_t}{\sigma_t}\Big)\, dt + z_t\, dW_t, \quad t \in [0, T) \\ y_T &= (S_T - K)^+. \end{cases} \tag{19}$$

For constant parameters (i.e. $r_t = r$, $\mu_t = \mu$, $\sigma_t = \sigma$, $\delta_t = \delta$), the analytic solution is

$$\begin{cases} y_t &= V(t, S_t) = S_t \exp\big(-\delta(T-t)\big)N(d_1) - K\exp\big(-r(T-t)\big)N(d_2), \\ z_t &= \frac{\partial V}{\partial S}\sigma = S_t \exp\big(-\delta(T-t)\big)N(d_1)\sigma, \\ d_{1/2} &= \frac{\ln\left(\frac{S_t}{K}\right) + \left(r \pm \frac{\sigma^2}{2}\right)(T-t)}{\sigma\sqrt{T-t}}, \end{cases} \tag{20}$$

where $N(\cdot)$ is the cumulative standard normal distribution function. In this example, we consider $T = 0.33$, $K = S_0 = 100$, $r = 0.03$, $\mu = 0.05$, $\delta = 0.04$, $\sigma = 0.2$, with the solution at (t_0, S_0) being $(y_0, z_0) \doteq (4.3671, 10.0950)$.

Note that the terminal condition has a non-smooth problem for the z_t process. Therefore, for discrete points near the strike price (also called at the money region), the initial value for the z_t process will cause large errors on the next time layers. To overcome this non-smoothness problem, we considered smoothing the initial conditions, cf. the approach of Hendricks [8]. For the forward part of (20), we have the analytic solution

$$S_t = S_0 \exp\Big(\big(\mu - \frac{\sigma^2}{2}\big)t + \sigma W_t\Big). \tag{21}$$

Discretizing (21), the exponential term will lead to a non-uniform grid. Therefore, instead of working in the stock price domain, we work in the log stock price domain. If we denote $X_t = \ln S_t$, then the analytic solution of X_t reads

$$X_t = X_0 + (\mu - \frac{\sigma^2}{2})t + \sigma W_t. \tag{22}$$

The backward part is the same as the (19). In Table 3 we show the importance of using the log stock price. Note that the speedup is relative to the serial case using a for loop.

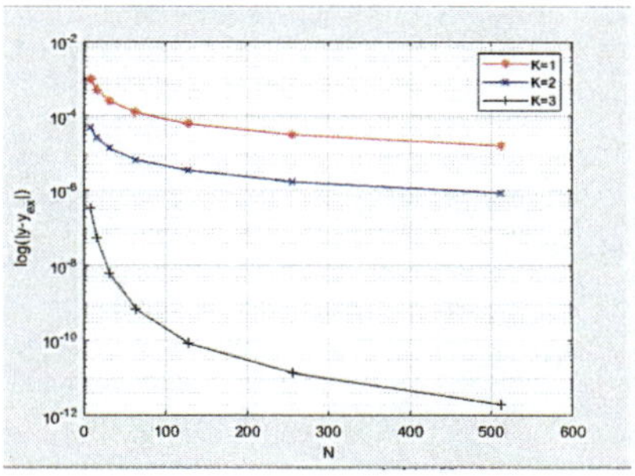

(a) Error for the y process.

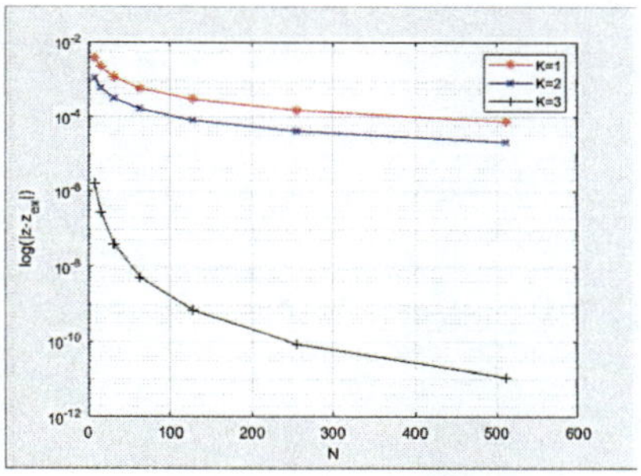

(b) Error for the z process.

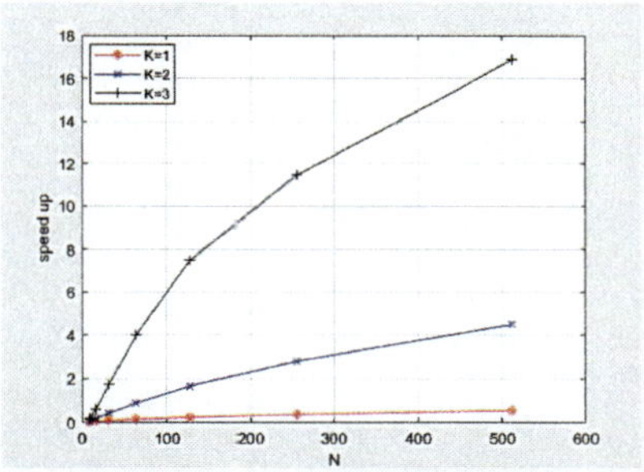

(c) Speedup

N	K	M	CPU time (sec)	GPU time (sec)	Speed up
8	1	158	0.01	0.35	0.02
16	1	224	0.02	0.39	0.05
32	1	316	0.05	0.60	0.09
64	1	446	0.15	1.01	0.15
128	1	632	0.43	1.81	0.24
256	1	892	1.23	3.40	0.36
512	1	1262	3.64	6.70	0.54
8	2	350	0.02	0.30	0.07
16	2	588	0.07	0.41	0.18
32	2	990	0.26	0.61	0.43
64	2	1664	0.95	1.04	0.91
128	2	2798	3.27	1.95	1.68
256	2	4704	10.76	3.82	2.81
512	2	7912	36.31	8.07	4.50
8	3	776	0.05	0.30	0.17
16	3	1552	0.25	0.41	0.62
32	3	3104	1.12	0.64	1.76
64	3	6208	4.62	1.16	4.00
128	3	12414	19.15	2.55	7.50
256	3	24826	76.94	6.72	11.44
512	3	49650	307.09	18.20	16.87

(d) Speedup results in table format.

Fig. 1. Results of naive parallelization for the Black-Scholes example.

The naive results using 256 threads per block are presented in Fig. 1 (note that $K_y = K_z = K$). It can be easily observed that the higher accuracy can be achieved when considering a 3-step scheme. Since we have more time layers to consider, more work can be assigned to the GPU and therefore increasing the speedup of the application. The highest speedup that we obtained for the Black-Scholes example is $17\times$.

Table 3. Comparison due to uniformity of the domain under log stock price transformation for the Black-Scholes model for $N = 256$, $K = K_y = K_z = 3$ and $M = 24826$.

Type	Time	Speedup
Serial (with for)	36825.27	
Serial (without for)	76.94	478.62
Parallel (with for)	237.57	155.01
Parallel (without for)	6.72	5476.19

```
==22320== Profiling result:
            Type  Time(%)      Time     Calls       Avg       Min       Max  Name
 GPU activities:   58.95%  7.24543s     85124  85.116us  15.520us  207.81us  void nrm2_kernel<double, dou
                    7.88%  968.66ms      1527  634.36us  595.40us  716.39us  sp_inter_non_grid_d_no_for(d
                    7.39%  908.40ms      1527  594.89us  520.64us  763.43us  gen_non_grid(double*, double
                    7.12%  875.58ms     42562  20.571us  19.040us  237.67us  void csrMv_kernel<double, dc
                    5.45%  670.20ms      1527  438.90us  388.77us  494.69us  calc_f_and_c_exp_d(double*,
                    3.55%  436.25ms     85124  5.1240us  3.5840us  8.8000us  void dot_kernel<double, doub
                    1.62%  199.67ms     42562  4.6910us  3.4560us  8.0000us  void axpy_kernel_val<double,
                    1.62%  199.27ms     85124  2.3400us  1.8880us  7.9040us  void reduce_1Block_kernel<dc
                    1.29%  159.04ms       509  312.45us  308.07us  333.60us  calc_c_exp_d(double*, double
                    1.19%  146.80ms     21569  6.8060us  4.4160us  8.7360us  step_3(double*, double*, dou
                    1.03%  126.59ms     42571  2.9730us     544ns  34.647ms  [CUDA memcpy HtoD]
                    0.68%  83.450ms    127687     653ns     608ns  7.6160us  [CUDA memcpy DtoH]
                    0.67%  82.697ms     21569  3.8340us  3.2960us  7.7440us  step_8(double*, double*, dou
                    0.42%  51.357ms     12192  4.2120us  3.4560us  6.2720us  copy_d(double*, double*, int
                    0.39%  48.323ms       509  94.936us  91.937us  103.94us  calc_y(double*, double*, dou
                    0.37%  45.056ms     20993  2.1460us  1.8880us  5.3760us  step_13(double*, double*, dc
                    0.12%  15.323ms      6144  2.4930us  1.5360us  5.8880us  void copy_kernel<double, int
                    0.12%  14.548ms       512  28.414us  26.560us  35.040us  spline_coeff(double*, double
                    0.07%  8.9152ms       509  17.515us  16.512us  19.680us  calc_z(double*, double*, dou
ouble*, int, double)
                    0.00%  15.168us         1  15.168us  15.168us  15.168us  CSR(double*, int*, int*, int
                    0.00%  4.0640us         1  4.0640us  4.0640us  4.0640us  csinterp_d(double*, double*,
                    0.00%     736ns         1     736ns     736ns     736ns  [CUDA memset]
```

(a) Performance of naive kernels

```
==22383== Profiling result:
            Type  Time(%)      Time     Calls       Avg       Min       Max  Name
 GPU activities:   18.57%  960.11ms      1527  628.75us  588.39us  899.21us  sp_inter_non_grid_d_no_for(
                   17.69%  914.31ms      1527  598.76us  521.16us  765.06us  gen_non_grid(double*, doubl
                   16.82%  869.32ms     42562  20.424us  18.880us  23.584us  void csrMv_kernel<double, d
                   12.76%  659.30ms      1527  431.76us  387.52us  487.91us  calc_f_and_c_exp_d(double*,
                   11.24%  580.94ms    127686  4.5490us  3.4240us  9.1200us  void dot_kernel<double, dou
                    5.27%  272.61ms    127686  2.1340us  1.8560us  114.79us  void reduce_1Block_kernel<d
                    3.42%  176.80ms     42562  4.1530us  2.9440us  7.7760us  void axpy_kernel_val<double
                    3.08%  159.16ms       509  312.69us  308.83us  334.11us  calc_c_exp_d(double*, doubl
                    2.72%  140.38ms     21569  6.5080us  4.4480us  8.0320us  step_3(double*, double*, do
                    1.86%  95.952ms         9  10.661ms     544ns  34.389ms  [CUDA memcpy HtoD]
                    1.53%  79.093ms    127687     619ns     576ns  7.5840us  [CUDA memcpy DtoH]
                    1.50%  77.376ms     21569  3.5870us  3.1360us  7.4240us  step_8(double*, double*, do
                    0.97%  50.224ms     12192  4.1190us  3.3600us  7.0080us  copy_d(double*, double*, in
                    0.93%  47.838ms       509  93.984us  87.809us  112.77us  calc_y(double*, double*, do
                    0.81%  41.774ms     20993  1.9890us  1.7920us  7.2960us  step_13(double*, double*, d
                    0.29%  15.097ms      6144  2.4570us  1.5040us  7.1040us  void copy_kernel<double, in
                    0.28%  14.292ms       512  27.914us  25.856us  35.840us  spline_coeff(double*, doubl
                    0.17%  8.7412ms       509  17.173us  16.352us  19.136us  calc_z(double*, double*, do
                    0.11%  5.5165ms      1024  5.3870us  4.6720us  6.1760us  rhs(double*, double*, int,
                    0.00%  14.048us         1  14.048us  14.048us  14.048us  CSR(double*, int*, int*, in
                    0.00%  3.9040us         1  3.9040us  3.9040us  3.9040us  csinterp_d(double*, double*
                    0.00%     736ns         1     736ns     736ns     736ns  [CUDA memset]
```

(b) Performance after first optimization iteration

```
==22446== Profiling result:
            Type  Time(%)      Time     Calls       Avg       Min       Max  Name
 GPU activities:   24.09%  910.16ms      1527  596.05us  552.26us  777.51us  sp_inter_non_grid_d_no_for(d
                   22.59%  853.73ms     42562  20.058us  18.624us  231.43us  void csrMv_kernel<double, dc
                   16.13%  609.41ms    127686  4.7720us  3.5840us  9.0240us  void dot_kernel<double, doub
                    7.92%  299.37ms    127686  2.3440us  1.8560us  8.0320us  void reduce_1Block_kernel<dc
                    5.90%  222.81ms      1527  145.91us  133.38us  167.65us  calc_f_and_c_exp_d(double*,
                    4.44%  167.65ms     42562  3.9380us  2.9440us  7.9680us  void axpy_kernel_val<double,
                    3.74%  141.18ms     21569  6.5450us  4.5120us  7.6800us  step_3(double*, double*, dou
                    2.75%  103.89ms      1527  68.035us  61.664us  92.384us  gen_non_grid(double*, double
                    2.55%  96.399ms         9  10.711ms     544ns  34.512ms  [CUDA memcpy HtoD]
                    2.10%  79.491ms    127687     622ns     576ns  7.5520us  [CUDA memcpy DtoH]
                    2.06%  77.984ms     21569  3.6150us  3.1680us  6.9120us  step_8(double*, double*, dou
                    1.29%  48.910ms     12192  4.0110us  2.7200us  7.2320us  copy_d(double*, double*, int
                    1.27%  47.908ms       509  94.122us  87.840us  108.10us  calc_y(double*, double*, dou
                    1.10%  41.495ms     20993  1.9760us  1.7600us  6.8160us  step_13(double*, double*, dc
                    0.90%  34.072ms       509  66.938us  65.377us  73.441us  calc_c_exp_d(double*, double
                    0.40%  15.041ms      6144  2.4480us  1.5040us  7.8400us  void copy_kernel<double, int
                    0.38%  14.304ms       512  27.937us  26.016us  35.040us  spline_coeff(double*, double
                    0.24%  8.9084ms       509  17.501us  16.768us  20.032us  calc_z(double*, double*, dou
                    0.16%  6.0294ms      1024  5.8880us  4.6400us  6.7840us  rhs(double*, double*, int, d
                    0.00%  13.856us         1  13.856us  13.856us  13.856us  CSR(double*, int*, int*, int
                    0.00%  4.0000us         1  4.0000us  4.0000us  4.0000us  csinterp_d(double*, double*,
                    0.00%     704ns         1     704ns     704ns     704ns  [CUDA memset]
```

(c) Performance after second optimization iteration

Fig. 2. Results of iterative parallelization for the Black-Scholes example.

Furthermore, we optimize the kernels created for the Black-Scholes BSDE for $N = 512$, $K_y = 3$ and $K_z = 3$. For this, we used the NVIDIA profiling tools (`nvprof` and `nvvp`) to gather information about performance bottlenecks and apply the proper optimization technique.

After applying `nvprof`, the main bottleneck of the application is the second norm kernel that calculates the errors in the BiCGSTAB algorithm as presented in Fig. 2a. Note that this kernel is already optimized by the NVIDIA developers. However, it is a kernel that serves for a general purpose. Instead of using second norm kernel, we used the dot kernel and later took the square root. This reduced the computation time from 7.2 s to 580.9 ms as presented in Fig. 2b. We consider it as the first iteration of the optimization process.

Moreover, we applied again `nvprof` and found that the next performance inhibitor is the kernel which calculates the non-grid values and another kernel that generates the non-grid values. This is due to the inefficient memory accesses. To overcome this problem, we considered loop interchanging and loop unrolling. As presented in Fig. 2c, the performance of above kernels is improved, from 960.1 ms to 910.1 ms and 914.3 ms to 103.9 ms respectively. Finally, we changed the thread configuration to 128 threads per block in order to increase parallelizm and we were able to achieve a 51× speedup. We present the speedups for each iteration of the optimization process in Table 4.

Table 4. Speedups of Black-Scholes model for $N = 256$, $K = K_y = K_z = 3$ and $M = 24826$.

Type	Time	Speedup
Serial	307.09	
Naive	18.2	16.87
First iteration	7.45	41.22
Second iteration	6.02	51.01

5 Conclusions and Outlook

In this work we parallelized the multistep method developed in [16] for the numerical approximation of BSDEs on GPU.

Firstly, we presented an optimal operation to find the location of the interpolated values. This was essential for the reduction of the computational time. The numerical results exhibited a high accuracy in very small computation times. Moreover, we optimized the application after finding the performance bottlenecks and applying optimization techniques. Using the *cuBLAS* kernel to calculate the error of the BiCGSTAB iterative method, loop interchange and loop unrolling provided us a 51× speedup for the Black-Scholes example.

Based on our results, the GPU architecture for the multistep scheme is well suited for the acceleration of BSDEs. For future work we will focus on parallelizing d-dimensional problems using more time steps as in [15] for a higher accuracy with financial applications such as multi-asset option pricing.

Acknowledgements. The authors were partially supported by the bilateral German-Portuguese Project *FRACTAL – FRActional models and CompuTationAL Finance*, the bilateral German-Hungarian Project *CSITI – Coupled Systems and Innovative Time Integrators* and the bilateral German-Slovakian Project *ENANEFA – Efficient Numerical Approximation of Nonlinear Equations in Financial Applications* all financed by the DAAD.

References

1. Bender, C., Steiner, J.: Least-squares Monte-Carlo for backward SDEs. In: Numerical methods in finance, pp. 257–289. Springer, Heidelberg (2012)
2. Bouchard, B., Touzi, N.: Discrete-time approximation and Monte-Carlo simulation of backward stochastic differential equations. Stochast. Process. Appl. **111**(2), 175–206 (2004)
3. Crisan, D., Manolarakis, K.: Solving backward stochastic differential equations using the cubature method: application to nonlinear pricing. SIAM J. Fin. Math. **3**(1), 534–571 (2012)
4. Dai, B., Peng, Y., Gong, B.: Parallel option pricing with BSDE method on GPU. In: 2010 Ninth International Conference on Grid and Cloud Computing, pp. 191–195. IEEE (2010)
5. El Karoui, N., Peng, S., Quenez, M.C.: Backward stochastic differential equations in finance. Math. Finan. **7**(1), 1–71 (1997)
6. Gobet, E., Lemor, J.P., Warin, X.: A regression-based Monte Carlo method to solve backward stochastic differential equations. Annal. Appl. Probab. **15**(3), 2172–2202 (2005)
7. Gobet, E., López-Salas, J.G., Turkedjiev, P., Vázquez, C.: Stratified regression Monte-Carlo scheme for semilinear PDEs and BSDEs with large scale parallelization on GPUs. SIAM J. Sci. Comput. **38**(6), C652–C677 (2016)
8. Hendricks, C.: High-Order Methods for Parabolic Equations in Multiple Space Dimensions for Option Pricing Problems, Dissertation, University of Wuppertal (2017)
9. Lemor, J.P., Gobet, E., Warin, X.: Rate of convergence of an empirical regression method for solving generalized backward stochastic differential equations. Bernoulli **12**(5), 889–916 (2006)
10. Pardoux, E., Peng, S.: Adapted solution of a backward stochastic differential equation. Syst. Control Lett. **14**(1), 55–61 (1990)
11. Peng, S.: Probabilistic interpretation for systems of quasilinear parabolic partial differential equations. Stochast. Stochast. Rep. **37**(1–2), 61–74 (1991)
12. Peng, Y., Gong, B., Liu, H., Dai, B.: Option pricing on the GPU with backward stochastic differential equation. In: 2011 Fourth International Symposium on Parallel Architectures, Algorithms and Programming, pp. 19–23. IEEE (2011)
13. Peng, Y., Liu, H., Yang, S., Gong, B.: Parallel algorithm for BSDEs based high dimensional American option pricing on the GPU. J. Comp. Inform. Syst. **10**(2), 763–771 (2014)
14. Teng, L.: A Review of tree-based approaches to solve forward-backward stochastic differential equations. arXiv preprint arXiv:1809.00325 (2018)
15. Teng, L., Lapitckii, A., Günther, M.: A Multi-step Scheme based on Cubic Spline for solving Backward Stochastic Differential Equations. arXiv preprint arXiv:1809.00324 (2018)
16. Zhao, W., Zhang, G., Ju, L.: A stable multistep scheme for solving backward stochastic differential equations. SIAM J. Numer. Anal. **48**(4), 1369–1394 (2010)
17. Zhao, W., Chen, L., Peng, S.: A new kind of accurate numerical method for backward stochastic differential equations. SIAM J. Sci. Comput. **28**(4), 1563–1581 (2006)

Two Epidemic Propagation Models and Their Properties

István Faragó(✉) and Fanni Dorner

Eötvös Loránd University Budapest, Pázmány Péter s. 1/C, Budapest 1117, Hungary
faragois@caesar.elte.hu, dornerfanni@gmail.com

Abstract. In order to have an adequate model, the continuous and the corresponding numerical models on some fixed mesh should preserve the basic qualitative properties of the original phenomenon. In this paper we focus our attention on some mathematical models of biology, namely, we consider different epidemic models. First we investigate the SIR model, then different models for malaria (Ross models). Special attention is paid to the investigation of the extended models with involving the demography of humans and malaria mosquitoes. We examine their qualitative properties, and prove their invariance wrt the data set. A numerical example demonstrates the theoretical results.

1 Introduction and Motivation

Most of the living populations have to cope with different diseases. Some of these diseases are communicable and can decrease the size of the population dramatically. This is why people are eager to understand the mechanism of epidemics and try to prevent their outbreak and propagation by efficient and affordable measures (e.g., hygiene, vaccination). One of the tools of the investigation of epidemics may be the construction of mathematical models and the analysis of the solutions of these models [1,5]. The mathematical modeling of infectious diseases is a tool which has been used to study the mechanisms by which diseases spread, to predict the future course of an outbreak and to evaluate strategies to control an epidemic.

The original epidemic propagation phenomenon has a number of essential characteristic properties that must be reflected not only by biological but also by mathematical models. Since the unknown functions describe the number or the density of the different individuals, obviously a reliable model could have only a non-negative solution. Therefore, one of the main requirements is the non-negativity property, which means that if the initial data are non-negative, then the solution of the models is also non-negative. In this paper we focus our attention to the analysis of this question for different mathematical models.

The paper is organised as follows. In Sect. 2 we introduce several mathematical models for direct and the indirect contacts. In Sect. 3 we investigate the qualitative properties of SIR models in detail. Section 4 investigates the malaria propagation models. Finally, in Sect. 5 we demonstrate the theoretical results on a numerical example.

I. Dimov and S. Fidanova (Eds.): HPC 2019, SCI 902, pp. 209–220, 2021.
https://doi.org/10.1007/978-3-030-55347-0_18

2 Mathematical Models of Epidemic Propagation

In 1927, Kermack and McKendrick [4] created an epidemic model (also known as SIR model) for the following three compartments:

- ***(S)*** susceptibles - who have yet to contract the disease and become infectious,
- ***(I)*** infectives - who can pass on the disease to others,
- ***(R)*** removed - who have been infected but cannot transmit the disease for some reason.

The model has the form of a system of ordinary differential equations

$$\begin{aligned}\frac{dS}{dt} &= -aS(t)I(t)\\ \frac{dI}{dt} &= aS(t)I(t) - bI(t)\\ \frac{dR}{dt} &= bI(t),\end{aligned} \tag{1}$$

where $S(t)$, $I(t)$ and $R(t)$ denote the number of susceptible, infective and recovered individuals as a function of time t. Contact rate a and recovery rate b are given positive parameters. It is easy to see that there is no vital dynamics in this model, and the movement between the three groups is one-way, which means that the recovered individuals do not become susceptible again. This model describes epidemic propagation of the direct type, i.e., when the disease spreads through direct physical connection.

The other one is the indirect type of epidemic propagation, i.e., when the disease spreads indirectly through an intermediary (so called vector). This model is usually called as host-vector-host model. For this type of disease the most typical example is malaria.

This kind of disease is caused by infection with single-celled (protozoan) parasites of genus Plasmodium and is characterized by paroxysms of chills, fever, headache, pain and vomiting. The parasites are transmitted to humans through the bites of infected female Anopheles mosquitoes (vectors) [5].

Malaria is widely spread in tropical and subtropical regions, including Africa, Asia, Latin America, the middle East and some parts of Europe. The most cases and deaths occur in sub-Saharan Africa. In particular, thirty countries in sub-Saharan Africa account for 90 % of global malaria deaths. Even though the disease has been investigated for hundreds of years, it still remains a major public health problem with 109 countries declared as endemic to the disease in 2008. There were 243 million malaria cases reported, and nearly a million deaths - primarily of children under 5 years [10,11]. Hence, malaria is responsible for the fifth greatest number of deaths due to infectious diseases and is the second leading cause of death in Africa behind HIV/ AIDS.

The so-called Ross-MacDonald model (see, [6,8,9]) is the earliest attempt to quantitatively describe the dynamics of malaria transmission at a population level. Ross introduced the deterministic differential equation model of the malaria by dividing the human population into susceptible (S_h) and infected (I_h) compartments, with the infected class returning to the susceptible class again, i.e., according to the SIS model. The mosquito population also has two compartments (S_m and I_m), and they are connected similarly to the human population. Assuming that the sizes of both populations are constant (N_h and N_m, respectively), the most typical model is given by the following system of ordinary differential equations:

$$\frac{dI_h}{dt} = qa(N_h - I_h)I_m(t) - \alpha I_h(t),$$
$$\frac{dI_m}{dt} = pa(N_m - I_m)I_h(t) - \mu I_m(t). \tag{2}$$

Here $I_h(t)$ and $I_m(t)$ represent the sizes of the populations in the infectious classes of human and female mosquitoes, respectively, at time t. Parameter a yields the number of bites per unit time, q and p are the probability of transmission of diseases from mosquitoes to humans and vice versa. The parameter α is the average recovery rate of humans, and μ is the death rate of the vectors. The model ignores the latencies within both human hosts and mosquitoes, and assumes no immunity of the recovered individuals. By analyzing this mathematical model, both Ross and McDonald found that it would be possible to eradicate the disease without killing all vector mosquitoes. This was in contrast to the traditional belief that malaria could be wiped out only by eradicating all vector mosquitoes, which would be impossible in practice.

We note that some other form of the Ross model, defined for the proportion of the human population infected $\mathbf{I_h}(t) = I_h(t)/N_h$ and for the proportion of the female mosquito population infected $\mathbf{I_m}(t) = I_m(t)/N_m$, can be written as follows

$$\frac{d\mathbf{I_h}}{dt}(t) = abm(1 - \mathbf{I_h}(t))\mathbf{I_m}(t) - \alpha\mathbf{I_h}(t),$$
$$\frac{d\mathbf{I_m}}{dt}(t) = ac(1 - \mathbf{I_m}(t))\mathbf{I_h}(t) - \mu_2\mathbf{I_m}(t). \tag{3}$$

Here b is the proportion of bites that produce infection in humans, c is the proportion of bites by which one susceptible mosquito becomes infected, m is the ratio of N_h and N_m, and μ_2 denotes the per capita rate of mosquito mortality.

However, the Ross model (3) has some shortcomings, too.

- Due to the constant population, there is no vital dynamics, i.e., birth and mortality are not taken into account. (At least, they are assumed to be equal.)
- The transition from the susceptible state to the infected state (from $S(t)$ to $I(t)$) is not direct. There is a transition state, called exposed state (in notation: $E(t)$) when the individuals who are infected are not able to pass on the infection to others.
- In the Ross model there is no recovery state ($R(t)$) for the humans, since it is based on the SIS model.

In the following we construct another model for the malaria propagation that eliminates the above-mentioned shortcomings, and in the sequel we call it extended model of malaria (cf. [7]).

In general, we divide the human population into four groups: susceptible humans, exposed humans, infectious humans and recovered humans. Unlike human population, the mosquito population is divided into three subclasses. (There is no recovered class due to their short life-cycle.) The model for the human is based on the SEIRS process, while for the mosquitoes on the SEI model, and is defined as follows.

$$
\begin{aligned}
\frac{dS_h}{dt} &= \Lambda_h - \frac{b\beta_h S_h(t) I_m(t)}{1+\nu_h I_m(t)} - \mu_h S_h(t) + \omega R_h(t)\\
\frac{dE_h}{dt} &= \frac{b\beta_h S_h(t) I_m(t)}{1+\nu_h I_m(t)} - (\alpha_h + \mu_h) E_h(t)\\
\frac{dI_h}{dt} &= \alpha_h E_h(t) - (r + \mu_h + \delta_h) I_h(t)\\
\frac{dR_h}{dt} &= r I_h(t) - (\mu_h + \omega) R_h(t)\\
\frac{dS_m}{dt} &= \Lambda_m - \frac{b\beta_m S_m(t) I_h(t)}{1+\nu_m I_h(t)} - \mu_m S_m(t)\\
\frac{dE_m}{dt} &= \frac{b\beta_m S_m(t) I_h(t)}{1+\nu_m I_h(t)} - (\alpha_m + \mu_m) E_m(t)\\
\frac{dI_m}{dt} &= \alpha_m E_m(t) - (\mu_m + \delta_m) I_m(t).
\end{aligned}
\tag{4}
$$

Adding the usual initial conditions

$$
\begin{aligned}
S_h(0) = S_{0h}, I_h(0) = I_{0h}, E_h(0) = E_{0h}, R_h(0) = R_{0h},\\
S_m(0) = S_{0m}, E_m(0) = E_{0m}, I_m(0) = I_{0m},
\end{aligned}
\tag{5}
$$

we arrive at the Cauchy problem for a seven-dimensional system of ordinary differential equations. The parameters of the model (4)–(5) are given in Table 1. In the sequel this model will be called extended Ross model.

In the following we examine the qualitative properties of the models discussed above.

3 Qualitative Analysis of the SIR Model

First we analyze the SIR model given by (1) and prove its basic property. The first equation can be written as

$$\frac{S'(t)}{S(t)} = -aI(t).$$

Table 1. Parameters in the extended model (4)

$S_h(t)$	Number of susceptible humans at time t
$E_h(t)$	Number of humans exposed to malaria infection at time t
$I_h(t)$	Number of infected humans at time t
$R_h(t)$	Number of recovered humans at time t
$S_m(t)$	Number of susceptible mosquitoes at time t
$E_m(t)$	Number of exposed mosquitoes at time t
$I_m(t)$	Number of infectious mosquitoes at time t
Λ_h	Birth number of humans
Λ_m	Birth number of mosquitoes
b	Biting rate of the mosquito
β_h	Probability that a bite by an infectious mosquito results in transmission of disease to human
β_m	Probability that a bite results in transmission of parasite to a susceptible mosquito
μ_h	Per capita rate of human mortality
μ_m	Per capita rate of mosquito mortality
δ_h	Disease induced death rate of human
δ_m	Disease induced death rate of mosquito
α_h	Per capital rate of progression of humans from the exposed state to the infectious state
α_m	Per capital rate of progression of mosquitoes from the exposed state to the infectious state
r	Recovery rate of humans
ω	Per capital rate of loss of immunity in humans
ν_h	Proportion of an antibody produced by human in response to the incidence of infection caused by mosquito
ν_m	Proportion of an antibody produced by mosquito in response to the incidence of infection caused by human

Hence its solution can be defined directly and it has the form

$$S(t) = S(0) \exp\left(-a \int_0^t I(s) ds\right). \tag{6}$$

Similarly, from the second equation in (1) we obtain

$$\frac{I'(t)}{I(t)} = aS(t) - b,$$

and hence

$$I(t) = I(0) \exp\left(\int_0^t (aS(s) - b) ds\right). \tag{7}$$

Finally, from the third equation in (1) we obtain the solution in the form

$$R(t) = R(0) + b \int_0^t I(s)ds. \tag{8}$$

Using the formulas (6)–(8) we have proved the following statement.

Theorem 1. *With non-negative initial conditions the Cauchy problem for the ODE system (1) has a non-negative solution.*

However, the above formulas mean the following, too. Since $S(t)$ and $I(t)$ are non-negative, the first and the third equations in (1) imply the relations $S'(t) \leq 0$ and $R'(t) \geq 0$. Therefore we have

Theorem 2. *With non-negative initial conditions the Cauchy problem for the ODE system (1) the solution $S(t)$ is a monotonically decreasing, while $R(t)$ is a monotonically increasing function.*

Remark 1. Since the sum of the equations in (1) results in $S'(t)+I'(t)+R'(t) = 0$, therefore $V(t) := S(t) + I(t) + R(t) = N = constant$ (where $constant = S(0) + I(0) + R(0) = V(0)$) holds, which is called mass-preservation property.

Since $S(t)$ is a monotonically decreasing function and it is bounded from below (by zero), therefore it has a limit as t tends to infinity, i.e., there exists $\lim_{t\to\infty} S(t) = S^\star$. Analogically, $R(t)$ is a monotonically increasing function and it is bounded from above (by N), and hence there exists $\lim_{t\to\infty} R(t) = R^\star$. Without loss of generality we may assume that at the initial time $t = 0$ there is no recovered individual, i.e., $R(0) = 0$, and hence $N = S(0) + I(0) = S^\star + R^\star$. By using (8) and the definition of $R^\star$, we have

$$R^\star = \lim_{t\to\infty} R(t) = b \int_0^\infty I(s)ds,$$

which means that $\lim_{t\to\infty} I(t) = 0$. Moreover, $\int_0^\infty I(s)ds = R^\star/b$.

On the other hand, by using (6) and the definition of $S^\star$, we have

$$S^\star = \lim_{t\to\infty} S(t) = S(0)\exp\left(-a\int_0^\infty I(s)ds\right) = S(0)\exp\left(-R^\star a/b\right). \tag{9}$$

Since $R^\star = N - S^\star = S(0) + I(0) - S^\star$, (9) can be written as

$$S^\star = S(0)\exp\left(-a(S(0) + I(0) - S^\star)/b\right). \tag{10}$$

From (10) it follows that

$$S_{rel} = \exp\left(-\kappa(1 + I(0)/S(0) - S_{rel})\right), \tag{11}$$

where $\kappa = aS(0)/b$ and $S_{rel} = S^\star/S(0)$. The dimensionless parameter κ is called Kermack–McKendrick parameter [3]. We note that the relation (11) yields an implicit nonlinear equation for the unknown number S_{rel}, which shows the relative change in the number of susceptible individuals.

Remark 2. The Kermack–McKendrick parameter κ has another, better known interpretation. The dynamics of the infectious class depends on the sign of the derivative $I'(t)$. From the second equation in (1) we can see that under the condition $aS(0) - b < 0$ it is negative at the initial state, i.e., the number of the infected decreases. In the opposite case ($aS(0) - b > 0$) the number of the infected increases. We introduce the notation $R_0 = aS(0)/b$ and it is the so-called basic reproduction number (also called basic reproduction ratio). In fact, this ratio is derived as the expected number of new (secondary) infections from a single infection in a population where all subjects are susceptible. One can see that the parameters R_0 and κ are equal. If $R_0 < 1$, then the modeled disease dies out, and if $R_0 > 1$, then the disease spreads in the population.

4 Qualitative Analysis of the Malaria Models

In the sequel we consider the mathematical models of malaria, formulated in Sect. 2, and we analyse their qualitative behavior. Our attention will be focused primarily on the non-negativity preservation property.

First we recall the Ross model in the form (3). Since in this model $\mathbf{I_h}$ and $\mathbf{I_m}$ denote the density of the human and vector populations, clearly they should be in the interval $[0, 1]$ at any time instant. This means that the model should possess the *density preservation property*, which means the following: if $I_h(0), I_m(0) \in [0, 1]$, then $I_h(t), I_m(t) \in [0, 1]$ for all $t > 0$, too.

In [2] the following statement is proved.

Theorem 3. *The Ross model has the density preservation property, which implies that this model has the non-negativity preservation property, too.*

Next we are moving on to the non-negativity analysis of the extended Ross malaria model. We prove the statement which demonstrates such a property of the model.

Theorem 4. *Assume that $\Lambda_h, \Lambda_m > 0$ and all initial values in (5) are positive. Then the solution of the model (4)–(5) is positive for all t.*

Proof. We will prove the above theorem indirectly. Suppose that there exists $t < +\infty$ where our theorem is not true, which means that there exists at least one component which is equal to zero in that point. Let $t^\star$ be the infimum of such values. Then every component is positive on $[0, t^\star)$, and there exists at least one component which is equal to zero in $t^\star$. (This is due to the continuity of the functions.) In the sequel we show that no one of components can have this property.

1. Suppose that S_h has this feature, i.e., $S_h(t^\star) = 0$.
   Then the other components are non-negative on $[0, t^\star]$, that is the functions $E_h(t)$, $I_h(t)$, $R_h(t)$, $S_m(t)$, $E_m(t)$, $I_m(t)$ are non-negative on this interval. Let us consider the first equation in (4) at the point $t = t^\star$. We get

$$S_h'(t^\star) = \Lambda_h - \frac{b\beta_h S_h(t^\star) I_m(t^\star)}{1 + \nu_h I_m(t^\star)} - \mu_h S_h(t^\star) + \omega R_h(t^\star) = \Lambda_h + \omega R_h(t^\star) > 0.$$

This means that $S_h(t)$ is strictly monotonically increasing at the point $t = t^\star$. Therefore $S_h(t) < S_h(t^\star)$ for all $t \in (t^\star - \varepsilon, t^\star)$ with some $\varepsilon > 0$. As $S_h(t^\star) = 0$ we get that $S_h(t) < 0$ on $(t^\star - \varepsilon, t^\star)$, which is a contradiction.

2. Assume that E_h has the above feature, i.e., $E_h(t^\star) = 0$. Multiplying both sides of the second equation in (4) with $e^{(\alpha_h+\mu_h)t}$, we obtain

$$e^{(\alpha_h+\mu_h)t}E_h'(t) + e^{(\alpha_h+\mu_h)t}(\alpha_h + \mu_h)E_h(t) = e^{(\alpha_h+\mu_h)t}\frac{b\beta_h S_h(t)I_m(t)}{1+\nu_h I_m(t)}.$$

Here on the left side stands the derivative of the function $e^{(\alpha_h+\mu_h)t}E_h(t)$. Hence, integrating this equality on the interval $[0, t]$ we get

$$e^{(\alpha_h+\mu_h)t}E_h(t) - E_h(0) = \int_0^t e^{(\alpha_h+\mu_h)s}\frac{b\beta_h S_h(s)I_m(s)}{1+\nu_h I_m(s)}ds.$$

Putting $t = t^\star$ this results in the relation

$$E_h(t^\star) = e^{-(\alpha_h+\mu_h)t^\star}E_h(0) + \int_0^{t^\star} e^{-(\alpha_h+\mu_h)(t^\star-s)}\frac{b\beta_h S_h(s)I_m(s)}{1+\nu_h I_m(s)}ds.$$

Hence $E_h(t^\star) > 0$, which yields a contradiction.

3. Assume now that I_h has the above feature, i.e., $I_h(t^\star) = 0$. Then the third equation in (4) at the point $t = t^\star$ results in the following:

$$I_h'(t^\star) = \alpha E_h(t^\star) - (r + \mu_h + \delta_h)I_h(t^\star) = \alpha E_h(t^\star) > 0,$$

which means that $I_h(t)$ is strictly monotonically increasing in $t^\star$. Hence, as in the first case, for S_h, we get again a contradiction.

4. Similarly we assume the above feature for the function $R_h(t)$ as well. Then

$$R_h'(t^\star) = rI_h(t^\star) - (\mu_h + \omega)R_h(t^\star) = rI_h(t^\star) > 0,$$

i.e., $R_h(t)$ is positive for all $t \in [0, t^\star]$, which contradicts the property $R_h(t^\star) = 0$.

For mosquitoes the proof is carried out analogously.

5 Suppose that $S_m(t)$ has the above feature, i.e., $S_m(t^\star) = 0$, and all the other components are non-negative at the point $t = t^\star$. Then the the fifth equation in (4) at the point $t = t^\star$ results in the relation

$$S_m'(t^\star) = \Lambda_m - \frac{b\beta_m S_m(t^\star)I_h(t^\star)}{1+\nu_m I_h(t^\star)} - \mu_m S_m(t^\star) = \Lambda_m > 0.$$

Hence, as in the case for $S_h(t)$, we got a contradiction.

6 Likewise for $E_m(t)$, by multiplication both sides of the corresponding equation by $e^{(\alpha_m+\mu_m)t}$, and then integrating on $[0, t]$, and substituting $t = t^\star$, we obtain

$$e^{(\alpha_m+\mu_m)t^\star}E_m(t^\star) - E_m(0) = \int_0^{t^\star} e^{(\alpha_m+\mu_m)t^\star}\frac{b\beta_m S_m(s)I_h(s)}{1+\nu_m I_h(s)}ds.$$

Hence

$$E_m(t^\star) = e^{-(\alpha_m+\mu_m)t^\star} E_m(0) + \int_0^{t^\star} e^{-(\alpha_m+\mu_m)(t^\star-s)} \frac{b\beta_m S_m(s) I_h(s)}{1+\nu_m I_h(s)} ds.$$

This means that $E'_m(t^\star) > 0$, which is the required contradiction.

7 Finally, for $I_m(t)$ we have

$$I'_m(t^\star) = \alpha E_m(t^\star) - (\mu_m + \delta_m) I_m(t^\star) = \alpha E_m(t^\star) > 0,$$

therefore $I_m(t^\star) > 0$, which is a contradiction.

These steps prove that there is no such $t^\star$ where any of the components turns into zero. Since at the initial point the components are positive, due to the continuity of the functions every component is positive for all t. □

In the Theorem 4 we have proved the positivity of the solution, i.e., its boundedness from below. In the following we consider the bound from above.

Let $V_h(t)$ denote the total number of the humans, i.e.,

$$V_h(t) = S_h(t) + E_h(t) + I_h(t) + R_h(t). \tag{12}$$

Similarly, for the mosquitoes we define

$$V_m(t) = S_m(t) + E_m(t) + I_m(t). \tag{13}$$

First we formulate the statement for the upper bound to V_h.

Theorem 5. *Suppose that at the initial time $t = 0$ the total human population is bounded by the number Λ_h/μ_h. Then at any time t the solution of the extended Ross model (4)–(5) is also bounded from above by Λ_h/μ_h.*

Proof. By the assumption we have

$$V_h(0) < \frac{\Lambda_h}{\mu_h}. \tag{14}$$

Summing up the equations in (4) and using the definition of $V_h(t)$, we get

$$V'_h(t) = \Lambda_h - \mu_h V_h(t) - \delta_h I_h(t).$$

Hence, rearranging and multiplying both sides by $e^{\mu_h t}$, we obtain

$$V'_h(t) \cdot e^{\mu_h t} + \mu_h e^{\mu_h t} V_h(t) = (\Lambda_h - \delta_h I_h(t)) e^{\mu_h t}.$$

We notice that on the left side stands the derivative of $V_h(t) \cdot e^{\mu_h t}$. Hence, integrating both sides from 0 to t, we get

$$V_h(t) e^{\mu_h t} - V_h(0) = \int_0^t (\Lambda_h - I_h(s)) e^{\mu_h s} ds.$$

Hence

$$V_h(t) = e^{-\mu_h t} V_h(0) + e^{-\mu_h t} \int_0^t (\Lambda_h - I_h(s)) e^{\mu_h s} ds. \tag{15}$$

Since

$$e^{-\mu_h t} \int_0^t \Lambda_h e^{\mu_h s} ds = e^{-\mu_h t} \Lambda_h \frac{1}{\mu_h} (e^{\mu_h t} - 1) = \frac{\Lambda_h}{\mu_h} (1 - e^{-\mu_h t}),$$

and according to the Theorem 4, the function $I_h(t)$ is positive, from (15) we get the estimate

$$V_h(t) \le e^{-\mu_h t} V_h(0) + \frac{\Lambda_h}{\mu_h} \left[1 - e^{-\mu_h t}\right] = e^{-\mu_h t} \left[V_h(0) - \frac{\Lambda_h}{\mu_h}\right] + \frac{\Lambda_h}{\mu_h}. \tag{16}$$

Introducing the notation $g(t) = e^{-\mu_h t} \left[V_h(0) - \frac{\Lambda_h}{\mu_h}\right] + \frac{\Lambda_h}{\mu_h}$ for the right side, an easy computation shows that, under the assumption (14), $g'(t) > 0$, which means that $g(t)$ is a monotonically increasing function. Therefore

$$g(t) \le \lim_{t \to \infty} g(t) = \frac{\Lambda_h}{\mu_h}. \tag{17}$$

Hence, (16) and (17) together imply the required estimate

$$V_h(t) < \frac{\Lambda_h}{\mu_h}$$

for any $t > 0$.

□

The statement for the $V_m(t)$ is as follows.

Theorem 6. *Suppose that at the initial time $t = 0$ the total population for the mosquitoes is bounded by the number Λ_m/μ_m. Then in the extended Ross model (4)–(5) the solution at any time t is also bounded from above by Λ_m/μ_m.*

Proof. Since the proof is carried out analogously to the proof of Theorem 5, for brevity we omit it. □

Theorems 4, Theorem 5 and Theorem 6 imply that each component in the extended Ross model is on the interval $(0, \Lambda_h/\mu_h)$ and $(0, \Lambda_m/\mu_m)$, respectively.

5 Numerical Example

In Sect. 4 we have shown the invariance of the solutions of the extended Ross model on the domains $(0, \Lambda_h/\mu_h)$ and $(0, \Lambda_m/\mu_m)$, respectively. In the following we demonstarte the results numerically.

Table 2. Values of the parameters in the extended model (4)

$S_h(0) = 900$	$E_h(0) = 30$	$I_h(0) = 50$	$R_h(0) = 20$
$S_m(0) = 900$	$E_m(0) = 330$	$I_m(0) = 270$	
$\Lambda_h = 20$	$\Lambda_m = 40$	$\beta_h = 0.3$	$\beta_m = 0.5$
$\mu_h = 0.015$	$\mu_m = 0.02$	$\delta_h = 0.05$	$\delta_m = 0.15$
$\alpha_h = 0.6$	$\alpha_m = 0.6$	$r = 0.05$	$\omega = 0.02$
$b = 0.075$	$\nu_h = 0.5$	$\nu_m = 0.5$	
$V_h(0) = 1000$	$\frac{\Lambda_h}{\mu_h} = 1333.3$	$V_m(0) = 1500$	$\frac{\Lambda_m}{\mu_m} = 2000$

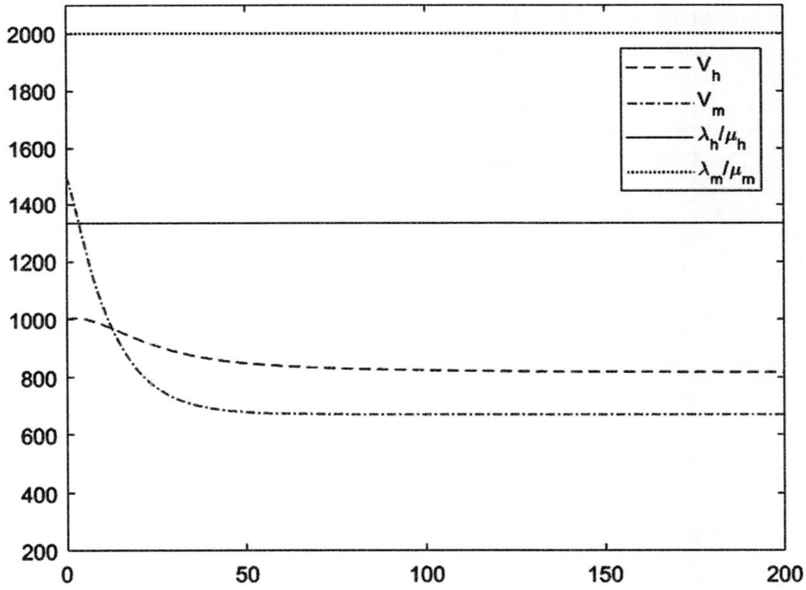

Fig. 1. Numerical solutions of the extended model (4) with the data from Table 2

For the numerical solution the simplest explicit method, the explicit Euler method has been used, and the parameters are set according to the Table 2.

The results are shown on the Fig. 1. As one can see, the solutions remain in the given domains as $t \to \infty$. Moreover, the curves tend to constant.

Acknowledgements. The project has been supported by the European Union, and co-financed by the European Social Fund (EFOP-3.6.3-VEKOP-16-2017-00002).

References

1. Capasso, V.: Mathematical Structures of Epidemic Systems. Springer, Heidelberg (1993)

2. Faragó, I., Mincsovics, M., Mosleh, R.: Reliable numerical modelling of malaria propagation. Appl. Math. **63**, 259–271 (2018)
3. Hoppenstadt, F.C.: Mathematical Methods for Analysis of a Complex Disease. AMS, New York, Courant Institute of Mathematical Sciences (2011)
4. Kermack, W.O., McKendrick, A.G.: A contribution to the mathematical theory of epidemics. I. Proc. Roy. Soc. Lond. Ser. A **115**, 700–721 (1927)
5. Mandal, S., Sarkar, R.R., Sinha, S.: Mathematival models of malaria – a review. Malaria J. **10**, 1–19 (2011)
6. McDonald, G.: The analysis of infection rates in diseases in which super-infection occurs. Trop. Dis. Bull. **47**, 907–915 (1950)
7. Olaniyi, S., Obabiyi, O.S.: Mathematival models for malaria transmiss dynamycs in human and mosquito populations with nonlinear forces of infection. Int. J. Pure and Appl. Math. **88**, 125–156 (2013)
8. Ross, R.:The Prevention of Malaria. J. Murray, London (1910)
9. Ross, R.: An application of the theory of probabilities to the study of a priori pathometry. I. Proc. Roy. Soc. Lond. Ser. A **47**, 204–230 (1916)
10. Xiao, Y.: Study of malaria transmission dynamics by mathematical models. Ph.D. thesis, The University of Western Ontario (2011)
11. World Health Organisation (WHO) and WHO Global Malaria Program. https://www.who.int/malaria/en/

HPC Simulations of the Atmospheric Composition Bulgaria's Climate (On the Example of Coarse Particulate Matter Pollution)

Georgi Gadzhev(✉), Kostadin Ganev, and Plamen Mukhtarov

National Institute of Geophysics, Geodesy and Geography, Bulgarian Academy of Sciences, Acad. G. Bonchev street, bl. 3, 1113 Sofia, Bulgaria
{ggadjev,kganev}@geophys.bas.bg, engpjm@abv.bg

Abstract. The present work aims to provide an overall view of the HPC facilities implementation for studying the regional atmospheric pollution transport and transformation processes of Bulgaria. The study aims at revealing the main features of the atmospheric composition of Bulgaria climate and at tracking and characterizing the main pathways and processes that lead to atmospheric composition formation in the country.

The US EPA Models–3 system is chosen as a modeling tool. As NCEP Global Analysis Data with 1° resolution is used as a meteorological background, the nesting capacities of WRF and CMAQ are used to reduce simulations over Bulgaria to 9 km. The Bulgarian national inventory is applied within the territory of the country and the TNO emission inventory is used as emission input outside of Bulgaria. Special pre-processing procedures are created for introducing temporal profiles and speciation of the emissions.

The study is based on a large number of numerical simulations carried out day by day between 2007–2014. The simulations were performed on the supercomputer Avitohol of IICT–BAS. The following atmospheric composition characteristics will be demonstrated, on the example of the coarse particulate matter, and discussed in the Paper:

1) Seasonal and annual concentration field's pattern, with their typical diurnal course;
2) Evaluation of the contribution of different dynamic and transformation processes to the formation of the atmospheric composition of the country's climate;
3) Vertical structure of the atmospheric composition fields, considered from a point of view of dynamic and transformation processes interaction.

Keywords: Vertical structure · Air pollution · Atmospheric composition · Numerical simulation · High performance computing

I. Dimov and S. Fidanova (Eds.): HPC 2019, SCI 902, pp. 221–233, 2021.
https://doi.org/10.1007/978-3-030-55347-0_19

1 Introduction

Air Quality (AQ) is a key element for the well-being and quality of life of European citizens, and Bulgaria also faces AQ problems [1–3]. It has been recently considered that the photo-oxidant and PM air pollution are the major environmental problems, but classic acidifying pollutants (SO_2, NO_x,), heavy metals (Hg, Cd, Pb) and persistent organic pollutants are still a serious problem and so the study of their environmental impact is also necessary.

Regional studies of the air pollution over the Balkans, including country–to–country pollution exchange, have been carried out for quite a long time - see for example [4–12]. HPC facilities, recently developed within the country provided opportunities to carry out more comprehensive studies with up-to-date modelling tools, detailed and reliable input data for long enough simulation periods and good spatial resolution.

Atmospheric composition modeling methods include statistical, deterministic and hybrid systems. The approach described in the present paper is deterministic - a combination of meteorological input, weather diagnosis/forecasting, additional meteorological pre-processing and chemical composition simulations. This is a fruitful approach for indicating excess limits and target values, assessing the effect of distinct emission sources on air pollution and formulating feasible long and short-term interventions to reduce air pollution. The deterministic approach could also help to better understand the function of different transport scales and events in the air pollution model formation, thereby adding to the validation of the model.

Among the many atmospheric composition features, the following will be proved on the example of coarse particulate matter addressed in the paper:

1) Seasonal and annual concentration fields pattern, with their typical diurnal course;
2) Evaluation of the contribution of different dynamic and transformation processes to the formation of the atmospheric composition of the country's climate: The air pollution transport is subject to different scale phenomena, each characterized by specific atmospheric dynamics mechanisms, chemical transformations, typical time scales etc. The air pollution pattern is formed as a result of interaction between different processes, so knowing the contribution of each of them to different meteorological conditions, a given emission spatial configuration and temporal behavior is by all means important;
3) Vertical structure of the atmospheric composition fields, considered from a point of view of dynamic and transformation processes interaction. The parameters of the atmosphere have key impact on the quality of life and human health. Because of this, it is quite natural, to study mostly the surface air quality. On the other hand, the incredible diversity of dynamic processes, the complex chemical transformations of the compounds and complex emission configuration together lead to the formation of a complex vertical structure of the atmospheric composition. The detailed analysis of this vertical structure with its temporal/spatial variability jointly with the atmospheric

dynamics characteristics can enrich significantly the knowledge about the processes and mechanisms which form air pollution, including those in the near to the earth surface.

2 Methods

The simulations have been performed using US EPA Model-3 system as modeling tools for 3D simulations:

- Meteorological model WRF [13];
- Atmosphere Composition Model CMAQ [14,15];
- Emission model SMOKE [16].

The NCEP Global Analysis Data meteorological background with $1° \times 1°$ resolution has been used. The models nesting capabilities were applied to downscale the simulations to 9 km for Bulgaria. TNO high resolution emission inventory [17,18] and National emission inventory have been used as emission input for Bulgaria. More detailed description of the experiments can be seen in [19,20].

The CMAQ "Integrated Process Rate Analysis" option has been applied to discriminate the role of different dynamic and chemical processes for the air pollution pattern formation. The procedure allows a concentration change of each compound for an hour to be presented as a sum of the contribution of the processes which determine the concentration. The processes that are considered are: advection, diffusion, mass adjustment, emissions, dry deposition, chemistry, aerosol processes and cloud processes/aqueous chemistry.

The WRF/CMAQ simulations have been performed day by day for 8 years – from 2007 to 2014. Thus, a quite extensive data base was created, which could be used for different studies and considerations of the main features and origins of the atmospheric composition in Bulgaria.

Performing extensive simulations of this kind with up to date highly sophisticated numerical models obviously requires large computer resources (for more details see: Table 1, 2 and [21]) and so the numerical experiments have been organized in effective HPC environment (Atanassov et al. [22]). The calculations have been implemented on the Supercomputer System Avitohol at IICT–BAS (Institute of Information and Communication Technologies–Bulgarian Academy of Sciences). The supercomputer consists of 150 HP Cluster Platform SL250S GEN8 servers, each one equipped with 2 Intel Xeon E5-2650 v2 8 C 2600 GHz CPUs and 64 GB RAM per server. The storage system is HP MSA 2040 SAN

Table 1. Computer resource requirements on 16 CPU-s for 1 Day simulation

	WRF	CMAQ and SMOKE	Total
Time (h)	3	2	5
HDD (GB)	0.5	1	1.5

Table 2. Computer resource requirements for 1 Day simulation for CMAQ at different number of CPU-s

		D2/27 km	D3/9 km	D4/3 km	D5/1 km	Total
	2 CPU	2 h 40 min	1 h 10 min	1 h 36 min	32 min	5 h 58 min
	4 CPU	1 h 20 min	35 min	45 min	15 min	2 h 55 min
	8 CPU	40 min	20 min	22 min	8 min	1 h 30 min
	16 CPU	30 min	18 min	12 min	5 min	1 h 5 min
1 day	HDD (MB)	255	420	70	145	890
7 Year	HDD (GB)	636	1048	175	362	2221

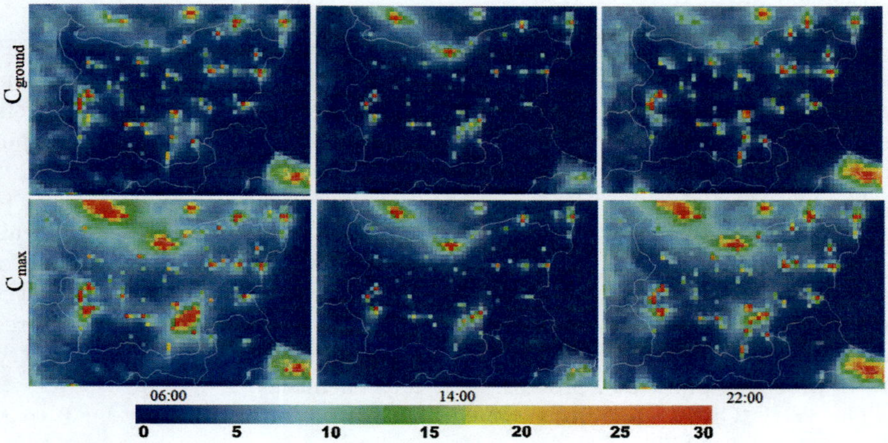

Fig. 1. Plots of the ensemble averaged surface and maximal CPRM summer concentrations [μg/m^3]

with a total of 96 TB of raw disk storage capacity. All the servers are interconnected with fully non-blocking FDR Infiniband, using a fat-tree topology [22]. The needed libraries and programs have been installed on the supercomputer for proper functioning and work of the models used in this study.

3 Results

The 8–year ensemble of computer simulation results is, hopefully, large and comprehensive enough to reflect the variety of meteorological conditions and to represent the main features of the atmospheric composition climate of the country–typical and extreme patterns with their typical recurrences.

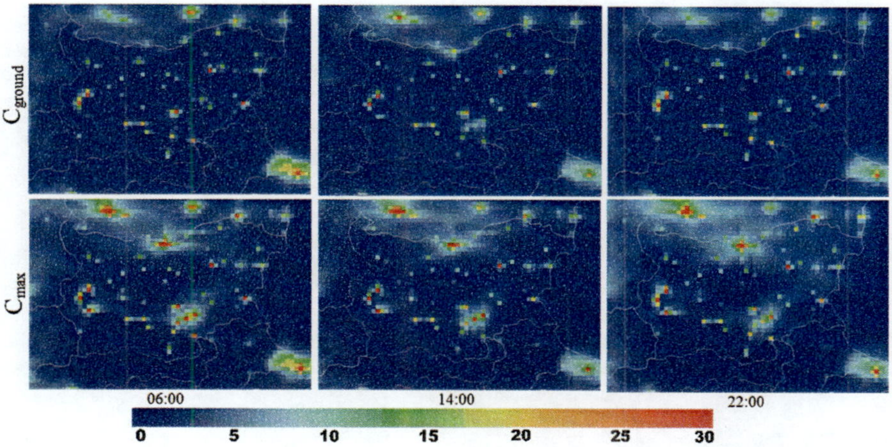

Fig. 2. Plots of the ensemble averaged surface and maximal CPRM winter concentrations [$\mu g/m^3$]

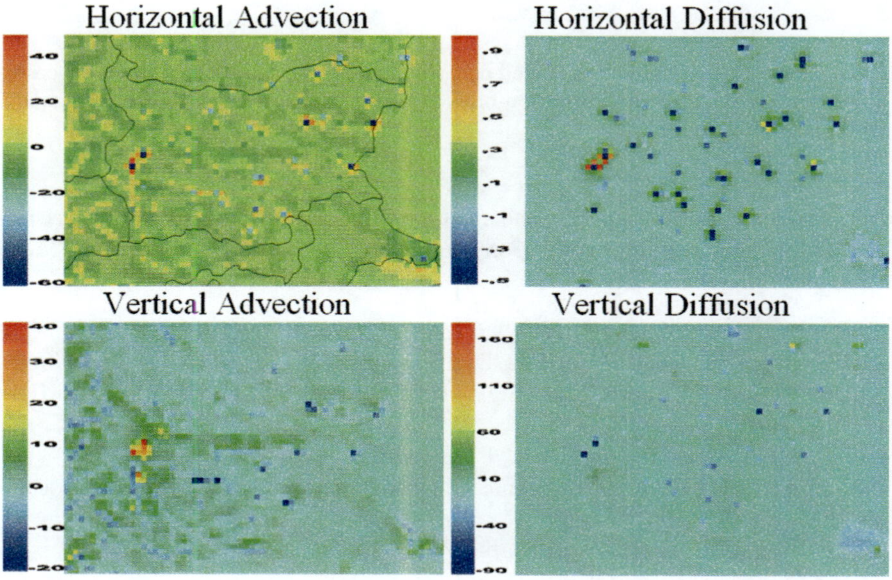

Fig. 3. Plots of the ensemble averaged contribution of horizontal and vertical advection and diffusion to the formation of the CPRM concentration hourly change in 06:00 UTC [$\mu g/m^3$]

The models output storage has been too large. Since not all of the output data has been that important for further air quality and environmental factors, post–processing procedures and respective software have been created in order to

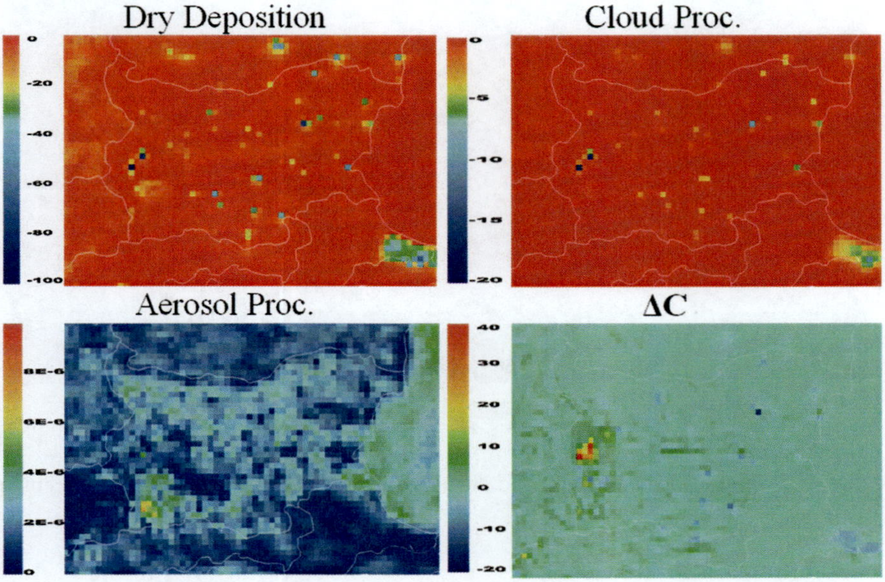

Fig. 4. Plots of the ensemble averaged contribution of dry deposition, cloud processes and aerosol processes to the formation of the CPRM concentration hourly change ΔC in 06:00 UTC [μg/m^3]

"filter" the output and keep only the most valuable data. Different characteristics of the following most important pollutants have been stored from the CMAQ output on hourly basis [19]:

- NO_2, NO, O_3, NO_3, OH, HO_2, $H2O_5$, HNO_3, HONO, PNA, H_2O_2, CO, FORM, ALD$_2$, C_2O_3, PAN, PACD, PAR, OLE, FACD, AACD, ETH, TOL, CRES, TO$_2$, XYL, MGLY, ISOP, ISPD, SO_2, SULF (H_2SO_4), UMHP, TERP, NH_3 (gases – 34);
  - PSO_4, PNH_4, PNO_3, POA, PEC (aerosol – 5);
  - SOAA, SOAB (Anthropogenic and Biogenic secondary organic aerosol – 2);
  - FPRM, CPRM (fine and coarse PM – 2).

The CPRM characteristics behaviour will be shown and discussed in the present paper as a demonstration of the HPC simulations of the atmospheric composition of Bulgaria climate. All the characteristics have been obtained by averaging/finding the mean value of the ensemble seasonally or annually, so they present the typical seasonal and diurnal variability of the CPRM behaviour.

The CPRM concentrations at ground level, together with the maximal CPRM concentrations (at whatever height they occur) are shown on Figs. 1 and 2 for summer and winter. The first thing that could be easily noticed from the figures is the well manifested seasonal and diurnal variability of the CPRM

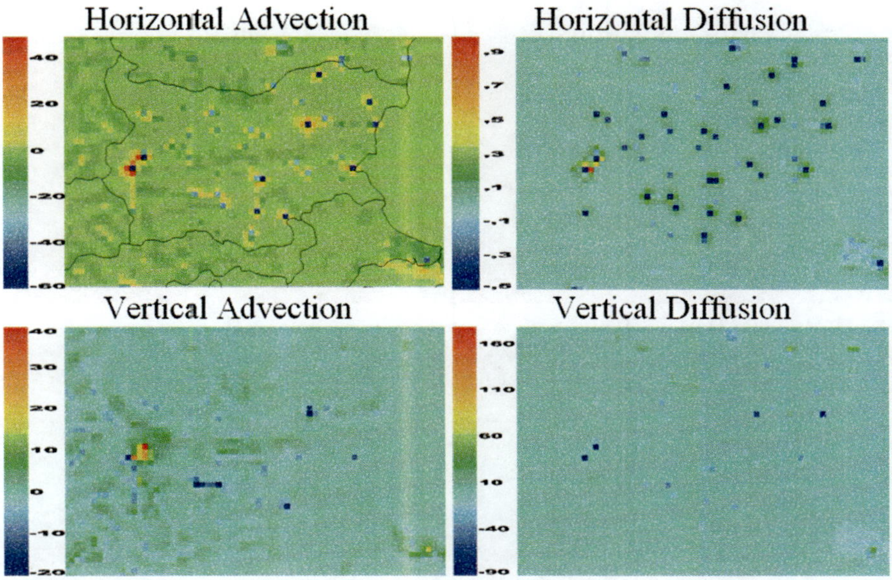

Fig. 5. Plots of the ensemble averaged contribution of horizontal and vertical advection and diffusion to the formation of the CPRM concentration hourly change in 18:00 UTC [μg/m^3]

concentrations. The concentration fields are very heterogeneous with the CPRM pollution concentrated around the pollution sources. This is most likely due to the CPRM gravity sedimentation because of which the CPRM is not so much issue of horizontal transport. During the summer, however, there is more intensive vertical turbulent transport which, to some extent, opposes the gravity sedimentation, the CPRM deposition is not so fast and so a larger CPRM amount is transported horizontally.

The annual fields of the hourly contribution of the different processes, which determine the CPRM formation in the first (ground) model layer, together with the resulting hourly concentration change ΔC are shown in Figs. 3 and 4 for 6 UTC and in Figs. 5 and 6 for 18 UTC. The horizontal heterogeneity and the diurnal course of the contribution of the different processes are again very well displayed. The major CPRM sources (mostly point sources) can be seen practically in all the plots, visualizing the contribution of the different processes.

The processes can have different signs, while the contribution of the dry deposition and the cloud processes (wash-out) are, apparently, always negative and the contribution of the aerosol processes (which lead to CPRM formation) is always positive.

The sign of the contribution of the other processes changes depending on the meteorological conditions and the emission sources configuration. For example,

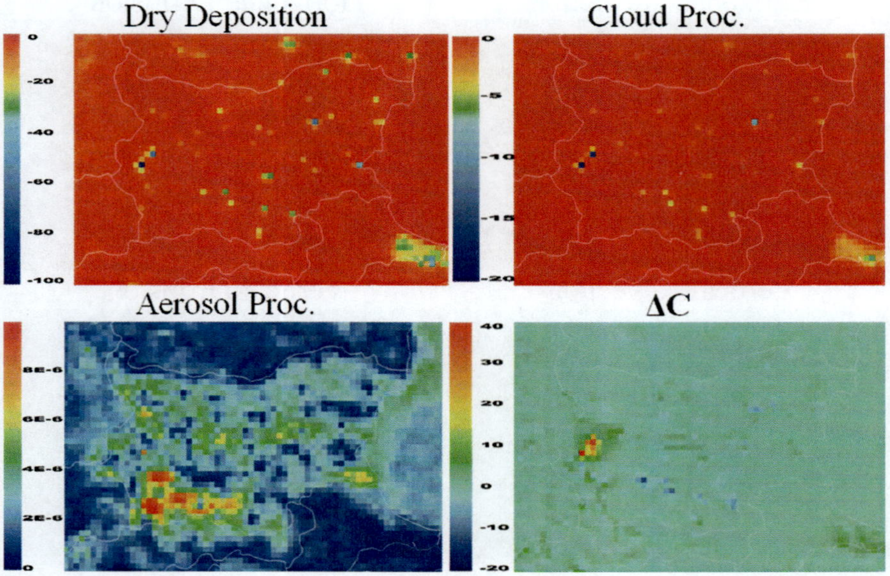

Fig. 6. Plots of the ensemble averaged contribution of dry deposition, cloud processes and aerosol processes to the formation of the CPRM concentration hourly change ΔC in 18:00 UTC [$\mu g/m^3$]

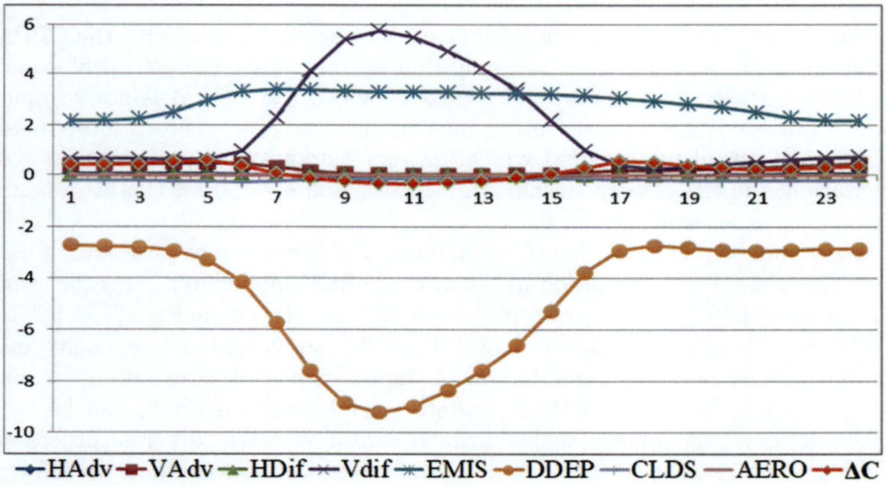

Fig. 7. Diurnal course of the averaged for the territory of Bulgaria annual contribution of different processes to the formation of the CPRM concentration hourly change ΔC [$\mu g/m^3$]

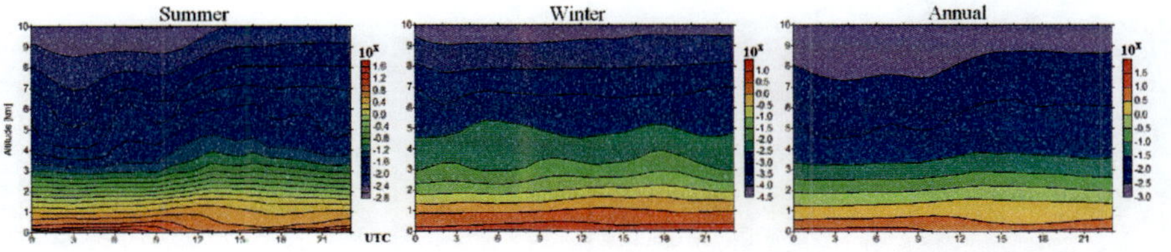

Fig. 8. Diurnal course of the summer, winter and annual vertical CPRM profiles, averaged for the territory of Bulgaria

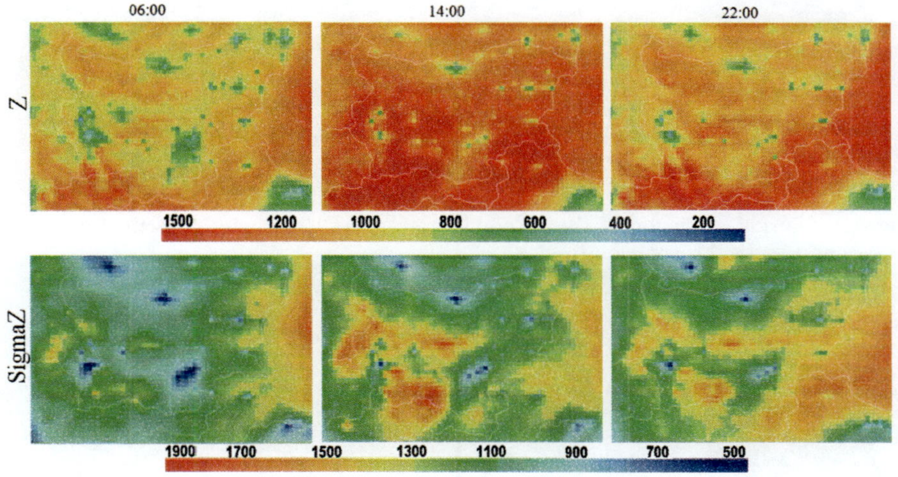

Fig. 9. Diurnal evolution of the ensemble averaged CPRM summer vertical center of masses and vertical dispersion [m]

the contribution of the vertical advection is mostly negative when the sources are within the first layer and positive, when they are above it.

The pattern of the contribution of the processes which determine the horizontal transport (advection and diffusion) is more complex and not so straightforward to explain, yet the emission sources are clearly visible even in this plots.

The diurnal course of the annual contribution of the different processes, averaged for the territory of Bulgaria is shown on Fig. 7. It can be seen that the major contributions are these of the – emissions, dry deposition and, during the day, the vertical diffusion. The contribution of the other processes, averaged for such a large territory, is much smaller, though locally these contributions could be significant. The resulting ΔC is also much smaller than the major processes and is slightly negative during the day, in spite of the intensive turbulent transport towards the ground layer.

The diurnal course of the summer, winter and annual vertical CPRM profiles, averaged for the territory of Bulgaria, is shown in Fig. 8. As it should be expected, the surface CPRM concentrations decrease during the day, due to the more

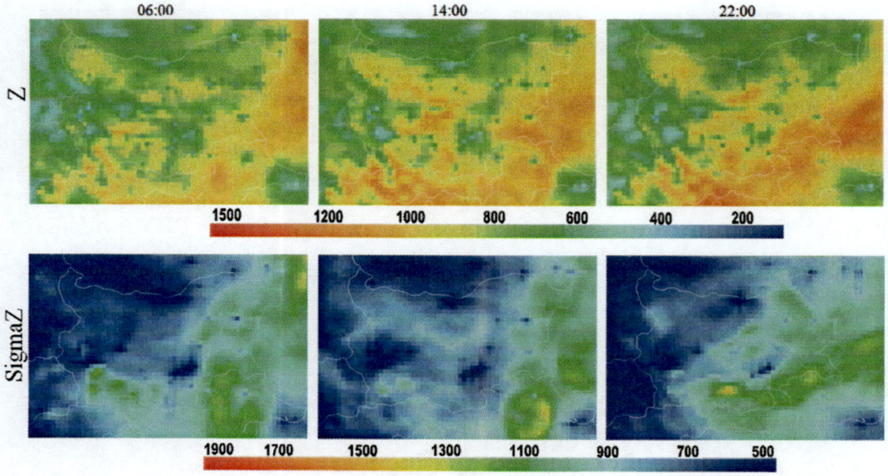

Fig. 10. Diurnal evolution of the ensemble averaged CPRM winter vertical center of masses and vertical dispersion [m]

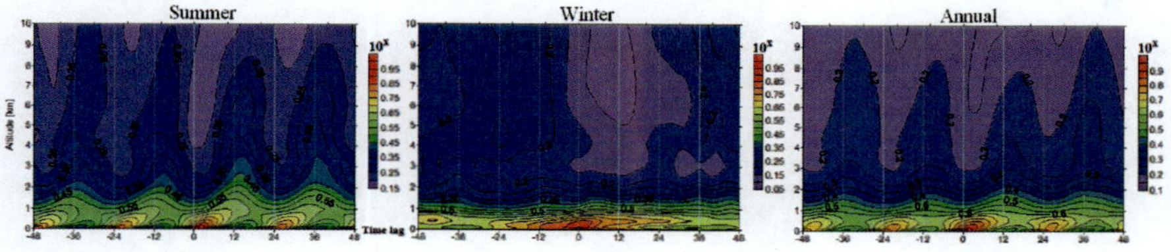

Fig. 11. Plots of the summer, winter and annual cross-correlations between the surface and higher level CPRM concentrations

intensive vertical diffusion transport. Naturally this effect is most prominent during the summer and almost non–existent during the winter.

Some impression of the very complex horizontal pattern of the CPRM vertical structure one can be obtained from Figs. 9 and 10, where the diurnal evolution of the CPRM vertical center of masses and vertical dispersion is demonstrated for the summer and winter periods. Both the horizontal heterogeneity and the diurnal and seasonal variability of these characteristics are very well displayed in the plots. As it should be expected both the centers of masses and the vertical dispersion have higher values during the summer and especially at noon. Again the location of the major sources can be detected at places with smaller values of the center of masses and the vertical dispersion.

Plots of the cross–correlations between the surface and higher level CPRM concentrations are shown in Fig. 11. The most remarkable thing in these plots is the 24-h diurnal cycle, very well displayed during the summer and to some extent during the year. It reflects, on one hand the diurnal courses of the emissions and on the other, the diurnal course of the atmospheric stability, hence the vertical diffusion transport. The effect cannot be observed in the winter period which probably means that the vertical diffusion is not so crucially important for the CPRM profiles formation during this period, or perhaps also that the vertical diffusion does not have such a prominent diurnal course.

4 Conclusions

This paper shows only some of the behavioral characteristics of one of the atmospheric compounds – the CPRM. Even the brief analysis, made in the paper demonstrates that the formation of the air pollution pattern is subject of the following factors:

– Interaction of processes of different natural, spatial and temporal scales;
– Significant spatial, seasonal and diurnal variability of the different processes;
– Temporal variability and spatial heterogeneity of the emission sources.

The numerical experiments performed, produced a huge volume of information, which has to be carefully analyzed and generalized so that some final conclusions could be made. The obtained ensemble of numerical simulation results is extensive enough to allow statistical treatment – calculating not only the mean concentrations, but also standard deviations, skewness, etc. with their dominant temporal modes (seasonal and/or diurnal variations). Some advanced and sophisticated methods for statistical treatment of the results should also be appropriately applied in order to:

– study the atmospheric pollution transport and transformation processes (accounting also for heterogeneous chemistry and the importance of aerosols for air quality and climate) from urban through local to regional (Balkan and country) scales;
– track and characterize the main pathways and processes that lead to atmospheric composition formation in different scales;
– provide high quality scientifically robust assessments of the atmospheric composition and its origin - the air pollution fields response to emission changes (model sensitivity to emission input) is obviously a task of great practical importance, obviously connected with formulaton of short–term (current) pollution mitigating decisions and long-term pollution abatement strategies.

Acknowledgements. This work has been accomplished with the financial support by the Grant No BG05M2OP001-1.001-0003, financed by the Science and Education for Smart Growth Operational Program (2014–2020) and co-financed by the European Union through the European structural and Investment funds.

Project “NATIONAL GEO-INFORMATION CENTER”, subject of the National Road Map for Scientific Infrastructure 2017–2023, funded by Contr. No D 1-161/28.08.2018.

The present work is supported by the Bulgarian National Science Fund (grant DN-04/2/13.12.2016).

References

1. Georgieva, I., Ivanov, V.: Computer simulations of the impact of air pollution on the quality of life and health risks in Bulgaria. Int. J. Environ. Pollut. **64**(1–3), 35–46 (2018)
2. Ivanov, V., Georgieva, I.: Air quality index evaluations for Sofia city. In: 17th IEEE International Conference on Smart Technologies, EUROCON 2017 - Conference Proceedings, art. no. 8011246, pp. 920-925 (2017)
3. Georgieva, I., Ivanov, V.: Impact of the air pollution on the quality of life and health risks in Bulgaria. In: HARMO 2017 - 18th International Conference on Harmonisation within Atmospheric Dispersion Modelling for Regulatory Purposes, Proceedings, October 2017, pp. 647-651 (2017)
4. Zerefos, C., Ganev, K., Kourtidis, K., Tzortziou, M., Vasaras, A., Syrakov, E.: On the origin of SO2 above Northern Greece. Geophys. Res. Lett. **27**(3), 365–368 (2000)
5. Syrakov, D., Prodanova, M., Ganev, K., Zerefos, Ch., Vasaras, A.: Exchange of sulfur pollution between Bulgaria and Greece. Environ. Sci. Pollut. Res. **9**(5), 321–326 (2002)
6. Ganev, K., Dimitrova, R., Syrakov, D., Zerefos, Ch.: Accounting for the mesoscale effects on the air pollution in some cases of large sulfur pollution in Bulgaria or Northern Greece. Environ. Fluid Mech. **3**, 41–533 (2003)
7. Chervenkov, H.: Estimation of the exchange of sulphur pollution in the Southeast Europe. J. Environ. Prot. Ecol. **7**(1), 10–18 (2006)
8. Zanis, P., Poupkou, A., Amiridis, V., Melas, D., Mihalopoulos, N., Zerefos, C., Katragkou, E., Markakis, K.: Effects on surface atmospheric photo-oxidants over Greece during the total solar eclipse event of 29 March 2006. Atmos. Chem. Phys. Discuss. **7**(4), 11399–11428 (2007)
9. Ganev, K., Prodanova, M., Syrakov, D., Miloshev, N.: Air pollution transport in the Balkan region and country-to-country pollution exchange between Romania, Bulgaria and Greece. Ecol. Model. **217**, 255–269 (2008)
10. Poupkou, A., Symeonidis, P., Lisaridis, I., Melas, D., Ziomas, I., Yay, O.D., Balis, D.: Effects of anthropogenic emission sources on maximum ozone concentrations over Greece. Atmos. Res. **89**(4), 374–381 (2008)
11. Symeonidis, P., Poupkou, A., Gkantou, A., Melas, D., Yay, O.D., Pouspourika, E., Balis, D.: Development of a computational system for estimating biogenic NMVOCs emissions based on GIS technology. Atmos. Environ. **42**(8), 1777–1789 (2008)
12. Marécal, V., et al.: A regional air quality forecasting system over Europe: The MACC-II daily ensemble production. Geosci. Model Dev. **8**, 2777–2813 (2015)
13. Shamarock, W.C., et al.: A description of the Advanced Research WRF Version 3 (2008). https://opensky.ucar.edu/islandora/object/technotes%3A500/datastream/PDF/view
14. Byun, D.: Dynamically consistent formulation in meteorological and air quality models for multi-scale atmospheric studies. J. Atm. Sci. (1998). in review. https://doi.org/10.1175/1520-0469(1999)056⟨3789:DCFIMA⟩2.0.CO;2
15. Byun, D., Ching J.K.S.: Science algorithms of the EPA Models 3 Community Multiscale Air Quality (CMAQ) Modeling System. United States Environmental Protection Agency, Office of Research and Development, Washington, DC 20460, EPA-600/R-99/030 (1999)

16. CEP Sparse Matrix Operator Kernel Emission (SMOKE) Modeling System. University of Carolina, Carolina Environmental Programs, Research Triangle Park, North Carolina (2003)
17. Visschedijk, A., Zandveld P., van der Gon, H.: A high resolution gridded European emission database for the EU integrated project GEMS, TNO report 2007-A-R0233/B, The Netherlands (2007)
18. Brunekreef, B., Holgate, S.: Air pollution and health. Lancet 2000 **360**, 1233–1242 (2007)
19. Gadzhev, G., Ganev, K., Miloshev, N., Syrakov, D., Prodanova, M.: Numerical study of the atmospheric composition in Bulgaria. Comput. Math. Appl. **65**(3), 402–422 (2013)
20. Gadzhev, G., Ganev, K., Syrakov, D., Prodanova, M., Miloshev, N.: Some statistical evaluations of numerically obtained atmospheric composition fields in Bulgaria. In: Proceedings of the 15th International Conference on Harmonisation within Atmospheric Dispersion Modelling for Regulatory Purposes, Madrid, Spain, 6–9 May, pp. 373–377 (2013)
21. Todorova, A., Syrakov, D., Gadjhev, G., Georgiev, G., Ganev, K.G., Prodanova, M., Miloshev, N., Spiridonov, V., Bogatchev, A., Slavov, K.: Grid computing for atmospheric composition studies in Bulgaria. Earth Sci. Inform. **3**(4), 259–282 (2010)
22. Atanassov, E., Gurov, T., Karaivanova, A., Ivanovska, S., Durchova, M., Dimitrov D.: On the Parallelization Approaches for Intel MIC Architecture. In: AIP Conference Proceedings, vol. 1773, p. 070001 (2016). https://doi.org/10.1063/1.4964983

HPC Simulations of the Present and Projected Future Climate of the Balkan Region

Georgi Gadzhev[1], Vladimir Ivanov[1], Rilka Valcheva[2], Kostadin Ganev[1], and Hristo Chervenkov[2](✉)

[1] National Institute of Geophysics, Geodesy and Geography, Bulgarian Academy of Sciences, Acad. G. Bonchev Street, bl. 3, 1113 Sofia, Bulgaria
{ggadjev,vivanov,kganev}@geophys.bas.bg

[2] National Institute of Meteorology and Hydrology, 66, Tsarigradsko Shose blvd., 1784 Sofia, Bulgaria
{Rilka.Valcheva,hristo.tchervenkov}@meteo.bg

Abstract. The forthcoming climate changes are the biggest challenge facing mankind today. They will exert influence on the ecosystems, on all branches of the international economy, and on the quality of life. The climate changes and their consequences have a great number of regional aspects, which global models cannot predict. That is why an operation plan for adaptation to climate changes has to be based on well-grounded scientific assessments, taking in consideration regional features of the climate changes and their consequences. The present work aims to give an overview of the high performance computing (HPC) facilities' implementation for studying of present and future regional climates of the Balkan region. Simulations were performed with the regional climate model RegCM on the supercomputer Avitohol of IICT – BAS. The global climate simulations, taken from HadGEM2 database were used as source for the initial and lateral boundary conditions. The simulations were performed for the following time slots: 1) 1975 – 2005 – reference period, 2) 2020 – 2050 – near future and 3) 2070 – 2099 – far future. The simulations for the future climate were carried out for 3 CMIP5 emission scenarios: RCP2.6 (optimistic scenario), RCP4.5 (realistic scenario) and RCP8.5 (pessimistic scenario). The simulation results for the reference period were compared with the independent, observation-based data set E-OBS. These preliminary simulation results reveal that the projected climate changes are strongest in the far future and RCP8.5. According to the temperature, the warming signal is spatially dominating with peak values the summer, when it is more than 4.5 °C. The projected precipitation change is more complex, both in time and space. It is worth emphasizing, however, that the expected summer months reduction, which could amplify the negative impact of the expected warmer climate.

I. Dimov and S. Fidanova (Eds.): HPC 2019, SCI 902, pp. 234–248, 2021.
https://doi.org/10.1007/978-3-030-55347-0_20

Keywords: RegCM · CMIP5 RCPs · Climate change signal · Balkan Peninsula

1 Introduction

The forthcoming climate changes are the biggest challenge facing mankind today. They will exert influence on the ecosystems, on all branches of the international economy, and on the quality of life. The climate changes and their consequences have a great number of regional aspects which the global models cannot predict. That is why an operational plan for adaptation to climate changes has to be based on well-grounded scientific assessments, taking into account regional features of the climate changes and their consequences.

There have been several collaborative projects in the last few decades, which support climate change studies on regional and global scales. One of the initiatives in that direction is the project CECILIA (Central and Eastern Europe Climate Change Impact and Vulnerability Assessment, http://www.cecilia-eu.org/) [8]. Its objective is to assess the impacts of climate change in the region by means of regional climate modelling of a broad range of topics as hydrology, water quality, urban air quality and agriculture. The aim of the more recent, fifth phase, of the Coupled Model Inter-comparison Project (CMIP5) is to study the climate and climate change in the past, present and future, using a set of simulations with different climate simulators in various spatial and temporal scales [14]. One of the types of CMIP5 simulations is a set of long-term simulations from the middle of the nineteenth century through the twenty-first century and beyond. It uses atmosphere-ocean global climate models (AOGCM), Earth system models of intermediate complexity (EMIC) and Earth System Models (ESM). AOGCM and EMIC represent the atmosphere, ocean, land and sea ice in an interactive way, and their response depends on time-varying concentrations of some atmospheric components. ESM are AOGCM with prescribed CO_2 concentrations and emissions, which account for the carbon cycle between the ocean, atmosphere and terrestrial biosphere. The CMIP5 experiment uses new emission scenarios called representative concentration pathways (RCP) [11] to assess the interactions between the human activities on one side and the environment on the other, and their evolution. They are based on the balance between the incoming and outgoing surface and atmospheric radiation which depends on the change of the atmospheric composition and provides a framework for modelling in the next stages of scenario-based research. Unlike the scenarios described in the Intergovernmental Panel on Climate Change (IPCC) "Special Report on Emissions Scenarios" (SRES) used for CMIP3, which did not include policy intervention, the RCPs are mitigation scenarios that assume policy actions will be taken to achieve certain emission targets. For CMIP5, four RCPs have been formulated that are based on a range of projections of future population growth, technological development, and societal responses. The labels for the RCPs provide a rough estimate of the radiative forcing in the year 2100 (relative to preindustrial conditions) [14]. In the current study, we have used three scenarios: RCP2.6,

RCP4.5 and RCP8.5. The RCP2.6 is the scenario with a peak of the radiative forcing at about 3 $W.m^{-2}$ and declining to 2.6 $W.m^{-2}$ by 2100. The RCP4.5 implies an original increase of the stabilizing radiative forcing at 4.5 $W.m^{-2}$ after 2100 year. The RCP8.5 scenario suggests rising pathway with 8.5 $W.m^{-2}$ in 2100.

A more recent initiative for studying the regional climate change in Europe is EURO–CORDEX ((http://www.euro-cordex.net/) [9, 10] which is the European part of the worldwide CORDEX (Coordinated Regional Downscaling Experiment) initiative. The project is focused on assessing the present-day and future climate with 11 regional climate models in various model configurations, which are involved in the project. The EURO–CORDEX provides climate projections for the European domain at about 12.5 km 50 km. The regional climate simulations uses initial and boundary conditions from the CMIP5 global climate projections and the emission scenarios of RCP2.6, RCP4.5 and RCP8.5.

The Earth system configuration of The Hadley Centre Global Einvironment Model version 2 (HadGEM2-ES) is a multicomponent model [4] which was designed for studying the earth climate system and its evolution for at least 100 years. It is a part of the long-term model set of the CMIP5 experiments. The model is composed of land-surface exchange scheme, large-scale hydrology module, river-modelling scheme, aerosols module, tropospheric chemistry module, terrestrial ecosystem, and ocean ecosystems. The two ecosystems simulate the interaction between the carbon cycle and the climate. The combination of all these procedures and the biogeochemical feedback increases the system's complexity which, in turn, increases uncertainty. However, that also makes the performance of the system more similar to the reality, which is more useful for our understanding of the current and future states of the climate system. The HadGEM2-ES model spin-up is until 1860 year, which is considered as "preindustrial" era. The model has horizontal resolution $1.875° \times 1.25°$ in the atmosphere and 1° horizontal resolution in the ocean. The time step of the model is 1800 s in the atmosphere and on the land, and 3600 s in the ocean.

The main aims of the present work are to study the present and projected future climate of the Balkan region in finer spatial and temporal scales with the latest emission scenario conception and to broaden the climate knowledge of the region under consideration in relation to the EURO – CORDEX and CECILIA. For that purpose, we perform simulations with regional climate model RegCM on the supercomputer Avitohol of the Institute of Information and Communication Technologies at the Bulgarian Academy of Sciences (IICT–BAS) with initial and boundary conditions taken from the HadGEM2-ES model. The regional climate model used is RegCM with configuration outlined in detail in [7]. The horizontal resolution and time step are 20 km and 20 sec respectively. We performed the simulations for the reference period ("control run", CR) from 1975 to 2005, which is representative for the present climate, for the climate in the near future (NF) from 2021 to 2050 and for the climate in the far future (FF) from 2071 to 2099. The study for the near future time slot is performed for three emission scenarios the described above and the simulations for the far future—only for the RCP2.6 and RCP8.5 thus far.

The article's structure is as follows. The model performance for the present climate is evaluated in Sect. 2. In Sect. 3 the projected future climate for the considered emission scenarios is discussed. Subsect. 3.1 is dedicated to the study of the seasonal and annual means of some meteorological parameters and Subsect. 3.2—on the obtained climate change signal. Concise concluding remarks as well as outlook for further research can be found in Sect. 4.

2 Simulations of the Present Climate

Central and Eastern Europe, and the Balkan Peninsula in particular, is a region in Europe, where most climate model validations show considerable problems. For instance, the coarser and finer version of the EURO-CORDEX ensemble tends to produce warm and dry summer bias for this region [10]. The bias ranges of the EURO-CORDEX ensemble mostly correspond to those of the ENSEMBLES simulations, but, among other achievements, less pronounced southern European warm summer bias can be identified. Some ensemble members overestimate winter temperatures over the area, but the cold biases are generally less expressed.

In the recent decades a lot of sensitivity analyses of RegCM have been completed regarding the selection of suitable integration domain, adequate horizontal resolution, potential driving models, applied physics schemes, or adaptation tools (see [3,12,15] and references therein). The main objective of the ongoing numerical experiment TVRegCM was to evaluate the simulation capabilities of 20 combinations of distinct parameterization schemes available in the model [3,7].

The validation of the current climate simulations in this research focuses on assessing the distributional similarity of the gridded time series of the minimum, mean and maximum temperature on daily basis, as well as the daily precipitation sum, which are subsequently denoted as *tn, tg, tx* and *rr*. This approach is different from that assumed in the beforementioned studies where the seasonal and multiyear means of the temperature and precipitaton were evaluated. The necessity of such initial step is to identify possible differences between observed and simulated climate variables, which could provide the basis for post-processing of both control and scenario RCM runs with some bias correction procedure [6]. The surface observation based data set of gridded time series E-OBS of the European Climate Assessment & data set (ECA&D) project is used as reference in the model validation as in [3,7,12] and many other studies. The version implemented in the current study is the newest, ensemble-based, one (19.0e) [5].

The Quantile-Quantile (Q-Q) plot, which is a plot of the quantiles of the first data set against the quantiles of the second one, is a commonly used technique for checking whether two data sets are distributed differently [6]. The method could be generalized for two dimensional samples, as in our case, narrowing the comparison for some key quantiles, for example, the 10th, 25th (lower quartile),

50th (median), 75th (upper quartile) and 90th percentile, traditionally noted as *X10, X25, X50, X75* and *X90*, correspondingly. The chosen evaluation concept has two main strengths. First and foremost, the quantiles are robust estimators: - they are significantly less sensitive to the presence of outliers in a sample, which are, more or less, typical for any time series of climate variables [2]. Second, since Q-Q plots compare distributions, instead of pairs of numbers, there is no need for the size of the two samples being compared to be equal. The latter is also an issue in our case: due to the driving model, the performed simulations are for 360-days years, in contrast to the standard calendar used in E-OBS. There are also, a relatively small number indeed, missing values for some of the selected variables in E-OBS.

Figure 1 shows the distribution of *X10, X25, X50, X75* and *X90* of the *tn*, denoted further for conciseness as *tnX10, tnX25, tnX50, tnX75* and *tnX90*, correspondingly from E-OBS and from RegCM output respectively as well as the bias (i.e. RegCM—E-OBS) between them. Two specific aspects of the fields of the biases are noticeable. The first one is the cold bias close to the northern border of the domain, most apparent in the NW corner, i.e. over the Alps, which is more extended for the first three quantiles and also notable for the last two quantiles in SE. There are also relatively lower adverse differences in Asia Minor. More significant, however, is the warm bias over the bigger part of the domain. It is distinguishable for all *tnX10*–*tnX90* with maximum values for *tnX10* over the western part of the Balkan Peninsula of 3–5 °C.

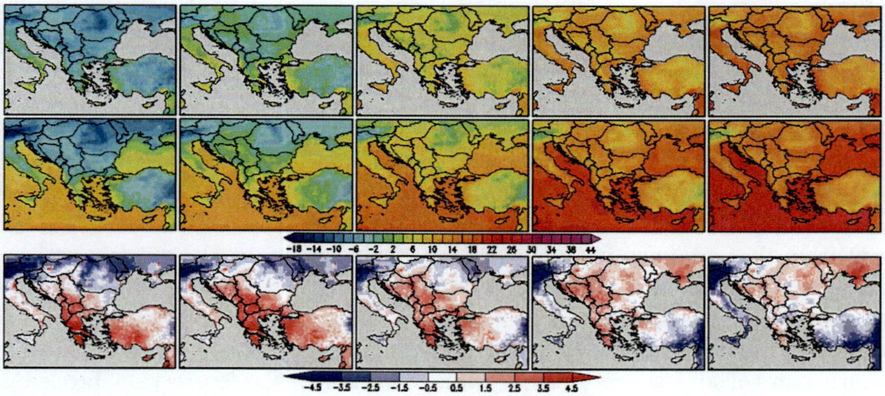

Fig. 1. *tnX10, tnX25, tnX50, tnX75* and *tnX90* from E-OBS (first row) and RegCM (second row), as well as the bias between them (third row). The units are °C.

Figure 2 is similar to Fig. 1 but for the mean temperature quantiles *tgX10*–*tgX90*. Although the warm bias over the Balkan Peninsula is still recognizable on the field of *tgX10* and *tgX25*, the spatially dominant deviation is negative, with lowest values below −4 °C over the Apennines and the Anatolian plateau. The differences between model result and reference over the latter region could be partially rooted in local weakness of E-OBS caused by the lack of (shared with the ECA&D) observations (G. van der Schrier from ECA&D, personal communication).

Figure 3 shows the fields of maximum temperature quantiles *txX10–txX90*. It can be expected from Fig. 2 and 3, the prevailing bias, is negative with values nearly below -5 °C, particularly for the upper margin, i.e. *txX50–txX90*.

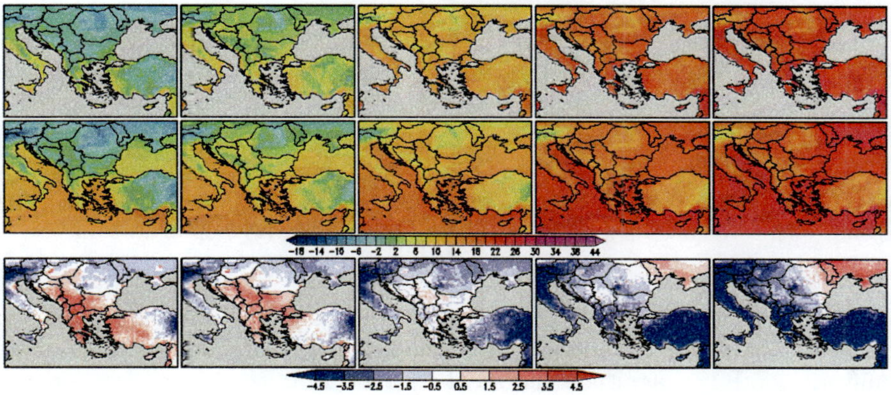

Fig. 2. Same as Fig. 1, but for the mean temperature quantiles *tgX10–tgX90*

Following the same idea for the temperature percentiles, Fig. 4 depicts the comparison between these statistical estimators for the daily precipitation sum which are noted as *rrX10, rr25, rr50, rr75* and *rrX90*, correspondingly. Most apparent on Fig. 4 seems the absence of any precipitation above 1 mm on the E-OBS-fields of rrX10, rrX25, rrX50. The problem could be, at least partially,

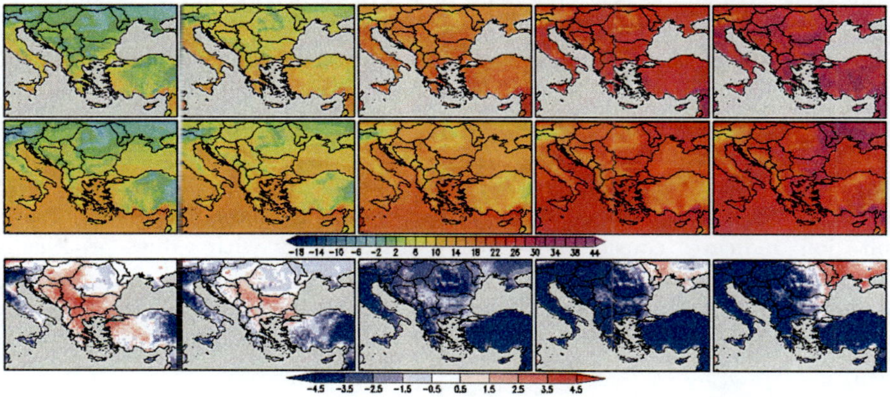

Fig. 3. Same as Fig. 1, but for the maximum temperature quantiles *txX10–txX90*

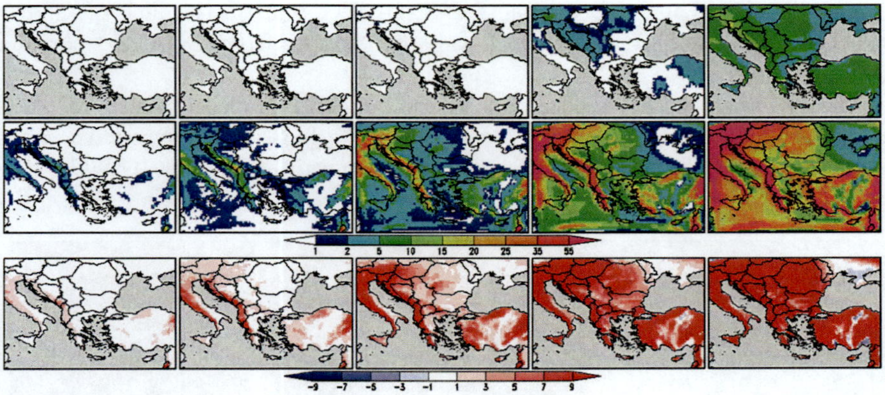

Fig. 4. *rrX10, rrX25, rrX50, rrX75* and *rrX90* from E-OBS (first row) and RegCM (second row), as well as the bias between them (third row). The units are mm.

rooted in the fact that reports of small rainfall amounts are subject to bad observer practice (see [5] and references therein). The discrepancy becomes more significant, both in magnitude and spatial extent, from the lower to the greater quantiles and is an overestimating direction. Since the E-OBS-weaknesses in reproducing precipitation extremes are known [2,5], this issue could not be regarded solely as a problem of the model RegCM.

Distributional dissimilarities are evaluated often in the climatology [1,17] by the statistic of the two-sample Kolmogorov-Smirnov (KS) test, which is defined as follows:

$$D_{n,m} = \sup_x |F_{1,n}(x) - F_{2,m}(x)|, \tag{1}$$

where $F_{1,n}$ and $F_{1,m}$ are the empirical (cumulative) distribution functions of the first and the second sample respectively, and sup is the supremum function. In our case the first sample is the reference time series (E-OBS) and the second—the model output from RegCM.

Figure 5 shows the distribution of the KS-statistic for the minimum, mean and maximum temperature as well as for the daily precipitation sum. Figure 5 shows, that the distribution patterns of the KS-statistic for the minimum, mean with some reservation for the maximum temperature are principally similar. Overall, the areas with comparatively large $D_{n,m}$, in Figs. 1, 2 and 3. Unlike to the first three subplots on 5, the last one, showing the KS-statistic for precipitation, indicates considerably greater dissimilarity with values practically everywhere above 0.4.

The null hypothesis H_0 of the KS- test is that the two samples originate from the same distribution. According to [16], the null hypothesis is rejected at level α if

$$D_s > \left[-\frac{1}{2}\left(\frac{1}{n}+\frac{1}{m}\right)\ln\left(\frac{\alpha}{2}\right)\right]^{1/2}, \tag{2}$$

where n is the size of the first sample (say, E-OBS time series length) and m—of the second (RegCM time series length) as in Eq. 1, and typically $\alpha = 0.05$. By absence of undefined values, both in E-OBS and RegCM, $n = 11323$ and $m = 11220$, which, according to Eq. 2, yields $D_s \lesssim 0.02$. This value is more than an order smaller than the depicted on Fig. 5 and therefore the null hypothesis has to be rejected for all considered parameters in all grid cells.

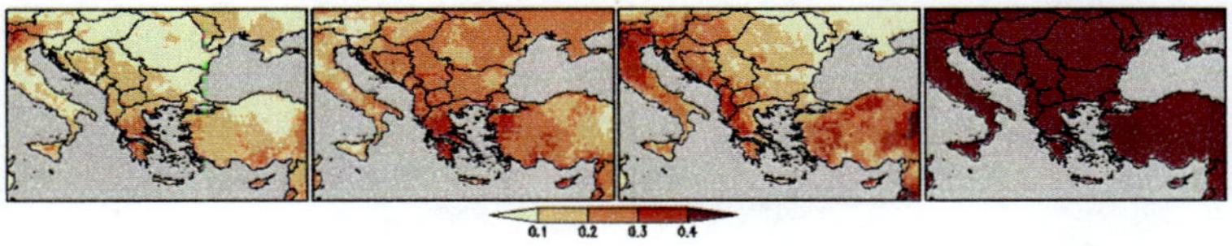

Fig. 5. KS-Statistic for *tn, tg, tx* and *rr* (from left to right). Color darkening from yellow to brown indicates increasing dissimilarity.

3 Projected Future Climate

3.1 Seasonal and Annual Means

The multiyear (i.e. over the whole period of interest) seasonal means of *tg* and *rr* are most frequently used as estimation metrics in the studies of the projected future climate. The raw model output consists of *tg* and *rr* on daily basis for the NF from the three considered emission scenarios and, thus far, for RCP2.6 and RCP8.5 for the FF. From these quantities, we have calculated the winter (December–February, noted further as DJF), spring (March–May, MAM), summer (June–August, JJA), autumn (September–November, SON) as well as the annual (January–December, ANN) multiyear means of *tg* and *rr*, as shown on Fig. 6 in the case of the *tg* for the NF.

The most apparent result of the analysis on Fig. 6 is that the distribution patterns of the seasonal and annual means of the mean temperature are practically identical, with relatively small graduate increase from RCP2.6 to RCP8.5 in direction warming over the bigger part of the domain.

Figure 7 is similar to Fig. 6 but for the daily precipitation sum *rr*. As could be expected, the field of the precipitation is more heterogeneous then this of the mean temperature. Similarly to Fig. 6, the fields of the seasonal and annual means are identical, in this case, however, without apparent difference from scenario to scenario.

Figure 8 shows the fields of the multiyear mean seasonal and annual mean temperature for the FF. The distribution of the *tg* is spatially coherent, as in the similar Fig. 6. Again, there is gradual increase of the temperature from RCP2.6 to RCP8.5 but in this case the difference between both scenarios seems more significant. The reason could be rooted in the longer period before the beginning of the FF, leading to a greater general effect of all cumulative factors.

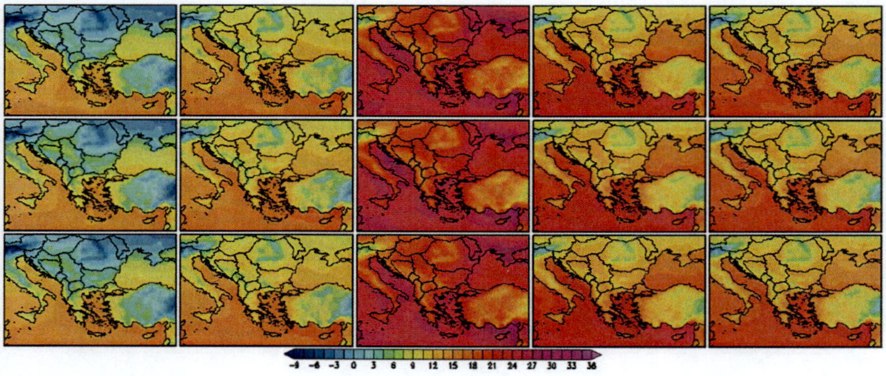

Fig. 6. Multiyear means for the NF of the *tg* (units: °C) for DJF, MAM, JJA and SON as well as ANN (from left to right) for RCP2.6 (first row), RCP4.5 (second row) and RCP8.5 (third row). Note that the color scale is different than this on Figs. 1, 2 and 3.

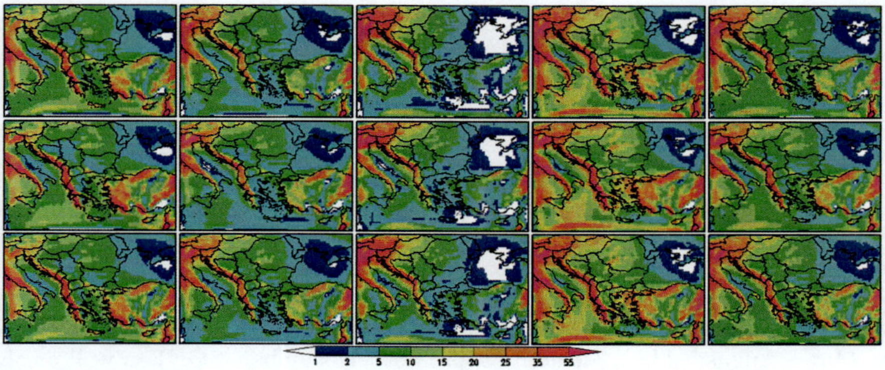

Fig. 7. Same as Fig. 6 but for the *rr*. The units are mm.

Figure 9 is analogous to Fig. 7 but for the FF. The spatial structure of the mean precipitation field is very similar, without principal changes from season to season and an annual basis; There are no apparent differences between both scenarios. The big and persistent dry spot over Black Sea, clearly distinguishable from all Fig. 8 subplots, seems larger here, particularly in the summer.

3.2 Climate Change Signal

The absolute change in the mean temperature $tg_{FT} - tg_{CR}$, respectively the relative change of the mean precipitation $(rr_{FT} - rr_{CR})/rr_{CR}$, where the subscript FT denotes the future period, NF or FF, are the most natural and widely used estimators of the projected climate change signal. Figure 10 shows the seasonal and annual fields of the multiyear mean temperature and the daily precipitation

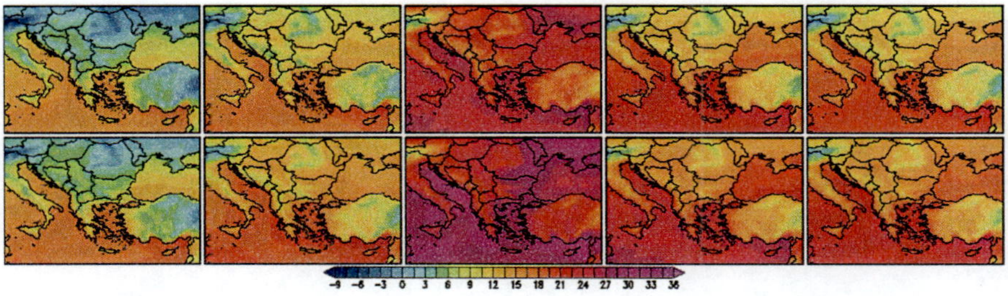

Fig. 8. Multiyear means for the FF of the *tg* (units: °C) for DJF, MAM, JJA and SON as well as ANN (from left to right) for RCP2.6 (first row) and RCP8.5 (second row).

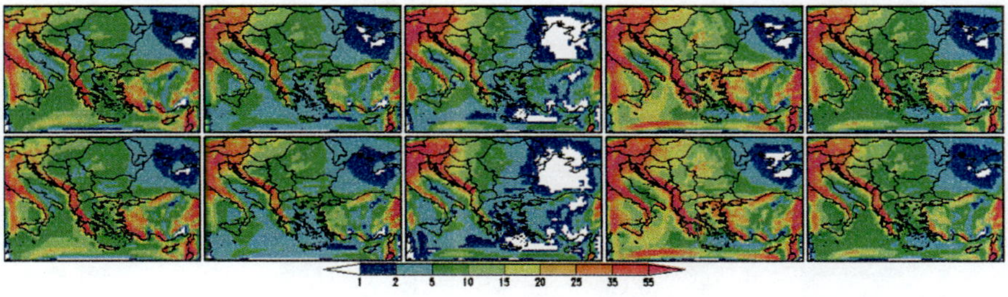

Fig. 9. Multiyear means for the FF of the *rr* (units: mm) for DJF, MAM, JJA and SON as well as ANN (from left to right) for RCP2.6 (first row) and RCP8.5 (second row).

sum for the CR (i.e. tg_{CR} and rr_{CR}), which are the basis for estimating the temporal changes. The absolute change of the multiyear means for the near future of the mean temperature is shown on Fig. 11.

The most obvious and significant result from the analysis of the temperature change is, that it is positive over the whole domain for all seasons and scenarios. The interseasonal changes, as well as these between the different parts of the domain, which are typically in the interval 1.5–2.5 °C, are almost negligible. Since the differences between the three scenarios seem also relatively small and not systematic, the fact that the signal over some regions, for example northern Italy and the northwestern part of the Balkan Peninsula, is stronger for RCP4.5 than RCP8.5, looks weird.

The relative change of the multiyear means of the mean daily precipitation sum is shown on Fig. 12.

The spatially prevailing signal for the winter and spring as well as this on annual basis is positive (i.e. the climate will become wetter) for all scenarios. The tendency for the summer and the autumn is opposite over the bigger part of the domain. The stronger saturation of the colors, both blue and red, of the subplots on the second row suggests more significant changes, in both directions, for RCP4.5. Keeping in mind that this scenario could be accepted as intermittent between the other two, this result is rather strange. It is worth to emphasize also

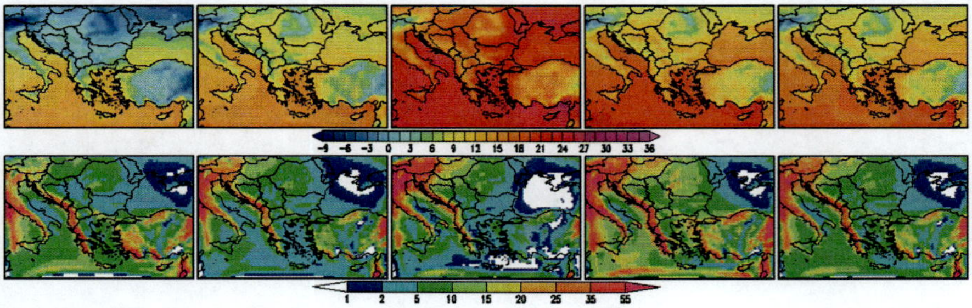

Fig. 10. Multiyear means for the CR of the the *tg* (units: °C, first row) and *rr* (units: mm, second row) for DJF, MAM, JJA and SON as well as ANN (from left to right).

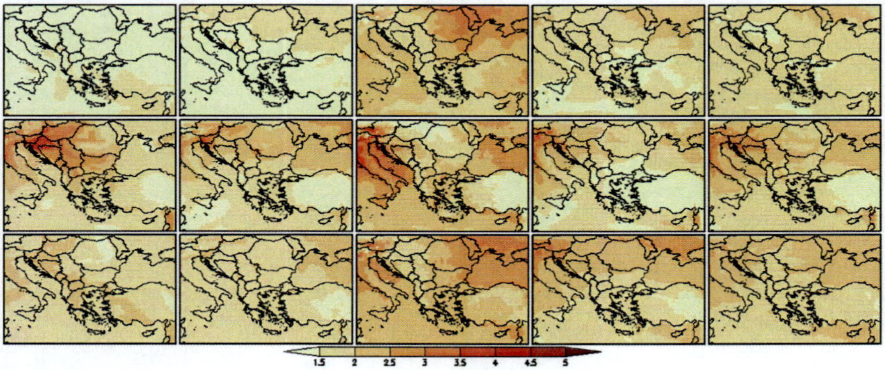

Fig. 11. Absolute change of the multiyear means for the NF of the *tg* (units: °C) for DJF, MAM, JJA and SON as well as ANN (from left to right) for RCP2.6 (first row), RCP4.5 (second row) and RCP8.5 (third row).

the relatively strong opposite tendencies over neighboring regions. The reason could be, partially, rooted in the very complex topography of the domain—fragmented coastlines and, especially near the Dinaric mountains and the Alps, steep change of the altitude.

The temperature and precipitation climate change signal is shown in the common plot Fig. 13 for the sake of brevity.

Similar to Fig. 11, the interseasonal temperature changes for both RCP2.6 and RCP8.5 for all seasons except the spring-summer and summer-autumn transition are low. The spatial distribution of the signal is also almost homogeneous, without clear horizontal gradient. The typical values of the temperature change for RCP2.6 for the NF are in the range 1.5–2.5 °C, which are close to the corresponding value for the NF. Unlike the signal for the NF, however, the difference between the two scenarios, RCP2.6 and RCP8.5, is remarkable. The maximum temperature rise is during the summer when the temperature changes above 4.5 °C over the larger portion of the domain. The spatial distribution of the pre-

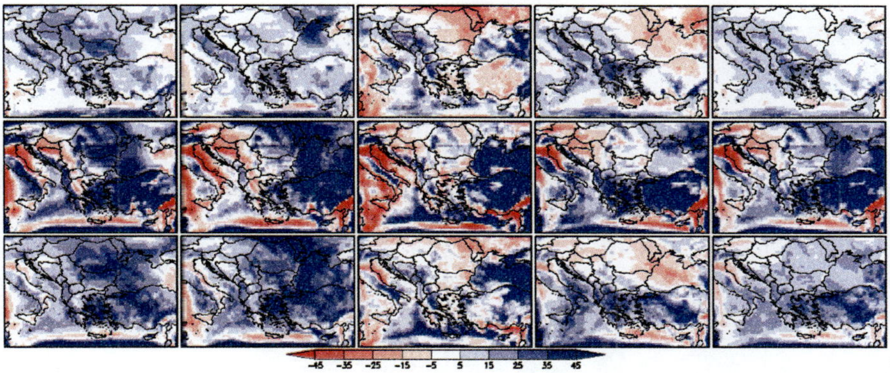

Fig. 12. Relative (in %) change of the multiyear means for the NF of the *rr* for DJF, MAM, JJA and SON as well as ANN (from left to right) for RCP2.6 (first row), RCP4.5 (second row) and RCP8.5 (third row).

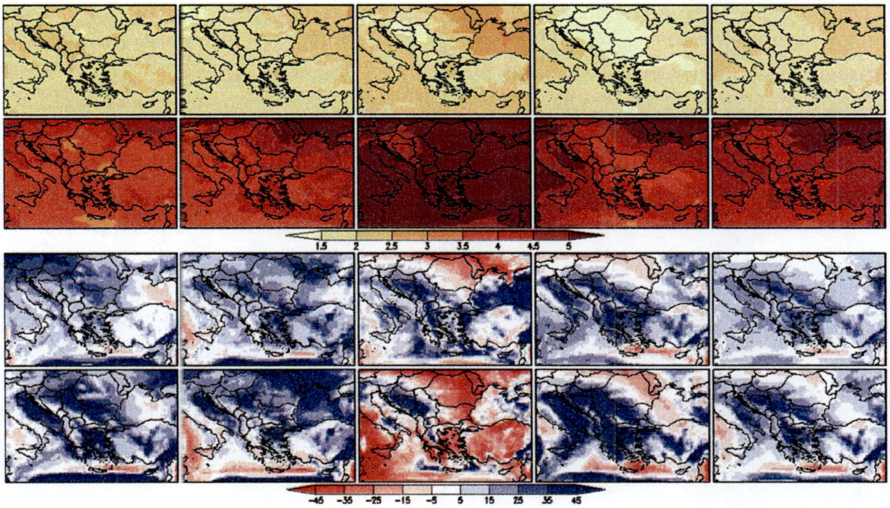

Fig. 13. Absolute change of the multiyear means for the FF of the *tg* (units: °C, upper section) for DJF, MAM, JJA and SON as well as ANN (from left to right) for RCP2.6 (first row) and RCP8.5 (second row). Relative (in %, lower section) change of the multiyear means for the FF of the *rr* for DJF, MAM, JJA and SON as well as ANN (from left to right) for RCP2.6 (first row) and RCP8.5 (second row).

cipitation change shows prevailing positive signal for all seasons except for the summer and on an annual basis. The projected increase is roughly 25–35% for RCP2.6 and 35–45% for RCP8.5. The expected precipitation reduction in the summer reaches values of 35–45% for RCP8.5 over Bulgaria and Romania.

4 Conclusion

The performed simulations of the present climate reveals that the mean temperatures, far from the tails of the distribution (i.e. close to the median), are relatively well reproduced, with acceptable low bias over the internal part of the domain. This is the most important conclusion of the outcomes from Sect. 2. However the extreme temperatures, as a whole, are not so well simulated—the minimum temperatures are overestimated and the maximum temperatures are underestimated. This is a common issue of most RCM, which shows similar weaknesses [10, 17]. In the present study, the bias for the upper tail of the distribution of the maximum temperatures is significantly bigger than the bias for the lower tail of the distribution of the minimum temperatures. This result confirm our previous findings [3, 7] that the model is generally colder. The spatial structure of the precipitation bias is more heterogeneous, but the model tends to overestimate the daily sums. This finding also agrees with the results of the studies mentioned above. The KS-test reveals that neither the temperatures (*tn, tg, tx*), nor the precipitation could originate from the same distributions as the corresponding references. Nevertheless, the distributional similarity for all these parameters is biggest in the regions, where the biases are minimal.

The simulation results for the estimated seasonal and annual average of mean temperature and precipitation reveal neither major changes in the spatial distribution of the two main parameters, nor interseasonal shifts.

The most significant result according to the projected temperature change which agrees with the prevailing number of modern evaluation studies (see [6, 13] and references therein) is the spatial dominating positive signal. Generally, it increases from RCP2.6 to RCP8.5, more clearly in the FF and is remarkably bigger for FF compared to NF for the RCP8.5 scenario. Despite all the uncertainties, it could be assumed from these facts, that the warming would be the general tendency in the 21th century. Such warming is the natural continuation of the tendency already detected by the analysis of the historical time series for the previous decades. The maximum values are obtained for the summer, which is prerequisite for more frequent and/or longer extremely hot episodes with all negative consequences.

As expected, the projected precipitation change is more complex, both in time and space. It is worth to emphasize, again, the expected summer months reduction, which could amplify the negative impact of the expected hotter climate.

The study could be continued in many directions, whereas the necessary next step is to finalize the simulations with RCP4.5 for the FF. More in-depth investigations of the future climate could also be performed, always keeping in mind the high ecological and social importance of the climatological research.

Acknowledgements. This work has been carried out in the framework of the National Science Program 'Environmental Protection and Reduction of Risks of Adverse Events and Natural Disasters', approved by the Resolution of the Council of Ministers №577/17.08.2018 and supported by the Ministry of Education and Science

(MES) of Bulgaria (Agreement №DO-230/06-12-2018). We acknowledge the provided access to the e-infrastructure of the Centre for Advanced Computing and Data Processing, with the financial support by the Grant No BG05M2OP001-1.001-0003, financed by the Science and Education for Smart Growth Operational Program (2014–2020) and co-financed by the European Union through the European structural and Investment funds. This work was partially supported by the Bulgarian Ministry of Education and Science under the National Research Programme 'Young scientists and postdoctoral students' approved by DCM 577/17.08.2018. Bulgarian National Science Fund (grant DN 14/3 13.12.2017). We acknowledge the E-OBS dataset from the EU-FP6 project UERRA http://www.uerra.eu and the data providers in the ECA&D project https://www.ecad.eu.

References

1. Brands, S., Gutiérrez, J.M., Herrera, S., Cofiño, A.S.: On the use of reanalysis data for downscaling. J. Clim. **25**, 2517–2526 (2012). https://doi.org/10.1175/JCLI-D-11-00251.1
2. Chervenkov, H., Slavov, K.: Theil-sen estimator for the parameters of the generalized extreme value distributions: demonstration for meteorological applications. C.R. Acad. Bulgare Sci. **70**(12), 1701–1708 (2017)
3. Chervenkov, H., Ivanov, V., Gadzev, G., Ganev, K.: Sensitivity study of different RegCM4.4 model set-ups–recent results from the TVRegCM experiment. Cybern. Inf. Technol. **17**(5), 17–26 (2017)
4. Collins, W.J., Bellouin, N., Doutriaux-Boucher, M., Gedney, N., Halloran, P., Hinton, T., Hughes, J., Jones, C.D., Joshi, M., Liddicoat, S., Martin, G., O'Connor, F., Rae, J., Senior, C., Sitch, S., Totterdell, I., Wiltshire, A., Woodward, S.: Development and evaluation of an earth-system model - HadGEM2. Geosci. Model Dev. **4**(4), 1051–1075 (2011). https://doi.org/10.5194/gmd-4-1051-2011
5. Cornes, R., van der Schrier, G., van den Besselaar, E.J.M., Jones, P.D.: An ensemble version of the E-OBS temperature and precipitation datasets. J. Geophys. Res. Atmos. **123** (2018). https://doi.org/10.1029/2017JD028200
6. Déqué, M.: Frequency of precipitation and temperature extremes over France in an anthropogenic scenario: model results and statistical correction according to observed values. Global Planet. Change **57**(1–2), 16–26 (2007). https://doi.org/10.1016/j.gloplacha.2006.11.030
7. Gadzhev, G., Georgieva, I., Ganev, K., Ivanov, V., Miloshev, N., Chervenkov, H., Syrakov, D.: Cimate applications in a virtual research environment platform. Scalable Comput. Pract. Experience **19**(2), 107–118 (2018). https://doi.org/10.12694/scpe.v19i2.134
8. Halenka, T.: Regional climate modeling activities in CECILIA project: introduction, Idöjárás 112, III–IX (2008)
9. Jacob, D., Petersen, J., Eggert, B., Alias, A., Christensen, O.B., Bouwer, L.M., Braun, A., Colette, A., Déqué, M., Georgievski, G., Georgopoulou, E., Gobiet, A., Menut, L., Nikulin, G., Haensler, A., Hempelmann, N., Jones, C., Keuler, K., Kovats, S., Kröner, N., Kotlarski, S., Kriegsmann, A., Martin, E., van Meijgaard, E., Moseley, C., Pfeifer, S., Preuschmann, S., Radermacher, C., Radtke, K., Rechid, D., Rounsevell, M., Samuelsson, P., Somot, S., Soussana, J., Teichmann, C., Valentini, R., Vautard, R., Weber, B., Yiou, P.: EURO-CORDEX: new high-resolution climate change projections for European impact research. Reg. Environ. Change **14**(2), 563–578 (2014). https://doi.org/10.1007/s10113-013-0499-2

10. Kotlarski, S., Keuler, K., Christensen, O.B., Colette, A., Dèquè, M., Gobiet, A., Goergen, K., Jacob, D., Lüthi, D., van Meijgaard, E., Nikulin, G., Schär, C., Teichmann, C., Vautard, R., Warrach-Sagi, K., Wulfmeyer, V.: Regional climate modeling on European scales: a joint standard evaluation of the EURO-CORDEX RCM ensemble. Geosci. Model Dev. **7**, 1297–1333 (2014). https://doi.org/10.5194/gmd-7-1297-2014
11. Moss, R.H., Edmonds, J.A., Hibbard, K.A., Manning, M.R., Rose, S.K., Van Vuuren, D.P., Carter, T.R., Emori, S., Kainuma, M., Kram, T., Meehl, G.A., Mitchell, J.F.B., Nakicenovic, N., Riahi, K., Smith, S.J., Stouffer, R.J., Thomson, A.M., Weyant, J.P., Wilbanks, T.J.: The next generation of scenarios for climate change research and assessment. Nature **463**(7282), 747–756 (2010). https://doi.org/10.1038/nature08823
12. Pieczka, I., Pongrácz, R., André, K.S., Kelemen, F.D., Bartholy, J.: Sensitivity analysis of different parameterization schemes using RegCM4.3 for the Carpathian region. Theor. Appl. Climatol. 1–14 (2016). https://doi.org/10.1007/s00704-016-1941-4
13. Spiridonov, V., Valcheva, R.: Stability of climate change at a given interval in a 30-year future period Example for the territory of Bulgaria (2020–2050). C. R. Acad. Bulgare Sci. **70**(3), 405–410 (2017)
14. Taylor, K.E., Stouffer, R.J., Meehl, G.A.: An overview of CMIP5 and the experiment design. Bull. Amer. Meteor. Soc. **93**, 485–498 (2012). https://doi.org/10.1175/BAMS-D-11-00094.1
15. Valcheva, R., Peneva, E.: Sensitivity to the parametrization of cumulus convection in the RegCM4.3 simulations focused on Balkan Peninsula and Bulgaria. Ann. Sofia Univ. "St. Kliment Ohridski" Fac. Phys. **107**, 113–131 (2014)
16. Wilks, D.S.: Statistical Methods in the Atmospheric Sciences, 3rd edn. International Geophysics Series, vol. 100, p. 154. Academic Press, Cambridge (2011)
17. Xue, Y., Janjic, Z., Dudhia, J., Vasic, R., De Sales, F.: A review on regional dynamical downscaling in intraseasonal to seasonal simulation/prediction and major factors that affect downscaling ability. Atmos. Res. **147–148**, 68–85 (2014). https://doi.org/10.1016/j.atmosres.2014.05.001

Numerical Identification of Time-Dependent Volatility in European Options with Two-Stage Regime-Switching

Slavi G. Georgiev(✉) and Lubin G. Vulkov

University of Ruse, 8 Studentska Street, 7017 Ruse, Bulgaria
{sggeorgiev,lvalkov}@uni-ruse.bg

Abstract. We develop numerical algorithms to solve inverse problems of determining time-dependent volatility according to point measurements inside of a truncated domain for regime-switching models of European options. An average linearization in time of the diffusion terms of the initial-boundary problems is used. Difference schemes on Tavella-Randall grids are derived. The numerical method is based on a decomposition of the difference solution with respect to the volatility for which the transition to the new time layer is carried out by solving two discrete elliptic system problems. Numerical experiments are performed to verify the effectiveness and robustness of the new algorithms.

1 Introduction

Despite the popularity of the Black-Scholes-Merton (BSM) model [4], it *fails* in various ways, such as the fact that *the implied volatility is not constant.* During the past few decades many extensions to the Black-Scholes-Merton model have been introduced to provide more realistic descriptions for asset price dynamics. In particular, the regime switching model is becoming quite popular since it proves to better capture the changing beliefs of investors towards the states of certain financial markets, cf. [6]. Therefore, it has been introduced to the area of financial pricing and extensively studied by a number of researchers. For example, under regime-switching models, Buffington and Elliot [6], Zhang [9] and Zhu [31] worked on the valuation problem of European options. Recently, the regime-switching mechanism has also been introduced intoclassical stochastic volatility models to form regime-switching stochastic volatility models [12,21]. Another example is Siu et. al. [22], where the currency options are evaluated under a two-factor stochastic volatility regime-switching model. Optimal portfolio indifference pricing utility regime-switching models are derived by Valdez and Vargiolu [25].

The regime-switching model as an alternative to the BSM model is computationally inexpensive, compared with the stochastic volatility and jump-diffusion models. Under this model, the market parameters such as interest rate, drift,

I. Dimov and S. Fidanova (Eds.): HPC 2019, SCI 902, pp. 249–261, 2021.
https://doi.org/10.1007/978-3-030-55347-0_21

volatility and dividend are allowed to take diverse values in a finite number of regimes. Numerical methods for regime switching models are developed in many papers, see e. g. [14,30]. There is a considerable support for the regime switching model from empirical studies, cf. [6,9,16,28].

In this paper we concentrate on numerical identification of time-dependent volatility from European options in two-stage regime-switching economy. For these problems, a closed-form solution does not exist. Hence, numerical methods are employed to solve them. This is a challenging task since the inverse problems are ill-posed and nonlinear. There are several existing methods of implied volatility calibration for Black–Scholes models, [1–3,7,8,10,11,32].

To the best of our knowledge, there are only a few papers on inverse regime-switching modelling. In the paper [16], the inverse problem of determining the local volatility functions is firstly established and then solved via Tikhonov regularization to obtain the optimal solution, which is achieved iteratively through a newly designed numerical algorithm. The problem of recovering the time-dependent volatilities in each state using an inverse Stieltjes moment method is studied in [28].

The advantages of our method is that it is simple to implement and independent of discretization schemes and meshes used. It also works for multi-factor problems, see the algorithm of [13].

The rest of the paper is organized as follows. In the next section the mathematical model of the two-stage regime-switching is discussed, while in Sect. 3 the inverse problem is formulated. Section 4 is devoted to the finite difference solution of the direct problem, while in Sect. 5 algorithms for solution to the inverse problems are developed. Numerical examples to show the accuracy and efficiency of the algorithms are presented in Sect. 6, while some conclusions are given in the last section.

2 Mathematical Model

In the classical Black-Scholes model, the volatility for both call and put options is constant across all possible strike prices K and maturity times T. However, the constant volatility assumptions is not consistent with the real behaviour of the options market, where the volatility implied by the market prices depends on both K and T. Regime-switching models capture better this feature of implied volatility.

Assuming the underlying economy switches among a finite number of states, which is modeled by a finite Markov Chain X_t. For simplicity, in this paper we consider the case that there are only two stages: $X_t = 1$ and $X_t = 2$. Let $r_i, i = 1,2$ and $\sigma_i, i = 1,2$ be a set of discrete risk-free interests rates and volatilities, respectively. Let the matrix

$$Q = \begin{pmatrix} -q_1 & q_1 \\ q_2 & -q_2 \end{pmatrix}$$

denote the generator of X_t , where q_1 and q_2 are positive constants. Under the risk-neutral measure the stochastic process for the underlying asset S is

$$dS = r_{X_t}(t)Sdt + \sigma_{X_t}(t)dW,$$

where dW is the increment of a Wiener process, r_{X_t} and σ_{X_t} can take different values depending on different regimes.

If we let $C_1(S,t)$ and $C_2(S,t)$ be option prices for state 1 and state 2, respectively a system of coupled Black-Scholes equations for the option prices can be derived according to Buffington and Elliot [6]:

$$\begin{aligned} &\tfrac{\partial C_1}{\partial t} + \tfrac{1}{2}\sigma_1^2(t)S^2\tfrac{\partial^2 C_1}{\partial S^2} + r_1(t)S\tfrac{\partial C_1}{\partial S} - r_1(t)C_1 - q_1(C_1 - C_2) = 0, \\ &\tfrac{\partial C_2}{\partial t} + \tfrac{1}{2}\sigma_2^2(t)S^2\tfrac{\partial^2 C_2}{\partial S^2} + r_2(t)S\tfrac{\partial C_2}{\partial S} - r_2(t)C_2 - q_2(C_2 - C_1) = 0. \end{aligned} \tag{1}$$

Once the contractually defined payoff of the derivative securities

$$C_k(S,T) = g_k(S), \quad k = 1,2 \tag{2}$$

is specified, system (1) can be solved backwards in time from the maturity date $t = T$ to the present time $t = 0$ in order to obtain the current value of the contract.

The original problem (1), (2) is posed on the domain $[0, \infty)$. The boundary condition at $S = 0$ is obtained simply by setting $S = 0$ in equations (1) which results in the dynamical boundary conditions

$$\begin{aligned} \tfrac{\partial C_1}{\partial t} &= (r_1(t) + q_1)C_1 - q_1 C_2, \\ \tfrac{\partial C_2}{\partial t} &= -q_2 C_1 + (r_2(t) + q_2)C_2. \end{aligned} \tag{3}$$

In case we use implicit numerical schemes, we must truncate the domain to $[0, S_{max}]$ and then to impose boundary condition at $S = S_{max}$. To avoid generating large errors in the solution due to the approximation of the boundary conditions, the truncated domain must be large enough, which results in a large cost. There are number of papers devoted to this problem, see e. g. [1,12,21,27].

One very simple way to handle the asymptotic behavior at infinity is to specify a Dirichlet condition at $S = S_{max}$ For instance, in [1,23,27] the condition

$$C_k(S_{max}, t) = C_{k0}(t), \quad k = 1,2 \tag{4}$$

is considered. For an European call option, this specification follows from the arguments when $S \gg K$, it is unlikely that the option would expire worthless, so that holding this option is roughly equivalent to owning the underlying asset, reduced with the strike price K. Such a boundary condition for a call option is

$$C_k(S_{max}, t) = S_{max} - Ke^{-rt}, \; k = 1,2. \tag{5}$$

A more universal right boundary condition is the linear boundary condition, see e. g. [27]:

$$\frac{\partial^2 C_k}{\partial S^2}(S_{max}, t) = 0, \; k = 1,2. \tag{6}$$

The transformed method [26] is the most effective and also can be applied to the regime-switching models.

For a call option the terminal condition is

$$g_k(S) = \max(S - K, 0), \quad k = 1, 2. \tag{7}$$

The problem (1)–(7) with unknown function $\{C_k\}_{k=1}^{M}$ is called *direct problem.*

All these methods *require some knowledge about the behaviour of the solution for large* S: either a known value for the solution or an assumption for linearity. Here, for a fixed in advance S_{max} with given *observations* at the points $S = S_1^\star,\ S = S_2^\star,\ 0 < S_k^\star < S_{max}$

$$C_k(S_k^\star, t) = \varphi_k(t), \quad 0 < t < T, \quad k = 1, 2 \tag{8}$$

we wish to obtain approximately $C_k(S_{max}, t),\ 0 < t < T,\ k = 1, 2.$

We note that finding the right boundary condition could also be formulated as an *inverse problem.* Such a problem we have solved in [14].

3 The Implied Volatility Inverse Problem

In this section we briefly recall the fundamental concept of *implied volatility.* The classical Black-Scholes model replicates *one regime* and is a particular case of (1) when one assume that the volatility $\sigma_{X_t}(t)$ and the interest rate $r_{X_t}(t)$ are equal in all the regimes, or the transition intensities for switching from one regime to another are zero. This yields the standard Black-Scholes formula for call price.

The analytical solution of the European call option for constant coefficients is given by the famous Black-Scholes formula [4]

$$C^{BS}(S, t) = Se^{-d(T-t)}N(d_1) - Ke^{-r(T-t)}N(d_2).$$

Here, $N(x)$ is the cumulative distribution function of the standard normal distribution, i. e. $N(x) := \dfrac{1}{2\pi}\displaystyle\int_{-\infty}^{x} e^{-t^2/2}\mathrm{d}t$, where

$$d_1 = \frac{\ln(S/K) + (r - d + \sigma^2/2)\,(T - t)}{\sigma\sqrt{T - t}}, \quad d_2 = d_1 - \sigma\sqrt{T - t}.$$

Note that C^{BS} is a strictly increasing function of $\sigma > 0$ with range in the interval bounded by the lower and upper no-arbitrage limits of the call. This implies that there is a 1-1 correspondence between Black-Scholes prices and volatilities, where the prices fall between no-arbitrage limits and $\sigma > 0$. We thus define the *implied volatility* σ^{imp} of a quoted (observed) price C^{obs} at time t and spot S as the unique solution σ^{imp} to the equation

$$C^{obs} = C^{BS}(S, t; K, t, r, \sigma^{imp}).$$

The implied volatility language is widely used by traders and practitioners since it gives a reference upon which option prices can be compared across different strikes, underlying and observation times [1,3,5,7,8,10].

In [15] we developed an algorithm for identifying the volatility in a regime–switching economy in the constant volatility case. Based on the formula of Naik [18], we propose a numerical and an analytical approaches to compute the implied volatility from point measurements.

It is well-known empirical remark that calls having different strikes, but otherwise identical, have different implied volatilities. This important observation is called the *smile effect*. It has been subject of intense research, see e. g [1–3,5,7,8].

Therefore, following this discussion, we consider the inverse problem of determining the volatilities $\sigma_1^2(t)$ and $\sigma_2^2(t)$ in (1)–(4) (or (5) or (6)) with additional conditions (8).

4 Solution of the Direct Problem

Since closed-form solutions in the general time-dependent case are not available for the system of equations (1), numerical solutions for the option prices C_1, C_2 are approximated in this section.

The results in this paper hold for arbitrary spatial grid, but we shall particularly consider grids that are obtained by Tavella–Randall like grid [23]. Let $\psi : [a,b] \to [L,S]$ be any given continuous function that is strictly increasing and satisfies $\psi(a) = L$, $\psi(b) = S$. Let integer $I \geq 3$ and

$$\xi = a + i\triangle\xi(0 \leq i \leq I+1) \text{ with } \triangle\xi = {}^{b-a}/_{I+1}. \tag{9}$$

Then a grid $L = S_0 < S_1 \cdots < S_I < S_{I+1} = S$ is defined by the transformation $S = \psi(\xi_i)$ $(0 \leq i \leq I+1)$ if the function ψ is sufficiently smooth, then also the grid is smooth in the sense that there exist real constants $\mathscr{C}_0, \mathscr{C}_1, \mathscr{C}_2 > 0$ (independent of i and I) such that the mesh width $h_i = S_i - S_{i-1}$ satisfy

$$\mathscr{C}_0\triangle\xi \leq h_i \leq \mathscr{C}_1\triangle\xi \text{ and } |h_{i+1} - h_i| \leq \mathscr{C}_2(\triangle\xi)^2.$$

We next formulate the FDS that we consider for the semidiscretization of (1). Let $f : [0, S_{max}] \to R$ by any given function and $1 \leq i \leq I$. Write $H_i = h_i + h_{i+1}$. For approximating the first derivative $f'(S_i)$ we deal with two central FDS:

$$f'(S_i) \approx -\frac{h_{i+1}}{h_i H_i} f(S_{i-1}) + \frac{h_{i+1} - h_i}{h_i h_{i+1}} f(S_i) + \frac{h_i}{h_{i+1} H_i} f(S_{i+1}), \tag{10}$$

$$f'(S_i) \approx (f(S_{i+1}) - f(S_{i-1}))/H_i \tag{11}$$

For approximating the second derivative $f''(S_i)$ we consider the central FDS

$$f''(S_i) \approx \frac{2}{h_i H_i} f(S_{i-1}) - \frac{2}{h_i h_{i+1}} f(S_i) + \frac{2}{h_{i+1} H_i} f(S_{i+1})$$

Letting $c_{k,i}^{j} \cong C_k(S_i, \tau^j)$, we arrive at the implicit scheme for Eq. (1) [14]:

$$\frac{c_{k,0}^{j+1} - c_{k,0}^{j}}{\triangle\tau} = -(r_k^{j+1} + q_k)c_{k,0}^{j+1} + q_k c_{3-k,0}^{j+1},$$

$$\frac{c_{k,i}^{j+1} - c_{k,i}^{j}}{\triangle\tau} = (\sigma_k^{j+1})^2 S_i^2 \left(\frac{c_{k,i-1}^{j+1}}{h_i H_i} - \frac{c_{k,i}^{j+1}}{h_i h_{i+1}} + \frac{c_{k,i+1}^{j+1}}{h_{i+1} H_i}\right) \tag{12}$$

$$+r_k^{j+1} S_i \left\{ \begin{array}{ll} \frac{c_{k,i+1}^{j+1} - c_{k,i-1}^{j+1}}{H_i}, & \text{for (11)}, \\ -\frac{h_{i+1}}{h_i H_i} c_{k,i-1}^{j+1} + \frac{h_{i+1} - h_i}{h_i h_{i+1}} c_{k,i}^{j+1} + \frac{h_i}{h_{i+1} H_i} c_{k,i+1}^{j+1} & \text{for (10)} \end{array} \right\}$$

$$-(r_k^{j+1} + q_k)c_{k,i}^{j+1} + q_k c_{3-k,i}^{j+1}, \quad (1 \le i < I),$$

$$\frac{c_{k,I}^{j+1} - c_{k,I}^{j}}{\triangle\tau} = r_k^{j+1} S_i \left(\frac{c_{k,I}^{j+1}}{h_I} - \frac{c_{k,I-1}^{j+1}}{h_I}\right)$$

$$-(r_k^{j+1} + q_k)c_{k,I}^{j+1} + q_k c_{3-k,I}^{j+1}, (0 \le j < J). \tag{13}$$

Then we define $c_k^j = \left[c_{k,1}^j, \dots, c_{k,I-1}^j\right]^\top$, $E_k = -\triangle\tau q_k \mathrm{I}$, where I is the $(I+1) \times (I+1)$ identity matrix, and two $(I+1) \times (I+1)$ matrices $M_k,\ k = 1, 2$ given by

$$M_k = \begin{bmatrix} C_{k,0} & B_{k,0} & & & \\ A_{k,1} & C_{k,1} & B_{k,1} & & \\ & \ddots & \ddots & \ddots & \\ & & A_{k,I-1} & C_{k,I-1} & B_{k,I-1} \\ & & & A_{k,I} & C_{k,I} \end{bmatrix}.$$

If $\mathbf{c}^j = \left[c_1^j, c_2^j\right]^\top$ and $\check{\mathbf{c}}^j = \left[c_2^j, c_1^j\right]^\top$, then the **implicit** scheme system could be rewritten in matrix form given below

$$\mathbf{M}\check{\mathbf{c}}^{j+1} = \mathbf{c}^j, \quad \text{where } \mathbf{M} = \begin{bmatrix} E_1 & M_1 \\ M_2 & E_2 \end{bmatrix}, \tag{14}$$

$$C_{k,0} = \triangle\tau(r_k^{j+1} + q_k) + 1, \quad B_{k,0} = 0, \quad A_{k,i} = -\frac{\triangle\tau(\sigma_k^{j+1})^2 S_i^2}{h_i H_i} + \frac{\triangle\tau r_k^{j+1} S_i h_{i+1}}{h_i H_i},$$

$$C_{k,i} = \frac{\triangle\tau(\sigma_k^{j+1})^2 S_i^2}{h_i h_{i+1}} - \frac{\triangle\tau r_k^{j+1} S_i (h_{i+1} - h_i)}{h_i h_{i+1}} + \triangle\tau(r_k^{j+1} + q_k) + 1,$$

$$B_{k,i} = -\frac{\triangle\tau(\sigma_k^{j+1})^2 S_i^2}{h_{i+1} H_i} - \frac{\triangle\tau r_k^{j+1} S_i h_i}{h_{i+1} H_i}, \ (1 \le i < I),$$

$$A_{k,I} = \frac{\triangle\tau r_k^{j+1} S_I}{h_I}, \qquad C_{k,I} = -\frac{\triangle\tau r_k^{j+1} S_I}{h_I} + \triangle\tau(r_k^{j+1} + q_k) + 1.$$

Solving the block diagonal system (14) is done making use of the *Schur complement* [29]. The Schur complement of E_2 is defined as $M/E_2 := E_1 - M_1 E_2^{-1} M_2$. Then, if we eliminate the second block equation for c_1^{j+1} we could solve for c_2^{j+1}:

$$(M/E_2) c_2^{j+1} = c_1^j - M_1 E_2^{-1} c_2^j.$$

Substituting the result gives an equation for c_1^{j+1}:

$$E_2 c_1^{j+1} = c_2^j - M_2 c_2^{j+1}.$$

Applying the Schur complement is useful here because we replace inverting the block diagonal $2(I+1) \times 2(I+1)$ matrix M with inverting the diagonal $(I+1) \times (I+1)$ matrix E_2, which is trivial. This yields a significant computational gain.

5 Solution of the Inverse Problem

For the classical Black-Scholes equation, i.e. in the case of one regime and constant coefficients, there are different approaches based on the famous Black-Scholes formula in determining the implied volatility for the options, see e.g. the recent review [19]. Here we present an algorithm considering the regime-switching economy with tho regimes. The inverse problem (12), (13) of determining the unknowns $\mathbf{c}^{j+1}$ and $(\boldsymbol{\sigma}^{j+1})^2 := \left((\sigma_1^{j+1})^2, (\sigma_2^{j+1})^2\right)$ is nonlinear. We use the approximation of the product ab at the mid-layer $\tau^{j+1/2}$, [13,17,20,24]:

$$a^{j+1/2} b^{j+1/2} = \frac{1}{2} a^{j+1} b^j + \frac{1}{2} a^j b^{j+1} + \mathscr{O}(\triangle\tau^2).$$

Now let us expose the algorithm for solving the inverse problem (12), (13):

5.1 Algorithm

We assume that a point observation of the form (8) $C_k(S_k^\star, \tau) = \varphi_k(\tau)$, $k = 1, 2$ is given for all τ, $(0 < \tau \leq T)$, where $i^\star$ is the spatial grid node of the measurement $S^\star$.

To solve the problem (12), (13) for $(\sigma_k^{j+1})^2$, $k = 1, 2$ at the new time layer, we employ the decomposition:

$$c_{k,i}^{j+1} = z_{k,i}^{j+1} + (\sigma_1^{j+1})^2 u_{k,i}^{j+1} + (\sigma_2^{j+1})^2 w_{k,i}^{j+1}, \quad (0 \leq i \leq I). \tag{15}$$

Step 1
Solve for the auxiliary function $\mathbf{z}^{j+1}$ the system

$$\frac{z_{k,0}^{j+1} - c_{k,0}^j}{\triangle\tau} = -(r_k^{j+1} + q_k) z_{k,0}^{j+1} + q_k z_{3-k,0}^{j+1},$$

$$\frac{z_{k,i}^{j+1} - c_{k,i}^{j}}{\triangle\tau} = \frac{(\sigma_k^j)^2 S_i^2}{2}\left(\frac{z_{k,i-1}^{j+1}}{h_i H_i} - \frac{z_{k,i}^{j+1}}{h_i h_{i+1}} + \frac{z_{k,i+1}^{j+1}}{h_{i+1} H_i}\right)$$

$$+r_k^{j+1} S_i \left\{\begin{array}{ll} \frac{z_{k,i+1}^{j+1} - z_{k,i-1}^{j+1}}{H_i}, & \text{for (11)}, \\ -\frac{h_{i+1}}{h_i H_i} z_{k,i-1}^{j+1} + \frac{h_{i+1}-h_i}{h_i h_{i+1}} z_{k,i}^{j+1} + \frac{h_i}{h_{i+1} H_i} z_{k,i+1}^{j+1} & \text{for (10)} \end{array}\right\}$$

$$-(r_k^{j+1} + q_k) z_{k,i}^{j+1} + q_k z_{3-k,i}^{j+1}, \ (1 \le i < I),$$

$$\frac{z_{k,I}^{j+1} - c_{k,I}^{j}}{\triangle\tau} = r_k^{j+1} S_i \left(\frac{z_{k,I}^{j+1}}{h_I} - \frac{z_{k,I-1}^{j+1}}{h_I}\right)$$

$$-(r_k^{j+1} + q_k) z_{k,I}^{j+1} + q_k z_{3-k,I}^{j+1}, (0 \le j < J).$$

Step 2.1
Solve for the auxiliary function $\mathbf{u}^{j+1}$ the system

$$\frac{u_{k,0}^{j+1}}{\triangle\tau} = -(r_k^{j+1} + q_k) u_{k,0}^{j+1} + q_k u_{3-k,0}^{j+1},$$

$$\frac{u_{k,i}^{j+1}}{\triangle\tau} + \left\{\begin{array}{ll} -\frac{S_i^2}{2}\left(\frac{c_{k,i-1}^j}{h_i H_i} - \frac{c_{k,i}^j}{h_i h_{i+1}} + \frac{c_{k,i+1}^j}{h_{i+1} H_i}\right) & \text{for } k = 1, \\ 0 & \text{for } k = 2 \end{array}\right\} =$$

$$\frac{(\sigma_k^j)^2 S_i^2}{2}\left(\frac{u_{k,i-1}^{j+1}}{h_i H_i} - \frac{u_{k,i}^{j+1}}{h_i h_{i+1}} + \frac{u_{k,i+1}^{j+1}}{h_{i+1} H_i}\right)$$

$$+r_k^{j+1} S_i \left\{\begin{array}{ll} \frac{u_{k,i+1}^{j+1} - u_{k,i-1}^{j+1}}{H_i}, & \text{for (11)}, \\ -\frac{h_{i+1}}{h_i H_i} u_{k,i-1}^{j+1} + \frac{h_{i+1}-h_i}{h_i h_{i+1}} u_{k,i}^{j+1} + \frac{h_i}{h_{i+1} H_i} u_{k,i+1}^{j+1} & \text{for (10)} \end{array}\right\}$$

$$-(r_k^{j+1} + q_k) u_{k,i}^{j+1} + q_k u_{3-k,i}^{j+1}, \ (1 \le i < I),$$

$$\frac{u_{k,I}^{j+1}}{\triangle\tau} = r_k^{j+1} S_i \left(\frac{u_{k,I}^{j+1}}{h_I} - \frac{u_{k,I-1}^{j+1}}{h_I}\right)$$

$$-(r_k^{j+1} + q_k) u_{k,I}^{j+1} + q_k u_{3-k,I}^{j+1}, (0 \le j < J).$$

Step 2.2
Solve for the auxiliary function $\mathbf{w}^{j+1}$ the system

$$\frac{w_{k,0}^{j+1}}{\triangle\tau} = -(r_k^{j+1} + q_k) w_{k,0}^{j+1} + q_k w_{3-k,0}^{j+1},$$

$$\frac{w_{k,i}^{j+1}}{\triangle\tau} + \left\{\begin{array}{ll} 0 & \text{for } k = 1, \\ -\frac{S_i^2}{2}\left(\frac{c_{k,i-1}^j}{h_i H_i} - \frac{c_{k,i}^j}{h_i h_{i+1}} + \frac{c_{k,i+1}^j}{h_{i+1} H_i}\right) & \text{for } k = 2 \end{array}\right\} =$$

$$\frac{(\sigma_k^j)^2 S_i^2}{2}\left(\frac{w_{k,i-1}^{j+1}}{h_i H_i} - \frac{w_{k,i}^{j+1}}{h_i h_{i+1}} + \frac{w_{k,i+1}^{j+1}}{h_{i+1} H_i}\right)$$

$$+r_k^{j+1} S_i \left\{ \begin{array}{ll} \frac{w_{k,i+1}^{j+1} - w_{k,i-1}^{j+1}}{H_i}, & \text{for (11)}, \\ -\frac{h_{i+1}}{h_i H_i} w_{k,i-1}^{j+1} + \frac{h_{i+1}-h_i}{h_i h_{i+1}} w_{k,i}^{j+1} + \frac{h_i}{h_{i+1} H_i} w_{k,i+1}^{j+1} & \text{for (10)} \end{array} \right\}$$

$$-(r_k^{j+1} + q_k) w_{k,i}^{j+1} + q_k w_{3-k,i}^{j+1}, \quad (1 \le i < I),$$

$$\frac{w_{k,I}^{j+1}}{\triangle\tau} = r_k^{j+1} S_i \left(\frac{w_{k,I}^{j+1}}{h_I} - \frac{w_{k,I-1}^{j+1}}{h_I}\right)$$

$$-(r_k^{j+1} + q_k) w_{k,I}^{j+1} + q_k w_{3-k,I}^{j+1}, (0 \le j < J).$$

Step 3
In order to compute $(\boldsymbol{\sigma}^{j+1})^2$, we substitute $\varphi_k^{j+1} := \varphi_k(\tau^{j+1}) = z_{k,i^\star}^{j+1} + (\sigma_1^{j+1})^2 u_{k,i^\star}^{j+1} + (\sigma_2^{j+1})^2 w_{k,i^\star}^{j+1}$ in (15) to obtain

$$\left(\boldsymbol{\sigma}^{j+1}\right)^2 = \left(\varphi_1^{j+1} - z_{1,i^\star}^{j+1}, \varphi_2^{j+1} - z_{2,i^\star}^{j+1}\right) \begin{pmatrix} u_{1,i^\star}^{j+1} & u_{2,i^\star}^{j+1} \\ w_{1,i^\star}^{j+1} & w_{2,i^\star}^{j+1} \end{pmatrix}^{-1}. \qquad (16)$$

To recover the option price $\mathbf{c}^{j+1}$, we put the computed values $(\boldsymbol{\sigma}^{j+1})^2$ back in (15).

6 Computational Results

In this section we present numerical results in order to illustrate the accuracy and efficiency of the proposed algorithm for the solution to the time-dependent implied volatility inverse problem.

6.1 Direct Problem

We define the direct problem as the system of PDEs (1), left boundary condition (3), linear right boundary condition (6) and initial condition (7). Here we take $S_{max} = 500$, $K = 100$, $\mathbf{r} = (5\%, 5\%)$, $\boldsymbol{\sigma} = (25\%, 15\%)$, $\mathbf{q} = (0.5, 0.5)$ and $T = 1$ year time to expiry.

For the computational experiments we use $I = 1001$ spatial nodes, unevenly distributed in the spatial domain by the transformation (9), where $\psi(\xi) = K + \alpha \sinh(\xi)$, $\xi \in \left[\sinh^{-1}\left(-{}^{K}\!/\!_{\alpha}\right), \sinh^{-1}\left({}^{S_{max}-K}\!/\!_{\alpha}\right)\right]$ and $\alpha = {}^{K}\!/\!_{5}$. The temporal step is constant $\triangle\tau = {}^{1}\!/\!_{100}$.

On Fig. 1 we see the solution to the formulated direct problem. The numerical results well agree with the analytical solution given in [18].

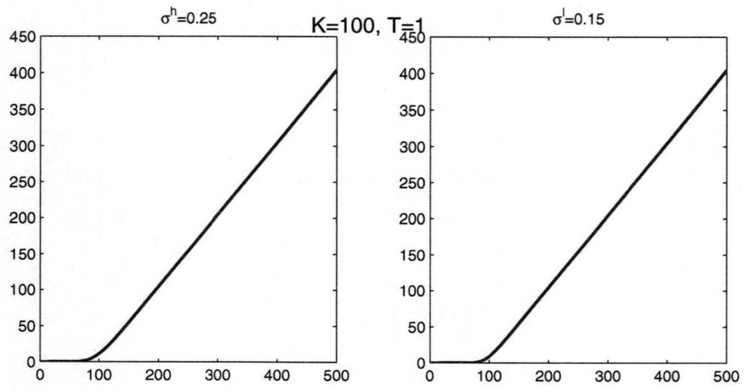

Fig. 1. Solution to the direct problem

6.2 Inverse Problem

Let us define an observation of type (8) and formulate the inverse problem as the system of PDEs (1), left boundary condition (3), linear right boundary condition (6) and initial condition (7). Here we use $I = 10001$ equidistant spatial nodes and time step $\triangle\tau = {}^1\!/_{12}$. In order to demonstrate the capabilities of the algorithm, we set $\sigma^h = \sigma^h(\tau) = 0.25 + 0.06 * \tau$ and $\sigma^l = \sigma^l(\tau) = 0.15 + 0.03 * \tau$.

To reconstruct the implied volatility $(\boldsymbol{\sigma}^{j+1})^2$, we invert the square matrix

$$\begin{pmatrix} u_{1,i^\star}^{j+1} & u_{2,i^\star}^{j+1} \\ w_{1,i^\star}^{j+1} & w_{2,i^\star}^{j+1} \end{pmatrix}.$$

For exploring the stability and convergence of the algorithm, we are interested in the determinant

$$D(i^\star, \mathscr{C}) := \begin{vmatrix} u_{1,i^\star}^{j+1} & u_{2,i^\star}^{j+1} \\ w_{1,i^\star}^{j+1} & w_{2,i^\star}^{j+1} \end{vmatrix} = u_{1,i^\star}^{j+1} w_{2,i^\star}^{j+1} - u_{2,i^\star}^{j+1} w_{1,i^\star}^{j+1}.$$

In Table 1 we present the values of D for different points of measurement $i^\star$ and different values of the discrete Courant number $\mathscr{C} := {}^{\triangle\tau}\!/_{2h^2}$ for the system (1). (The symbol ∞ in the table represents a very big number; the actual value is `Inf` in the MATLAB environment.)

Table 1. Values of $D(i^\star, \mathscr{C})$ for various $i^\star$ and $\mathscr{C}$ at the last time-stepping

$i^\star \setminus \mathscr{C}$	10,526316	13,333333	18,181818	28,571429	66,666667
96	−9,88949e+27	11,678957	22,142781	69,523105	750,25615
98	1,02425e+34	11,113759	31,435989	76,523371	554,64821
100	−121,80508	−106034677	−1919,0966	−322,05632	143,62992
102	−7,82416e+32	1,8913180	∞	73097,288	730,74394
104	6,98620e+17	5,44634e+58	19,017990	42,322066	887,09828
106	−33628201	2064752977	4,28523e+14	−1,24194e+42	839,65047

As for the inverse problems in general, the bigger the Courant number, the more stable the inverse problem computations become [14].

Another interesting implication is that the quality of the reconstructed volatility $\boldsymbol{\sigma}^{j+1}$ gets higher when the measurement $S^\star$ is taken around the strike price, and decreases with pushing $S^\star$ away from K. However, if the observation is taken *exactly* at the strike level, the results are not acceptable due to the discontinuity of the payoff function in this point. In Table 1 we can analyze the behaviour of $D(i^\star, \mathscr{C})$ according to some reasonable values of its arguments.

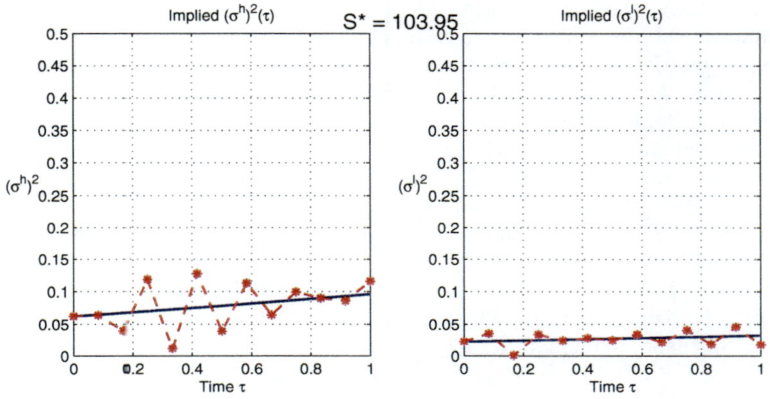

Fig. 2. Solution to the inverse problem

In Fig. 2 we can observe the implied volatility (16) for $S^\star = 103.95$ and $\mathscr{C} = 15,384615$. The recovered values slightly oscillates around the true ones. In the case of σ^h, the fluctuations are more pronounced, but they gradually fade away. The reconstructed option premia (15) match exactly their fair values.

7 Conclusion

In this paper the problem of finding the time-dependent volatility in a regime-switching economy is considered. We implement an algorithm based on a special decomposition of the numerical solution to compute the implied volatility if given point observations.

Our numerical experiments confirm the algorithm is capable of calculating the option prices and volatility levels in an efficient way.

In spite of the fact that we consider European call options, it is possible to extend and generalize the algorithm for computing the implied volatility of more complex options in a regime-switching model.

Acknowledgements. The authors are thankful to the reviewers for their constructive comments and suggestions, which significantly improved the quality of the paper.

This paper contains results of the work on project No 2019 - FNSE - 05, financed by "Scientific Research" Fund of Ruse University. The research is also supported by the Bulgarian National Science Fund under Project DN 12/4 "Advanced analytical and numerical methods for nonlinear differential equations with application in finance and environmental pollution" from 2017, and project No 2019 - FNSE - 03.

References

1. Achdou, Y., Pironneau, O.: Computational methods for option pricing. In: SIAM Frontiers in Applied Mathematics (2005)
2. Avellaneda, M., Friedman, C., Holmes, R., Samperi, D.: Calibrating volatility surfaces via relative entropy minimization. Appl. Math. Finance **4**, 37–64 (1997)
3. Baumeister, J.: Inverse problems in finance. In: Gestner, T., Kloeden, P. (eds.) Recent Developments in Computational Finance, Interdisciplinary Mathematical Sciences, vol. 14, pp. 81-159 (2013)
4. Black, F., Scholes, M.: The pricing of options and corporate liabilities. J. Polit. Econ. **61**, 637–654 (1973)
5. Bouchouev, I., Isakov, V.: The inverse problem of option pricing. Inverse Prob. **13**, 7–11 (1997)
6. Buffington, J., Elliott, R.J.: Regime switching and European options. In: Pasik-Duncan, B. (ed.) Stochastic Theory and Control. Proceedings of a Workshop, pp. 73–82. Springer, Heidelberg (2002)
7. Coleman, T.F., Li, Y., Verma, A.: Reconstructing the unknown local volatility function. World Sci. Book Chapters **2**, 77–102 (2015)
8. Crépey, S.: Calibration of the local volatility in a generalized Black-Scholes models using Tikhonov regularization. SIAM J. Math. Anal. **34**, 1183–1206 (2003)
9. Dai, M., Zhang, Q., Zhu, J.Q.: Trend following trading under regime switching models. SIAM J. Finance Math. **1**(1), 780–810 (2010)
10. Deng, Z.-C., Hon, Y.C., Isakov, V.: Recovery of the time-dependent volatility in optionpricing model. Inverse Prob. **32**, 115010 (2016)
11. Dupire, B.: Pricing with a smile. Risk **7**(2), 16–20 (1994)
12. Ehrhardt, M., Günther, M., ter Maten, E.J.W. (eds.) Novel Methods in Computational Finance. Springer (2017)
13. Georgiev, S.G., Vulkov, L.G.: Computational recovery of time-dependent volatility from integral observations in option pricing. J. Comput. Sci. **39**, 101054 (2019)
14. Georgiev, S.G., Vulkov, L.G.: Numerical determination of the right boundary condition for regime-switching models of European options from point observations. In: AIP Conference Proceedings, vol. 2048, p. 030003 (2018)
15. Georgiev, S.G.: Numerical and analytical computation of the implied volatility from option price measurements under regime–switching. In: AIP Conference Proceedings, vol. 2172, p. 070007 (2019)
16. He, X.-J., Zhu, S.-P.: On full calibration of hybrid local volatility and regime-switching models. J. Future Mark. **38**(5), 586–606 (2018)
17. Isakov, V.: Recovery of time-dependent volatility coefficient by linearization. Evol. Equ. Control Theory **3**(1), 119–134 (2014)
18. Naik, V.: Option valuation and hedging strategies with jumps in the volatility of asset returns. J. Finance **48**(5), 1969–1984 (1993)
19. Orlando, G., Taglilatela, G.: A review on implied volatility calculation. J. Comput Appl. Math. **320**, 202–220 (2017)

20. Samarskii, A.A., Vabishchevich, P.: Numerical Methods for Solving Inverse Problems of Mathematical Physics. Walter de Gruter, Berlin (2007)
21. Ševčovič, D., Stehlíková, B., Mikula, K.: Analytical and Numerical Methods for Pricing Financial Derivatives. Nova Science, Hauppauge (2011)
22. Siu, T.K., Yang, H., Liu, J.W.: Pricing currency option under two-factor Markov-modulated stochastic volatility models. Insur. Math. Econ. **43**(3), 295–302 (2008)
23. Tavella, D., Randall, C.: Pricing Financial Instruments: The Finite Difference Method. Wiley, Hoboken (2000)
24. Vabishchevich P.N., Klibanov, M.V.: Numerical identification of the leading coefficient of a parabolic equation. Differ. Equ. **52**(7), 896–953 (2016)
25. Valdez, A.R.L., Vargiolu, T.: Optimal portfolio in the regime-switching model, In: Dalang, R.C., Dozzi, M., Russo, F. (eds.) Proceedings of the Ascona'11 Seminar on Stochastic Analysis, Random Fields and Applications, pp. 435-449 (2013)
26. Valkov, R.: Fitted finite volume method for generalized Black-Scholes equation transformed on finite interval. Numer. Algorithm **65**, 195–220 (2014)
27. Windcliff, H., Forsyth, P.A., Vetzal, K.R.: Analysis of the stability of linear boundary condition for the Black-Sholes equation. J. Comp. Finance **8**, 65–92 (2004)
28. Xi, X., Rodrigo, M.R., Mamon, R.S.: Parameter estimation of a regime-switching model using an inverse Stieltjes moment approach. In: Cohen, S.N., Madan, D., Siu, T.K., Yang, H. (eds.) Stochastic Processes, Finance And Control: A Festschrift in Honor of Robert J. Elliott, pp. 549-569 World scientific, Singapore (2012)
29. Zhang, F.: Matrix Theory - Basic Results and Techniques, 2nd edn. Springer, Heidelberg (2011)
30. Zhang, K., Teo, K.L., Swarts, M.: A robust numerical scheme to pricing American options under regime-switching based on penalty method. Comput. Econ. **43**, 463–483 (2013)
31. Zhu, S.-P., Badzan, A., Lu, X.: A new exact solution for pricing European options in a two-state regime-switching economy. Comp. Math. Appl. **64**(8), 2744–2755 (2014)
32. Zubelli, J.P.: Inverse problems in finance, a short survey of calibration techniques. In: Kolev, N., Morettin, P. (eds.) Proceedings of the Second Brazilian Conference on Statistical Modelling in Insurance and Finance, pp. 64-76 (2005)

Multiple Richardson Extrapolation Applied to Explicit Runge–Kutta Methods

Teshome Bayleyegn[1] and Ágnes Havasi[2(✉)]

[1] Eötvös Loránd University, Pázmány Péter s. 1/C, Budapest 1117, Hungary
sbayleyegn130@gmail.com

[2] Eötvös Loránd University and MTA-ELTE Numerical Analysis and Large Networks Research Group, Pázmány Péter s. 1/C, Budapest 1117, Hungary
havasia@cs.elte.hu

Abstract. Richardson extrapolation has long been used to enhance the accuracy of time integration methods for solving differential equations. The original version of Richardson extrapolation is based on a suitable linear combination of numerical solutions obtained by the same numerical method with two different time-step sizes. This procedure can be extended to more than two step sizes in a natural way, and the resulting method is called repeated Richardson extrapolation. In this talk we investigate another possible generalization of the idea of Richardson extrapolation, where the extrapolation is applied to the combination of some underlying method and the classical Richardson extrapolation. The procedure obtained in this way, called multiple Richardson extrapolation, is analysed for accuracy and absolute stability when combined with some explicit Runge–Kutta methods.

1 Introduction and Motivation

Richardson extrapolation is a sequence acceleration method introduced by L. F. Richardson [3–5]. The original algorithm (which we will call Classical Richardson extrapolation and abbreviate as CRE) consists in the linear combination of two numerical solutions obtained by the same numerical method, but two different discretization parameters. It is primarily applied during the numerical solution of ordinary differential equations, where the two solutions are calculated by using time steps h and $h/2$ in the following manner.

Consider the Cauchy problem for a system of ODE's

$$\begin{cases} y'(t) = f(t, y), & t \in [0, T] \\ y(0) = y_0, \end{cases} \tag{1}$$

where the unknown function y is of the type $\mathbb{R} \to \mathbb{R}^d$ and $y_0 \in \mathbb{R}^d$. We define two meshes on $[0, T]$:

$$\Omega_h := \{t_n = nh : n = 0, 1, \ldots, N_t\}, \tag{2}$$

I. Dimov and S. Fidanova (Eds.): HPC 2019, SCI 902, pp. 262–270, 2021.
https://doi.org/10.1007/978-3-030-55347-0_22

where $N_t = T/h$, and

$$\Omega_{h/2} := \{t_k = kh/2 : k = 0, 1, \ldots, 2N_t\}. \tag{3}$$

We solve the problem on both meshes with the same convergent one-step numerical method of order p. The corresponding numerical solutions at time t_n of the coarse mesh will be denoted by z_n and w_n, respectively. If the exact solution $y(t)$ is $p+1$ times continuously differentiable, then

$$y(t_n) - z_n = K \cdot h^p + \mathcal{O}(h^{p+1}), \tag{4}$$

and

$$y(t_n) - w_n = K \cdot (h/2)^p + \mathcal{O}(h^{p+1}), \tag{5}$$

where the quantity K is independent of the chosen time step size. By eliminating the p-th order terms, we get the following linear combination, defining the CRE method:

$$y_{\text{CRE},n} := \frac{2^p w_n - z_n}{2^p - 1}, \tag{6}$$

which approximates the exact solution to the order $p+1$.

The above procedure can be applied in passive or active mode. In the passive CRE the combined solution is never used in the further computations, while in the active CRE from every point t_n of the coarse mesh we take the next time step from the combined solution on both meshes. When the CRE is applied, **three** time steps are performed during one large time step, which means roughly 1.5 times more computations than solving the problem with $h/2$ (i.e., by $2N_t$ steps) using the underlying method. However, if **we have already calculated** the solution with time step h (N_t steps), then the passive CRE hardly requires more time than performing $2N_t$ steps with the underlying method. When parallelized, the active RE does not require much more time, either, than taking $2N_t$ steps with the underlying method. In this paper we will only deal with active Richardson extrapolation.

It is natural to raise the question of how to obtain an even higher order of approximation to the exact solution. One possibility is to combine numerical solutions obtained by more than two different time-step sizes, e.g., by h, $h/2$ and $h/4$. Denoting by z_n, w_n and q_n the obtained numerical solutions, we have

$$y(t_n) - z_n = K_1 \cdot h^p + K_2 \cdot h^{p+1} + \mathcal{O}(h^{p+2}), \tag{7}$$

$$y(t_n) - w_n = K_1 \cdot (h/2)^p + K_2 \cdot (h/2)^{p+1} + \mathcal{O}(h^{p+2}), \tag{8}$$

$$y(t_n) - q_n = K_1 \cdot (h/4)^p + K_2 \cdot (h/4)^{p+1} + \mathcal{O}(h^{p+2}). \tag{9}$$

Eliminating K_1 and K_2, we get a method of order $p+2$. This procedure is called repeated Richardson extrapolation (RRE), and has been extensively studied (up to seven repetitions) by Zlatev et al. [6] regarding accuracy and absolute stability. Another approach could be to apply Richardson extrapolation to the combined method (underlying method + Richardson extrapolation), possibly several times. (The well-known Rombergs's rule of integral calculus is based on this idea.) We will refer to this procedure as multiple Richardson extrapolation (MRE).

In this paper we present the method of multiple Richardson extrapolation, and investigate its accuracy and absolute stability by different choices of the underlying explicit Runge–Kutta method.

2 Multiple Richardson Extrapolation (MRE)

Assume that a p-th order underlying numerical method has been selected, and we have combined it with the classical Richardson extrapolation using the meshes (2) and (3). Denote the combined solution (calculated with the underlying method combined with CRE) at time t_n of the coarse mesh by $z_{CRE}(t_n)$. If we now define an even finer grid

$$\Omega_{h/4} := \{t_l = lh/4 : l = 0, 1, \ldots, 4N_t\}, \tag{10}$$

we can again combine the numerical solutions calculated on the meshes (3) and (10). (This means that the combined method is applied with $h/2$ being the long time step.) The new numerical solution will be denoted as $w_{CRE,n}$. Since the CRE yields a numerical approximation of order $p+1$, $z_{CRE,n}$ and $w_{CRE,n}$ are to be combined according to the general formula (6), but with values $p+1$ instead of p in the exponents. This results in the formula of the multiple Richardson extrapolation:

$$y_{\mathrm{MRE}}(t_n) := \frac{2^{p+1} w_{CRE,n} - z_{CRE,n}}{2^{p+1} - 1}. \tag{11}$$

Obviously, this procedure can be applied several times, however, we will only deal with the simplest case where formula (11) is used.

We tested the increase of the accuracy on the following examples.

Example 1. We considered the scalar problem

$$\begin{cases} y' = -ty^2, & t \in [2,3] \\ y(2) = 1. \end{cases} \tag{12}$$

The exact solution is $y(t) = \frac{2}{t^2-2}$. The global errors were calculated in absolute value at the end of the time interval $[2,3]$. The expected increase of the convergence order from 1 to 2 when the CRE is used and to 3 when the MRE is applied is seen from Table 1.

Example 2. We applied the above methods to the linear, constant coefficient system of ODE's

$$\begin{cases} y'(t) = Ay(t), & t \in [0,10] \\ \\ y(0) = (1,0,2)^T \end{cases} \tag{13}$$

Table 1. Global errors at $t = 3$ in absolute value in Example 1

h	EE	EE + CRE	EE + MRE
0.1	0.0217	5.1060e−04	7.9827e−05
0.01	0.0021	4.4863e−06	7.0195e−08
0.001	2.0785e−04	4.4412e−08	6.9197e−11
0.0001	2.0777e−05	4.4368e−10	5.0354e−13

with $y : \mathbb{R} \to \mathbb{R}^3$ and

$$A = \begin{bmatrix} -1 & 0 & 0 \\ -0.2 & -0.8 & 0 \\ -0.2 & -0.2 & -0.6 \end{bmatrix}$$

The eigenvalues of A are all real, and the three solution components tend to zero as t is increased. Table 2 shows the expected increase of the accuracy from RE to CRE and from CRE to MRE again.

Table 2. Maximal global errors on $[0, 10]$ in Example 2

h	EE	EE + CRE	IE + MRE
1	0.3679	0.1321	0.0137
0.1	0.0192	6.6154e−04	8.1369e−06
0.01	0.0018	6.1775e−06	7.7102e−09
0.001	1.8402e−04	6.1359e−08	7.6597e−12

It is easily seen that MRE is a slightly more costly way to obtain order $p + 2$ than RRE by using the same step size. By RRE, **seven** time steps are performed during one large time step, while this number is **eight** for the MRE. However, the computational time is only negligibly longer for the MRE when parallel computers are used. From the viewpoint of computational efficiency it is crucial, however, that we can choose a large enough time step while ensuring absolute stability. Therefore, in the next section we discuss the issue of absolute stability and examine how the MRE method affects the stability region of certain underlying numerical methods in comparison with the other versions of Richardson extrapolation (CRE and RRE).

3 Absolute Stability Analysis

The concept of absolute stability is related to Dahlquist's test problem [1]

$$\begin{cases} y'(t) = \lambda y(t), & t \in [0, \infty) \\ y(0) = y_0, \end{cases} \tag{14}$$

where $\lambda \in \mathbb{C}$, $\mathrm{Re}\lambda \leq 0$. It is known that the exact solution is $y(t) = y_0 \exp(\lambda t)$, which is bounded for $\mathrm{Re}\lambda \leq 0$. We expect that the numerical solution should also remain bounded on a mesh with a fixed time step as $t_n \to \infty$.

Let $\mu := h\lambda$. Then the time-stepping operator of any one-step numerical method when applied to the problem (14) can be expressed by μ:

$$y_n = R(\mu)y_{n-1} \quad \Rightarrow \quad y_n = (R(\mu))^n y_0 \qquad (n = 1, 2, \ldots). \tag{15}$$

The function $R(\mu)$ is called stability function of the method. For explicit methods it is a polynomial, while for implicit methods it is a rational function.

Example: for the explicit Euler (EE) method ($y_n = y_{n-1} + hf(t_{n-1}, y_{n-1})$)

$$y_n = (1 + h\lambda)y_{n-1}, \tag{16}$$

so the stability function of the explicit Euler method is $R_{EE}(\mu) = 1 + \mu$.

Clearly, a one-step method is bounded at the grid points of $[0, \infty)$ iff $|R(\mu)| \leq 1$.

Definition 1. The set $\{\mu \in \mathbb{C} : |R(\mu)| \leq 1\}$ is called stability region of the one-step numerical method with stability function $R(\mu)$.

Definition 2. A one-step numerical method with stability function $R(\mu)$ is called A-stable if

$$|R(\mu)| \leq 1 \quad \forall \mu \in \mathbb{C}^-, \tag{17}$$

where $\mathbb{C}^-$ denotes the left half plane of the complex plane including the imaginary axis.

The stability function of any one-step underlying method combined with MRE can be derived as follows. Denote by $R(\mu)$ the stability function of the selected underlying method of order p, and assume that y_{n-1} has been calculated during the numerical solution of (14).

Step 1: Perform one large time-step by the combined method (underlying method + CRE) with a time-step size h by using a starting value y_{n-1} to calculate:

$$z_{CRE,n} = \frac{2^p R(\frac{\mu}{2})^2 - R(\mu)}{2^p - 1} y_{n-1}. \tag{18}$$

We introduce the notation

$$R_{CRE}(\mu) := \frac{2^p R(\frac{\mu}{2})^2 - R(\mu)}{2^p - 1} \tag{19}$$

for the stability function of the combined underlying method + CRE.

Step 2: Perform two small time-steps by the combined method with a time-step size $h/2$ by using y_{n-1} as a starting value in the first small time-step:

$$w_{CRE,n} = \left[R_{CRE}\left(\frac{\mu}{2}\right)\right]^2 y_{n-1} = \left[\frac{2^p R(\frac{\mu}{4})^2 - R(\frac{\mu}{2})}{2^p - 1}\right]^2 y_{n-1} \tag{20}$$

Step 3: Improve the solution by applying the basic formula (11). This leads to the stability function

$$R_{MRE}(\mu) = \frac{2^{p+1}[R_{CRE}(\frac{\mu}{2})]^2 - R_{CRE}(\mu)}{2^{p+1}-1}. \tag{21}$$

Particularly, if the underlying method is the explicit Euler method ($p = 1$), then we have

$$R_{EE}(\mu) = 1 + \mu; \tag{22}$$

$$R_{EE+CRE}(\mu) = \frac{2^1[R_{EE}(\frac{\mu}{2})]^2 - R_{EE}(\mu)}{2^1 - 1} = 1 + \mu + \frac{\mu^2}{2}; \tag{23}$$

$$R_{EE+MRE}(\mu) = \frac{2^2[R_{EE+CRE}(\frac{\mu}{2})]^2 - R_{EE+CRE}(\mu)}{2^2 - 1} =$$

$$\frac{4}{3}\left(1 + \frac{\mu}{2} + \frac{\mu^2}{8}\right)^2 - \frac{1}{3}\left(1 + \mu + \frac{\mu^2}{2}\right) = 1 + \mu + \frac{1}{2}\mu^2 + \frac{1}{6}\mu^3 + \frac{1}{48}\mu^4. \tag{24}$$

Note that since any p-th order, m-stage explicit RK method for $p = m$ ($m \leq 4$) has the stability function $R(\mu) = 1+\mu+\frac{1}{2}\mu^2+\ldots+\frac{1}{p!}\mu^p$ (see [2], p. 202), therefore the stability function of EE + CRE is the same as the stability function of any second-order two-stage RK (2RK) method. From this also follows the fact that the stability function of any 2RK method + CRE is the same as that of EE + MRE. However, the stability function of 2RK + CRE is different from the stability function of any 3RK method. (The latter being only a third-degree polynomial.) Moreover, the joint stability polynomial of 2RK + CRE and EE + MRE is different from the stability polynomial of 4RK, since the coefficient of the fourth-degree term is not the same.

Figure 1 shows that the stability regions are increasingly larger for the following methods: i) the EE method, ii) the EE + CRE method, iii) the EE + RRE method and iv) the EE + MRE method. In Fig. 2 third-order methods are compared. The stability region of the EE + RRE method is somewhat larger than that of any third-order, three-stage Runge–Kutta (3RK) method. Both are contained by the even larger joint stability region of the 2RK + CRE and EE + MRE methods. Figure 3 compares the following fourth-order methods: i) RK4; ii) 3RK + CRE; iii) 2RK + RRE and iv) 2RK + MRE. The last one has the largest stability region. Note the change of the scales when making comparisons with Figs. 1 and 2.

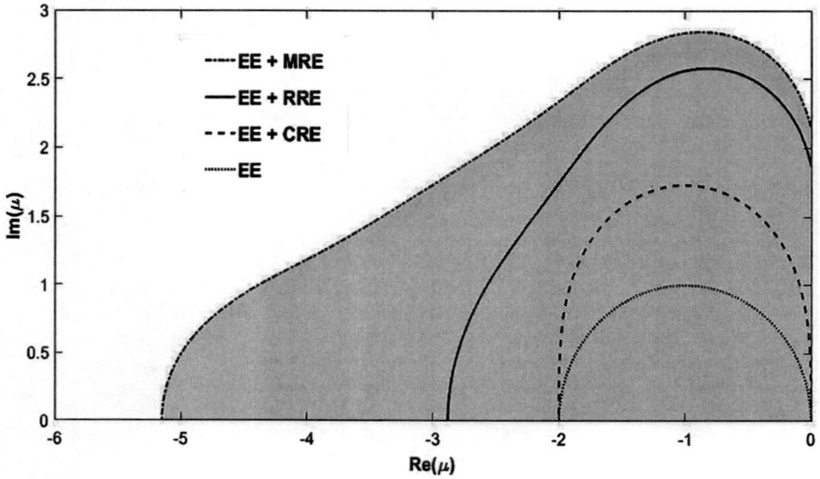

Fig. 1. Stability regions (in grey) for the EE with different RE methods

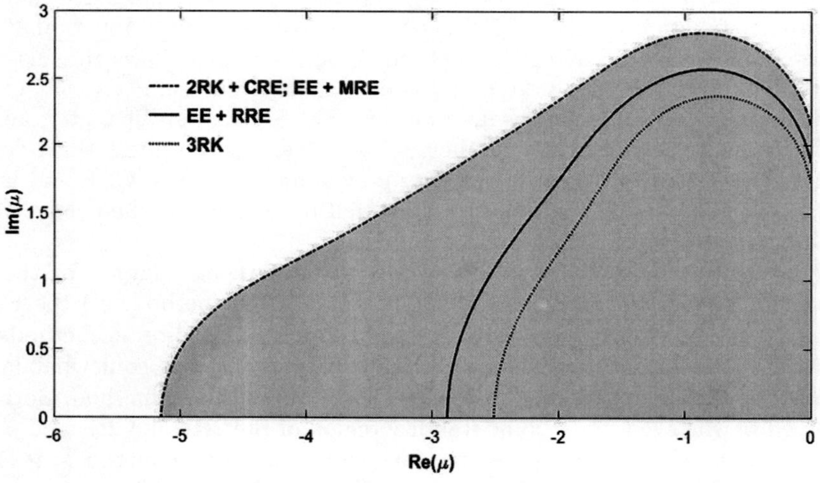

Fig. 2. Stability regions (in grey) for certain 3rd order methods

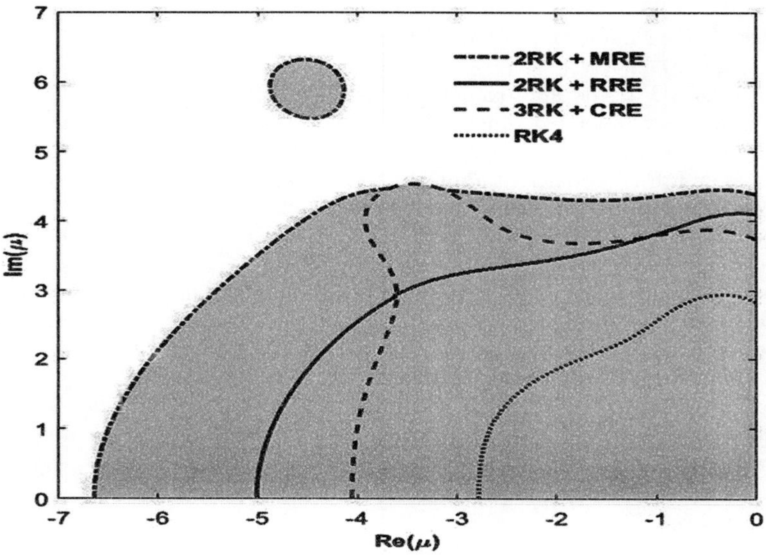

Fig. 3. Stability regions (in grey) for certain 4th order methods

4 Conclusion and Further Plans

Multiple Richardson extrapolation (MRE) is a simple tool by which the accuracy of a one-step numerical method can be increased from p to $p+2$. In this paper we examined the absolute stability of some explicit Runge–Kutta schemes when applied together with MRE. It is known that the classical Richardson extrapolation enhances the stability region for explicit RK methods [5]. Zlatev et al. [7] have recently shown that each repetition during RRE increases the stability region for ERK methods. From the stability analysis presented in this paper we can conclude that the stability region remains bounded when any of the studied ERK methods (EE, 2RK, 3RK) are combined with MRE. Moreover, MRE has an even larger stability region than RRE for all the studied explicit RK schemes.

As a next step, we plan to examine the absolute stability properties of implicit Runge–Kutta schemes combined with multiple Richardson extrapolation.

Acknowledgement. Project no. ED_18-1-2019-0030 (Application-specific highly reliable IT solutions) has been implemented with the support provided from the National Research, Development and Innovation Fund of Hungary, financed under the Thematic Excellence Programme funding scheme.

References

1. Dahlquist, G.: A special stability problem for linear multistep methods. BIT **3**, 27–43 (1963)

2. Lambert, J.D.: Numerical Methods for Ordinary Differential Equations. Wiley, New York (1991)
3. Richardson, L.F.: The approximate arithmetical solution by finite differences of physical problems including differential equations, with an application to the stresses in a masonry dam. Philos. Trans. R. Soc. Lond. Ser. A **210**, 307–357 (1911)
4. Richardson, L.F.: The deferred approach to the limit. I. Single Lattice Philos. Trans. Roy. Soc. London Ser. A **226**, 299–349 (1927)
5. Zlatev, Z., Dimov, I., Faragó, I., Havasi, Á.: Richardson Extrapolation - Practical Aspects and Applications. De Gruyter, Berlin (2017)
6. Zlatev, Z., Dimov, I., Faragó, I, Havasi, A.: absolute stability and implementation of the two-times repeated richardson extrapolation together with explicit Runge-Kutta methods. In: 7th International Conference, FDM 2018, Lozenetz, Bulgaria, 11–16 June 2018, Revised Selected Papers (2018)
7. Zlatev, Z., Dimov, I., Faragó, I., Georgiev, K.,Havasi, Á.: Explicit Runge-Kutta methods combined with advanced versions of the richardson extrapolation, CMAM (submitted)

A Stochastic Analysis of RC Structures Under Progressive Environmental Collapse Considering Uncertainty and Strengthening by Ties

A. Liolios[1(✉)], G. Skodras[2], K. Liolios[3], K. Georgiev[3], and I. Georgiev[3,4]

[1] Department of Civil Engineering, Democritus University of Thrace, Xanthi, Greece
alioliosc@civil.duth.gr, angeloslioliosc@gmail.com
[2] Department of Mechanical Engineering, School of Engineering, University of Western Macedonia, Kozani, Greece
gskodras@uowm.gr
[3] Institute of Information and Communication Technologies, Bulgarian Academy of Sciences, Sofia, Bulgaria
kliolios@env.duth.gr, {georgiev,ivan.georgiev}@parallel.bas.bg
[4] Institute of Mathematics and Informatics, Bulgarian Academy of Sciences, Sofia, Bulgaria

Abstract. In Civil Engineering praxis, existing reinforced concrete (RC) structures are sometimes subjected to removal of structural elements degraded by environmental effects. These structures can be strengthened by cable elements (tension-ties) in order to avoid a progressive collapse. The probabilistic analysis of this problem is herewith numerically investigated. Emphasis is given to uncertainties for the estimation of structural input parameters, common to old RC structures. Monte Carlo techniques are used and the unilateral behaviour of the cable-elements, which can undertake tension stresses only, is strictly taken into account.

Keywords: Computational structural mechanics · Removal of structural elements · Upgrading by ties · Input parameters uncertainties · Monte Carlo techniques

1 Introduction

Environmental actions, e.g. corrosion, earthquakes etc., can often cause significant damages to Civil Engineering reinforced concrete (RC) structures [1–3]. A main such defect is the strength degradation, resulting into a reduction of the loads bearing capacity of some structural elements. For some of such degraded elements is sometimes obligatory to be removed, and so a further reduction of the whole structure capacity is caused, which can lead to progressive collapse [23, 24].

I. Dimov and S. Fidanova (Eds.): HPC 2019, SCI 902, pp. 271–278, 2021.
https://doi.org/10.1007/978-3-030-55347-0_23

To overcome such defects, strengthening of the remaining structure is suggested. Among the available strengthening methods [4], cable-like members (tension-only tie-elements) can be used as a first strengthening and repairing procedure [1, 2, 5]. Cables can undertake tension but buckle and become slack and structurally ineffective when subjected to a sufficiently large compressive force. Thus, the governing conditions take an equality as well as an inequality form and the problem becomes a high non-linear one. Therefore, the problem of structures containing as above cable-like members belongs to the so-called Inequality Problems of Mechanics, as their governing conditions are of both, equality and inequality type [6–11]. A realistic numerical treatment of such problems can be obtained by mathematical programming methods (optimization algorithms).

Moreover, for the analysis of such old RC structures, many uncertainties for input parameters must be taking into account [12–15]. These mainly concern the holding properties of the old materials that had been used for the building of such structures, e.g. the remaining strength of the concrete and steel, as well as the cracking effects etc. Therefore, a probabilistic estimation of the input parameters must be performed [16–18].

The aim of this paper is to deal with a stochastic numerical approach for the analysis of existing old industrial framed RC buildings, which are strengthened by cable elements in order to overcome capacity reduction after the obligatory removal of some degraded structural elements. The computational approach is based on Monte Carlo methods [16–18] and on an incremental problem formulation. Finally, an application is presented for a simple typical example of an industrial RC frame strengthened by bracing ties after the removal of some ground floor columns.

2 Method of Analysis

The probabilistic approach for the analysis of existing RC structures under input-parameters uncertainties can be obtained by using Monte Carlo simulations. As well known, see e.g. [16–18], Monte Carlo simulation is simply a repeated process of generating deterministic solutions to a given problem. Each solution corresponds to a set of deterministic input values of the underlying random input variables. A statistical analysis of the so obtained simulated solutions is then performed. Thus, the computational methodology consists of solving first the deterministic problem for each set of the random input variables and finally realizing a statistical analysis.

2.1 Numerical Treatment of the Deterministic Problem

A reinforced concrete (RC) framed structure containing cable-like members is considered. For the general analysis of such a structure, a double discretization is applied: in space by finite elements and in time by a direct time-integration method [1, 2, 5]. The RC structure is discretized to frame elements with generally non-linear behavior. For the cables, pin-jointed bar elements with unilateral behavior are used. The rigorous mathematical investigation of the problem can be obtained by using the variational or hemivariational inequality concept, see Panagiotopoulos [9, 10]. So, the behavior of the cables and the generally non-linear behavior of RC elements, including loosening, elastoplastic or/and elastoplastic-softening-fracturing and unloading - reloading effects, can be expressed mathematically by the subdifferential relation:

$$s_i(d_i) \in \hat{\partial} S_i(d_i) \tag{1}$$

Here, for the example case of a typical i-th cable element, s_i and d_i are the (tensile) force and the deformation (elongation), respectively, $\hat{\partial}$ is the generalized gradient and S_i is the superpotential function [9, 10]. By definition, relation (1) is equivalent to the following hemivariational inequality, expressing the Virtual Work Principle for inequality problems:

$$S_i^{\uparrow}(d_i, e_i - d_i) \geq s_i(d_i) \cdot (e_i - d_i) \tag{2}$$

Here $S_i^{\uparrow}$ denotes the subderivative of S_i, and e_i, d_i are kinematically admissible (virtual) deformations. For the numerical treatment of practical inequality problems, a piece-wise linearization is usually applied to relations (1) and (2), see e.g. [5–8]. So, for the case of cables, the unilateral behavior of the i-th cable-element ($i = 1, \ldots, N$) is expressed by the following relations [5]:

$$e_i = F_{0i} \,.\, s_i + e_{i0} - v_i, \tag{3a}$$

$$s_i \geq 0, \quad v_i \geq 0, \quad s_i v_i = 0. \tag{3b}$$

Here e_i, F_{0i}, s_i, e_{i0} and v_i denote the strain (elongation), "natural" flexibility constant, stress (tension), initial strain and slackness, respectively. Relations (3b) consist the *Linear Complementarity Conditions* (LCC) and express that either a non-negative stress (tension) or a non- negative slackness exists on cables at every time-moment. The above considerations lead to formulate the problem and to solve it at every time-moment as a *Linear Complementarity Problem* (LCP). The numerical treatment of this LCP is obtained by using optimization methods [5–11].

In an alternative approach, the incremental dynamic equilibrium for the assembled structural system with cables is expressed in matrix form by the equation:

$$\underline{M}.\Delta\underline{\ddot{u}}(t) + \underline{C}.\Delta\underline{\dot{u}}(t) + \underline{K}_T.\Delta\underline{u}(t) = \Delta\underline{p}(t) + \underline{T}.\Delta\underline{s}(t) \tag{4}$$

Here $\underline{u}(t)$ and $\underline{p}(t)$ are the time dependent displacement and the given load vectors, respectively. $\underline{C}(\underline{\dot{u}})$ and $\underline{K_T}(\underline{u})$, are the damping and the time dependent tangent stiffness matrix, respectively. Dots over symbols denote derivatives with respect to time. $\underline{T}$ is a transformation matrix. By $\underline{s}(t)$ is denoted the time dependent cable stress vector with elements satisfying the relations (1), (2), (3a) and (3b).

The above matrix equation combined with the initial conditions consist the problem formulation, where, for given $\underline{p}(t)$, the vectors $\underline{u}(t)$ and $\underline{s}(t)$ are to be computed. For the numerical treatment of the above problem, the structural analysis software Ruaumoko [19] is herewith used. When the static case of the problem is only to be investigated, a Dynamic Relaxation approach [20] is appropriately used.

2.2 The Probabilistic Approach Using Monte Carlo Simulation

In order to calculate the random characteristics of the response of the considered RC buildings, the Monte Carlo simulation is used [14–18]. As mentioned, the main element of a Monte Carlo simulation procedure is the generation of random numbers from a specified distribution. The random variable simulation is implemented by using the technique of *Latin Hypercube Sampling* (LHS), [14–18]. The LHS is a selective sample technique by which, for a desirable accuracy level, the number of the sample size is significantly smaller than the direct Monte Carlo simulation.

Random design variables for the herein considered RC buildings are the uncertain quantities describing the plastic-hinges behavior and the spatial variation of input old materials parameters. According to *Joint Committee Structural Safety* (JCSS), see [21], concrete strength and elasticity modulus follow the Normal distribution, whereas the steel strength follows the *Lognormal* distribution.

3 Numerical Example

The RC plane frame structure of Fig. 1 had been initially constructed with two more internal columns in the ground floor, which have been removed due to degradation caused by environmental actions. The axial loads, which were initially undertaken by these two columns, are now shown as the two applied vertical concentrated loads of 180 kN and 220 kN.

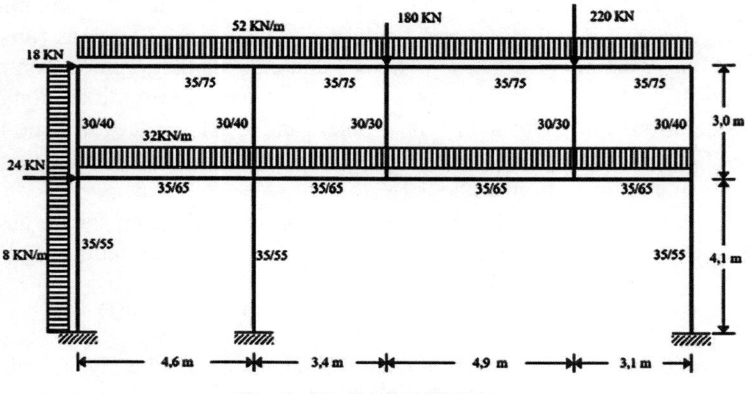

Fig. 1. The initial RC frame.

Due to removal of the above two columns, and in order to prevent a progressive collapse, the initial RC frame of Fig. 1 is strengthened by ten (10) steel cables (tension-only tie-elements) as shown in Fig. 2. In the so formulated system, it is wanted to be computed which of the cables are activated and which are not, under the considered static loading of Fig. 1.

The estimated concrete class of initial frame is C12/15 and the steel class is S220. As mentioned, according to JCSS (Joint Committee Structural Safety), see [21], concrete strength and elasticity modulus follow a Normal distribution and the steel strength

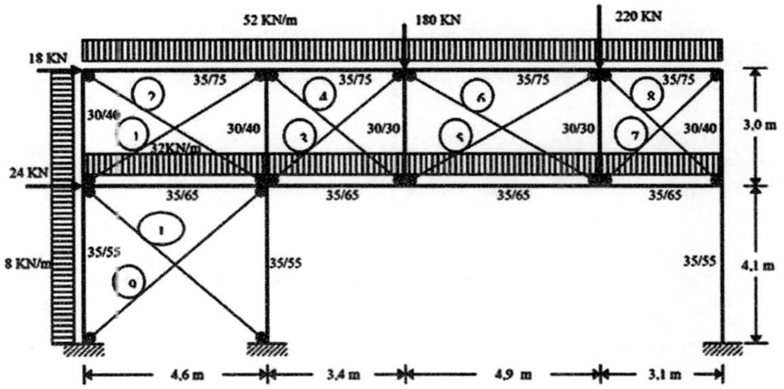

Fig. 2. The RC frame strengthened by 10 cables.

follows the Lognormal distribution. So the statistical characteristics of the input random variables concerning the building materials are estimated to be as shown in Table 1. By COV is denoted the coefficient of variation.

Table 1. Statistical data for the building materials treated as random variables

	Distribution	Mean	COV
Compressive strength of concrete	Normal	12.0 MPa	15%
Yield strength of steel	Lognormal	191.3 MPa	10%
Initial elasticity modulus, concrete	Normal	26.0 GPA	8%
Initial elasticity modulus, steel	Normal	200 GPA	4%

Using Ruaumoko software [19], the columns and the beams of the frame are modeled by prismatic frame RC elements. The effects of cracking on columns and beams are estimated in a probabilistic way by applying the guidelines of [22]. So, the stiffness reduction due to cracking results to effective stiffness with mean values of 0.60 I_g for the external columns, 0.80 I_g for the internal columns and 0.40 I_g for the beams, where I_g is the gross inertia moment of their cross-section. The relevant coefficient of variation COV is estimated to be 10%.

Nonlinearity at the two ends of the RC frame structural elements is idealized by using one-component plastic hinge models, following the Takeda hysteresis rule [19]. Concerning the constitutive diagrammes of plastic hinges, a typical normalized moment-normalized rotation backbone is shown in Fig. 3.

The strengthening cable members have a cross-sectional area $F_r = 8\ cm^2$ and are of steel class S1400/1600 with elasticity modulus $E_s = 210$ GPa. The cable constitutive law concerning the unilateral (slackness), hysteretic, fracturing, unloading-reloading etc. behavior, has the diagram depicted in Fig. 4.

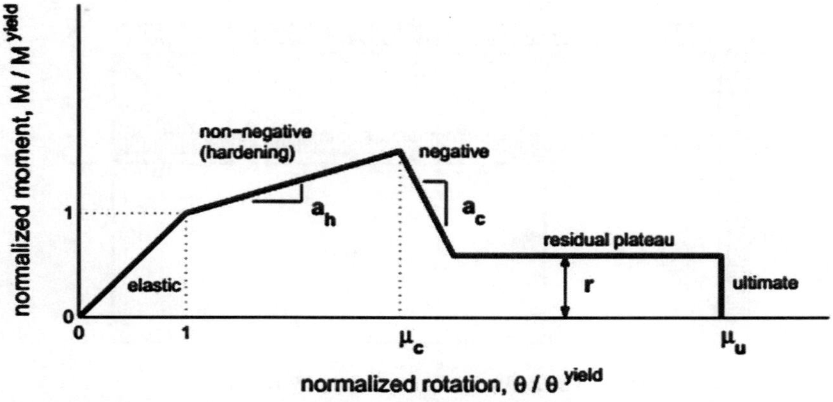

Fig. 3. Constitutive backbone diagramme for normalized moment - normalized rotation in plastic hinges [13].

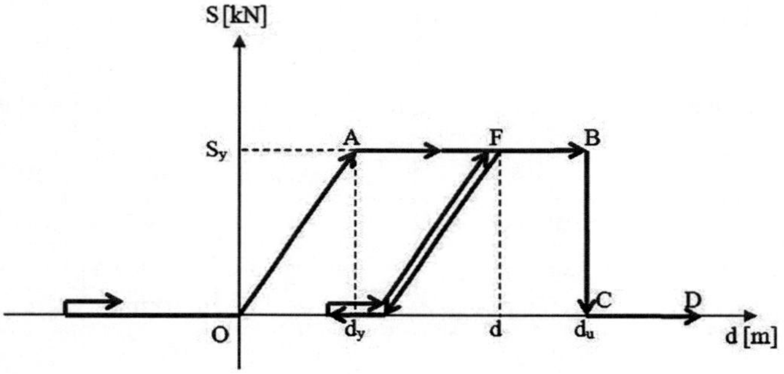

Fig. 4. Constitutive law of the cable-elements.

The application of the proposed numerical procedure using 200 Monte Carlo samples [14–18] and applying a pseudo Dynamic Relaxation approach [20] gives the following results for the cable-elements:

a. The mean values of the slackness of the no activated cable-elements are:
   $v_1 = 0.848*10^{-3}$ m, $v_3 = 10.321*10^{-3}$ m, $v_5 = 1.082*10^{-3}$ m,
   $v_8 = 9.564*10^{-3}$ m, $v_{10} = 1.652*10^{-3}$ m.
   The relevant mean coefficient of variation is COV = 18.27%.
b. The elements of the stress vector $\underline{s}$, where: $\underline{s} = [S_1, S_2, \ldots, S_{10}]^T$, are computed to have the following values (in kN) for the non-active cables:
   $S_1 = S_3 = S_5 = S_8 = S_{10} = 0.0$,
   whereas for the active cables, the mean values are:
   $S_2 = 10.17$ kN, $S_4 = 346.04$ kN, $S_6 = 18.84$ kN,
   $S_7 = 342.08$ kN, $S_9 = 25.81$ kN.
   The relevant mean coefficient of variation is COV = 24.48%.

Thus, cables 2, 4, 6, 7 and 9 are the only ones which have been activated, appearing zero slackness. The other cables 1, 3, 5, 8 and 10 cannot contribute to the system resistance under the given loads of Fig. 1.

4 Concluding Remarks

The proposed computational approach can be effectively used for the stochastic numerical investigation concerning the inelastic behaviour of existing RC framed-structures, which are subjected to removal of some degraded structural elements and strengthened by cable elements. As in a numerical application has been shown, the probabilistic estimation of the uncertain input parameters can be effectively realized by using Monte Carlo simulation.

Acknowledgement. This work is accomplished with the support by the Grant No BG05M2OP001-1.001-0003, financed by the Science and Education for Smart Growth Operational Program (2014-2020) and co-financed by the European Union through the European structural and Investment funds.

References

1. Liolios, A., Liolios, K., Moropoulou, A., Georgiev, K., Georgiev, I.: Cultural heritage RC structures environmentally degradated: optimal seismic upgrading by tension-ties under shear effects. In: Lirkov, I., Margenov, S. (eds.) LSSC 2017. LNCS 10665, 2018, pp. 516–526. Springer (2017)
2. Liolios, A., Chalioris, K., Liolios, A., Radev, S., Liolios, K.: A computational approach for the earthquake response of cable-braced reinforced concrete structures under environmental actions. In: Lirkov, I., Margenov, S., Waśniewski, J. (eds.) LSSC 2011. LNCS, vol. 7116, pp. 590–597. Springer, Heidelberg (2012). https://doi.org/10.1007/978-3-642-29843-1_67
3. Moncmanova, A.: Environmental Deterioration of Materials. WIT Press, Southampton (2007)
4. Dritsos, S.E.: Repair and Strengthening of Reinforced Concrete Structures (in Greek). University of Patras, Greece (2017)
5. Liolios, A., Moropoulou, A., Liolios, A., Georgiev, K., Georgiev, I.: A computational approach for the seismic sequences induced response of cultural heritage structures upgraded by ties. In: Margenov, S., Angelova, G., Agre, G. (eds.) Innovative Approaches and Solutions in Advanced Intelligent Systems. SCI, vol. 648, pp. 47–58. Springer, Cham (2016). https://doi.org/10.1007/978-3-319-32207-0_4
6. Nitsiotas, G.: Die Berechnung statisch unbestimmter Tragwerke mit einseitigen Bindungen. Ingenieur-Archiv **41**, 46–60 (1971)
7. Maier, G.: A quadratic programming approach for certain classes of non-linear structural problems. Meccanica **3**, 121–130 (1968)
8. Cottle, R.W.: Fundamentals of quadratic programming and linear complementarity. In: Cohn, M.Z., Grierson, D.E., Maier, G. (eds.) Engineering Plasticity by Mathematical Programming. Pergamon Press, Oxford (1979)
9. Panagiotopoulos, P.D.: Inequality Problems in Mechanics and Applications. Convex and Nonconvex Energy Functions. Birkhäuser Verlag, Basel (1985)
10. Panagiotopoulos, P.: Hemivariational Inequalities. Applications in Mechanics and Engineering. Springer, Berlin (1993)

11. Mistakidis, E.S., Stavroulakis, G.E.: Nonconvex Optimization in Mechanics. Smooth and Nonsmooth Algorithmes, Heuristic and Engineering Applications. Kluwer, London (1998)
12. Strauss, A., Hoffmann, S., Wendner, R., Bergmeister, K.: Structural assessment and reliability analysis for existing engineering structures, applications for real structures. Struct. Infrastruct. Eng. **5**(4), 277–286 (2009)
13. Vamvatsikos, D., Fragiadakis, M.: Incremental dynamic analysis for estimating seismic performance sensitivity and uncertainty. Earthq. Eng. Struct. Dyn. **39**(2), 141–163 (2010)
14. Papadrakakis, M., Stefanou, G. (eds.): Multiscale Modeling and Uncertainty Quantification of Materials and Structures. Springer, Berlin (2014)
15. Thomos, G., Trezos, C.: Examination of the probabilistic response of reinforced concrete structures under static non-linear analysis. Eng. Struct. **28**(2006), 120–133 (2006)
16. Dimov, I.T.: Monte Carlo Methods for Applied Scientists. World Scientific (2008)
17. Ang, A.H., Tang, W.H.: Probability Concepts in Engineering Planning and Design, Vol. 2: Decision, Risk, and Reliability. Wiley, New York (1984)
18. Kottegoda, N., Rosso, R.: Statistics, Probability and Reliability for Civil and Environmental Engineers. McGraw-Hill, London (2000)
19. Carr, A.J.: RUAUMOKO - Inelastic Dynamic Analysis Program. Department of Civil Engineering, University of Canterbury, Christchurch, New Zealand (2008)
20. Papadrakakis, M.: A method for the automatic evaluation of the dynamic relaxation parameters. Comput. Methods Appl. Mech. Eng. **25**(1), 35–48 (1981)
21. JCSS: Joint Committee on Structural Safety. Probabilistic Model Code-Part 1: Basis of Design (12th draft), March 2001. http://www.jcss.ethz.ch/
22. Pauley, T., Priestley, M.J.N.: Seismic Design of Reinforced Concrete and Masonry Buildings. Wiley, New York (1992)
23. Starossek, U.: Progressive Collapse of Structures, 2nd edn. Thomas Telford Ltd., London (2017)
24. Penelis, G., Penelis, Gr.: Concrete Buildings in Seismic Regions, 2nd edn. CRC Press, Taylor and Francis Group Ltd., London (2019)

Numerical Simulation of Thermoelastic Nonlinear Waves in Fluid Saturated Porous Media with Non-local Darcy Law

Miglena N. Koleva[1(✉)], Yuri Poveschenko[2], and Lubin G. Vulkov[1]

[1] University of Ruse, 8 Studentska str., 7017 Ruse, Bulgaria
{mkoleva,lvalkov}@uni-ruse.bg
[2] Keldysh Institute of Applied Mathematics of Russian Academy of Sciences, Miusskaya sq., 4, Moscow 125047, Russia
hecon@mail.ru

Abstract. In this work we present a numerical study of a non-linear space-fractional parabolic system, describing thermoelastic waves propagation in fluid saturated porous media with non-local Darcy law. We use implicit cell-centered finite difference method for numerical discretizations, combined with $L1-2$ formula for the fractional derivative approximation. We apply Newton's method to compute the finite difference solution. We solve a linear system of algebraic equation in block-matrix form on each iteration, using Schur complement. Numerical experiments attest good attributes of the proposed numerical method.

1 Introduction

In the recent years fractional calculus approach has been used to handle deviations from the classical diffusion to model particle transport in porous media [4,5,8,13,15,16]. The usage of fractional derivatives to describe fluid motion inporous media is not new. In the paper [9], the author studied seepage flow in porous media and observed that the fractional derivatives can be used to describe deviations from Darcy's law behavior.

In this paper we investigate numerically the propagation of non-linear waves in fluid saturated anomalous porous media, governed by the parabolic space-fractional system, suggested by Garra and Salusti [6]. The key role for the modeling of this process is played by a non-local Modified Darcy Law (MDL) by considering a fractional derivative generalization of the classical case. The starting point for the derivation of the model are Natale and Salusti equations [12] of transient solutions for temperature and pressure waves in fluid-saturated porous rock. Using the formalism of the fractional calculus, Garra and Salusti [6] introduce the MDL:

$$\phi U(z,t) = -\frac{K_f}{\mu}\frac{\partial^\gamma P}{\partial z^\gamma} = -\frac{K_f}{\mu}\frac{1}{\Gamma(1-\gamma)}\int_0^z \frac{P'(\xi)d\xi}{(z-\xi)^\gamma}, \qquad 0<\gamma<1, \quad z>0,$$

I. Dimov and S. Fidanova (Eds.): HPC 2019, SCI 902, pp. 279–289, 2021.
https://doi.org/10.1007/978-3-030-55347-0_24

where $\frac{\partial^\gamma P}{\partial z^\gamma}$ is the Caputo fractional partial derivative, see e.g. [3,14], U is the velocity of the fluid content within the porous media, ϕ is the porosity, K_f is the medium permeability, μ is the fluid viscosity and $\Gamma(\cdot)$ is the Gamma function.

Solving new non-linear problem of continuous mechanics requires advances in quality, accuracy, and robustness of the numerical algorithms.

There is a vast literature on numerical investigation of time-fractional or space-fractional $(1 < \gamma < 2)$ problems [1,17].

In the first equation of the parabolic system, considered in the paper, the space-fractional derivative $(0 < \gamma < 1)$ is inherent for the physical process. The results for numerical investigation of convection sub-diffusion problems, especially non-linear systems, are scarce.

In this paper we construct weighted finite difference method, realized by Newton's iteration process. We develop iteration process based on the Schur complement of the coefficient matrix with respect to the diagonal matrix in order to reduce the computational time.

The remaining part of the paper is organized as follows. In the next section, we formulate the model problem, involving scaling of the parameters. In Sect. 3 we develop weighted finite difference scheme and the corresponding iteration procedures. Numerical examples are presented and discussed in Sect. 4. Finally we give some concluding remarks.

2 Model Problem Formulation

We consider the fractional-order non-linear system suggested in [6], which describes dynamics of the temperature $T(z,t)$ and the pressure $P(z,t)$ at depth z and time $t:\ 0 < t \leq T_{fin}$. We investigate the temperature and the pressure in the upper layer, i.e. $0 < z < \infty$. In this case the model reads as follows

$$\frac{\partial T}{\partial t} - \eta \frac{\partial^2 T}{\partial z^2} = \beta \frac{\partial T}{\partial z} \frac{\partial^\gamma P}{\partial z^\gamma} + \chi \frac{\partial^\gamma P}{\partial z^\gamma} \frac{\partial P}{\partial z}, \tag{1}$$

$$\frac{\partial P}{\partial t} - \lambda \frac{\partial^2 P}{\partial z^2} = \alpha \frac{\partial T}{\partial t}, \tag{2}$$

$$T(z,0) = T_0^+, \quad P(z,0) = P_0^+, \ 0 < z < \infty, \tag{3}$$

$$\frac{\partial P}{\partial z}(z,t) = 0, \ T(z,t) = 0, \ z \to \infty, \quad P(0,t) = P_\circ, \ T(0,t) = T_\circ. \tag{4}$$

Here η is the thermal diffusivity due to the diffusion, β is the thermal diffusivity due to the convection, χ is the matrix-fluid friction coefficient, λ is the fluid diffusivity coefficients, α is a source term coefficient and the Caputo fractional partial derivative is defined by

$$\frac{\partial^\gamma P}{\partial z^\gamma} = \frac{1}{\Gamma(1-\gamma)} \int_0^z \frac{P'(\xi) d\xi}{(z-\xi)^\gamma}, \qquad 0 < \gamma < 1, \ \ z > 0.$$

We solve the problem (1)–(4) on large enough finite computational interval for z variable, namely $z \in [0, L]$.

As the model parameters vary widely [2,6,11], the computational time-space domain is large, the values of the pressure (Pa) are quite greater in comparison with the temperature ($^\circ C$) (see [2,11]), we introduce the following scaling

$$P' = \frac{P}{p^*}, \quad T' = \frac{T}{\mathbf{t}^*}, \quad t' = \frac{t}{t^*}, \quad z' = \frac{z}{z^*}. \tag{5}$$

Thus, we obtain a problem defined on the truncated domain $[0, L'] \times [0, T'_{fin}]$, which has the same structure as (1)–(4), but all quantities are replaced by the corresponding primed notation

$$\eta' = \frac{\eta t^*}{(z^*)^2}, \ \beta' = \frac{\beta p^* t^*}{(z^*)^2}, \ \chi' = \frac{\chi (p^*)^2 t^*}{(z^*)^2 \mathbf{t}^*}, \ \lambda' = \frac{\lambda t^*}{(z^*)^2}, \ \alpha^* = \frac{\alpha \mathbf{t}^*}{p^*},$$

$$P'_\circ = \frac{P_\circ}{p^*}, \ T'_\circ = \frac{T_\circ}{\mathbf{t}^*}, \ (P_0^+)' = \frac{P_0^+}{p^*}, \ (T_0^+)' = \frac{T_0^+}{\mathbf{t}^*}, \ T'_{fin} = \frac{T_{fin}}{t^*}, \ L' = \frac{L}{z^*}.$$

Further, for simplicity, we preserve the previous notation, omitting the primes.

3 Numerical Method

We define uniform mesh $\overline{\omega} = \overline{\omega}_\tau \times \overline{\omega}_h$ in the space-time computational domain:

$$\overline{\omega}_\tau = \{t_j = j\tau, \ j = 0, 1, \ldots, J, \ \tau = T_{fin}/J\}, \ \ \overline{\omega}_h = \{z_i = ih, \ i = 0, 1, \ldots, N, \ h = L/N\}.$$

Numerical solution at grid node (z_i, t_j) is denoted by $P_i^j = P(z_i, t_j)$ and $T_i^j = T(z_i, t_j)$.

Further, we construct θ-weighted ($0 \le \theta \le 1$) finite difference scheme:

$$\begin{aligned}
&\frac{T_i^{j+1} - T_i^j}{\tau} - \theta\eta \frac{T_{i-1}^{j+1} - 2T_i^{j+1} + T_{i+1}^{j+1}}{h^2} - \theta L_i^{j+1} \frac{\partial_h^\gamma P_i^{j+1}}{\partial z^\gamma} \\
&\qquad = (1-\theta)\eta \frac{T_{i-1}^j - 2T_i^j + T_{i+1}^j}{h^2} + (1-\theta) L_i^j \frac{\partial_h^\gamma P_i^j}{\partial z^\gamma}, \ i = 1, 2, \ldots, N-1, \\
&\frac{P_i^{j+1} - P_i^j}{\tau} - \theta\lambda \frac{P_{i-1}^{j+1} - 2P_i^{j+1} + P_{i+1}^{j+1}}{h^2} \\
&\qquad = (1-\theta)\lambda \frac{P_{i-1}^j - 2P_i^j + P_{i+1}^j}{h^2} + \alpha \frac{T_i^{j+1} - T_i^j}{\tau}, \qquad i = 1, 2, \ldots, N-1, \\
&\frac{P_N^{j+1} - P_N^j}{\tau} - 2\theta\lambda \frac{P_{N-1}^{j+1} - P_N^{j+1}}{h^2} = 2(1-\theta)\lambda \frac{P_{N-1}^{j+1} - P_N^{j+1}}{h^2} + \alpha \frac{T_N^{j+1} - T_N^j}{\tau}, \\
&T_0^{j+1} = T_\circ, \ T_N^{j+1} = 0, \ P_0^{j+1} = P_\circ, \\
&T_i^0 = T_0^+, \ P_i^0 = P_0^+, \ i = 0, 1, \ldots, N,
\end{aligned} \tag{6}$$

where

$$L_i^j = \beta \frac{T_{i+1}^j - T_{i-1}^j}{2h} + \chi \frac{P_{i+1}^j - P_{i-1}^j}{2h}$$

and the fractional derivative is approximated by $L1-2$ formula, developed in [7]

$$\frac{\partial^\gamma P_i^j}{\partial z^\gamma} \approx \frac{\partial_h^\gamma P_i^j}{\partial z^\gamma} := \frac{h^{-\gamma}}{\Gamma(2-\gamma)}\left[c_0^{(\gamma)} P_i^j - \sum_{s=1}^{i-1}\left(c_{i-s-1}^{(\gamma)} - c_{i-s}^{(\gamma)}\right) P_s^j - c_{i-1}^{(\gamma)} P_0^j\right]. \quad (7)$$

The coefficients in (7) are given as follows. For $i=1$, $c_0^{(\gamma)} = a_0^{(\gamma)} = 1$ and for $i \geq 2$:

$$c_s^{(\gamma)} = \begin{cases} a_0^{(\gamma)} + b_0^{(\gamma)}, & s=0, \\ a_s^{(\gamma)} + b_s^{(\gamma)} - b_{s-1}^{(\gamma)}, & 1 \leq s \leq i-1, \\ a_s^{(\gamma)} - b_{s-1}^{(\gamma)}, & s=i, \end{cases}$$

$$a_s^{(\gamma)} = (1+s)^{1-\gamma} - s^{1-\gamma}, \quad 0 \leq s \leq i-1,$$

$$b_s^{(\gamma)} = \frac{1}{2-\gamma}[(1+s)^{2-\gamma} - s^{2-\gamma}] - \frac{1}{2}[(l+s)^{1-\gamma} + s^{1-\gamma}], \ s \geq 0.$$

It is proved in [7] that the discretization (7), $0<\gamma<1$ for $P(z,t^j) \in C^3[0,z_i]$ has a local truncation error $O(h^{2-\gamma})$ for $z_0 < z < z_1$ and $O(h^{3-\gamma})$ for $z_0 < z < z_i$, $i \geq 2$.

The formula (7) has been constructed and applied for solving time-fractional differential problem. Here we implement this approximation to discretizy a space-fractional derivative. We implement Newton's method to compute the solutions of (6). Setting $T^{(0)} := T^j$, $P^{(0)} := P^j$, for $k=0,1,\dots$, the resulting linearized system (6) is

$$\begin{aligned}
&\left(-\frac{\theta\beta}{2h}\frac{\partial_h^\gamma P_i^{(k)}}{\partial z^\gamma} - \frac{\theta\eta}{h^2}\right) T_{i+1}^{(k+1)} + \left(\frac{1}{\tau} + \frac{2\theta\eta}{h^2}\right) T_i^{(k+1)} + \left(\frac{\theta\beta}{2h}\frac{\partial_h^\gamma P_i^{(k)}}{\partial z^\gamma} - \frac{\theta\eta}{h^2}\right) T_{i-1}^{(k+1)} \\
&+ \left(-\frac{\theta\chi}{2h}\frac{\partial_h^\gamma P_i^{(k)}}{\partial z^\gamma}\right) P_{i+1}^{(k+1)} + \left(-c_0^{(\gamma)}\theta\frac{h^{-\gamma}}{\Gamma(2-\gamma)} L_i^{(k)}\right) P_i^{(k+1)} \\
&+ \left(\frac{\theta\chi}{2h}\frac{\partial_h^\gamma P_i^{(k)}}{\partial z^\gamma} + (c_0^{(\gamma)} - c_1^{(\gamma)})\theta\frac{h^{-\gamma}}{\Gamma(2-\gamma)} L_i^{(k)}\right) P_{i-1}^{(k+1)} \\
&+ \theta\frac{h^{-\gamma}}{\Gamma(2-\gamma)} L_i^{(k)}\left[\sum_{s=1}^{i-2}(c_{i-s-1}^{(\gamma)} - c_{i-s}^{(\gamma)}) P_s^{(k+1)} - c_{i-1}^{(\gamma)} P_0^{(k+1)}\right] = \frac{T_i^j}{\tau} \\
&+ (1-\theta)\eta\frac{T_{i-1}^j - 2T_i^j + T_{i+1}^j}{h^2} + (1-\theta)L_i^j\frac{\partial_h^\gamma P_i^j}{\partial z^\gamma} - \theta L_i^{(k)}\frac{\partial_h^\gamma P_i^{(k)}}{\partial z^\gamma}, \quad i=1,2,\dots,N-1, \\
&T_0^{(k+1)} = T_\circ, \quad T_N^{(k+1)} = 0, \quad T_i^0 = T_0^+, \quad i=0,1,\dots,N,
\end{aligned} \quad (8)$$

$$
\begin{aligned}
&-\frac{\theta\lambda}{h^2}P_{i-1}^{(k+1)} + \left(\frac{1}{\tau} + \frac{2\theta\lambda}{h^2}\right)P_i^{(k+1)} - \frac{\theta\lambda}{h^2}P_{i+1}^{(k+1)} - \frac{\alpha}{\tau}T_i^{(k+1)} \\
&= \frac{P_i^j - \alpha T_i^j}{\tau} + (1-\theta)\lambda\frac{P_{i-1}^j - 2P_i^j + P_{i+1}^j}{h^2}, \quad i = 1, 2, \ldots, N-1, \\
&\left(\frac{1}{\tau} + \frac{2\theta\lambda}{h^2}\right)P_N^{(k+1)} - \frac{2\theta\lambda}{h^2}P_{N-1}^{(k+1)} - \frac{\alpha}{\tau}T_N^{(k+1)} = \frac{P_N^j - \alpha T_N^j}{\tau} + 2(1-\theta)\lambda\frac{P_{N-1}^j - P_N^j}{h^2}, \\
&P_0^{(k+1)} = P_\circ, \quad P_i^0 = P_0^+, \quad i = 0, 1, \ldots, N.
\end{aligned} \tag{9}
$$

When the desired accuracy ε (in maximum discrete norm $\|\cdot\|$) is reached

$$
\varepsilon = \max\left\{\|T^{(k+1)} - T^{(k)}\|, \|P^{(k+1)} - P^{(k)}\|\right\},
$$

we set $T^{j+1} := T^{(k+1)}$, $P^{j+1} := P^{(k+1)}$.

For further investigations we slightly modify the left boundary condition for P in (6), namely $P_0^{j+1} = P_\circ$ is replaced by

$$
P_0^{j+1} - \frac{\alpha}{\tau}T_0^{j+1} = P_\circ - \frac{\alpha}{\tau}T_\circ
$$

and consequently in the iteration process (9) we obtain

$$
P_0^{(k+1)} - \frac{\alpha}{\tau}T_0^{(k+1)} = P_\circ - \frac{\alpha}{\tau}T_\circ. \tag{10}
$$

Consider the equivalent matrix-vector form of (8)–(9) with the modified boundary condition (10)

$$
\mathbf{M_1}^{(k)}T^{(k+1)} + \mathbf{Q}^{(k)}P^{(k+1)} = F_1^{(k)}, \tag{11}
$$

$$
\mathbf{D}T^{(k+1)} + \mathbf{M_2}^{(k)}P^{(k+1)} = F_2^{(k)}, \tag{12}
$$

where

- $P^{(k+1)} = \left(P_0^{(k+1)}, P_1^{(k+1)}, \ldots, P_N^{(k+1)}\right)$ and $T^{(k+1)} = \left(T_0^{(k+1)}, T_1^{(k+1)}, \ldots, T_N^{(k+1)}\right)$ are column vectors;
- $\mathbf{M_1}^{(k)}$ and $\mathbf{M_2}^{(k)}$ are tridiagonal $(N+1)\times(N+1)$ matrices, built from the coefficients of $T_{i-1}^{(k+1)}$, $T_i^{(k+1)}$, $T_{i+1}^{(k+1)}$ in (8) and $P_{i-1}^{(k+1)}$, $P_i^{(k+1)}$, $P_{i+1}^{(k+1)}$ in (9), respectively;
- $\mathbf{D} = -\tau\mathbf{E}/\alpha$, where $\mathbf{E}$ is the unit $(N+1)\times(N+1)$ matrix;
- $\mathbf{Q}^{(k)}$ is generated as a result of the fractional derivative approximation and the contribution of the convection term (i.e the coefficients of $P_{i-1}^{(k+1)}$, $P_i^{(k+1)}$, $P_{i+1}^{(k+1)}$ in (8)), see Fig. 2;
- $F_1^{(k)}$ and $F_2^{(k)}$ are column vectors with length $N+1$, corresponding to the right-hand side in (8) and (9), respectively.

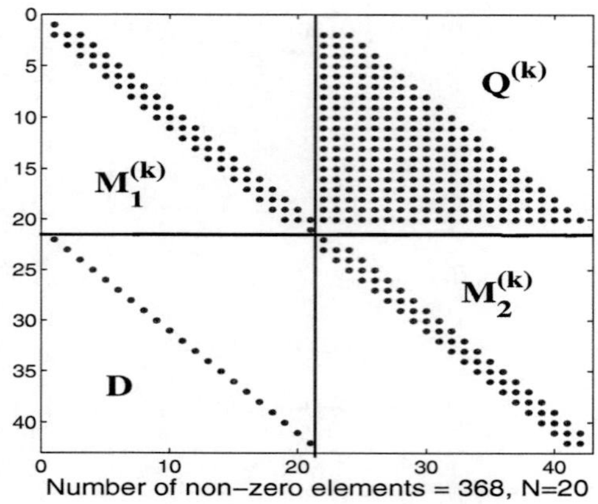

Fig. 1. Structure of the coefficient matrix in (11), (12), $N = 20$

From (12) we express the temperature $T^{(k+1)} = \mathbf{D}^{-1}(F_2^{(k)} - \mathbf{M_2}^{(k)}P^{(k+1)})$ and substitute in (11). Taking into account that $D^{-1} = -\tau\mathbf{E}/\alpha$, we derive

$$\begin{aligned} P^{(k+1)} &= \left(\mathbf{Q}^{(k)} + \frac{\tau}{\alpha}\mathbf{M_1}^{(k)}\mathbf{M_2}^{(k)}\right)^{-1}\left(F_1^{(k)} + \frac{\tau}{\alpha}\mathbf{M_1}^{(k)}F_2^{(k)}\right), \\ T^{(k+1)} &= -\frac{\tau}{\alpha}\left(F_2^{(k)} - \mathbf{M_2}^{(k)}P^{(k+1)}\right). \end{aligned} \tag{13}$$

Note that to compute the solution with the iteration scheme (13), at each time level and at each iteration we have to invert a $(N+1) \times (N+1)$ matrix, i.e. the Schur complement

$$S_D = \mathbf{Q}^{(k)} + \frac{\tau}{\alpha}\mathbf{M_1}^{(k)}\mathbf{M_2}^{(k)},$$

while with (8), (9), we need to invert a $2(N+1) \times 2(N+1)$ coefficient matrix.

Theorem 1. *The Schur complement S_D is nonsingular, if*

$$\begin{aligned} \tau &\le \frac{h\Gamma(2-\gamma)}{\theta\chi\Gamma(2-\gamma)\left\|\frac{\partial_h^\gamma P_i^{(k)}}{\partial z^\gamma}\right\| + 2\theta|c_0^{(\gamma)}|\|L^{(k)}\|h^{1-\gamma}}, \\ h &\le \frac{2\eta}{\beta}\left\|\frac{\partial_h^\gamma P_i^{(k)}}{\partial z^\gamma}\right\|^{-1}, \qquad \alpha \le 1. \end{aligned} \tag{14}$$

Proof (outline). Taking into account the properties of the coefficients $c_i^{(\gamma)}$ [7], we prove that the block matrix $\begin{pmatrix} \mathbf{M_1}^{(k)} & \mathbf{Q}^{(k)} \\ \mathbf{D} & \mathbf{M_2}^{(k)} \end{pmatrix}$ is strictly diagonal dominant, if the conditions (14) are fulfilled. Since $\mathbf{D}$ is a diagonal matrix, we conclude that S_D is nonsingular [10,18].

Note that, the second condition in (14) is not difficult to be satisfied due to the scaling (5).

4 Computational Results

In this section we provide and discuss results from the numerical tests. We compute the solution of the problem (1)–(4), (5) by the iteration processes (8)–(9) and (13). First, we consider model parameters corresponding to the *clay rock* [2,6]:

$$\alpha = 6.10^2, \quad \beta = 2.10^{-13}, \quad \chi = 10^{-19}, \quad \eta = 3.10^{-7}, \quad \lambda = 4.10^{-6}$$

and scaling parameters $p^* = 10^7$, $\mathbf{t}^* = 10$, $t^* = 10^8$, $z^* = 10^3$.

To test the order of convergence, we add residual function in the right hand side of the Eq. (1)–(2), such that $P(z,t) = e^{t/2}(4L^3 z - z^4)$, $T(z,t) = e^{-2t}(L^3 z - z^4)$ to be the exact solution of the modified system (1), (4) with initial and left Dirichlet boundary condition, chosen regarding to the exact solution.

We give the errors (E_P^N, E_T^N) in maximum discrete norm and the corresponding order of convergence (CR_P^N, CR_T^N) at final time T_{fin}:

$$E_P^N = \max_{0 \le i \le N} |P(z_i, T_{fin}) - P_i^J|, \quad CR_P^N = \log_2 E_P^{N/2}/E_P^N,$$

$$E_T^N = \max_{0 \le i \le N} |T(z_i, T_{fin}) - T_i^J|, \quad CR_T^N = \log_2 E_T^{N/2}/E_T^N.$$

The computations are performed for $\theta = 0.5$, $\tau = h$ and accuracy of the iteration process $\varepsilon = 10^{-6}$. We denote by AvIt the average number of iterations at each time level and CPU is the computer time (in seconds) for the whole computational process.

Let $L' = 1$, $T'_{fin} = 1$ (i.e. $L = 1$ km , $T_{fin} \approx 39$ months). In Tables 1, 2 and 3 we give the results from the computations with teration schemes (8)–(9) and (13) for $\gamma = 0.25$, $\gamma = 0.5$ and $\gamma = 0.75$, respectively.

Table 1. Results for $\gamma = 0.25$, $\tau = h$

N	*Iteration scheme* (8)–(9)						*Iteration scheme* (13)					
	E_P^N	CR_P^N	E_T^N	CR_T^N	$AvIt$	CPU	E_P^N	CR_P^N	E_T^N	CR_T^N	$AvIt$	CPU
20	5.225e−3		1.473e−3		2.85	0.09	5.225e−3		1.473e−3		2.85	0.08
40	1.335e−3	1.969	3.758e−4	1.970	2.93	0.16	1.335e−3	1.969	3.758e−04	1.970	2.93	0.14
80	3.406e−4	1.971	9.527e−5	1.980	2.00	0.43	3.406e−4	1.971	9.528e−5	1.978	2.00	0.30
160	8.516e−5	2.000	2.391e−5	1.995	2.00	2.31	8.516e−5	2.000	2.390e−05	1.995	2.00	1.02
320	2.124e−5	2.004	5.985e−6	1.998	1.99	13.48	2.124e−5	2.004	5.997e−06	1.995	1.99	4.91
640	5.316e−6	1.998	1.499e−6	1.997	2.00	116.27	5.316e−6	1.998	1.620e−06	1.888	2.00	49.37
1280	1.331e−6	1.998	3.752e−7	1.998	2.00	1257.27	1.335e−6	1.994	6.821e−07	1.248	2.71	657.67

The conditions in Theorem 1 are fulfilled for all computations. We observe that for $\theta = 0.5$ the order of convergence is $O(\tau^2 + h^2)$, the iteration process (13) performs faster than (8)–(9), but for fine meshes the second order of convergence of (13) is destroyed and the number of iterations is increased. The reason is

Table 2. Results for $\gamma = 0.5$, $\tau = h$

N	*Iteration scheme* (8)–(9)						*Iteration scheme* (13)					
	E_P^N	CR_P^N	E_T^N	CR_T^N	*AvIt*	CPU	E_P^N	CR_P^N	E_T^N	CR_T^N	*AvIt*	CPU
20	5.225e−3		1.423e−3		2.85	0.09	5.225e−3		1.423e−3		2.85	0.08
40	1.335e−3	1.969	3.691e−4	1.947	2.93	0.17	1.335e−3	1.969	3.691e−4	1.947	2.93	0.13
80	3.406e−4	1.971	9.497e−5	1.958	2.00	0.42	3.406e−4	1.971	9.497e−5	1.958	2.00	0.31
160	8.516e−5	2.000	2.414e−5	1.976	2.00	2.31	8.516e−5	2.000	2.413e−5	1.976	2.00	0.99
320	2.124e−5	2.004	6.103−6	1.984	1.99	13.56	2.124e−5	2.004	6.122e−6	1.979	1.99	4.75
640	5.316e−6	1.998	1.540e−6	1.987	2.00	115.06	5.315e−6	1.998	1.662e−6	1.881	2.00	50.41
1280	1.331e−6	1.998	3.875e−7	1.990	2.00	1258.11	1.328e−6	2.001	6.790e−7	1.292	2.62	649.57

Table 3. Results for $\gamma = 0.75$, $\tau = h$

N	*Iteration scheme* (8)–(9)						*Iteration scheme* (13)					
	E_P^N	CR_P^N	E_T^N	CR_T^N	*AvIt*	CPU	E_P^N	CR_P^N	E_T^N	CR_T^N	*AvIt*	CPU
20	5.225e−3		9.748e−4		2.85	0.09	5.225e−3		9.748e−4		2.85	0.08
40	1.335e−3	1.969	2.600e−4	1.906	2.93	0.17	1.335e−3	1.969	2.600e−4	1.906	2.93	0.14
80	3.406e−4	1.971	6.954e−5	1.903	2.00	0.43	3.406e−4	1.972	6.954e−5	1.903	2.00	0.31
160	8.516e−5	2.000	1.847e−5	1.913	2.00	2.33	8.516e−5	2.000	1.846e−5	1.913	2.00	1.02
320	2.124e−5	2.004	4.855e−6	1.928	1.99	13.45	2.124e−5	2.004	4.874e−6	1.921	1.99	4.96
640	5.316e−6	1.998	1.266e−6	1.940	2.00	115.20	5.316e−6	1.998	1.383e−6	1.817	2.00	49.45
1280	1.331e−6	1.998	3.275e−7	1.950	2.00	1263.89	1.332e−6	1.998	6.631e−7	1.061	2.65	650.93

that the condition number $\kappa(S_D)$ of the Schur complement for fine meshes is extremely large, i.e. $1/\kappa(S_D)$ is close to the computer zero. To restore the second order of convergence of (13), we apply the simple and fast *Jacobi (diagonal) preconditioning* of S_D.

Let us denote by d_{pq}, $p, q = 1, 2, \ldots N + 1$ the elements of S_D. The first equation in (13) is replaced by the equivalent one

$$P^{(k+1)} = \left[\mathbf{S}^{-1}\left(\mathbf{Q}^{(k)} + \frac{\tau}{\alpha}\mathbf{M_1}^{(k)}\mathbf{M_2}^{(k)}\right)\right]^{-1}\left[\mathbf{S}^{-1}\left(F_1^{(k)} + \frac{\tau}{\alpha}\mathbf{M_1}^{(k)}F_2^{(k)}\right)\right],$$

where $\mathbf{S}^{-1}$ is the $(N+1) \times (N+1)$ diagonal matrix

$$S^{-1} = \begin{pmatrix} 1/d_{11} & 0 & \ldots & 0 \\ 0 & 1/d_{22} & \ldots & 0 \\ \ldots & \ldots & \ldots & \ldots \\ 0 & 0 & \ldots & 1/d_{N+1N+1} \end{pmatrix}.$$

The efficiency of this simple procedure is illustrated in Table 4. Now, a second order of convergence is attained and the average number of iterations is optimal. Moreover, we observe better performance of the iteration process (13) with Jacobi preconditioning, than (8), (9).

Consider a new set of parameters, corresponding to *rocksalt* [6,11]

$$\alpha = 10^5, \quad \beta = 10^{-18}, \quad \chi = 10^{-25}, \quad \eta = 3.10^{-6}, \quad \lambda = 16.10^{-8}$$

Table 4. Results for iteration process (13) with Jacobi preconditioning, $\tau = h$

N	$\gamma = 0.25$						$\gamma = 0.75$					
	E_P^N	CR_P^N	E_T^N	CR_T^N	$AvIt$	CPU	E_P^N	CR_P^N	E_T^N	CR_T^N	$AvIt$	CPU
20	5.225e−3		1.472e−3		2.85	0.08	5.225e−3		9.748e−4		2.85	0.08
40	1.335e−3	1.969	3.758e−4	1.970	2.93	0.14	1.335e−3	1.969	2.600e−4	1.906	2.93	0.15
80	3.406e−4	1.971	9.528e−5	1.980	2.00	0.32	3.406e−4	1.971	6.954e−5	1.903	2.00	0.34
160	8.516e−5	2.000	2.389e−5	1.995	2.00	1.03	8.516e−5	2.000	1.846e−5	1.913	2.00	1.21
320	2.124e−5	2.004	5.989e−6	1.996	1.99	5.15	2.124e−5	2.004	4.867e−6	1.924	1.99	5.25
640	5.316e−6	1.998	1.572e−6	1.930	2.00	55.74	5.316e−6	1.998	1.335e−6	1.866	2.00	55.95
1280	1.331e−6	1.998	3.779e−7	2.056	2.00	648.50	1.331e−6	1.998	3.456e−7	1.950	2.00	647.33

and scaling parameters $p^* = 10^7$, $\mathbf{t}^* = 10$, $t^* = 10^8$, $z^* = 10^3$. The computational domain is similar as for the previous example and $\tau = h$.

On Fig. 2 we illustrate the second order of convergence of the numerical solution for $\gamma = 0.25$ and $\gamma = 0.75$, computed by both iteration schemes (8)–(9) and (13) with Jacobi preconditioning.

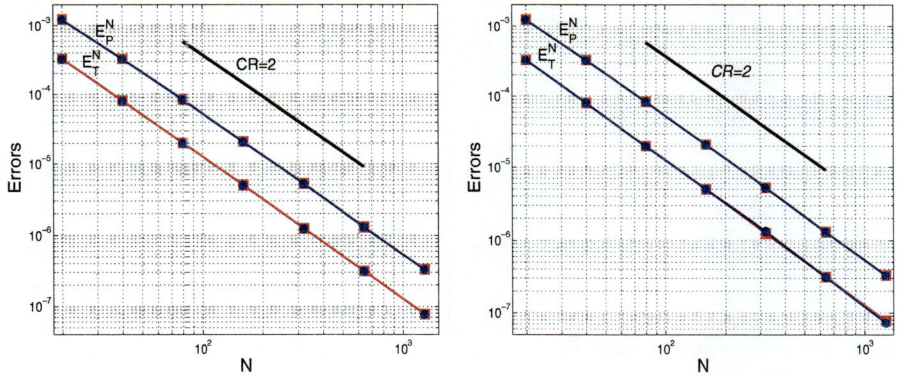

Fig. 2. Error in maximum discrete norm vs. N (in logarithmic scale) of the numerical values of P and T, computed by (8)–(9) (*line with circles*) and (13) with Jacobi preconditioning (*line with squares*) for $\gamma = 0.25$ (*left*) and $\gamma = 0.75$ (*right*); comparison line (*solid line*), corresponding to the exact second order of convergence

On Figs. 3 and 4, we depict the computational efficiency of both schemes (8)–(9) and (13) with Jacobi preconditioning for $\gamma = 0.25$ and $\gamma = 0.75$, respectively.

We observe better efficiency for the considered examples of the scheme (13) with Jacobi preconditioning in comparison with (8)–(9) - for one and the same CPU time, the solution obtained by (13) is more precise.

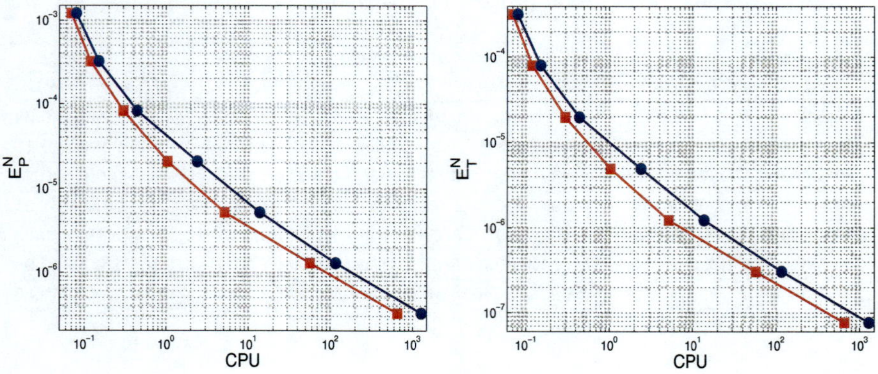

Fig. 3. Error in maximum discrete norm vs. CPU time (in logarithmic scale) of the numerical values of P (*left*) and T (*right*), computed by (8)–(9) (*line with circles*) and (13) with Jacobi preconditioning (*line with squares*) for $\gamma = 0.25$

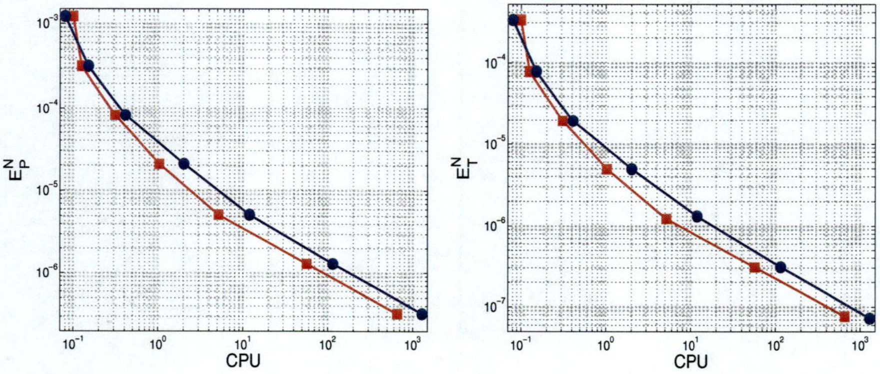

Fig. 4. Error in maximum discrete norm vs. CPU time (in logarithmic scale) of the numerical values of P (*left*) and T (*right*), computed by (8)–(9) (*line with circles*) and (13) with Jacobi preconditioning (*line with squares*) for $\gamma = 0.75$

5 Conclusions

In this work, we constructed and investigated finite difference scheme for a parabolic system describing non-local propagation of non-linear waves conducting by Modified Darcy Law. The space sub-diffusion derivative is approximated by a L1-2 formula. The non-linear system of difference equations is solved by Newton's iteration method. Also, we use Schur complement in order to accelerate the computational process for solving the linear system of algebraic equations.

Test examples show second order of convergence (in maximum norm) and fast performance of the algorithm, when Schur complement is implemented and combined with preconditioning technique.

Our next investigations will concern the sign-preserving property and convergence of the numerical scheme.

Acknowledgements. This work is supported by the bilateral Project "Development and investigation of finite-difference schemes of higher order of accuracy for solving applied problems of fluid and gas mechanics, and ecology" of Bulgarian National Science Fund under Project DNTS/Russia 02/12 and Russian Foundation for Basic Research under Project 18-51-18004-bolg_a.

References

1. Baleanu, D., Diethelm, K., Scalas, E., Trujillo, J.: Fractional Calculus: Models and Numerical Methods. World Scientific (2012). ISBN 978-981-4355-20-9
2. Bonafede, M.: Hot fluid migration: an efficient source of ground deformation to the 1982–1985 crisis at Campi Flegrei-Italy. J. Volcanol. Geotherm. Res. **48**, 187–198 (1991)
3. Caputo, M.: Models of flux in porous media with memory. Water Resour. Res. **36**, 693–705 (2000)
4. Fellah, M., Fellah, Z.E.A., Depollier, C.: Transient wave propagation in inhomogeneous porous matirials; application of fractional derivatives. Signal Process **86**, 2658–2667 (2006)
5. Garra, R.: Fractional-calculus model for temperature and pressure waves in fluid-saturated porous rocks. Phys. Rev. E **84**, 036605 (2011)
6. Garra, R., Salusti, E.: Application of the nonlocal Darcy law to propagation of nonlinear thermoelastic waves in fluid saturated porous media. Phys. D **250**, 52–57 (2013)
7. Gao, G.H., Sun, Z.Z., Zhang, H.W.: A new fractional numerical differentiation formula to approximate caputo fractional derivative and its application. J. Comput. Phys. **259**, 33–50 (2014)
8. Di Giuseppe, E., Moroni, M., Caputo, M.: Flux in porous media with memory models and experiments. Transp. Porous Med. **83**(3), 479–500 (2010)
9. He, J.-H.: Approximate analytical solution for seepage flow with fractional derivatives in porous media. Comput. Methods Appl. Mech. Eng. **167**(1–2), 57–68 (1998)
10. Horn, R., Zhang, F.: Basic properties of the Schur complement. In: Zhang, F. (ed.) The Schur Complement and its Applications, pp. 17–47. Springer (2005)
11. McTigue, D.F.: Thermoelastic response of fluid-saturated rock. J. Geophs. Res. **91**, 9533–9542 (1986)
12. Natale, G., Salusti, E.: Transient solutions for temperature and pressure waves in fluid-saturated porous rocks. Geophys. J. Int. **124**, 649–656 (1996)
13. Ochoa-Tapia, J.A., Valdes-Parada, F.J., Alvarez-Ramirez, J.: A fractional-order Darcy's law. Phys. A: Stat. Mech. Appl. **374**(1), 1–14 (2007)
14. Podlubny, I.: Fractional Differential Equations. Academic Press, New York (1999)
15. Sen, M., Ramos, E.: A spatially non-local model for flow in porous media. Transp. Porous Med. **92**(1), 29–39 (2012)
16. Schumer, R., Benson, D.A., Meerschaert, M.M., Wheatcraft, S.W.: Eulerian derivation of the fractional advection-dispersion equation. J. Contam. Hydrol. **48**, 69–88 (2001)
17. Sukale, Y., Gejji, V.: A review of the numerical methods to solve fractional differential equations. In: National conference on Modeling optimization and controlAt, VIT, Pune (2015)
18. Tain, Y., Takane, Y.: The inverse of any two-by-two nonsingular partitioned matrix and three matrix inverse completion problems. Comput. Math. Appl. **57**, 1294–1304 (2009)

Note on Weakly and Strongly Stable Linear Multistep Methods

M. E. Mincsovics[1,2(✉)]

[1] MTA-ELTE Numerical Analysis and Large Networks Research Group, Pázmány Péter sétány 1/C, Budapest 1117, Hungary
mincso@math.bme.hu
[2] Department of Differential Equations, Budapest University of Technology and Economics, Building H, Egry József utca 1, Budapest 1111, Hungary

Abstract. Our goal is to give explanations on the differences between weakly and strongly stable linear multistep methods from different perspectives.

1 Introduction

Consider the initial value problem

$$\begin{cases} u(0) = u^0\,, \\ u'(t) = f(u(t))\,, \end{cases} \tag{1}$$

where $t \in [0,T], \mathbf{u}^0 \in \mathbb{R}$ is the initial value, $u : [0,T] \to \mathbb{R}$ is the unknown function and f is Lipschitz continuous.

Linear multistep methods (LMMs) serve as a popular tool to approximate the solution of the initial value problem (1). LMMs can be given in the following way:

$$\begin{cases} u_i = c^i\,, & i = 0,\ldots,k-1 \\ \frac{1}{h}\Sigma_{j=0}^k \alpha_j u_{i-j} = \Sigma_{j=0}^k \beta_j f(u_{i-j})\,, & i = k,\ldots,n+k-1\,, \end{cases} \tag{2}$$

where k denotes the number of steps. LMMs can be scaled differently, we will prefer the normalisation $\Sigma_{j=0}^k \beta_j = 1$. 2 defines a recursion which can be explicit ($\beta_0 = 0$) or implicit, in both cases it requires k previous values to determine the next.

The *first* and *second characteristic polynomial* associated to (2) are

$$\rho(\xi) = \Sigma_{j=0}^k \alpha_j \xi^{k-j}\,, \qquad \sigma(\xi) = \Sigma_{j=0}^k \beta_j \xi^{k-j}\,. \tag{3}$$

It is well known that for (at least first order) consistency the LMM must satisfy $\rho(1) = 0$ and $\rho'(1) = \sigma(1) = 1$, where $\sigma(1) = 1$ means the normalisation. Moreover, a method is 0-stable, if for every root of the first characteristic polynomial

I. Dimov and S. Fidanova (Eds.): HPC 2019, SCI 902, pp. 290–297, 2021.
https://doi.org/10.1007/978-3-030-55347-0_25

$|\xi_i| \le 1$ holds and if $|\xi_i| = 1$ then it is a simple root (this is usually called the root-condition). The two subclasses of consistent and 0-stable LMMs are the strongly and weakly stable methods:

Definition 1. The method is said to be *strongly stable* if for every root $\xi_i \in \mathbb{C}$ of the first characteristic polynomial $|\xi_i| < 1$ holds except $\xi_1 = 1$, which is a simple root.

A not strongly stable method is said to be *weakly stable* if for every root $\xi_i \in \mathbb{C}$ of the first characteristic polynomial $|\xi_i| \le 1$ holds and if $|\xi_i| = 1$ then it is a simple root, moreover $\xi_1 = 1$.

Example 1. The *Adams–Bashforth two step method* (AB2)

$$\frac{1}{h}\left(y_n - y_{n-1}\right) = \frac{3}{2} f_{n-1} - \frac{1}{2} f_{n-2} \tag{4}$$

is a strongly stable LMM, since its first characteristic polynomial is $\xi(\xi - 1)$ with roots 1 and 0, while the explicit midpoint method (EM) - sometimes called leapfrog method, especially in the context of numerical solution of PDEs -

$$\frac{1}{h}\left(\frac{1}{2} y_n - \frac{1}{2} y_{n-2}\right) = f_{n-1} \tag{5}$$

is only a weakly stable method, since its first characteristic polynomial is $\frac{1}{2}\left(\xi^2 - 1\right)$ with roots 1 and -1. We also note that these are second order explicit methods.

Why should we distinguish these two types? The literature of LMMs is extremely large but this question is usually illustrated only with an example and sometimes in an incomplete form, from which the Reader might draw an inappropriate conclusion. In this paper we will also start with a simple example to illustrate some difference between these two types and at the same time we try to explain the consequences of it to settle this example in a satisfactory way. But this serves just as the starting point of the further investigation. There are so many types of stability concepts in this topic that it is not clear what is the relation of the instability which will turn up in this first example and the general definition of stability. The remaining part of the paper is devoted to clarify this relation. Up to the recent years the only contribution to this question was the so-called Spijker's example. Here we sum up and explain two new results from the recent years which developed the theory further.

2 Differences Between Weakly and Strongly Stable LMMs

2.1 A Simple Example

Similar investigations can be found in many textbooks, see e.g. [1, Example 5.7] [2, Subsection 423], [4, Example 6.12].

We consider the Dahlquist test equation

$$\begin{cases} \dot{y}(t) = \lambda y(t) \\ y(0) = 1 \end{cases} \tag{6}$$

which solution is $y(t) = e^{\lambda t}$. We will compare the performance of the AB2 and EM methods. We assume that $\Re\lambda < 0$, and $|h\lambda| \ll 1$ to eliminate the possible side-effect of their explicitness.

The approximating solution is in the form

$$y_n = c_1\xi_1^n + c_2\xi_2^n\,, \tag{7}$$

where ξ_1, ξ_2 are the roots of the characteristic test-equation

$$\rho(\xi) - h\lambda\sigma(\xi) = 0\,.$$

- In the case of the Adams–Bashforth method the roots of the characteristic test-equation

  $$\xi^2 = \xi\left(1 + \frac{3}{2}h\lambda\right) - \frac{1}{2}h\lambda$$

  are

  $$\begin{cases} \xi_1 = e^{h\lambda} + \mathcal{O}\left(h^3\lambda^3\right) \\ \xi_2 = \dfrac{1}{2}\left(h\lambda - h^2\lambda^2\right) + \mathcal{O}\left(h^3\lambda^3\right)\,, \end{cases}$$

  where ξ_1 is called the *principal root* since ξ_1^n approximates the solution well, while ξ_2 is called the *parasitic root.* Fortunately, for the AB2 $|\xi_2| \ll 1$, which means $\xi_2^n \to 0$ rapidly, making the parasitic root harmless.
- In case of the EM the characteristic test-equation is

  $$\xi^2 = 1 + 2h\lambda\xi$$

  which roots are

  $$\xi_{1,2} = h\lambda \pm \sqrt{h^2\lambda^2 + 1} = \begin{cases} e^{h\lambda} + \mathcal{O}\left(h^3\lambda^3\right) = \xi_1 \\ -e^{-h\lambda} + \mathcal{O}\left(h^3\lambda^3\right) = \xi_2\,. \end{cases}$$

  This is completely different from what we obtained in the case of the AB2 method, since

  $$|\xi_1| < 1 < |\xi_2|$$

  thanks to that $\lambda < 0$. This means that the absolute value of the error caused by the parasitic root is approximately $c_2 e^{-\lambda T}$ at the endpoint of the integration interval T. Thus, the error grows exponentially with T and oscillates (since $\xi_2 < 0$), which sounds very badly. Fortunately, the coefficients in 7 are

  $$c_1 = 1 + \delta \quad \text{and} \quad c_2 = -\delta\,, \quad \text{where} \quad \delta = \mathcal{O}\left(h^3\lambda^3\right)\,,$$

if we assume that we approximated y_1 with at least some second order method. This clears things up, the coefficient of the harmful term is small. However, if T is large enough (or h is not small enough, it depends on from which direction we look at it), then it can ruin the approximation completely, as we can see on the Fig. 1. It is worth to mention that c_2 is also small in the case of AB2.

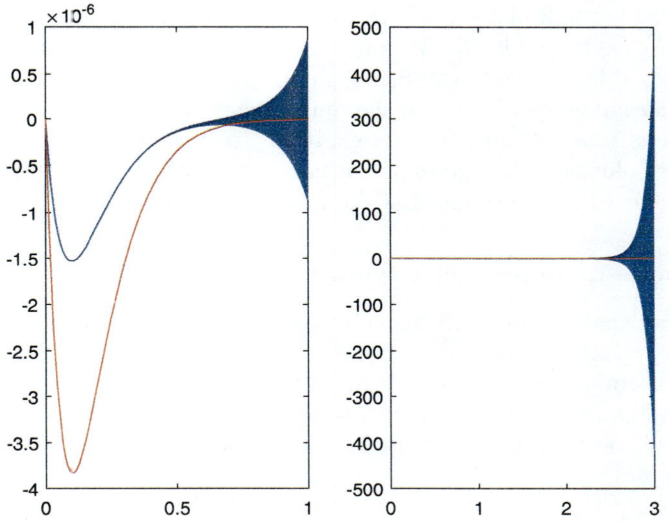

Fig. 1. Errors of the two methods. $h = 5 \cdot 10^{-4}$, $\lambda = -10$, y_1 was determined by the classical Runge-Kutta second order method. Left panel: $T = 1$, the midpoint method (blue) performs better at the initial part of the integration interval than the Adams-Bashforth second order method (red). At the end part the parasitic root starts to dominate, but it is not problematic yet. Right panel: $T = 3$, the midpoint method (blue) produces a completely useless approximation, since the integration interval is long enough to the parasitic root to show its effect.

The characteristic test equation can be considered as a perturbation of the first characteristic polynomial. We know that the roots of a polynomial depend continuously on the coefficients. In case of the AB2 if $h\lambda$ is small enough - as we assumed - then the 1, the first root of σ will move to $e^{h\lambda} + \mathcal{O}\left(h^3\lambda^3\right)$ which is inside of the unit circle and the 0 will move to $\frac{1}{2}\left(h\lambda - h^2\lambda^2\right) + \mathcal{O}\left(h^3\lambda^3\right)$ remaining small. In the case of the EM we can tell the same about the first root, while the second root which was already at the boundary of the unit circle, gets outside. Generally, for a strongly stable method, if h is small enough, all roots of the characteristic test equation will be inside of the unit circle, since the only one root of σ, which might be questionable, will move to some approximation of $e^{h\lambda}$.

This suggests that being weakly or strongly stable has some connection with the notion of stability region. We recall that the stability region of some method consists of the complex numbers $z = h\lambda$ for which all the roots of the characteristic test equation are inside of the (closed) unit circle. The best methods from this point of view are the A-stable methods, which stability region contains the left half plane. However, it is known that the order of A-stable LMMs can not exceed two (second Dahlquist barrier). The EM belongs to the worst methods from this point of view, since its absolute stability region does not contain any point from the left half plane at all. However, it would be too restrictive to avoid the usage of weakly stable LMMs, since for a k-step method, with k even, the maximum possible order is $k+2$ (first Dahlquist barrier). For methods with this maximal order all zeros of σ lie on the unit circle.

The EM is unstable in some sense. If we fix h and increase T, then the approximate solution blows up exponentially, and oscillates. At the other hand it does not affect the convergence of the EM method. Or does it?

2.2 Spijker's Example and Beyond

To investigate the stability and the convergence of some method we summarize the necessary theory of the Lax-Stetter framework in a nutshell. First we rewrite (2) into the form which fits into this framework.

A method can be represented with a sequence of operators $F_N : \mathscr{X}_N \to \mathscr{Y}_N$, where $\mathscr{X}_N$, $\mathscr{Y}_N$ are $k+n$ dimensional normed spaces with norms $\|\cdot\|_{\mathscr{X}_N}$, $\|\cdot\|_{\mathscr{Y}_N}$ respectively and

$$(F_N(\mathbf{u}_N))_i = \begin{cases} u_i - c^i\,, & i = 0, \dots, k-1 \\ \frac{1}{h}\Sigma_{j=0}^k \alpha_j u_{i-j} - \Sigma_{j=0}^k \beta_j f(u_{i-j})\,, & i = k, \dots, n+k-1 = N\,. \end{cases}$$

Finding the approximating solution means that we have to solve the non-linear system of equations $F_N(\mathbf{u}_N) = \mathbf{0}$.

Definition 2. We call a method *stable in the norm pair* $\left(\|\cdot\|_{\mathscr{X}_n}, \|\cdot\|_{\mathscr{Y}_n}\right)$ if for all IVPs (1) $\exists S \in \mathbb{R}_0^+$ and $\exists N_0 \in \mathbb{N}$ such that $\forall N \geq N_0\,, \forall \mathbf{u}_N, \mathbf{v}_N \in \mathbb{R}^{k+n}$ the estimate

$$\|\mathbf{u}_N - \mathbf{v}_N\|_{\mathscr{X}_N} \leq S\,\|F_N(\mathbf{u}_N) - F_N(\mathbf{v}_N)\|_{\mathscr{Y}_N} \tag{8}$$

holds.

To define stability in this way has a definite profit. It is general in the sense that it works for almost every type of numerical method approximating the solution of ODEs and PDEs as well. Convergence can be proved by the recipe "consistency + stability = convergence"

$$\|\varphi_{\mathbf{N}}(\bar{\mathbf{u}}) - \bar{\mathbf{u}}_{\mathbf{N}}\|_{\mathscr{X}_N} \leq S\,\|F_N\left(\varphi_{\mathbf{N}}(\bar{\mathbf{u}})\right) - F_N(\bar{\mathbf{u}}_N)\|_{\mathscr{Y}_N} = S\,\|F_N\left(\varphi_{\mathbf{N}}(\bar{\mathbf{u}})\right)\|_{\mathscr{Y}_N} \to 0\,,$$

where $\bar{u}$, $\bar{\mathbf{u}}_N$ denote the solution of the original problem (1) and the approximating problem $F_N(\mathbf{u}_N) = \mathbf{0}$ respectively, $\varphi_N : \mathscr{X} \to \mathscr{X}_N$ are projections from the normed space where the original problem is set, thus $\varphi_{\mathbf{N}}(\bar{\mathbf{u}}) - \bar{\mathbf{u}}_{\mathbf{N}}$ represents the error (measured in $\mathscr{X}_N$). Finally, in this framework $\|F_N(\varphi_{\mathbf{N}}(\bar{\mathbf{u}}))\|_{\mathscr{Y}_N} \to 0$ is exactly the definition of consistency. We note that the existence of $\bar{\mathbf{u}}_N$ (from some index) is also the consequence of stability, see [3, Lemma 24. and 25.], cf. [10, Lemma 1.2.1].

Note, that the popular recipe "0-stability + consistency = convergence" is norm dependent in this context. We know that

- weakly and strongly stable methods are stable in the norm pair $(\|\cdot\|_{k\max}, \|\cdot\|_{k\max})$ see e.g. [5], where for $k \in \mathbb{N}$ fixed, $\mathbf{u}_n \in \mathbb{R}^{k+n}$ the $k\max$ norm is defined as $\|\mathbf{u}_n\|_{k\max} = \max_{0\le i\le k-1}|u_i| + \max_{k\le i\le k+n-1}|u_i|$;
- strongly stable methods are stable in the norm pair $\left(\|\cdot\|_{k\max}, \|\cdot\|_{k\$}\right)$, as well, see [6], where $\|\mathbf{u}_n\|_{k\$} = \max_{0\le i\le k-1}|u_i| + h\max_{k\le l\le k+n-1}\left|\Sigma_{i=k}^{l} u_i\right|$ is the k–Spijker-norm;
- at the other hand, weakly stable methods are not stable in the norm pair $\left(\|\cdot\|_{k\max}, \|\cdot\|_{k\$}\right)$, see [7]. This is the extension of Spijker's example, see [9] and [10, Example 2 in Sect. 2.2.4] for a more available reference, which showed that the EM is not stable in this norm pair.

This makes it clear that the stability in the norm pair $\left(\|\cdot\|_{k\max}, \|\cdot\|_{k\$}\right)$ separates the weakly and strongly stable LMMs. However, this is a theoretical result, it would be much better to find some difference on the level of convergence. This can not be archived using this difference, since a weakly stable method is stable in the norm pair $(\|\cdot\|_{k\max}, \|\cdot\|_{k\max})$ and if it is consistent in the norm $\|\cdot\|_{k\max}$ with order m then it is convergent in the norm $\|\cdot\|_{k\max}$ with the same order. A strongly stable method can not do better using the stability of the norm pair $\left(\|\cdot\|_{k\max}, \|\cdot\|_{k\$}\right)$, since it will imply convergence in the same norm $\|\cdot\|_{k\max}$ as in the weakly stable case. The only gain is that in some of the cases, it is easier to prove consistency in the norm $\|\cdot\|_{k\$}$, see [10, Example 1 in Sect. 2.2.4].

To improve this approach the kC1 norm $\|\mathbf{u}_N\|_{kC1} = \max_{0\le i<k}|u_i| + \max_{k\le i<N}\dfrac{1}{h}|u_i - u_{i-1}|$. was introduced in [8]. This norm can capture the possible spurious oscillations of the numerical approximation which are typical for the weakly stable LMMs. The name refers to its continuous counterpart $\|u\|_{C^1} = |u(0)| + \max|\dot{u}(t)|$.

Based on the paper [8] we know the following.

- Consider a strongly stable k-step LMM whose order of convergence is γ with respect to the $k\max$ norm. Assume that the starting values are calculated with a one-step method with the same order. Then it is convergent with respect to the kC1 norm and its order is also γ.
- Unfortunately, it is still an open question for weakly stable LMMs. Moreover, the numerical experiments suggest that it might be true for weakly stable LMMs, as well.

We note that using a one-step method with the same order is rare in practice. Instead of this it is usual to use a one-step method with order of only $\gamma - 1$ to

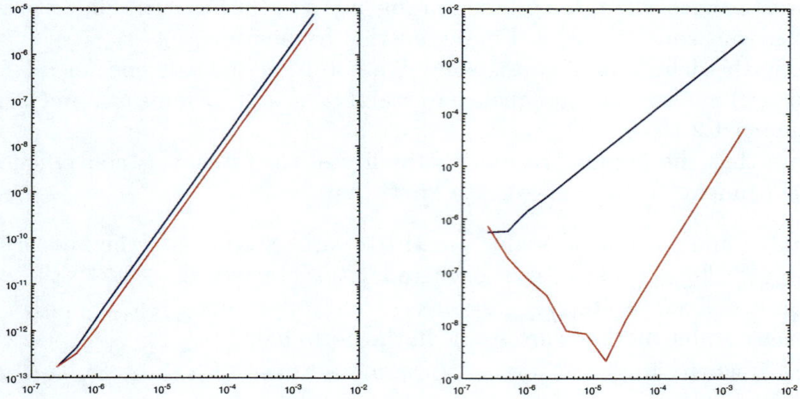

Fig. 2. Different errors of the EM. We approximated the IVP $\dot{y}(t) = 3\cos(3t) - (y(t) - \sin(3t))$, $y(0) = 1$, which solution is $\sin(3t) + e^{-t}$. y_1 was determined in two different ways: the classical explicit second order Runge-Kutta method (red) and the explicit Euler first order method (blue). We plotted the errors vs. the stepsize h on a log-log scale. Left panel: $k$max norm errors. Right panel: kC1 norm errors. It is clear, that the choice which determined y_1 does not really affected the convergence in the $k$max norm, while we lost one order in the kC1 norm (the slope of the left blue curve is approximately 1). We note that the rounding errors are responsible for the strange behaviour of the right red curve, showing that this norm is more sensitive from this point of view.

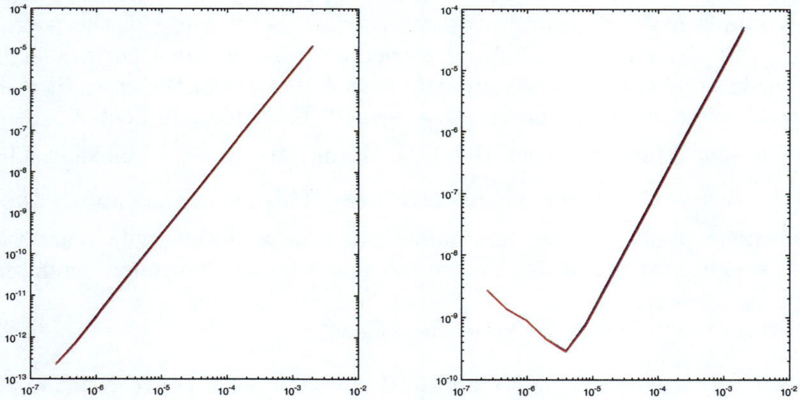

Fig. 3. The same numerical experiment for the AB2 method. The difference is negligible in both norms.

produce the starting values, or recursively by lower order members from the same family of the LMM we want to use. It is well known that these choices of the starting values will result in the required order of convergence if we use the k max norm. However, if we use the kC1 norm, it has a remarkable effect on weakly stable methods, see Fig. 2, while have to be Fig. 3.

If we go back to the test-equation, this phenomena can be easily understood. Using a first order method to get y_1 will result in a larger coefficient $c_2 = \mathcal{O}\left(h^2\lambda^2\right)$. Calculating

$$\left| c_2 \frac{(-1)^{n+1} e^{-\lambda h(n+1)} - (-1)^n e^{-\lambda hn}}{h} \right| = \mathcal{O}\left(h\lambda^2\right) e^{-\lambda hn}\left(e^{-\lambda h} + 1\right) = \mathcal{O}\left(h\lambda^2\right)$$

shows that the Fig. 3 is superfluous, drop it.

Summarizing, we tried to find the differences between weakly and strongly stable LMMs. We started with an example which showed that the EM is unstable from some sense. On the level of the general stability definition we are able to characterize the difference between these methods, the k-Spijker norm plays the key role in this comparison. On the level of convergence, the kC1 norm and also the precision of the first k steps are making the difference between weakly and strongly stable LMMs.

References

1. Ascher, U.M., Petzold, L.R.: Computer methods for ordinary differential equations and differential-algebraic equations (1997)
2. Butcher, J.C.: Numerical Methods for Ordinary Differential Equations, 3rd edn. Wiley, Hoboken (2016)
3. Faragó, I., Mincsovics, M.E., Fekete, I.: Notes on the basic notions in nonlinear numerical analysis. In: Proceedings 9'th Colloquium QTDE, 2011, no. 6, pp. 1–22 (2012). E. J. of qualitative theory of differential equations
4. Griffiths, D.F., Higham, D.J.: Numerical Methods for Ordinary Differential Equations. Springer, London (2010)
5. Mincsovics, M.E.: Stability of one-step and linear multistep methods – a matrix technique approach. In: Colloquium on the Qualitative Theory of Differential Equations, no. 15, pp. 1–10 (2016). https://doi.org/10.14232/ejqtde.2016.8.15
6. Mincsovics, M.E.: Note on the stability of strongly stable linear multistep methods. AIP Conf. Proc. **1895**(1), 110006 (2017)
7. Mincsovics, M.E.: Spijkers example and its extension, In: Dimov I., Faragó I., Vulkov L. (eds.) Finite Difference Methods. Theory and Applications. FDM 2018. Lecture Notes in Computer Science, vol. 11386, pp. 354–361. Springer, Cham (2018)
8. Mincsovics, M.E.: Discrete C^1 convergence of linear multistep methods. J. Comput. Appl. Math. **363**, 234–240 (2020)
9. Spijker, M.N.: Stability and convergence of finite-difference methods. University of Leiden, Thesis (1968)
10. Stetter, H.J.: Analysis of Discretization Methods for Ordinary Differential Equations. Springer, Heidelberg (1973)

Estimating the Statistical Power of the Benjamini-Hochberg Procedure

Dean Palejev(✉) and Mladen Savov

Institute of Mathematics and Informatics, Bulgarian Academy of Sciences, Georgi Bonchev Str., bl. 8, 1113 Sofia, Bulgaria
{palejev,mladensavov}@math.bas.bg

Abstract. In this work we investigate the properties of the statistical power of the Benjamini-Hochberg procedure. That procedure provides a widely used method for controlling the False Discovery Rate in a multiple comparison setup. We show that in many cases the statistical power of the p-values adjusted by the Benjamini-Hochberg procedure could be approximated by a normal distribution with both its mean and standard deviation having an exponential fit and convergence when the number of tests increase. As a result one could estimate the probability that the power belongs in a predetermined interval in a very computationally efficient way using only simulations for several values of the number of tests parameter. The fit for the mean also supports the conjecture that the power of the test decreases to the limit power, which is known to exist, when increasing the number of tests. The latter is a very favourable observation from practical perspective and we try to offer partially rigorous explanation why the monotonicity is present.

1 Introduction

The multiple comparison problem occurs in experiments where simultaneous tests of large number of hypotheses need to be implemented. Due to the large number of tests by the law of randomness one is bound to encounter many cases whereby the p-value of a certain hypothesis is below the predetermined threshold level α even if the null hypothesis is in fact true. The natural way to amend this is to change the rule by which a null hypothesis is rejected, whilst preserving the false discovery rate. The most popular approach for solving the multiple comparison problem is the Benjamini-Hochberg procedure, see [2], which can be described as follows: for a given significance level α and a set of m p-values, the tests corresponding to the l smallest p-values are rejected, where l is the largest integer such that $p_l \leq \alpha l/m$, where p_l is the l-th smallest p-value. We note that this is quite different from the classical Bonferroni correction which would reject a hypothesis if $p \leq \alpha/m$. The Benjamini-Hochberg method is widely used, through software implementation, in many applied areas of science where thousands and sometimes millions hypotheses are tested for a particular experiment, most often

I. Dimov and S. Fidanova (Eds.): HPC 2019, SCI 902, pp. 298–308, 2021.
https://doi.org/10.1007/978-3-030-55347-0_26

in the field of Genetics. Most of the applications are in the analysis of large number of Genetic features, for example genes or SNPs, see [5]. This has led to its prominence in science, being the 59th most cited scientific paper as of 2014, see [7]. There are various successful attempts to relax the assumptions of the Benjamini-Hochberg procedure, see [3,6], but in this work we focus on the classical method.

For practical applications it is important to know what proportion of the declared statistically significant p-values is in fact due to random fluctuations or in other words wrongly declared so. Therefore, it is important to determine/estimate apriori the power of the Benjamini-Hochberg procedure, that is the proportion of significant hypotheses that have been rightly declared as such. Under fairly general conditions, it is well-known that when the number of tests grows the power converges to an explicit number and this partially answers the question above. However, in practical terms, when the number of tests is fixed, albeit large, it is not easy to determine how close and how likely the power of the Benjamini-Hochberg procedure is to the limiting one. The aim of this paper, in specific cases, is to propose a model which fits the mean and the variance of the proportion in terms of the number of the tests and other intrinsic to the model quantities. We shall also discuss the empirical phenomenon that with the increase of the number of the tests the power decreases to the limiting one. This is a very favourable observation from practical perspective. Before going into details we will introduce rigorously the Benjamini-Hochberg procedure.

2 The Benjamini-Hochberg Method

2.1 Description and Notation

Let $X_1, X_2, \cdots, X_m$, $m \in \mathbf{N}^+$, with $\mathbf{N}^+ = \{1, 2, \cdots\}$, be independent random variables that are defined on a common probability space $(\Omega, \mathscr{F}, \mathbf{P})$. For some $0 \leq m_0 \leq m$ the random variables $(X_j)_{j \leq m_0}$ are assumed to have a common cumulative distribution function (c.d.f.) of the uniform random variable on $(0, 1)$, say F, whereas $(X_j)_{m_0 < j \leq m}$ are assumed to have a common c.d.f. of some Beta distribution, say G. We emphasize that solely the latter is a minor, but plausible restriction. These random variables describe the p-values of the tests, where the p-values for all valid null hypotheses must be random variables that are uniformly distributed on $(0, 1)$. In our case we have m_0 of those. We set $m_0/m = \gamma \in [0, 1]$ and denote $m_1 := m - m_0 \in [0, m]$. The parameter γ can be assumed larger than $1/2$ since the first m_0 random variables represent the non-significant tests which typically exceed the significant ones by a large margin. If $X_m^{(i)}$, $i \leq m$, is i-th order statistics of $X_1, X_2, \cdots, X_m$ then for a fixed level of rejection $\alpha \in (0, 1)$, the Benjamini-Hochberg procedure declares significant the first R_m order statistics, where

$$R_m = \max \left\{ m \geq i \geq 1 : X_m^{(i)} \leq \alpha \frac{i}{m} \right\}. \tag{1}$$

The non-significant discoveries are all null hypotheses declared as significant from the Benjamini-Hochberg procedure despite having c.d.f F. The significant discoveries by assumption possess c.d.f. G. According to [4, Theorem 3.1] the limit in probability, that is $\lim_{m\to\infty} R_m/m = x_{\alpha,\gamma}$ exists, provided the equation

$$(1-\gamma) \sup_{t\in[\alpha x,\alpha]} \frac{G(t)-t}{t} = \frac{1}{\alpha} - 1 \tag{2}$$

has a unique solution for $x = x_{\alpha,\gamma}$. The latter is certainly the case if $\{G(t)-t\}/t$ is decreasing on $[0,\alpha]$ which will be implicitly guaranteed in our paper by the choice of the parameters of the Beta distributions that we are going to utilize. The permissible ones have precisely the structure a significant p-value ought to have, that is to have more mass close to zero. We proceed by studying the question about the power of the test. It is clear that if we denote

$$S_m = \sum_{j=m_0+1}^{m} \mathbf{I}_{\{X_j \le \alpha \frac{R_m}{m}\}}, \tag{3}$$

where $(X_j)_{m_0<j\le m}$ are the aforementioned i.i.d. with c.d.f G then

$$S_m^* = \frac{S_m}{m_1} \tag{4}$$

is the proportion of tested elements that have been rightly identified as significant or simply the empirical power of the test. It is known from [4, Corollary 3.3] that under the assumptions we have in this paper, see the discussion concerning the limit of R_m/m above,

$$\lim_{m\to\infty} S_m^* = x_{\alpha,\gamma}\frac{1-\alpha\gamma}{1-\gamma}. \tag{5}$$

We emphasize that this limit requires that $m_0/m = \gamma$ is kept constant as m increases but it is non-binding as what happens in practice is that researchers know m, have a some approximate guess for m_0 and thus γ is known within certain bounds. Then working with γ within those bounds, our model gives the mean and variance of S_m^*.

2.2 Heuristics of the Monotone Decrease of the Power with the Number of Tests

In simulations, see Sect. 3.1, we observe that the empirical power of the Benjamini-Hochberg procedure decreases with the number of the tests to the quantity in (5). This in practical terms is very beneficial as $x_{\alpha,\gamma}\frac{1-\alpha\gamma}{1-\gamma}$ may serve as a lower bound to the empirical power. Though not entirely mathematically rigorous, we shall attempt to offer some heuristics why this monotonicity is valid.

For this purpose we introduce the empirical cumulative distribution function H

$$H_m(t) := \frac{1}{m}\sum_{j=1}^{m} \mathbf{I}_{\{X_j \le t\}} = \gamma F_{m_0}(t) + (1-\gamma)G_{m_1}(t), \tag{6}$$

where F_{m_0}, G_{m_1} are the respective empirical cumulative distribution functions of $(X_j)_{j=1}^{m_0}$ and $(X_j)_{j=m_0+1}^{m}$. Then it is known, see [4, (1.2)], that

$$\mathbf{P}\left(\frac{R_m}{m} \ge x\right) = \mathbf{P}\left(\max_{A_\alpha(m,x)} \frac{H_m(t)-t}{t} \ge \frac{1}{\alpha} - 1\right), \tag{7}$$

where

$$A_\alpha(m,x) = \left\{\frac{k}{m}\alpha : \lceil mx \rceil \le k \le m\right\}.$$

Also the limit in (7), as m increases, is 0 if $x > x_{\alpha,\gamma}$ and has a value of 1 if $x < x_{\alpha,\gamma}$, see [4, Theorem 3.1]. Assume now that we change m to $2m$. Then clearly

$$A_\alpha(m,x) \subseteq A_\alpha(2m,x)$$

and

$$\max_{A_\alpha(2m,x)} \frac{H_m(t)-t}{t} \ge \max_{A_\alpha(m,x)} \frac{H_m(t)-t}{t}$$

since we take maxima on a larger domain. This means that the quantities in (7) increase. Since (5) contains $x_{\alpha,\gamma}$ and (3) depends on R_m/m this monotonicity explains loosely the monotonicity observed in Sect. 3.1.

3 Estimating the Distribution of the Statistical Power

3.1 Numerical Simulations

In order to numerically estimate the statistical power in different scenarios, we consider a set of m tests with $m_1 = m - m_0$ of them being significant. The p-values of the significant ones are assumed to have a $B(a,b)$ distribution and of course, the non-significant have a $U(0,1)$ distribution. The choice of Beta distribution covers different possible shapes of the distribution of the p-values of the significant tests. Recall that the Beta distribution has density

$$f(x) = \frac{\Gamma(a+b)}{\Gamma(a)\Gamma(b)} x^{a-1}(1-x)^{b-1},\ x \in (0,1),$$

where Γ is the celebrated Euler's Gamma function. Then when $a < 1$ we have that $\lim_{x \to 0} f(x) = \infty$ which means that more mass is concentrated at zero, which befits a model of significant p-values, as they will have to cluster more at zero. The smaller the a with $b \geq 1$ fixed, the more mass is located in the vicinity of 0. The parameter b plays a similar role but in this case the larger it is, the more mass is concentrated around zero. Also when $a < 1, b \geq 1$, it is easily checked that for the c.d.f. G it holds true that $\{G(t) - t\}/t$ decreases.

The simulations were done 1 million times for all combinations of the following values of the parameters: $a \in \{0.1, 0.2, 0.25, 0.333, 0.5\}$, $b \in \{2, 5, 10, 20, 50, 100\}$, m taking various values between 50 and 20000 and the proportion of significant tests $\gamma_1 = m_1/m \in \{0.02, 0.05, 0.1, 0.15, 0.2, 0.3, 0.4, 0.5\}$. In all cases we use the standard significance level α of 0.05.

For each set of values of a, b, m and γ_1 we calculate $m_1 = \gamma_1 m$ and $m_0 = m - m_1$. If $\gamma_1 m$ is not an integer value, then we round it to take m_1. Then for each of the 1 million iterations we simulate m_1 values from $B(a, b)$ and m_0 values from the $U(0, 1)$ distribution. For each iteration we then apply the Benjamini-Hochberg procedure to the set of m p-values and calculate the empirical statistical power being the proportion of the adjusted significant ones that are less or equal to the significance level α.

3.2 Estimating the Mean and the Standard Deviation of the Distribution of the Statistical Power

When the expected number of significant tests m_1 was moderate or large, the empirical distribution of the power was close to normal, e.g. as in Fig. 1. Therefore in this case it is sufficient to estimate the mean and the standard deviation of that normal distribution.

Some results of these estimates are given in Table 1. Note that for the table to fit in this article, we show only portion of the results. However, to show that the estimates are consistently good, for each parameter we have chosen the smallest and the largest value and at least one value between them. The first 2 columns of the table are the parameters of the Beta distribution and the third is the proportion of the significant tests. The fourth column shows the error when estimating the means of the statistical power over the different values of m as a function of $\log_{10}(m)$. The next two columns are respectively the minimum and the maximum of the mean statistical power over the values of m. The next column being the difference of the previous 2 values. The last column is the error when estimating the standard deviation as a function of $\log_{10}(m)$.

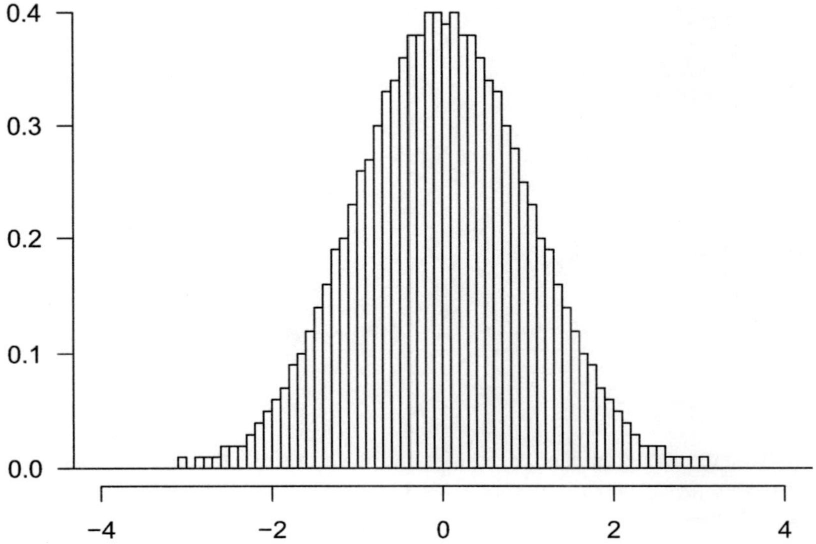

Fig. 1. Histogram of the distribution of the statistical power for $m = 10000$, $a = 0.1$, $b = 10$, $\gamma_1 = 0.1$

When the range of the values of the mean over the different values of m is large enough, say the difference between the largest and the smallest is at least 0.0025, we estimate the mean as an exponential function of the base-10 logarithm of the number of tests using the following formula: $y \sim A + Be^{-Cx}$, where y is that mean and $x = \log_{10}(m)$. When the number of tests increases to ∞, according to the above model, the power should converge to A. At the same time, the limit is the solution of a particular equation, see (5), where $x_{\alpha,\gamma}$ can be easily calculated numerically, see (2); for more properties, see article [4].

The fit works surprisingly well in most cases as shown in a particular example in Fig. 2. The fit was done using the R function `nls`. However when the range is small, practically all means are very close to each other. In this case we can estimate the fit as being a horizontal line. Such cases are denoted by an asterisk after the error in the fourth column.

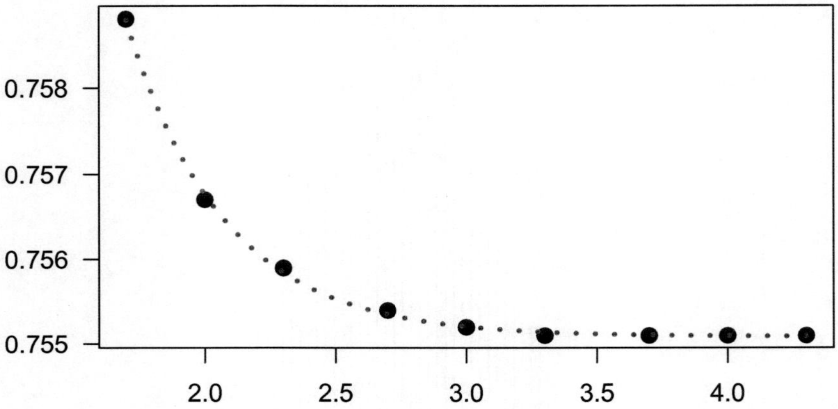

Fig. 2. Estimate of the mean of the statistical power for $a = 0.1$, $b = 10$, $\gamma_1 = 0.1$. The x-axis is the $\log_{10}(m)$ and the y axis is the empirical mean of the statistical power for that value of m

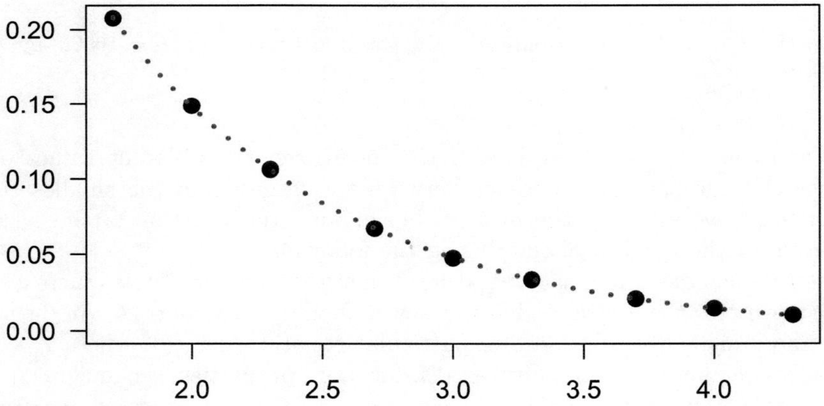

Fig. 3. Estimate of the standard deviation of the statistical power for $a = 0.1$, $b = 10$, $\gamma_1 = 0.1$. The x-axis is the $\log_{10}(m)$ and the y axis is the empirical standard deviation of the statistical power for that value of m

We estimate the standard deviation of the distribution as an exponential function of the base-10 logarithm of the number of tests using again the formula $y \sim A + Be^{-Cx}$. The result for particular set of values is shown in Fig. 3. As the standard deviation decreases faster to its limit, the ranges were wider and there is no need to use a linear fit. At the same time, as the standard deviation converges to 0 when the number of tests increases, the parameter A is equal to 0.

Table 1. Results from the numerical simulations

a	b	γ_1	Error of the mean fit	Min power	Max power	Power difference	Error of the standard deviation fit
0.1	2	0.02	0.00035759608	0.5189	0.5554	0.036500	0.001589205960
0.1	2	0.1	0.00003592543	0.62	0.6257	0.005700	0.000438660775
0.1	2	0.2	0.0008016*	0.669	0.6714	0.002400	0.000132286252
0.1	2	0.5	0.00003294532	0.7385	0.7392	0.000700	0.000031737943
0.1	10	0.02	0.00025035455	0.6337	0.6641	0.030400	0.002562905008
0.1	10	0.1	0.00004302625	0.7551	0.7588	0.003700	0.000329261948
0.1	10	0.2	0.0004126*	0.8118	0.813	0.001200	0.000086364367
0.1	10	0.5	0.00006009*	0.8862	0.8864	0.000200	0.000011405705
0.1	100	0.02	0.00005737354	0.8157	0.8325	0.016800	0.003579576672
0.1	100	0.1	0.0001394*	0.9404	0.9408	0.000400	0.000054475340
0.1	100	0.2	0.0000441*	0.9765	0.9766	0.000100	0.000007391393
0.1	100	0.5	0.0000371*	0.9973	0.9973	0.000000	0.000001664355
0.25	2	0.02	0.00182590018	0.1374	0.2264	0.089000	0.002026492616
0.25	2	0.1	0.00027781730	0.2338	0.2682	0.034400	0.000659909047
0.25	2	0.2	0.00003439259	0.2939	0.3112	0.017300	0.000680690726
0.25	2	0.5	0.00003059273	0.396	0.4015	0.005500	0.000185761790
0.25	10	0.02	0.00209049414	0.2469	0.3515	0.104600	0.001345550814
0.25	10	0.1	0.00003799672	0.419	0.447	0.028000	0.001482339334
0.25	10	0.2	0.00005739530	0.5229	0.5337	0.010800	0.000653759195
0.25	10	0.5	0.00001947062	0.6863	0.6884	0.002100	0.000096245222
0.25	100	0.02	0.00136270288	0.5299	0.6181	0.088200	0.003957810590
0.25	100	0.1	0.00006885447	0.8254	0.8297	0.004300	0.000745211001
0.25	100	0.2	0.00003333*	0.9288	0.9289	0.000100	0.000012030032
0.25	100	0.5	0.00002731*	0.9915	0.9915	0.000000	0.000003411931
0.5	2	0.02	0.00004366615	0.004032	0.0494	0.045368	0.000051766937
0.5	2	0.1	0.00028106199	0.0123	0.05487	0.042570	0.000440700688
0.5	2	0.2	0.00043295430	0.02369	0.06204	0.038350	0.000628575281
0.5	2	0.5	0.00028780522	0.05773	0.08542	0.027690	0.000220344079
0.5	10	0.02	0.00038832360	0.01546	0.1158	0.100340	0.000246166491
0.5	10	0.1	0.00105673847	0.06537	0.1466	0.081230	0.001086221264
0.5	10	0.2	0.00073056962	0.1285	0.1887	0.060200	0.000324766014
0.5	10	0.5	0.00004562724	0.304	0.327	0.023000	0.001009663506
0.5	100	0.02	0.00343608051	0.1345	0.358	0.223500	0.000838449480
0.5	100	0.1	0.00030172055	0.5532	0.5986	0.045400	0.003818030428
0.5	100	0.2	0.00001424068	0.8051	0.8072	0.002100	0.000535288694
0.5	100	0.5	0.00003531*	0.9753	0.9754	0.000100	0.000010077466

In the cases denoted by asterisk, a linear fit is used as described in Subsect. 3.2.

3.3 Binomial Consistency

When the expected number of significant tests m_1 was less than 30, we estimated the number of significant tests as the binomial $B(m_1, p)$ distribution with a known m_1 and therefore we only needed to estimate the parameter p. This was done by using MLE, although the parameter estimate is the same using other estimators. In Table 2 we compare the estimates of the mean and the standard deviation using the normal and the binomial model. To save space in this article, the table shows the results only for some values of the parameters. The first column shows the number of tests, the next two are the parameters of the Beta distribution, then we show the proportion and the number of significant tests. Then we show the fit of the mean when using the normal model and then its estimate using the binomial model, as well those of the standard deviation. As we can see both values for the means are very close (within few percent), whereas there is somehow larger difference between the standard deviations. In practice, in this case we recommend to use the estimates based on a binomial model as it matches closer the actual distribution.

4 Summary and Practical Considerations

4.1 Conclusions

We have developed a computationally efficient method in order to estimate the parameters of the distribution of the statistical power of the Benjamini-Hochberg procedure in a wide variety of cases. Practically with this estimation, one would only need to simulate the power for 3 small or moderate values of m, then estimate the parameters of the exponential models for both the mean and the standard deviation of the distribution, or the linear model for the mean. Having these estimates one could calculate the probability that the power is greater than a pre-specified number, e.g. the commonly used value of 80%.

4.2 Workflow

Here we describe the suggested workflow for given values of the parameters of the Beta distribution with parameters a and b and the proportion of significant tests γ_1 and a pre-specified threshold typically $\alpha = 0.05$:

1. Choose few values for the number of tests m. To save computational time, these could be small or moderate. For each of them:
   a. Calculate the number of significant tests $m_1 = \gamma_1 m$ and if necessarily round this number.
   b. Repeat large number of times, 10000 iterations should be enough in many cases, although here we used 1 million
      i. Generate m_1 values from $Beta(a, b)$ and $m_0 = m - m_1$ values from the standard uniform $U(0, 1)$ distribution.

Table 2. Binomial consistency

m	a	b	γ_1	m_1	Normal mean fit	Binomial mean fit	Normal st. dev. est	Binomial st. dev. est
1000	0.1	100	0.02	20	0.79640049	0.81624398	0.10377610	0.08659958
1000	0.1	10	0.02	20	0.64386461	0.63493272	0.13022474	0.10765527
1000	0.25	100	0.02	20	0.53718364	0.53479461	0.17504669	0.11153236
1000	0.25	10	0.02	20	0.24900938	0.25571167	0.14856053	0.09755081
1000	0.5	100	0.02	20	0.16650437	0.16868670	0.16986597	0.08373515
1000	0.5	10	0.02	20	0.03356742	0.03369923	0.05741249	0.04035071
200	0.1	100	0.05	10	0.89730902	0.89109884	0.11499517	0.09850975
200	0.1	100	0.1	20	0.95073175	0.94046426	0.06010027	0.05291089
200	0.1	10	0.05	10	0.70218534	0.70279840	0.17310597	0.14452433
200	0.1	10	0.1	20	0.75704814	0.75586084	0.11592679	0.09605603
200	0.25	100	0.05	10	0.69967676	0.70111490	0.21423991	0.14475939
200	0.25	100	0.1	20	0.82883425	0.82628910	0.12038845	0.08471583
200	0.25	10	0.05	10	0.34465376	0.35024926	0.21998592	0.15085580
200	0.25	10	0.1	20	0.42512241	0.42581692	0.17346249	0.11056601
200	0.5	100	0.05	10	0.36566948	0.36675064	0.29594556	0.15239574
200	0.5	100	0.1	20	0.57034921	0.56424612	0.24217557	0.11087661
200	0.5	10	0.05	10	0.07536366	0.07616842	0.12032861	0.08388492
200	0.5	10	0.1	20	0.09811173	0.09851102	0.12033170	0.06663580
50	0.1	100	0.2	10	0.96993666	0.97655398	0.05186600	0.04785008
50	0.1	100	0.3	15	0.99163268	0.98930332	0.02788459	0.02656095
50	0.1	100	0.4	20	0.98154753	0.99478124	0.01660976	0.01611138
50	0.1	100	0.5	25	1.00110315	0.99731349	0.01048302	0.01035238
50	0.1	10	0.2	10	0.80722288	0.81298798	0.14669523	0.12330390
50	0.1	10	0.3	15	0.84996728	0.84581546	0.11078488	0.09324222
50	0.1	10	0.4	20	0.86872813	0.86893435	0.08939984	0.07546106
50	0.1	10	0.5	25	0.87679293	0.88644419	0.07468727	0.06345420
50	0.25	100	0.2	10	0.93674985	0.92886214	0.10225027	0.08128792
50	0.25	100	0.3	15	0.95781473	0.96687826	0.05358244	0.04620584
50	0.25	100	0.4	20	0.98244848	0.98354551	0.03124976	0.02844622
50	0.25	100	0.5	25	0.98862112	0.99144285	0.01946534	0.01842164
50	0.25	10	0.2	10	0.53763569	0.53365900	0.23715156	0.15775521
50	0.25	10	0.3	15	0.60029298	0.59775859	0.19519717	0.12660786
50	0.25	10	0.4	20	0.66209261	0.64785002	0.16518294	0.10680365
50	0.25	10	0.5	25	0.69242788	0.68836107	0.14242372	0.09263263
50	0.5	100	0.2	10	0.79251435	0.80723292	0.21414563	0.12474291
50	0.5	100	0.3	15	0.90724174	0.90740458	0.10968857	0.07484273
50	0.5	100	0.4	20	0.96071594	0.95318664	0.06066389	0.04723445
50	0.5	100	0.5	25	0.99096815	0.97530897	0.03640890	0.03103636
50	0.5	10	0.2	10	0.18683575	0.18869185	0.21492370	0.12372843
50	0.5	10	0.3	15	0.23166455	0.23377617	0.21922714	0.10927789
50	0.5	10	0.4	20	0.27931694	0.28025935	0.22108908	0.10042760
50	0.5	10	0.5	25	0.32906692	0.32695925	0.21822561	0.09382045

ii. Adjust all these m values using the Benjamini-Hochberg procedure and calculate the power for that iteration, that is the number of the values from the Beta distribution that have adjusted p-values less or equal than the pre-specified threshold α.
iii. Take the sample mean and standard deviation over these iterations.

2. Create a model for estimating the mean of the statistical power for these values of m using either the exponential or linear model from Subsect. 3.2. Calculate numerically the limit of the power using the equations mentioned there. After moving that limit to the left hand side of the equations and taking logarithms of both sides, this yields a linear model. Note that when the number of significant tests is relatively small, one should use the binomial model.
3. Create a model for estimating the standard deviation of the statistical power for these values of m using the exponential model from Subsect. 3.2. As mentioned there A should be 0, and similarly to above, after taking logarithms of both sides of the equations, we only need to fit a linear model.
4. Now we have the underlying distribution for all values of m and from there we could calculate any properties of it. Note that in the binomial case the power would have a stepwise function, therefore we may not necessarily be able to achieve exactly a pre-specified value.

Acknowledgements. This work has been supported by grant DN-02/13 (Modern mathematical methods for Big data) with Bulgarian NSF.

The extensive numerical calculations were performed on the HPC facility Avitohol of IICT-BAS, described in [1].

References

1. Atanassov, E., Gurov, T., Ivanovska, S., Karaivanova, A.: Parallel monte carlo on intel MIC architecture. Procedia Comput. Sci. **108**, 1803–1810 (2017)
2. Benjamini, Y., Hochberg, Y.: Controlling the false discovery rate: A practical and powerful approach to multiple testing. J. R. Stat. Soc. Ser. B (Methodol.) **57**(1), 289–300 (1995)
3. Benjamini, Y., Yekutieli, D.: The control of the false discovery rate in multiple testing under dependency. The Annals of Statistics **29**(4), 1165–1188 (2001)
4. Ferreira, J.A., Zwinderman, A.H.: On the Benjamini - Hochberg method. Ann. Statist. **34**(4), 1827–1849 (2006)
5. Li, J., Witten, D.M., Johnstone, I.M., Tibshirani, R.: Normalization, testing, and false discovery rate estimation for RNA-sequencing data. Biostatistics **13**(3), 523–538 (2012)
6. Storey, J.D., Taylor, J.E., Siegmund, D.: Strong control, conservative point estimation and simultaneous conservative consistency of false discovery rates: A unified approach. J. R. Stat. Soc. Ser. B (Stat. Methodol.) **66**(1), 187–205 (2004)
7. Van Noorden, R., Maher, B., Nuzzo, R.: The top 100 papers. Nature **514**(7524), 550–553 (2014)

HPC Simulations of the Extreme Thermal Conditions in the Balkan Region with RegCM4

Vladimir Ivanov[1(✉)], Rilka Valcheva[2], and Georgi Gadzhev[1]

[1] National Institute of Geophysics, Geodesy and Geography – Bulgarian Academy of Sciences, Acad. G. Bonchev Street, bl. 3, 1113 Sofia, Bulgaria
{vivanov,ggadjev}@geophys.bas.bg

[2] National Institute of Meteorology and Hydrology, 66, Tsarigradsko Shose blvd, 1784 Sofia, Bulgaria
Rilka.Valcheva@meteo.bg

Abstract. The outdoor environment influences human beings differently. If it is too cold for the organism, that could cause frostbite, freezing and other dangerous health conditions. On the other hand, if it is too hot, the heat stress could become too high, and consequences such as heat stroke, fainting, etc. could emerge. In either case, the outdoor environment could become deadly for humans. The climate changes impose an adaptation of the human individuals to the current and the future thermal conditions. The current research aims to study the risk of extreme heat and cold thermal stress on the Balkan Peninsula and some adjacent territories. For that purpose, we performed numerical simulations with the regional climate model RegCM4 for the period from 1975 to 2005. The initial and boundary conditions were taken from the HadGEM2 Earth System Model. The simulations were performed on the supercomputer Avitohol of IICT—BAS. We studied the extreme thermal conditions in the cold and the warm seasons with two different biometeorological indexes. The first one describes how cold the human individual feels in low temperatures and windy environment. The second one describes how hot the human being feels in high temperatures and humid conditions. The simulation results for the influence of the current climate on human beings are demonstrated in the paper.

Keywords: RegCM · Heat stress · Wind chill · Heat wave · HPC simulation · Climate

1 Introduction

The heat stress for human beings is one of the most important problems in modern society. There is a lot of research work addressing that problem and the need for mitigation activities towards decreasing the harmful impacts on the people. According to IPCC reports, the temperature is increasing at many locations in

I. Dimov and S. Fidanova (Eds.): HPC 2019, SCI 902, pp. 309–324, 2021.
https://doi.org/10.1007/978-3-030-55347-0_27

the world. At some places like the Mediterranean, the climate projections show that the changes in the future are towards drought and temperature increase on the surface. A study of the simulations from 20 CMIP3 models with SRES emission scenarios [1] shows that one of the most sensitive areas to the change of the global climate is the Mediterranean with a decrease of precipitation and a following increase of the precipitation and temperature variability from April to September. Another research [2] deals with climate change hotspots from the CMIP5 model ensemble. This one refers to the comparison between the historical, near future and far-future climate. Using RCP4.5 and RCP8.5 emission scenarios, it stresses out that the climate change in the Mediterranean will exert massive intensification of the extreme events in the hot season in 2080–2099 years for the RCP8.5 scenario. According to another research with the regional climate model RegCM3 [3], the increasing greenhouse gas concentrations, when comparing A2 and B2 emission scenarios, worsen the heat risk in the Mediterranean region. The inclusion of humidity also shows that the conditions become a danger when there is a steeper change in topography. The model, however, shows a warm bias, especially in areas with higher maximum temperatures. According to the [4], the heat stress mortality in the European countries under emission scenarios RCP4.5 and RCP8.5 will keep rising from nowadays to 2099, in particular in the Mediterranean and Eastern Europe, and will be amplified by high emission scenarios. Some countries will suffer a projected increase in heat-related mortality more than twice per day. Another study of the temperature-related mortality across 23 countries for different regions with model results from General Circulations Models for different time slots from 1984 to 2099 [5], has reached similar conclusions. It has found that the temperature-related mortality under four emission scenarios of the Representative Concentration Pathways will show a steady increase of the heat-related mortality and a decrease of the cold-related one for RCP4.5 and RCP8.5 scenarios in Southern Europe. The climate conditions under RCP2.6 in that region are milder, with negative or zero heat- and cold-related mortality trends after 2050s. The Central European region, where some of the locations under investigation are at the border of the Balkan Peninsula, has similar although not that well expressed trends. The models with their different geographical boundaries, spatial resolution, and temporal resolution add additional uncertainty to that caused by physical and dynamical features. The present work aims to study the biometeorological conditions in a human-friendly manner for the area comprising only the Eastern Mediterranean and the Balkan Peninsula. For that purpose, we used a 3-hour output data from the regional climate model RegCM4. The output parameters were used to calculate two indexes for heat and cold stress conditions warnings.

2 Methods

We perform simulations with the hydrostatic regional climate model RegCM4 for a 31-year period from 1974 to 2004 with one year for spin-up, representative for the present climate with initial and boundary conditions from HadGEM2-ES model [6] with a horizontal resolution of 1.875° × 1.25° in the atmosphere, with 38 vertical levels. The computer resource requirements of the modeling system were very big Table 1. The numerical experiments were organized in effective HPC environment [7]. The calculations were performed on the Supercomputer System "Avitohcl" at IICT-BAS.

Table 1. Computer resource requirements 128 CPU-s on Avitohol

	1 Month	1 Year	Case
Time (h)	0.22	2.64	79.2
HDD output (GB)	1.9	22.8	684
HDD input (GB)	1.1	12.8	397

Some processes are too complex for direct modelling by the RegCM, and are represented in a simpler way. The land surface processes are represented by the BATS scheme [8]. The boundary layer physics is simulated by the Holtslag scheme [9,10]. The large-scale moisture processes are parameterized by the Subgrid Explicit Moisture Scheme [11]. The cumulus convection is parameterized by the Grell scheme [12], with Arakawa and Schubert closure [13]. We evaluate the human comfort conditions by numerical values resulting from calculation of two biometeorological indices categorizing the human sensation to the environmental temperature conditions. As temperature changes significantly throughout the year, we choose an index called heat index for high temperatures [14,15], and another one called the new wind chill index for low temperatures [16]. The heat index is defined as the temperature a human individual situated in a real environment, would feel in a base environment with effective wind speed 2.5 m/s, vapor pressure 1600 Pa, barometric pressure 1013 hPa, and zero extra radiation. It depends on the air temperature and relative humidity, and categorizes the thermal sensation according to the heat stress for the human organism [17]. When the temperature and the relative humidity become very high, the evaporation from the human body starts to decrease, which could increase the core body temperature to dangerous levels. The heat index is applied for air temperatures above 26 °C, and relative humidity above 40% Table 2. For temperatures higher or equal to 20 °C we used an ncl algorithm for the heat index [18,19]. For air temperatures lower than or equal to 4.4 °C, we use the new wind chill index (or just 'wind chill index'), which indicates the influence of cooling power of the wind on the thermal sensation in low air temperatures and wind speed equal or greater than 1.34 m/s. The wind increases the convection from the human body, and when the temperatures are low enough, the human could feel surface or deep frostbite, hypothermia, or even die [20]. It is categorized in seven grades according to the sensation and the rate of skin freezing in Table 3.

Table 2. Heat index categories description

Interval	Category name	Physiological response
Below 26.7 °C	No Stress	No
26.7 to 32.2 °C	Caution	Fatigue is possible with prolonged exposure and activity. Continuing activity could result in heat cramps
32.22 to 40.5 °C	Extreme Caution	Heat cramps and heat exhaustion are possible. Continuing activity could result in heat stroke
40.5 to 54.4 °C	Danger	Heat cramps and heat exhaustion are likely; heat stroke is probable with continued activity
54.4 °C and above	Extreme Danger	Heat stroke is imminent

Table 3. Wind chill index categories description

Interval	Category name	Physiological response
0 to −9 °C	Low Risk	Slight increase in discomfort
−10 to −27 °C	Moderate risk	Uncomfortable. Risk of hypothermia and frostbite if outside for long periods without adequate protection
−28 to −39 °C	High Risk	High risk of frostnip or frostbite.Exposed skin can freeze in 10 to 30 min. Check face and extremities for numbness or whiteness.High risk of hypothermia if outside for long periods without adequate clothing or shelter from wind and cold
−40 to −47 °C	Very High Risk	Very high risk of frostbite. Exposed skin can freeze in 5 to 10 min. Check face and extremities for numbness or whiteness. Very high risk of hypothermia if outside for long periods without adequate clothing or shelter from wind and cold
−48 to −54 °C	Severe Risk	Severe risk of frostbite. Exposed skin can freeze in 2 to 5 min. Check face and extremities frequently for numbness or whiteness. Severe risk of hypothermia if outside for long periods without adequate clothing or shelter from wind and cold
−55 °C and below	Extreme risk	DANGER! Exposed skin can freeze in less than 2 min. Outdoor conditions are hazardous

To assess some important temperature extreme events, we use two procedures developed in the project STAtistical and Regional dynamical Downscaling of EXtremes for European regions [21]. It has a lot of aims for inter-comparison of statistical and dynamical downscaling methods, extreme events study, and so on. One of its objectives was to develop a diagnostic software tool for calculating a standard set of extreme event statistics from daily temperature and precipitation data. Two characteristics of that set are the heat wave duration and the cold wave duration [22].

We studied the heat and cold waves for each season and for the whole year. To reflect the influence of the relative humidity and the wind speed over the heat wave day and cold wave day calculation, we replaced the temperature with the new wind chill index or the heat index if the conditions for the one or the other index were met. When the temperature was between 4.4 °C and 20 °C, the air temperature was used.

3 Results

3.1 Thermal Stress Modelling

The results showing the geographical distribution of the percentages of occurrence of the two heat index categories – Caution and Extreme Caution in three seasons except for the winter are given in the Fig. 1. We can see that only 1% to 2% of the cases in the spring are with Caution conditions in the northern part of the domain in Moldova, parts of Romania and Ukraine. There are no cases with Extreme Caution. The autumn is also with a small number of Caution cases, between 1% and 2% at some places, and unlike the spring, there are places with 1% to 10% at the most southwestern and southwestern parts of the domain.

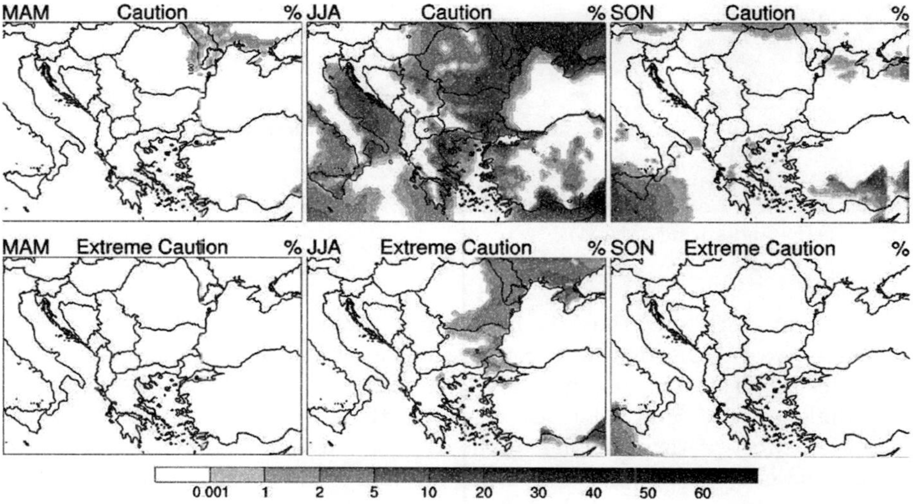

Fig. 1. Percentage of cases with heat index categories for the spring (MAM), summer (JJA) and autumn (SON)

The summer is characterized by a large number of Caution cases with a more complex distribution pattern. The Caution cases are between 1% and 2% in parts of Pannonian Basin, Transylvanian Plateau, Tyrrhenian Sea, the coastal areas, and in some of the western parts of the Balkan Peninsula. The places with Caution cases between 5% and 10% are located in the low terrain forms of the Balkan Peninsula, Adriatic Sea, Ukraine and the most southern parts of Asia Minor. There are also locations where the Caution is 20% to 30% of all cases: in the Ukraine and the southeastern parts of the domain, where the percentages go beyond 30 at some places. The percentage of Extreme Caution cases at the eastern parts of the Balkan Peninsula and Ukraine are mostly between 1% and 5% of all cases, with some places between 5% and 10%. Some spots in the most southern parts of Asia Minor account for 1% and 20% Extreme Caution cases. Other parts of the domain have no Extreme Caution cases.

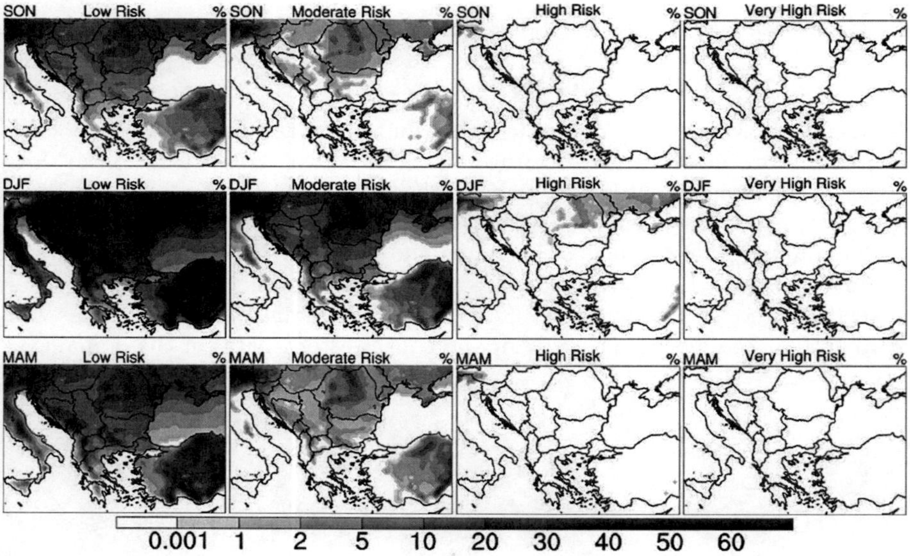

Fig. 2. Percentage of cases with wind chill index categories for the autumn (SON), winter (DJF) and spring (MAM)

It was interesting to study the biometeorological conditions under cold temperatures. We did that by the percentages of cases of the wind chill index (WCI) categories for all seasons but the summer (Fig. 2). We can see that the cases with wind chill from 0 °C and below are mainly in the Low Risk and Moderate Risk categories. Most of them belong to the first of the aforementioned categories. The percentages of Low Risk cases in the autumn are between 1% and 5% in the southern low terrain forms and in the northern parts of the Black Sea, between 5% and 10% in Bulgaria and on the Western Balkan Peninsula, and between 10%

and 20% in Romania, Hungary, Ukraine and Austria. There are locations with percentages above 20 in the most northern parts and in the Alps in particular. The Low Risk category percentages in Asia Minor are from 1 to 20 and some places with 20 to 30. The middle and north parts of Italy display percentages of Low Risk cases between 2 and 5 to above 50 in the Alps. The percentages of Moderate Risk cases are mostly between 1 and 5 in the middle and northern areas of the domain, with 5 to 20, and even above 60 in the Carpathians and the Alps. Concerning the High Risk category, there is a place in the northwestern corner of the domain with percentages between 1 and 5. The percentages of Low Risk in the winter are above 20 in the land areas of the domain, and above 60 in the highest mountains. Only the coastal area and the adjacent territories are with LR cases below 20%. The percentage of Moderate Risk cases in the winter are mostly between 10% and 20% in Bulgaria, Serbia, Bosnia and Herzegovina, and Croatia. It is above 20% in Romania, Ukraine, and the Alps, and from 10% to 30% in Hungary. The percentage with Moderate Risk conditions in Italy is from 1 to 20 in the middle parts, to above 60% in the Alps. The winter High Risk is between 1% to 5% in the most northern parts, and up to 30% in the Alps. The distribution of the Low Risk percentages in the spring is similar to that in the autumn. However, the percentages are higher since the same areas with 2% to 5% are now from 5% to 10%, others with latter percentages of wind chill index categories are already from 10% to 20%, and so on. The distribution of Moderate Risk category in the spring is more similar to the one in the autumn, but still with patterns pointed out in the previous sentence. Therefore, it appears that the springs from 1975 to 2004 are a bit colder than the autumns, in particular concerning the wind chill index between 0 °C and −9 °C, which is visible especially in the Asia Minor Peninsula.

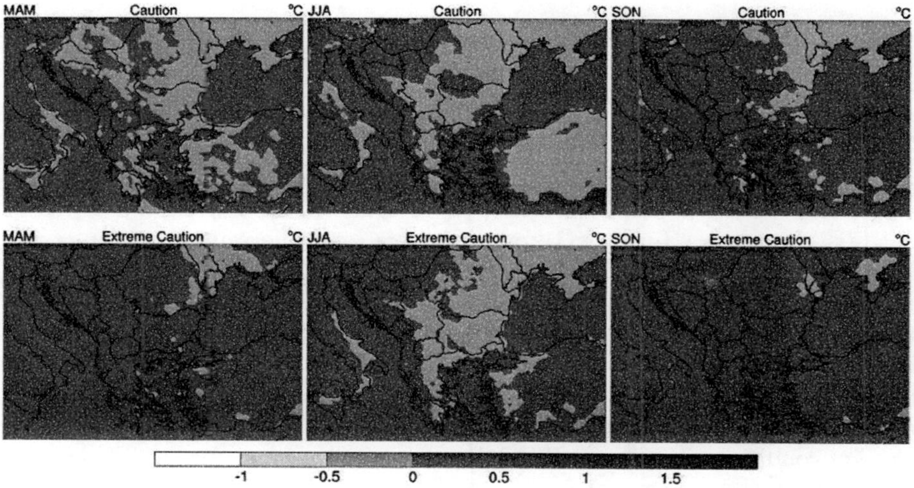

Fig. 3. Average difference between the heat index and the temperature for the spring (MAM), summer (JJA) and autumn (SON).

The heat index and the wind chill index comprise one factor in addition to the temperature. We can see the influence of that additional factor on the difference between the respective index, and the temperature. The results show that the average differences between the heat index (Fig. 3) and the temperature are −0.5 °C mostly in the land parts of the domain, and 0.5 °C at marine ones. For the cases with Extreme Caution in the spring and autumn the average difference is mainly 0.5 °C in comparison to the other combination of season and category. These results show that, the model simulations are too dry for the sensitivity of the heat index to the moisture.

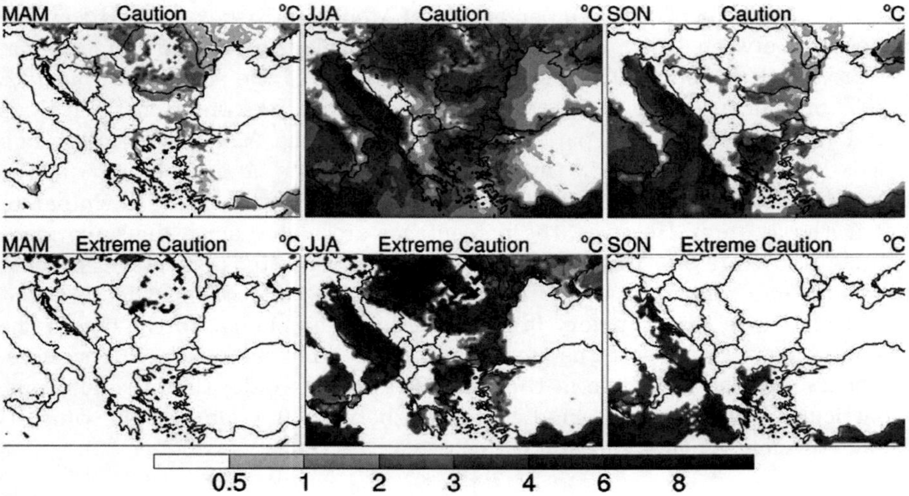

Fig. 4. The maximum difference between the heat index and the temperature for the spring (MAM), summer (JJA) and autumn (SON).

The maximums of these differences increase above 8 °C (Fig. 4). The maximum difference in the spring for the Caution cases is between 0.5 °C and 3 °C with some spots between 3 °C and 4 °C in some of the low terrain forms of Bulgaria, Romania, Hungary, Ukraine, Greece, and the southern part of the Asia Minor Peninsula. For the Extreme Caution there are just some spots with maximum difference from 3 °C to 6 °C. Other parts are with maximum difference below 0.5 °C. The Caution cases in summer go from 0.5 °C to 3 °C in the coastal, marine, and high terrain parts to above 3 °C in the low terrain land ones, and below 0.5 °C in the highest terrain parts. There are more Extreme Caution cases in summer with maximum differences of less than 0.5 °C and above 6 °C. The Caution conditions in autumn are characterized with maximum a difference of up to 3 °C in the eastern parts of the Balkan Peninsula, and as in the summer of up to 6 °C in the marine areas. The maximum differences for the Extreme Caution are mainly in the marine areas, except the Black Sea. Therefore, the humidity becomes high enough for a significant change of the heat index value, but that is probably a very rare event.

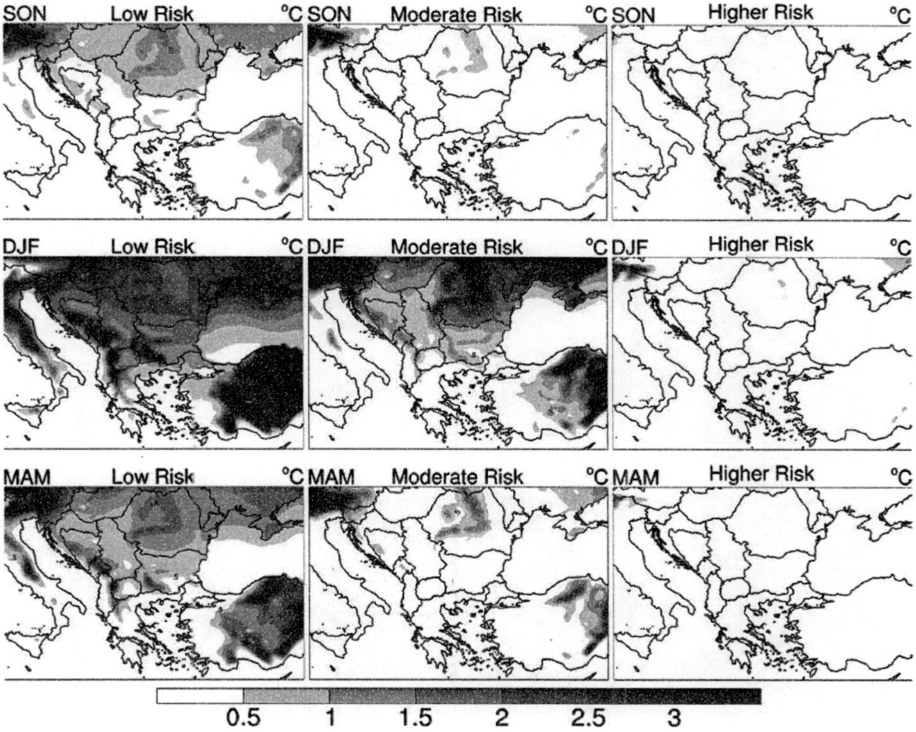

Fig. 5. Average difference between the temperature and the new wind chill index for the autumn (SON), winter (DJF) and spring (MAM).

The average difference between the temperature and the new wind chill index also suggests some patterns. In autumn, the difference in Low Risk cases is mainly between 0.5 °C and 2 °C, but in the northern part of the domain. In the case of Moderate Risk, the difference is above zero only in the Carpathians and some locations near the eastern border of the model domain. It goes to above 2.5 °C in the mountainous parts. The winter cases of Low Risk have an average difference from 0.5 °C in the southern coastal areas to between 1.5 °C and 2.5 °C in Bulgaria, Romania, Serbia and other countries. There are areas from 0.5 °C to 3 °C in the Black Sea. The winter Moderate Risk cases are with lower average differences and more locations with one below 0.5 °C. The spring Low Risk conditions have more places with average difference between the temperature and the new wind chill index above 0.5 °C in comparison to the average difference distribution in the autumn. The same applies to the Moderate Risk cases. The average difference in cases with Higher Risk and more severe categories is above 0.5 °C in some places, and especially in the Alps during the winter (Fig. 5).

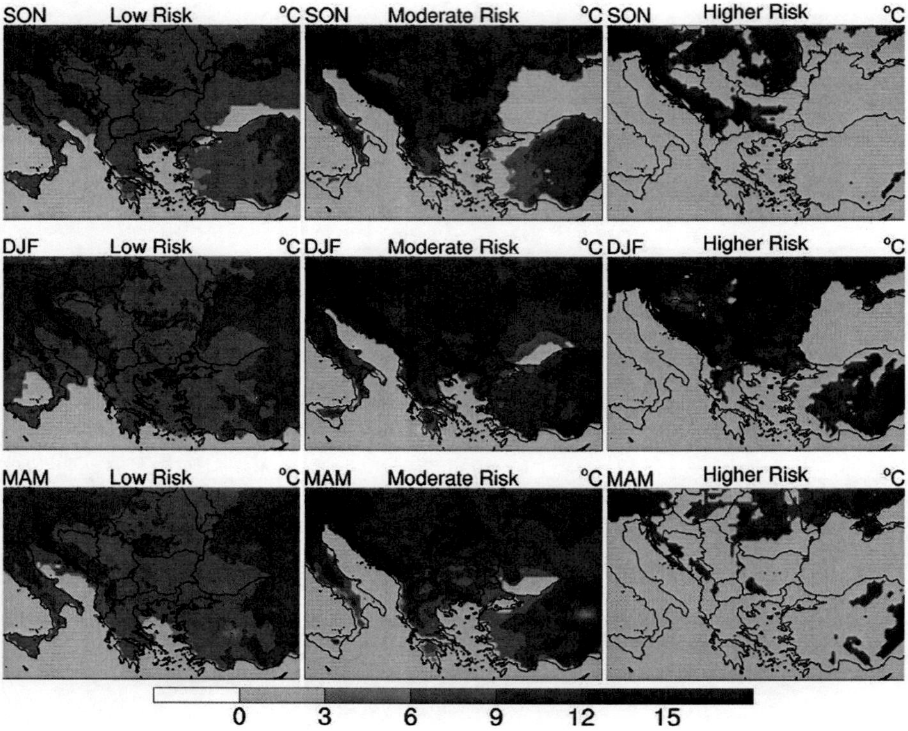

Fig. 6. Maximum difference between the temperature and the new wind chill index for the spring (SON), winter (DJF) and autumn (MAM).

As to the maximum difference between the air temperature and new wind chill index (Fig. 6), the Low Risk cases in the autumn, winter and spring have a very similar spatial distribution from 6 °C to 12 °C in most of the areas, and between 0 °C and 3 °C in the Mediterranean Sea. The Moderate Risk cases are also with less similar distribution in the three aforementioned seasons. There are locations with a maximum difference between 12 °C and 15 °C and above 15 °C in Alps and some mountainous parts of Asia Minor. When the wind chill category is High Risk or more severe, there are more areas with maximum difference between 0 °C and 3 °C, and above 15 °C especially during the winter.

The active hours for the human activity are mainly during the day. That is why it is interesting to see the frequency distributions of the indexes categories in the most active times of the day, which we chose to be between 06 UTC and 18 UTC. As we see for the heat index categories during the active hours (Fig. 7), the spatial distribution of percentages is similar to the one for all-day

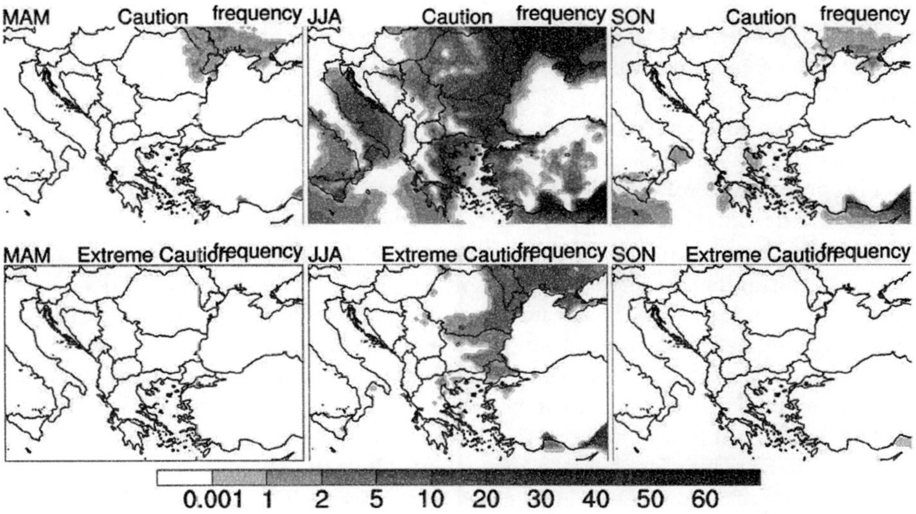

Fig. 7. Percentage of cases with the heat index categories for the spring (MAM), summer (JJA) and autumn (SON), for the time of day between 06 UTC and 18 UTC.

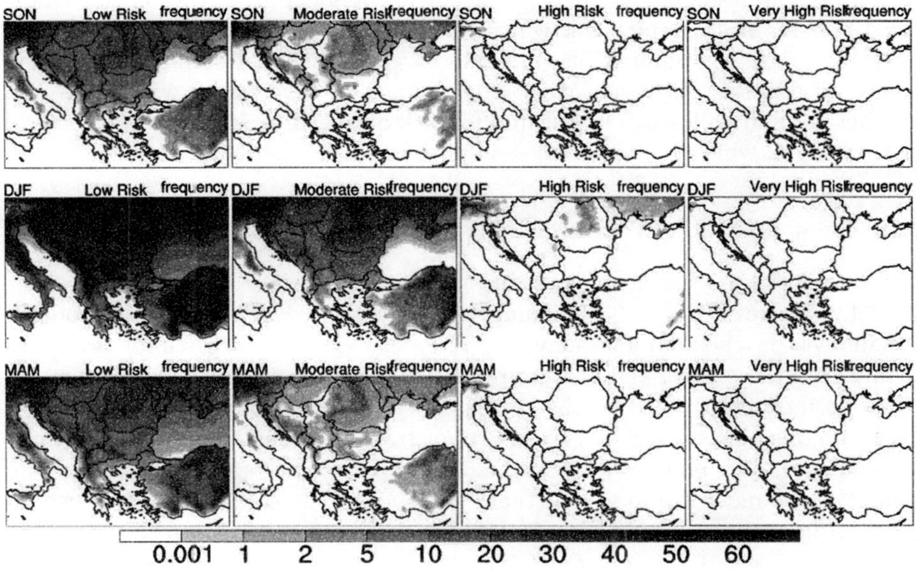

Fig. 8. Percentage of cases with the new wind chill index categories for the autumn (SON), winter (DJF) and spring (MAM), for the time of day between 06 UTC and 18 UTC.

(Fig. 1) but with some differences. There are more areas with a percentage of cases with Caution heat index between 20% and 30%, and there are even cases with 30% to 40%. That pattern is presented in the eastern parts of Bulgaria and

Romania, Ukraine, and the Aegean Sea coast. There are also more locations with percentages of cases with Extreme Caution between 5% and 10% in the eastern part of Romania, and between 2% and 5% in the Eastern Bulgaria. These results are due to the fact that the heat stress during daytime is heavier than during nighttime.

The percentages of wind chill index categories in the active hours are given at Fig. 8. The differences with the percentages of the wind chill categories for all times of the day are more subtle than those for the heat index. The main difference is the increased percentages of Low Risk at the expense of decreased percentages for Moderate Risk and High Risk. That change in the index distribution can be seen more notably in the northern part of the domain. Therefore, the change of the heat stress from night-time to daytime is much more notable than the change of the cold stress between these periods of the day.

The model simulations show that the occurrence of heat index categories Danger and Extreme Danger are rare events. For the spring season, there are only 56 cases of Danger and zero cases for Extreme Danger. In the summer, there are 6725 cases with Danger heat index, and 14 cases with Extreme Danger heat index. There are no cases of Extreme Danger category in the autumn, but there are 432 cases of Danger. Concerning the wind chill, the two most extreme cold categories also have a rare occurrence. The model output during the winter has 134 cases of Extreme Risk wind chill and 6077 cases of Severe Risk wind chill. In spring, the number of wind chill cases for risk of freezing categories is 0 cases of Extreme Risk wind chill and 115 cases of Severe Risk wind chill. The number of wind chill cases in autumn for risk of freezing categories is 0 cases of Extreme Risk wind chill and 12 cases of Severe Risk wind chill.

3.2 Heat and Cold Waves

The calculation of the heat wave duration and cold wave duration days gives the following results (Fig. 9). The number of the heat wave days depends on the specific season chosen for evaluation. The heat waves in the winter are more than 220 in the central and northern parts of the Balkan Peninsula, from under 20 to at most about 220 to the south, and more than 620 in the Black Sea. There are also heat waves in some parts of Italy. The heat waves in the winter are a much higher number than in the other seasons. The spring heat waves are from 20 to at most 220 in the northern parts and to at most 20 in the southern parts of the domain. In northern Asia Minor, the heat waves go up to about 220. The summer heat waves are more than those in the spring in South Romania, Western Asia Minor, Bulgaria and Greece. The autumn heat waves are up to about 260 in Romania, a bigger number than in spring and summer. Most of the other locations in the domain are zero. The heat wave days concerning the whole year, not just one season, have a spatial structure similar to that in the winter.

The winter cold waves are at least 220 in the middle and north Balkan Peninsula and Ukraine and from 80 to 220 in the other parts except Italy and marine areas without the Black Sea. The spring cold waves in the Northern and Southern Central Bulgaria, Romania, some parts of Serbia and Croatia, Hungary and the Black Sea decrease to at most 220 with some spots up to 260. In the other locations, they are below 140, and even below 80 in Greece, Italy and Western Asia Minor. The summer is characterized by almost no cold waves except for the Black Sea and around the Alps. The autumn has a greater number of cold wave days than spring. The year-based cold waves are above 460 in all areas except Greece and Asia Minor where their number decreases to below 20 at some points, but is generally above 60. Italy is more diverse in the number of cold waves which increase from south to the north.

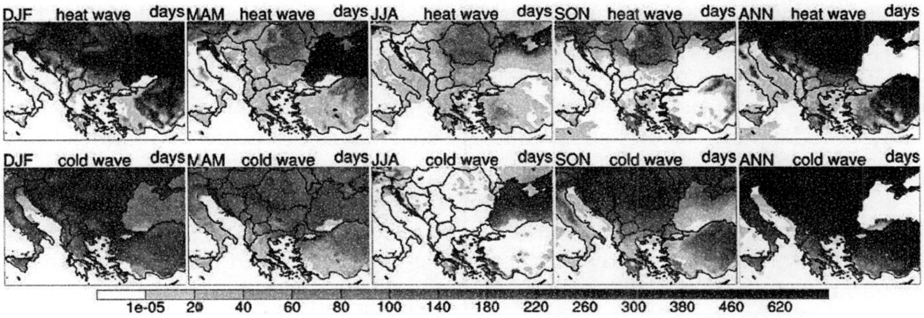

Fig. 9. Number of heat and cold waves for the winter (DJF), spring (MAM), summer (JJA) autumn (SON), and annually (ANN).

4 Summary

The study shows that the RegCM4 model simulations with HadGEM-ES initial and boundary conditions for the years 1975—2004 years contain Caution and Extreme Caution heat stress categories. Caution is a big issue in the eastern parts of the domain. Thus people in the eastern parts of Bulgaria, Romania and Moldova, as well as in the most southern parts of Ukraine are threatened mostly from fatigue from physical activity and long exposure in high-temperature conditions in the summer. The existence of Extreme Caution conditions suggest that people mainly in Romania, Moldova, and parts of the Ukraine which are in the model domain, have to pay additional attention to the bigger possibility for heat cramps, heat exhaustion and sunstroke. The influence of the humidity on the heat index calculation is generally small but in some cases, it could reach more than 4 °C in the low terrain forms and nearby rivers and marine areas, especially in Extreme Caution conditions during the summer. Therefore, we have to be very careful in the summer, in particular when temperatures are above 32 °C. Concerning the cold stress conditions, there are rare occurence of conditions with risk of skin freezing in the geographical area of the domain, mostly in the northern parts and the Alps in the winter. There is a possibility for slight cold discomfort in spring, winter, and autumn. There are also conditions for hypothermia due to prolonged exposure in all these seasons, but mostly in the winter and northern areas. The influence of the wind in wind chill index calculation is not as small as the humidity in high temperatures, and is generally increasing from south to the north. The average differences reach 3 °C and above. It is bigger mostly in the Low Risk conditions in the three seasons and Moderate Risk ones during the winter. The results for the maximum difference between the temperature and the wind chill index show that there is a very small number of cases with risk of skin freezing in the land areas. Generally, the influence of the wind on the thermal sensation calculated with the wind chill in the winter is higher than the one of the humidity on the temperature during the summer, calculated with the heat index. We suggest that the model simulations are too dry when the temperature is very high in the summer months for that period, due to the weak summer atmospheric circulation over that region which does not supply enough moisture. That explains the weak humidity contribution to the heat index. The results show that the cold waves are a higher number than the heat waves in the transition seasons. The heat wave days are more than cold wave days in the summer and on some low terrain forms in the winter. Therefore, we suggest that the transition seasons are characterized mainly by cold events, and the winter and the summer by heat events. The RegCM4 model has positive temperature biases, except for the spring [23]. The biggest one (1.9 °C) is for the summer. The RMSE for all seasons is above 2 °C and the biggest one is for he winter (3.6 °C). Thus, the model probably overestimate the health risk in the summer and to some extent underestimate the cold stress in the winter.

Acknowledgments. This work has been accomplished with the financial support by the Grant No BG05M2OP001-1.001-0003, financed by the Science and Education for Smart Growth Operational Program (2014–2020) and co-financed by the European Union through the European structural and Investment funds.

This work has been carried out in the framework of the National Science Program 'Environmental Protection and Reduction of Risks of Adverse Events and Natural Disasters' approved by the Resolution of the Council of Ministers 577/17.08.2018 and supported by the Ministry of Education and Science (MES) of Bulgaria (Agreement D01–230/06.12.2018).

This work was partially supported by the Bulgarian Ministry of Education and Science under the National Research Programme 'Young scientists and postdoctoral students' approved by DCM 577 / 17.08.2018.

References

1. Giorgi, F.: Climate change hot-spotsClimate change hot-spots. Geophys. Res. Lett. **33**, L08707 (2006). https://doi.org/10.1029/2006GL025734
2. Diffenbaugh, N.S., Giorgi, F.: Climate change hotspots in the CMIP5 global climate model ensemble. Clim. Change **114**, 813–822 (2012). https://doi.org/10.1007/s10584-012-0570-x
3. Diffenbaugh, N.S., Pal, J.S., Giorgi, F., Gao, X.: Heat stress intensification in the Mediterranean climate change hotspot. Geophys. Res. Lett. **34**, L11706 (2007). https://doi.org/10.1029/2007GL030000

4. Kendrovski, V., Baccini, M., Martinez, G.S., Wolf, T., Paunovic, E., Menne, B.: Quantifying projected heat mortality impacts under 21st-centurywarming conditions for selected European countries. Int. J. Environ. Res. Pub. Health **14**(7), 729 (2017). https://doi.org/10.3390/ijerph14070729
5. Gasparrini, A., Guo, Y., Sera, F., Vicedo-Cabrera, A.M., Huber, V., Tong, S., de Sousa Zanotti Stagliorio Coelho, M., Nascimento Saldiva, P.H., Lavigne, E., Matus Correa, P., Valdes Ortega, N., Kan, H., Osorio, S., Kyselý, J., Urban, A., Jaakkola, J.J.K., Ryti, N.R.I., Pascal, M., Goodman, P.G., Zeka, A., Michelozzi, P., Scortichini, M., Hashizume, M., Honda, Y., Hurtado-Diaz, M., Cesar Cruz, J., Seposo, X., Kim, H., Tobias, A., Iñiguez, C., Forsberg, B., Åström, D.O., Ragettli, M.S., Guo, Y.L., Wu, C., Zanobetti, A., Schwartz, J., Bell, M.L., Dang, T.N., Van, D.D., Heaviside, C., Vardoulakis, S., Hajat, S., Haines, A., Armstrong, B.: Projections of temperature-related excess mortality under climate change scenarios. Lancet Planet. Health **1**(9), 360–367 (2017)
6. Collins, W.J., Bellouin, N., Doutriaux-Boucher, M., Gedney, N., Halloran, P., Hinton, T., Hughes, J., Jones, C.D., Joshi, M., Liddicoat, S., Martin, G., O'Connor, F., Rae, J., Senior, C., Sitch, S., Totterdell, I., Wiltshire, A., Woodward, S.: Development and evaluation of an earth-system model - HadGEM2. Geosci. Model Dev. **4**(4), 1051–1075 (2011). https://doi.org/10.5194/gmd-4-1051-2011
7. Atanassov, E., Gurov, T., Karaivanova, A., Ivanovska, S., Durchova, M., Dimitrov, D.: On the parallelization approaches for intel MIC architecture. AIP Conf. Proc. **1773**, 070001 (2016). https://doi.org/10.1063/1.4964983
8. Dickinson, R.E., Henderson-Sellers, A., Kennedy, P.J.: Biosphere-atmosphere transfer scheme (BATS) version 1e as coupled to the NCAR community climate model. Tech. Rep, National Center for Atmospheric Research (1993)
9. Holtslag, A.A.M., de Bruijn, E.I.F., Pan, H.: A high resolution air mass transformation model for short-range weather forecasting. Mon. Weather Rev. **118**(8), 1561–1575 (1990). https://doi.org/10.1175/1520-0493(1990)118⟨1561:AHRAMT⟩2.0.CO;2
10. Holtslag, A.A.M., Boville, B.A.: Local versus nonlocal boundary-layer diffusion in a global climate model. J. Climate **6**(10), 1825–1842 (1993). https://doi.org/10.1175/1520-0442(1993)006⟨1825:LVNBLD⟩2.0.CO;2
11. Pal, J.S., Small, E.E., Eltahir, E.A.B.: Simulation of regional-scale water and energy budgets: Representation of subgrid cloud and precipitation processes within RegCM. J. Geophys. Res. Atmos. **105**(D24), 29579–29594 (2000). https://doi.org/10.1029/2000JD900415
12. Grell, G.A.: Prognostic evaluation of assumptions used by cumulus parameterizations. Mon. Weather Rev. **121**(3), 764–787 (1993). https://doi.org/10.1175/1520-0493(1993)121⟨0764:PEOAUB⟩2.0.CO;2
13. Arakawa, A., Schubert, W.H.: Interaction of a cumulus cloud ensemble with the large scale environment. Part I, J. Atmos. Sci. **31**, 674–701 (1974)
14. Steadman, R.G.: The assessment of sultriness, Part I: A temperature-humidity index based on human physiology and clothing science. J. Appl. Meteorol. **18**(7), 861–873 (1979). https://doi.org/10.1175/1520-0450(1979)018⟨0861:TAOSPI⟩2.0.CO;2
15. Rothfusz, L.P.: NWS Technical Attachment (SR 90-23) (1990)
16. Osczevski, R., Bluestein, M.: The new wind chill equivalent temperature chart. Bull. Am. Meteorol. Soc. **86**(10), 1453–1458 (2005). https://doi.org/10.1175/BAMS-86-10-1453
17. Heat Index. https://www.weather.gov/jetstream/hi. Accessed 20 May 2019

18. The Heat Index Calculation. https://www.wpc.ncep.noaa.gov/html/heatindex_equation.shtml. Accessed 20 May 2019
19. The NCAR Command Language (Version 6.4.0) [Software]. Boulder, Colorado: UCAR/NCAR/CISL/TDD (2019). https://doi.org/10.5065/D6WD3XH5
20. Wind chill index. https://www.canada.ca/en/environment-climate-change/services/weather-health/wind-chill-cold-weather/wind-chill-index.html. Accessed 20 May 2019
21. STARDEX (STAtistical and Regional dynamical Downscaling of Extremes for European regions). http://www.cru.uea.ac.uk/projects/stardex. Accessed 20 May 2019
22. STARDEX Diagnostic Extremes Indices Software. https://crudata.uea.ac.uk/projects/stardex/deis/Diagnostic_tool.pdf. Accessed 20 May 2019
23. Temperature and Precipitation Projections According to RCP45 Scenario For Bulgaria and Balkan Peninsula Regional Climate Models, Trieste, Italy, 28 May–08 June 2018

Modeling, Simulation; Optimization in a Data-rich Environment

BiqBin: Moving Boundaries for NP-hard Problems by HPC

Timotej Hrga[1], Borut Lužar[2], and Janez Povh[1(✉)] and Angelika Wiegele[3]

[1] Faculty of Mechanical Engineering, University of Ljubljana, Ljubljana, Slovenia
{timotej.hrga,janez.povh}@lecad.fs.uni-lj.si
[2] Faculty of Information Studies, Novo mesto, Slovenia
borut.luzar@gmail.com
[3] Alpen-Adria Universitaet Klagenfurt, Klagenfurt, Austria
angelika.wiegele@aau.at

Abstract. In this paper we present a parallel Branch and Bound (B&B) algorithm to solve the Stable Set Problem, which is a well-known combinatorial optimization problem. The algorithm is based on tight semidefinite programming bounds. Numerical results, based on using up to 192 CPU cores, show that this algorithm scales well.

This algorithm is available as a part of the online BiqBin solver, which enables online submissions of problem instances. After submission, it automatically generates computational jobs and runs them using the high-performance computer available at University of Ljubljana. BiqBin demonstrates how to bring HPC closer to specific user community – in our case the mathematical optimization community.

Keywords: High performance computing · Stable Set Problem · Online solver · Mathematical optimization

1 Introduction

1.1 Mathematical Optimization

Mathematical optimization is a very vibrant area within applied mathematics dealing mainly with searching for an element x in a given set $\mathcal{F}$ such that some objective function f attains the largest value in this x. The function f and the set of feasible solutions $\mathcal{F}$ represent specifics and constraints implied by the underlying real-life problem.

There are only few classes of such problems that can be efficiently solved to optimality, which means that there exist algorithms with good (i.e., polynomial) theoretical and/or practical performance. Linear programming, convex quadratic programming and semidefinite programming problems are three families where the so-called interior point methods (IPMs) have very good theoretical and practical performance.

Most of the problems in the domain of mathematical optimization are very hard, in fact, in most of the cases one can prove that they are $\mathbb{NP}$-hard. Thus,

I. Dimov and S. Fidanova (Eds.): HPC 2019, SCI 902, pp. 327–339, 2021.
https://doi.org/10.1007/978-3-030-55347-0_28

one is satisfied with finding good (not optimum) solutions obtained by solving some simplifications of the problems and/or employing algorithms, which can find solutions in reasonable time, with or without estimates on how good these solutions are (we call such algorithms *heuristics*). For more about mathematical optimization, we recommend [5,20].

1.2 High Performance Computing in Mathematical Optimization

High-Performance Computing (HPC) is—with its state-of-the-art computing, storage infrastructure and with related knowledge—an ecosystem that is essential for scientific research and industrial development. A recent document from the European Commission (EC)[1] points out the opportunities and challenges at the interface of Big Data, High-Performance Computing and Mathematics. In particular, this document stress out that Modelling, simulation and optimization (MSO) is one of the three main pillars of for scientific progress and innovation. HiPEAC Vision 2017[2] clearly states that Mathematics and Algorithms for extreme scale HPC systems is one out of seven current EU research priorities related to HPC. Nevertheless, the most recent results of the PRACE call[3] for using TIER-0 HPC infrastructure reveals that 47 computationally intensive projects were awarded, but none of them from the area of mathematical optimization.

This situation can be observed as being more general in the field of mathematical optimization. When browsing literature, we can see that very few researchers working on hard optimization problems also try to implement their solutions using HPC. For example, at the International Symposium on Mathematical programming 2018[4] in Bordeaux (the largest mathematical optimization conference in 2018) only 2 sessions out of approx. 480 sessions were devoted to using high-performance computers in optimization, although almost all of them were talking about hard mathematical optimization problems and methods to solve them. Similarly, at SIAM Conference on Optimization 2017[5] in Vancouver there were only 2 out of 150 mini-symposia devoted to using supercomputers for solving hard-optimization problems.

1.3 Motivation and Contribution

The main motivation for the research presented in this paper is twofold:

- to check how much high-performance computers can contribute to solving specific mathematical optimization problems;

[1] http://www.euro-math-soc.eu/system/files/news/MathematicsforDigitalScience.pdf.
[2] https://www.hipeac.net/v17.
[3] http://www.prace-ri.eu/17th-call-for-proposals-for-project-access/.
[4] https://ismp2018.sciencesconf.org/.
[5] https://www.siam.org/meetings/op17/.

- to bring an HPC based solver closer to mathematical programming users by providing a web interface.

In the paper, we present general binary quadratic problems and focus to a special instance, the Stable Set Problem (SSP). In particular, we

- explain how to develop an efficient parallel solver for the Stable Set Problem (SSP) based on parallelization of Branch and Bound algorithm;
- present a web application BiqBin, where the new parallel solver for (SSP) is available as an on-line solver;
- present numerical results for (SSP) obtained by the new solver.

2 Binary Quadratic Problems

2.1 General Binary Quadratic Problems

General binary quadratic problem (BQP) is an optimization problem in binary variables, where the objective function is quadratic, while the constraints are linear. It can be formulated as follows

$$\begin{aligned} &\max \mathbf{x}^\top Q\mathbf{x} + \mathbf{p}^\top \mathbf{x} + q \\ &\text{s.t. } A\mathbf{x} = \mathbf{b},\ \mathbf{x} \in \{0,1\}^n, \end{aligned} \tag{BQP}$$

where Q is a symmetric matrix and $\mathbf{p}, \mathbf{b}$ are real number vectors of appropriate size. This problem on one hand encompasses large variety of hard optimization problems, including the k-Clustering Problem [7, Chap. 10.3], [19], the Graph Partitioning Problem [12,13], the Quadratic Assignment Problem [14], etc. On the other hand, (BQP) can be transformed into the Max-Cut Problem via so-called exact penalty approach [10]. Therefore, a general approach to solve (BQP) can be utilized based on this transformation and by using efficient solvers for the Max-Cut Problem, like BiqMac [16,17] and BiqCrunch [9].

2.2 The Stable Set Problem

Suppose we have a list of project activities and a list of workers involved in this activities. Two activities that do not share any worker can be performed simultaneously. We call them *independent*. The problem of finding the largest subset of independent activities is the Stable Set Problem.

We can define it formally as an optimization problem in the language of graph theory. Suppose we have a graph G with vertex set V (representing the list of activities from the motivating example above) and the set of (undirected) edges $E \subset V \times V$ between the vertices from V. Two vertices $u, v \in V$ are connected (i.e., the corresponding activities share at least one worker) if and only if $\{u, v\} \in E$. The Stable Set Problem therefore asks for the largest subset of $S \subset V$ such that for all $v_i, v_j \in S$ it holds $\{v_i, v_j\} \notin E$. The size of a maximum stable set is called the *stability number* of G and denoted by $\alpha(G)$.

We can formulate the Stable Set Problem as a binary quadratic problem with linear constraints:

$$\begin{aligned} \alpha(G) &= \max\ \mathbf{x}^\top \mathbf{x} \\ &\text{s.t. } \mathbf{x} \in \{0,1\}^n,\ x_i + x_j \le 1,\ \forall\{i,j\} \in E \end{aligned} \tag{SSP}$$

Indeed, every $\mathbf{x}$ feasible for (SSP) defines a stable set $S(\mathbf{x}) := \{v_i \in V \mid x_i = 1\}$ with cardinality equal to $\sum_i x_i$. The optimum of (SSP) is always attained, since the feasible set is finite and any optimum $\mathbf{x}$ gives a stable set of maximum cardinality. We demonstrate definition of the Stale set problem on Fig. 1.

One can prove that the stability number of a graph can also be computed as the optimum value of the following optimization problem:

$$\begin{aligned} \alpha(G) &= \max\ \langle \mathbf{J}, Y \rangle \\ &\text{s.t. } Y = \mathbf{x}\mathbf{x}^\top,\ \operatorname{tr}(Y) = 1 \\ &\quad Y_{ij} = 0,\ \forall\{i,j\} \in E\,, \end{aligned} \tag{SSP'}$$

where $\mathbf{J}$ is the matrix of all ones. The hardness of the Stable Set Problem is embedded into the non-convex constraint $Y = \mathbf{x}\mathbf{x}^\top$. If we relax it to $Y \succeq 0$ (i.e., Y must be positive semidefinite), we obtain the well-known ϑ-function as an upper bound for $\alpha(G)$:

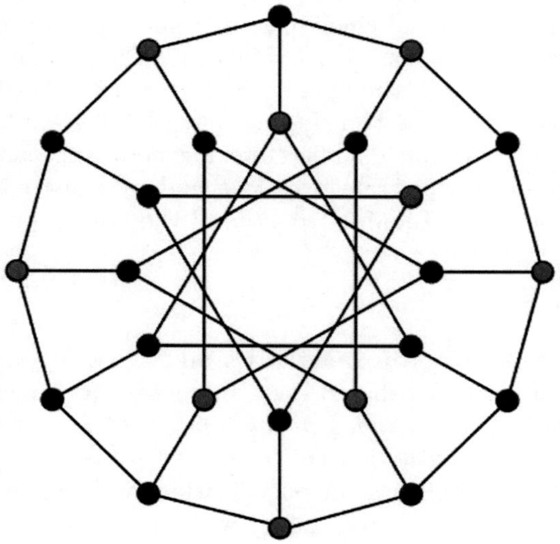

Fig. 1. This figure, taken from Wikipedia, depicts the generalized Petersen graph. The blue dots represent a stable set on 9 vertices (https://en.wikipedia.org/wiki/Independent_set_(graph_theory)).

$$\vartheta(G) = \max \langle \mathbf{J}, Y \rangle$$
$$\text{s.t. } Y \succeq 0,\ \operatorname{tr}(Y) = 1,\ Y_{ij} = 0,\ \forall \{i,j\} \in E$$

The ϑ-function received great attention for several reasons. One of them is that it is the optimum value of a semidefinite programming problem, for which there exists a solving method that runs efficiently both in theory and in practice, see e.g. [2,11,15].

2.3 Branch and Bound Algorithm for (SSP')

The Branch and Bound algorithm (B&B) is one of the standard methods to compute the global optimum of an optimization problem [3,4]. The main ingredients are (i) the branching procedure, which splits current problem into more problems of smaller dimensions, and (ii) the bounding procedure, which provides for each instance of a problem (also for the smaller problems obtained via branching) a lower and an upper bound for the optimum value.

In the case of (SSP), the ϑ-function problem provides a natural and relatively cheap upper bound. For the lower bounds, we need to find efficiently a stable set being as large as possible. We usually use the following heuristics: if Y is the optimum value for ϑ-function, we extract its diagonal as $\mathbf{y} = \operatorname{diag}(Y)$. It has values between 0 and 1. We order $\mathbf{y}$ in a descending order obtaining $\tilde{\mathbf{y}}$ and construct a stable set as follows: for each $\tilde{y}_j$, we check if the corresponding vertex $\tilde{v}_j \in V$ is adjacent with at least one of the vertices $\tilde{v}_i$ $(1 \leq i < j)$. If it does not, we add it to the stable set.

Using the ingredients described above, we formulate the B&B algorithm for (SSP') in Algorithm 2.

In practice we have several strategies how to select (G', ub') in the WHILE loop and how to choose v in the second IF sentence. In our case, we use priority queue of subproblems sorted according to upper bound. The pair with the largest value ub' is selected since this ensures that, when we have $LB = ub$, we are at the optimum.

As a bounding routine, we use ϑ-function. Depending on the underlying sparsity of the subgraph, the combination of two semidefinite solvers is used (the interior-point method tailored for SST and the boundary point method [15]).

We can also select v in the last IF sentence according to different strategies. There are several natural choices for a branching vertex. We implemented the following rules: for the first rule R1 we choose the vertex with minimum or maximum degree in the subgraph G'. The second rule R2 is based on the optimal solution matrix Y of ϑ-function. We extract its diagonal as $\mathbf{y} = \operatorname{diag}(Y)$ and for the next branching variable we select the index i for which the value y_i is closest to 1. Depending on the class of problems, either R1 or R2 was more efficient (Fig. 2).

INPUT: $G = (V, E)$.
INITIALIZATION
$G_0 = G$; Compute $\vartheta(G)$ to obtain Y_0 and initial upper bound $ub_0 = \vartheta(G)$.
S_0 starting stable set obtained from Y_0.
$LB = |S_0|$ starting lower bound.
$\mathcal{Q} = \{(G_0, \emptyset)\}$.
WHILE $\mathcal{Q} \neq \emptyset$
Select some $(G', ub') \in \mathcal{Q}$.
Compute

$$\begin{aligned} ub &= \vartheta(G') \\ S &= \text{ stable set in } G' \\ lb &= |S|; \end{aligned}$$

IF $lb > LB$
$LB := lb$;
Remove from $\mathcal{Q}$ all (G', ub') with $ub' < LB$;
IF $LB < ub$
choose $v \in V(G')$ for branching. Create N_v - neighborhood of v in G', which includes v.
Let G'_1 and G'_2 be the induced subgraphs on $V(G') \setminus v$ and $V(G') \setminus N_v$.
Add to $\mathcal{Q}$: (G'_1, ub) and (G'_2, ub);

Fig. 2. A B&B algorithm for (SSP')

3 Parallel B&B Algorithm for (SSP)

In this section, we describe how the algorithmic ingredients from serial B&B algorithm are combined into a parallel solver running on a high-performance computer. Local coordinator/worker paradigm with distributive work pools is applied in which different workers concurrently explore branches of the B&B tree. Each worker has its own local queue of subproblems and work is shared when one of them becomes idle. The master node knows the status of each worker (active or idle) and acts as a load coordinator receiving messages and based on their content replies in appropriate manner.

At the beginning, the load coordinator reads and broadcasts the original graph to the workers. It is important that every process has the knowledge of the original graph, since construction of subproblems via branching and encoding the MPI messages is done based on this information. Next, initial branching is applied (see Sect. 3.2), number of problems in queue is increased and each of them is sent to a different worker. Throughout the algorithm the master node manages bool array containing statuses of workers. They specify whether the process is active or idle. The load coordinator uses this information to reply to the requests received from workers and distributes information specifying which workers are free and sends best lower bound. During execution of the algorithm, working processes compute multiple candidates for optimal solutions. The master node is keeping the track of currently best optimal value and solution. When new solution is received, the value is compared, updated if necessary and distributed back during communication phase.

After a worker computes lower and upper bound, it compares these values to see if this branch of enumeration tree can be safely pruned or further branching is needed and construction of new subproblems takes place. In the latter, request message is sent to the load coordinator, asking for idle processes to share one of the new generated subproblems or subproblems left in the queue from previous branching processes. If no idle worker is available, the generated subproblems are placed in worker's own queue and work continues locally. Otherwise, subproblems are encoded and sent to available idle workers.

In order to efficiently construct a subproblem from the received data, we only send global labels of vertices in the related subgraph via MPI_Send. As from the beginning each process has the knowledge of whole graph, the subproblem can be efficiently constructed using this information. The algorithms for load coordinator and workers are summarized in Algorithms 3 and 4.

3.1 Implementation in C++

The parallel algorithm for SSP is implemented in C++. Distributed parallelization is applied in order to search the B&B enumeration tree in parallel. We used the routines from message-passing libray MPI [6] to conduct communication between independent computational nodes. In addition, linear algebra library Armadillo [18] is linked with Intel Math Kernel Library [1] to perform computationally intensive matrix operations.

3.2 Initial Branching Strategy

Before load coordinator sends problems to workers, initial branching strategy is used to ensure parallelism from start and to fully exploit all the cores. At the beginning of the algorithm only the original problem is in the queue of problems that need to be explored. If the load coordinator would send the initial problem to the first computing node, other processes woud be idle and resources would be wasted.

Our strategy is to repeatedly take problems from the queue, branch them as long as the number of subproblems equals the number of workers, and then send one subproblem to each worker. With this approach, we achieve the advantage that no computing node is idle and processes start working immediately in parallel. Secondly, computing the ϑ-function that is used in the bounding routine is the most costly operation for a worker. In order to increase the performance of the solver, the number of times we need to compute it should be kept to minimum and evaluated on graphs with smaller number of vertices. With the proposed initial branching strategy the size of the original graph gets reduced and we omit computing semidefinite bounds for bigger graphs (Fig. 3).

INPUT: $G = (V, E)$.
INITIALIZATION
  Broadcast graph G to the workers.
  Compute $\vartheta(G)$ to obtain Y_0.
  Compute starting stable set S_0 obtained from Y_0.
  Set starting lower bound $LB = |S_0|$ and distribute the value.
  Put original problem into queue and apply initial branching strategy.
  Set status of every worker to busy.
WHILE there are busy workers
  Receive message from worker.
  IF *IDLE*
    Mark worker as free.
  ELSE IF *NEW_VALUE*
    Worker found better LB. Compare, update and distribute new value.
  ELSE IF *SEND_WORKERS*
    Send ranks of the idle workers to the sender.
Send *FINISH* message to the workers.

Fig. 3. Algorithm for load coordinator

4 BiqBin Solver

The parallel solver for (SSP) described above is a part of the BiqBin web-based solver aimed to provide different solvers for several special subproblems of binary quadratic problems (BQP). The BiqBin will consist of several well adapted solvers for specific versions of (BQP), beside the code described above also the parallel code for the Max-Cut problem and its extension to cover (BQP) based on the exact penalty approach from [10]. A pilot version of BiqBin is already available to the research community as an online solver running on HPC at University of Ljubljana. The code will soon be available also under the GNU GPL (Fig. 4).

INITIALIZATION
Receive graph G and best lower bound LB from the load coordinator.
DO
Receive subproblem from the load coordinator or other worker and put the problem into local queue.
WHILE queue is not emty
Select some subproblem and compute lower bound lb and upper bound ub.
IF $lb > LB$
Send *NEW_VALUE* message to the load coordinator and receive updated LB.
IF $LB < ub$
Generate new subproblems and place them in queue.
Send request message *SEND_WORKERS* and receive ranks of idle workers.
Encode subproblems and send them to available processes.
Send *IDLE* message to the load coordinator and wait for new tasks.
IF *FINISH*, stop

Fig. 4. Algorithm for worker process

4.1 Online BiqBin

The online BiqBin solver is available as a web application.[6] It allocates on one hand dynamically optimal available computing power and storage and on the other hand provides to the problem contributors necessary statistics and information about the progress in solving.

It enables (registered) users to test the solvers against their own graph data. The web application contains main information about the BiqBin project and benchmark results which are available for everybody. Upon registration, the user can access the core functionalities of the application. In short, the user uploads her problem data files and decides which algorithm/solver she wants to use to solve the problem. The application then sends these data to HPC, where the selected algorithm is executed against the given data, and after some interval it returns the solution (optimal or approximate) back to the web application, which finally notifies the user by an email that the result is ready.

Let us describe the web application system in more details. The initial step for the user is to prepare the problem graph data in an appropriate format, which is described in the description of each function. Then, she creates a new **instance**, which is the entity determining the problem data. The user determines instance's name, data (by uploading her problem graph data), and description. In this way, we enable the user to reuse the problem data later, e.g. for running other solvers against it.

[6] http://biqbin.fis.unm.si.

Next, the user creates a new **task**, which is comprised by an instance and a solver (function) the user wants to run with the instance's data. The user also determines the tasks' name, and specifies if the result and the instance's data can be listed publicly for benchmarking purposes. When all the properties are specified the task can either be started automatically, or the user can decide when she will start it.

Starting a task, means that the system sends a demand with the execution data to the HPC where the solvers are installed. For each sent task, a job is started on a specified number of cores. Execution of a solver is being regularly monitored by saving intermediate results per every given time interval. After a specified maximum execution time, or after finding an optimal solution, the solver's execution is stopped, and the result files are sent back to the web application together with tasks' metadata. The result is then saved in the database, and the user is notified by email that the job is completed. If the task has been marked as publicly available, a benchmark record is also created and immediately visible on the Benchmarks page.

The system is designed in such a way that additional solvers or algorithms can be easily added; in fact even automatically if no additional packages are required for HPC execution. The number of cores used for a single taks is fixed, so that the users are limited, however, the users with special roles can specify the number of cores themselves. In this way, the architecture of the system is highly extendible in all main directions. Scalability depends on the HPC facilities, but this is also considered in the web application, as tasks for every solver can connect to a special HPC (Fig. 5).

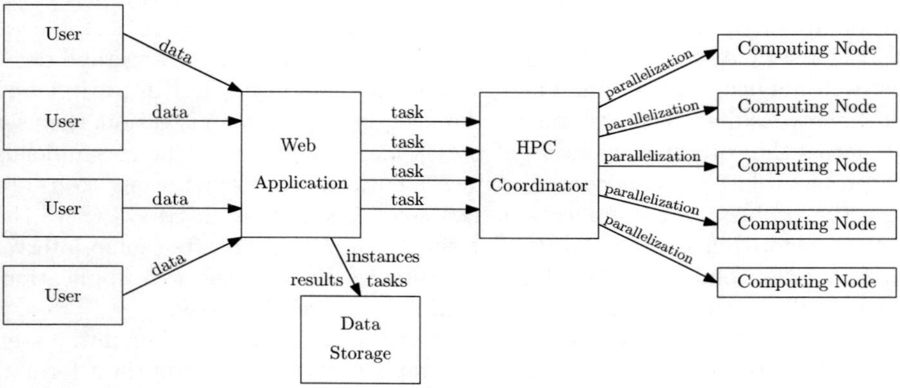

Fig. 5. A workflow diagram with involved system entities.

5 Numerical Results

The perfomance of the proposed parallel algorithm was evaluated in terms of execution time and speedups, and tested on random instances as well as on

DIMACS graphs (publicly available collection of benchmark graphs for solving various graph-related problems) [8]. In Table 1, we consider random graphs of specified edge density and different number of vertices n and provide computation times (in seconds). The results for DIMACS graphs are summarized in Table 2. All experiments were repeated 10 times and average results are presented. Different number of nodes were used to see how the algorithm scales. The cluster with different number of computational nodes which consist of a dual CPU Intel 2,93 GHz Xeon X5670, each with six physical cores was used.

Table 1. Computing the stability number on random graphs. Computation times in seconds, averaged over 10 instances. Serial specifies the runtime of sequential algorithm, other parameters specify the number of CPU cores used in parallel algorithm.

n	Density	Serial	12	24	48	96	192
100	0.1	4.15	0.69	0.69	0.75	1.14	–
100	0.2	5.46	1.50	0.87	0.68	1.15	–
100	0.3	5.66	2.03	1.41	0.88	1.20	–
100	0.5	18.43	7.23	2.54	1.04	1.15	–
150	0.1	143.90	42.36	29.22	10.20	5.38	4.61
150	0.2	120.02	20.14	14.79	7.34	5.43	4.30
150	0.3	222.49	24.35	18.44	13.90	9.89	6.83
150	0.5	62.58	15.39	10.83	7.49	3.64	2.80
200	0.2	6431.33	1463.36	669.71	297.75	145.58	67.41
200	0.3	355.70	92.49	56.87	30.87	12.40	11.80
200	0.5	447.91	153.83	88.70	68.58	42.82	35.28

Table 2. Computing the stability number on DIMACS graphs. Computation times are in seconds.

Graph	n	Density	24	48	96	192
brock200_1	200	0.25	82.54	38.77	22.06	16.21
brock400_1	400	0.75	1005.20	839.06	486.96	151.40
C250.9	250	0.90	32.55	26.85	21.44	17.23
keller4	171	0.35	15.29	15.08	10.80	9.76
p-hat300-3	300	0.25	287.52	155.97	59.43	39.78

6 Conclusion

In this paper we presented a parallel Branch and Bound (B&B) algorithm to solve the Stable Set Problem. In each node of the B&B tree, we solve an instance of semidefinite programming problem to obtain better upper bounds for the original problem. This solution also gives lower bounds via good stable sets.

We tested the algorithm on a high-performance cluster available at University of Ljubljana. By using up to 192 CPU cores, we show that this algorithm scales well.

We integrated the algorithm into an on-line BiqBin solver, which is a result of our effort to bring high-performance computing closer to mathematical optimization society. It enables on-line submissions of problem instances and, when it finds a solution, it automatically notifies the contributor about the result. We believe that our approach, where we offer a service and not only HPC adopted source code or HPC infrastructure, is the way how HPC community should continue to decrease the gap between HPC ecosystems and users.

Acknowledgment. The authors would like to thank to Slovenian Research Agency (ARRS) and to Austrian Science Foundation (FWF) for support via joint Austrian-Slovenian project nr. N1-0057 and via programs P2-0256 and P1-0383.

References

1. Intel Math Kernel Library. Reference Manual. Intel Corporation (2009)
2. Anjos, M.F., Lasserre, J.B. (eds.): Handbook on Semidefinite, Conic and Polynomial Optimization. International Series in Operations Research & Management Science, vol. 166. Springer, New York (2012)
3. Boyd, S., Mattingley, J.: Branch and Bound Methods (2007)
4. Clausen, J.: Branch and Bound Algorithms - Principles and Examples, pp. 1–30. Department of Computer Science, University of Copenhagen (1999)
5. Cottle, R.W., Thapa, M.N.: Linear and Nonlinear Optimization. International Series in Operations Research & Management Science. Springer, New York (2017)
6. Message Passing Interface Forum: MPI: A message-passing interface standard. Technical report, Knoxville, TN, USA (1994)
7. James, G., Witten, D., Hastie, T., Tibshirani, R.: An Introduction to Statistical Learning: with Applications in R. Springer Texts in Statistics, vol. 103. Springer, New York (2013)
8. Johnson, D.S., Trick, M.A.: Cliques, Coloring, and Satisfiability: Second DIMACS Implementation Challenge, October 11–13, 1993, vol. 26. American Mathematical Soc. (1996)
9. Krislock, N., Malick, J., Roupin, F.: BiqCrunch User Guide (2014)
10. Lasserre, J.B.: A max-cut formulation of 0/1 programs. Oper. Res. Lett. **44**(2), 158–164 (2016)
11. Malick, J., Povh, J., Rendl, F., Wiegele, A.: Regularization methods for semidefinite programming. SIAM J. Optim. **20**, 336–356 (2009)
12. Povh, J.: Semidefinite approximations for quadratic programs over orthogonal matrices. J. Global Optim. **48**(3), 447–463 (2009)
13. Povh, J.: Contribution of copositive formulations to the graph partitioning problem. Optimization **62**(1), 71–83 (2013)
14. Povh, J., Rendl, F.: Copositive and semidefinite relaxations of the quadratic assignment problem. Discret. Optim. **6**(3), 231–241 (2009)

15. Povh, J., Rendl, F., Wiegele, A.: A boundary point method to solve semidefinite programs. Computing **78**, 277–286 (2006)
16. Rendl, F., Rinaldi, G., Wiegele, A.: BiqMac – a solver for *bi*nary *q*uadratic and *ma*x-*c*ut problems (2006). http://BiqMac.aau.at/
17. Rendl, F., Rinaldi, G., Wiegele, A.: Solving max-cut to optimality by intersecting semidefinite and polyhedral relaxations. Math. Program. **121**(2, Ser. A), 307–335 (2010)
18. Sanderson, C., Curtin, R.: Armadillo: a template-based c++ library for linear algebra. J. Open Source Soft. **1**(2), 26 (2016)
19. Rui, X., Donald Wunsch, I.I.: Survey of clustering algorithms. IEEE Trans. Neural Netw. **16**(3), 645–678 (2005)
20. Yang, X.S.: Introduction to Mathematical Optimization: From Linear Programming to Metaheuristics. Cambridge International Science Publishing (2008)

Modeling and Assessment of Financial Investments by Portfolio Optimization on Stock Exchange

Todor Stoilov[1,2](✉), Krasimira Stoilova[1,2], and Miroslav Vladimirov[3]

[1] Institute of Information and Communication Technologies, Bulgarian Academy of Sciences, Sofia, Bulgaria
{t.stoilov,k.stoilova}@hsi.iccs.bas.bg
[2] Nikola Vaptsarov Naval Academy, Varna, Bulgaria
[3] Varna University of Economics, Varna Administration, Varna, Bulgaria
vladimirov@ue-varna.bg

Abstract. The portfolio optimization is a complex task, addressing decision making of financial investments. A requirement for the portfolio management is to deal with a lot of securities, traded on the stock exchange, frequently containing hundreds of stocks and securities. The modern portfolio theory insists to be defined a portfolio optimization problem, which parameters have to cope historical trends of the assets' returns, subjective views of experts and additional noisy data, generated from practical and political decisions. The paper addresses the actual model for portfolio optimization based on Black-Litterman approach for integration of historical data and expert views about posterior, future levels of assets' returns. A special formalism is derived, adding additional expert assessment based on the historical trend of the asset returns, which modify the classical Black-Litterman portfolio model. Risk and Return parameters of the portfolios are evaluated in mathware environments, applying classical methods of the mean variance optimization and the new modified model on Black Litterman optimization. By mathware tools it is demonstrated the ability to evaluate and to assess the portfolio performance. Real data from the world stock exchanges are taken for numerically illustration of the implementation of a modified Black Litterman portfolio model. This research can serve business applications and the topics of portfolio modeling, optimization, decision making, data driven analysis.

Keywords: Portfolio modeling · Optimization · Decision making · Data driven analysis

1 Introduction

The outcome of the portfolio theory is quite simple: how to allocate an amount of investments, buying a set of securities for an initial moment t_0 and after period of time T

T. Stoilov—Executive Manager of Investment Consult ICO Company.
Special Session: Modeling, Simulation; Optimization in a Data-rich Environment – (MSO-DE).

I. Dimov and S. Fidanova (Eds.): HPC 2019, SCI 902, pp. 340–356, 2021.
https://doi.org/10.1007/978-3-030-55347-0_29

to sell the securities and to make positive return from this investment. Despite that this task looks simple for understanding the portfolio investment implemented according to the requirements of the Modern Portfolio Theory (MPT) is a very complicated and risky job. The drawback for the portfolio management concerns the difference between the times t_0 and $t_0 + T$. The investor has to make decisions of the allocation of its investment at initial time t_0. But the portfolio return is evaluated and achieving in a final, future moment $t_0 + T$. The portfolio return in a future moment $t_0 + T$ depends on the assets returns, participating in the portfolio. Thus, the investor has to forecast the assets returns by means to assess its portfolio return. For the initial time t_0 the investor can estimate the current returns of the assets. But for the future moment t_0 + T he has to make realistic assumptions about the behavior of the asset returns. Because the assumption of precise value for the future return even of one asset is very ambitious task, the MPT implements an additional parameter for the assets returns, which is the risk. This parameter is assumed as a diapason of values, between which the future asset return will obtain its real value. This case of having information of the portfolio assets in time t_0 and the need for forecasting the values of returns and risks of assets is formally presented in Fig. 1.

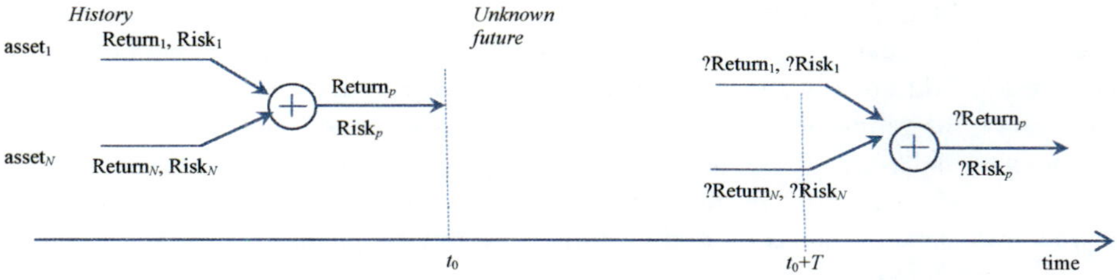

Fig. 1. The portfolio investment must forecast risk and return of the assets

Because the asset parameters about their returns and risks are unknown and have to be predicts (?Return$_i$, ?Risk$_i$) where i = 1, N – number of types of assets in the portfolio, the corresponding values of the portfolio Return and Risk (Return$_p$, Risk$_p$) are also strongly influenced by the predictions (?Return$_i$, ?Risk$_i$), i = 1, N.

2 Quantitative Assessment of the Portfolio Characteristics

The quantification of the portfolio theory is assumed to be performed initially by the works of Markowitz [9]. The assets returns are assumed to be the mean value of the returns, evaluated for a predefined historical period. The corresponding asset risk is quantified as the volatility of the asset return for the same historical period. Assuming that the returns and risks for all assets will keep their values from the initial moment t_0 till the future one $t_0 + T$, the Markowitz formalization of the portfolio problem is given by the optimization problems

$$\underset{\mathbf{w}}{max}\begin{bmatrix} \mathbf{E}^{\mathrm{T}}w \\ \mathbf{w}^{\mathrm{T}}\mathbf{\Sigma}\mathbf{w} \leq \sigma_{\max}^2 \end{bmatrix} \quad \text{or} \quad \underset{\mathbf{w}}{min}\begin{bmatrix} \mathbf{w}^{\mathrm{T}}\Sigma\mathbf{w} \\ \mathbf{E}^{\mathrm{T}}\mathbf{w} \geq E_{\min} \end{bmatrix} \tag{1}$$

where $\mathbf{E}^T = (E_1, \ldots, E_N)$ is a vector with the returns of the assets for the historical period;

N - number of assets in the portfolio;

$\mathbf{\Sigma}$ - the covariance $N \times N$ matrix between the asset returns for the historical period, which is the quantitative assessment of the risk and the correlations between the returns of the assets;

$\mathbf{w}^T = (w_1, \ldots, w_N)$, w_i – is a component of the vector of the weights, which presents the relative part of the investment, allocated to asset *i*. The vector **w** is a solution of the portfolio problem;

$\sigma^2_{\max}$ - the maximal allowed risk for the portfolio;

E_{min} - the minimal requested return by the portfolio.

The vector **E** and the covariance matrix $\mathbf{\Sigma}$ can change their values during the future period $(t_0, t_0 + T)$. This results in strong sensitivity of the solutions **w** of the portfolio problems (1) according to the initial data **E** and $\mathbf{\Sigma}$. This is the general reason the modern portfolio theory not to be widely applied in real case of portfolio investments and many investors to consider that the portfolio problems (1) are impractical for use.

The portfolio investments, based on problems (1) suffers from the constraint set of data, which is used for the definition of the portfolio characteristics **E** and . The last are estimated based on historical data. But the length of the historical period is not recommended by the portfolio theory and it is defined by the subjective assumptions of the investor. Additionally, the future investment period T is also defined in a subjective way. Hence, the definition of the portfolio problem insists strong inclusion of subjective decisions for the estimation of the problem parameters **E** and $\mathbf{\Sigma}$.

Problems (1) define the portfolio problem as an vector optimization one. To reduce the inconvenience of the vector optimization, currently problems (1) are aggregated in a simpler problem [13]

$$\max_{w} \left[\mathbf{E}^T\mathbf{w} - \lambda \mathbf{w}^T \Sigma \mathbf{w}\right] \tag{2}$$

$$\mathbf{w}^T|\mathbf{1}| = 1, \mathrm{w}^T \geq 0$$

$|\mathbf{1}|$ is $N \times 1$ vector with ones, λ is a risk aversion parameter. This parameter can take values from the set $[0, \infty]$. This parameter numerically estimates the level of risk, which the particular investor can accept, making his investment. For the case $\lambda = 0$ the investor is not care about the risk and his goal is to maximize the portfolio return. For the case $\lambda = \infty$ the investor is targeting minimum of the portfolio's risk. The parameter λ is also an input for the portfolio problem, which has to be estimated before solving (2). Such estimation subjectively influences the definition and respectively the solution of the portfolio problem (2). By changing λ from 0 to ∞ the portfolio problem (2) gives deferent solutions for the weights $\mathbf{w}(\lambda)$. Respectively, the values for the different portfolios of returns are $E_p(\lambda) = \mathbf{E}^T\ \mathbf{w}(\boldsymbol{\lambda})$ and the portfolio risk is quantified with the volatility $V_p\ (\lambda) = \mathbf{w}^T\ (\lambda)\mathbf{\Sigma}\mathbf{w}(\lambda)$. The couple $(Ep(\lambda), Vp(\lambda))$ gives one point in the space Risk(Return). Each asset and portfolio is represented as a point in this space. By changing the values of λ and solving (2) the corresponding portfolios lays on a curve, named "efficient frontier", Fig. 2.

Fig. 2. Efficient frontier of the portfolio optimization

The MPT recommends that the portfolio investment has to be represented as a point belonging to the "efficient frontier". We would like to conclude that even for problem (2) an additional parameter λ must be estimated in subjective way, which quantifies the ability of the investor to undertake risk for his investment.

Using the Markowitz portfolio formalism of problems (1) and (2) an additional contribution to the MPT was given by Sharpe by the definition of the Capital Market Theory [13]. As a component of this theory is the Capital Asset Pricing Model. This model deals with a special linear relation, named "Capital Market Line" (CML). This line results from the introduction of a special portfolio, named "market portfolio". The market portfolio weight $\mathbf{w_M}$ contains the weights of all assets, traded on this market, and their values are equal to the market capitalization weights of all assets. The geometrical interpretation of the market portfolio is given in Fig. 3. The point $(r_f,0)$ represents the risk free asset. It has 0 risk and constant return r_f. The market portfolio (E_M, V_M) is the tangent point of line t towards the "efficient frontier", which passes through the point $(r_f,0)$. The line t is the CML. The troubles in finding analytical description of the CML are due to the lack of analytical relation for the "efficient frontier". This line is smooth for the case of unconstrained optimization problem (2) [10]. But having problem (2), which is a constraint optimization one, the "efficient frontier" is not a smooth line. In [4] it is assessed that the "efficient frontier" is a combination of parabolic quadratic curves. Hence, an analytical description of the "efficient frontier" as a quadratic curve, is not possible. The "efficient frontier" can be found as a set of points by solving problem (2) with different values for $\lambda = 0,\ldots,\infty$. In practice, the market characteristics (E_M, V_M) are substituted by the mean return and volatility of the market index (Nasdaq, Dow Jones, FTSE etc.). But for the case of limited number N of types of assets, chosen for inclusion in the portfolio by the investor, the estimation of the market characteristics for such particular markets is quite complicated task.

The MPT tries to decrease the subjective cases for definition and evaluation of the assets and portfolio characteristics. The short analysis till now illustrated that the subjective actions are obligatory for the definition and finding final portfolio's solution. A solution to incorporate subjective influence for the definition of the portfolio problem is given by the Black Litterman (BL) model [3]. This model is assumed as a next step in the development of MPT. The idea of this model is to integrate both historical characteristics of the assets and additional recommendations and views, given by experts. Thus, the final estimation of the assets returns and risks is based both on objective, historical data and subjective expert views and assessments. Thus, the forecasting of the assets and portfolio parameters is based on additional experts' views [1, 5, 6]. The added value of

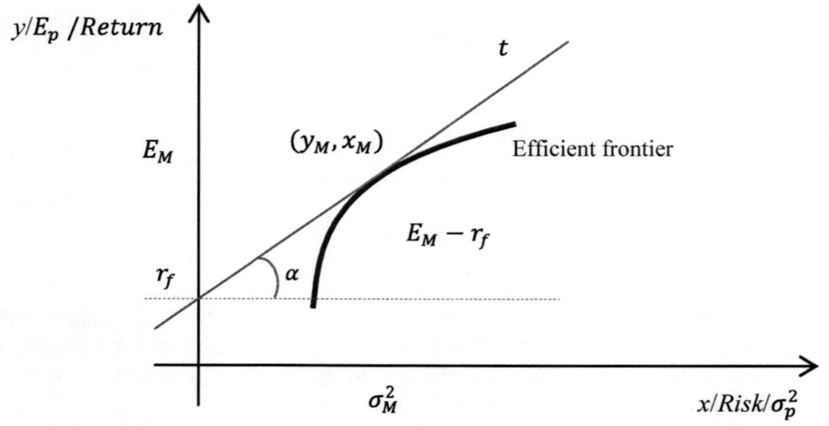

Fig. 3. Graphical interpretation of the Capital Market Line and Market portfolio

such an integration of two sources of information for the assets characteristics results in lack of sharp cut of assets in the content of the portfolio. In classical Markowitz problems (1) or (2), the content of the portfolio is limited to small types of securities and the corresponding values of their weights **w** is high, close to the relative 1. Small changes in the assets characteristics results in a sharp change of the content of the portfolio. Such changes are not accepted by the investors and the investment process contains high level of risk and lack of diversification. In the case of BL model, such sharp changes of the portfolio content is not met, which is a prerequisite for implementation of diversification of the investment. Nevertheless of this positive future, the BL model is not accepted without remarks [11].The general concern to this model is the question about its efficiency. But the given portfolio solutions by the BL model are assumed as more acceptable from the investors' community.

3 Black-Litterman Model for Assessing the Assets Characteristics

The BL model evaluates new values about the assets characteristics: assets returns, covariation between the assets returns. This new values are found by integration of objective, historical data about the assets returns and the expert forecasts about these assets returns. A graphical interpretation of the BL estimation of the assets characteristics is given in Fig. 4. The estimation of the vector of mean assets returns **E** and the covariance matrix between the returns $\mathbf{\Sigma}$ is done according the available historical data. These assets characteristics are evaluated from objective historical data about the dynamics of the assets returns. But the values of **E** and $\mathbf{\Sigma}$ are modified by the expert views, formalized by the matrices **P** and **Q.** The formal presentation of the BL relations is shortly described below.

The BL model introduces new assets characteristics, named "implied returns" Π_i, $i = 1,N$. These assets returns differ from the historical real mean values E_i, $i = 1,N$. The reason that these two types of assets returns are different is the existence of noise on the market, which influences the market dynamics of the assets returns. Thus, the BL model

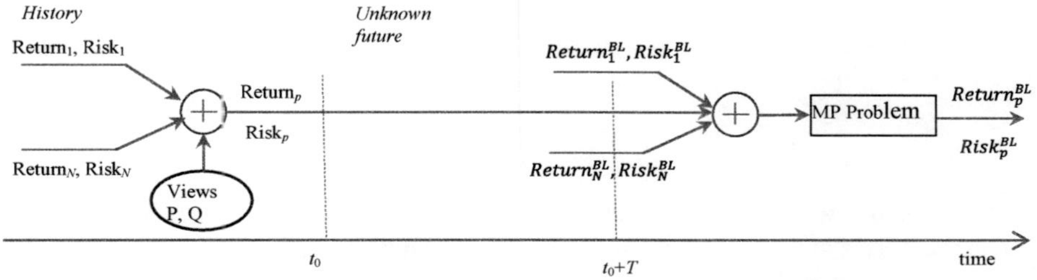

Fig. 4. Graphical interpretation of the BL assessment of assets characteristics

operates with the "implied returns" Π_i instead of the mean historical returns E_i, which are used in the classical mean-variance portfolio theory.

A key point in the BL model is the evaluation of the assets "implied excess returns" Π_i, $i = 1, N$. Following [15], the implied excess returns of the assets have to satisfy the market portfolio. Respectively, the goal function of problem (2) becomes in a form (3) where the market capitalization weights $\mathbf{w_M}$ are known from the market portfolio

$$\max_{\mathbf{w_M}} \left\{ \mathbf{w_M^T \Pi} - \frac{1}{2}\lambda \mathbf{w_M^T \Sigma w_M} \right\}. \tag{3}$$

The covariance matrix $\mathbf{\Sigma}$ is assumed to be the same, which is evaluated only with historical data of the assets returns. The only unknown parameters in (3) are the "implied excess returns" $\Pi^{\mathbf{T}} = (\Pi_1, \ldots, \Pi_N)$. The evaluation of Π is performed by the so called "inverse optimization", which is based on unconstrained optimization of the portfolio goal function or

$$\frac{d}{dw}\left(\mathbf{w_M^T \Pi} - \frac{1}{2}\lambda \mathbf{w_M^T \Sigma w_M} \right) = 0$$

and finally it follows

$$\mathbf{\Pi} - \lambda \mathbf{\Sigma w_M} = 0. \tag{4}$$

In this relation the risk aversion parameter λ is not known. For its evaluation the market characteristics (E_M, V_M, $\mathbf{w_M}$) are used again. Both sides of relation (4) are multiplied by the market capitalization weights $\mathbf{w_M}$ or

$$\mathbf{w_M^T}\Pi = \lambda \mathbf{w_M^T \Sigma w_M}.$$

The left side of this equality gives the market excess return over the risk free return r_f, $E_M - r_f$. The value $\mathbf{w_M^T \Sigma w_M}$ is the market risk (volatility) σ_M^2. Hence, the unknown risk aversion parameter λ can be evaluated by the market characteristics or

$$\lambda = \frac{E_M - r_f}{\sigma_M^2}. \tag{5}$$

Finally, the "implied excess returns" can be evaluated from (4) and (5)

$$\mathbf{\Pi} = \frac{E_M - r_f}{\sigma_M^2} \mathbf{\Sigma w_M}. \tag{6}$$

The "implied returns" Π^* of the assets are found by taking into account the level of riskless asset return r_f

$$\prod^* = \prod + r_f \tag{7}$$

These values will be used for the evaluation of the unknown BL asset returns $\mathbf{E_{BL}}$. Due to the noises in the market the difference between the "implied returns" and the BL returns $\mathbf{E_{BL}}$ is given by the matrix relation

$$\mathbf{\Pi}^* = \mathbf{E_{BL}} + \boldsymbol{\varepsilon},\ \mathbf{E}_{\mathrm{BL}} = \begin{vmatrix} E_{BL1} \\ \vdots \\ E_{BLN} \end{vmatrix}, \tag{8}$$

where the vector of the noise $\boldsymbol{\varepsilon}$ has zero mean value, independent components, and volatility proportional to the assets volatilities $\varepsilon \sim N(0, \tau\mathbf{\Sigma})$. The coefficient $\boldsymbol{\tau}$ has value less than 1 and determines that the noise volatility has the same character as the volatility of the assets returns $\mathbf{\Sigma}$, but with less intensity. Because the noise components are independent each other, only the diagonal elements are used from $\mathbf{\Sigma}$, which corresponds to the individual risks of the assets.

The BL model introduces a new relation to (8) for the evaluations of the returns $\mathbf{E_{BL}}$. It is derived from the requirement for integration of subjective experts views for the forecast of the assets returns $\mathbf{E_{BL}}$.

The subjective expert views are formalized by the equation

$$\mathbf{Q} = \mathbf{PE}_{\mathrm{BL}} + \boldsymbol{\eta}, \tag{9}$$

where the matrices $\mathbf{Q}$ and $\mathbf{P}$ are defined in appropriate way. The noise $\boldsymbol{\eta}$ of the expert views is assumed again with zero mean value with independent components, because the experts are chosen to have independent views, $\boldsymbol{\eta} \sim \mathbf{N}(0, \mathbf{\Omega})$. The variation of the noise is noted $\mathbf{\Omega}$,. Due to the independence of the expert views the matrix of noise variation $\mathbf{\Omega}$ has only diagonal elements. Each element of $\mathbf{\Omega}$, following [7], is evaluated as

$$\mathbf{\Omega} = \tau diag\left(\mathbf{P\Sigma P^T}\right). \tag{10}$$

The matrix $\mathbf{P}$ and vector $\mathbf{Q}$ are defined according to several assumptions [12]. Assuming that k experts give independent views, the vector $\mathbf{Q}$ has kx1dimension. The component q_j, $j = 1,k$ has value about the future level of return, which an asset will have. The number of the asset is given by matrix $\mathbf{P}$ with dimensions $k \times N$. The component $p_{j,i}, j = 1, k; i = 1, N$ identifies that the expert j assess that asset i will increase or decrease its return in the future.

4 Definition of Expert Views

Currently, the BL model applies two types of experts' views: absolute and relative [12]. An example below clears the meaning of this kind of definition of the experts' views.

Assuming portfolio with 5 assets, $N = 5$, and two experts, $k = 2$ the absolute and relative views are explained by examples.

Absolute form of views: The notations $\mathbf{P^T} = \begin{vmatrix} 01000 \\ 000-10 \end{vmatrix}$, $\mathbf{Q} = \begin{vmatrix} 3,5 \\ 2 \end{vmatrix}$ means that the first expert forecast that asset $N = 2$ will increase its return with 3,5%. The second experts gives decrease of asset $N = 4$ with 2%. Because the absolute form of views is more difficult to be provided by the experts, more practical is to make relative forecasts.

Relative form of views: $\mathbf{P^T} = \begin{vmatrix} 0100-1 \\ 100-10 \end{vmatrix}$, $\mathbf{Q} = \begin{vmatrix} 3,5 \\ 2 \end{vmatrix}$.

For this case the values of the matrix $\mathbf{P}$ and matrix $\mathbf{Q}$ are given in the way which presents that the first expert forecasts that the return of asset $N = 2$ will outperform the return of asset $N = 5$ with 3,5%. The second expert makes his assessment that the return of asset $N = 1$ will outperform with 2% the return of asset $N = 4$. For both forms for the views the matrix $\mathbf{P}$ contains values of $+1$ or -1. The sum of a row of $\mathbf{P}$ gives always zero value.

In this research we introduce new weighted form of the expert views, which contains components of $\mathbf{P}$, which are different from $+1$ or -1.

5 Weighted Form of Views

The idea behind this special formalization of the experts' views is to use additional information, which can be derived from the historical data of the assets returns. The MPT currently uses only two general characteristics of the numerical series of asset returns, recorded for historical period of time: mean values $\mathbf{E}$ and the correlations $\mathbf{\Sigma}$ between the assets returns. The diagonal components of the correlation matrix are the volatilities (risks) of the assets returns. But additional assets characteristics are also available and introduced for the portfolio optimization like the market return $\mathbf{E_M}$ and risk $\mathbf{V_M}$, "implied returns" Π. For the portfolio analysis the investor can take into consideration additional parameters and dynamical features of the assets returns. For the case of BL model experts' views are considered. The definition of the weighted form of the expert views targets integration of new data about the behavior of the assets returns. As a result the need to have experts in the BL model is overcome and their role is substituted by new economics characteristics of the market. Such a substitution allows to make comparisons between the portfolio solutions, evaluated by classical Markowitz portfolio problems and these obtained by BL models. Currently the BL model gives portfolio solutions, which are strongly influenced by the number and professional experience of the experts. Hence the solutions of portfolios, based on BL model cannot be compared and assessed on a common base. The usage of common data, derived from objective characteristics of the market gives opportunity to be compared different portfolio problems based on classical or BL or other models.

The weighted form of expert views was introduced initially in [14]. Here it is additionally extended and motivated. The additional assets characteristics are based on the difference between the historical average asset returns $\mathbf{E}$ and the evaluated implied returns Π, $(\mathbf{E} - \Pi)$. The component i with maximal value will be the ith element of the matrix row of **P**

$$\alpha_i ==> \overset{max}{\underset{i}{}} (E_i - \Pi_i), i = 1, N. \tag{8}$$

It means that asset i is underestimated and it is expected that in the future its return will increase. For the case

$$\alpha_j ==> \overset{min}{\underset{j}{}} \left(E_j - \Pi_j\right) \tag{9}$$

it means that asset j is overestimated and is expected in the feature its return to decrease. This view make a row s, $s = 1,k$ in matrix **P** $\mathbf{p}_s = \left| 0 \cdots \alpha_i \; 0 \ldots \; 0 - \alpha_j \; \ldots 0 \right|$. The corresponding value q_s in matrix **Q,** is used with value

$$q_s = \overset{max}{\underset{i}{}} (E_i - \Pi_i), i = 1, N. \tag{10}$$

6 Black Litterman Modified Asset Parameters

Following [2, 7, 8, 15] the BL relations for the new average assets returns and covariance matrix are given by the relations

$$\mathbf{E}_{\text{BL}} = \left[(\tau\boldsymbol{\Sigma})^{-1} + \mathbf{P}^T\boldsymbol{\Omega}^{-1}\mathbf{P}\right]^{-1}\left[(\tau\boldsymbol{\Sigma})^{-1}\Pi + \mathbf{P}^T\boldsymbol{\Omega}^{-1}\mathbf{Q}\right] + r_f$$

$$\boldsymbol{\Sigma}_{BL} = \boldsymbol{\Sigma} + \left[(\tau\boldsymbol{\Sigma})^{-1} + \mathbf{P}^T\boldsymbol{\Omega}^{-1}\mathbf{P}\right]^{-1}. \tag{11}$$

These new assets parameters are used to solve the portfolio optimization problem (2). By changing the parameter λ the efficient frontier with the BL data can be found, which is required from the investor to make its final investment decision.

7 Numerical Example for Comparison of Markowitz and BL Portfolio Solutions

The numerical example is based on real market with data from 2018. Since 2017 the riskless deposits in the banks in Bulgaria have very low return, close to zero and/or even with negative values. That's the reason why investors started to pay attention to real form of portfolio management, applying combinations of risky assets. Such risky investments have been operated by many mutual funds, which in Bulgaria are about 125. These mutual funds are managed by 28 business entities. These mutual funds perform asset management and their results give higher return in comparison with the riskless form of investment. The financial instruments, which participate in their portfolios, contain shares, bonds, funds on the currency markets. These mutual funds can be assessed also by their historical returns and risks.

The numerical example chooses 8 mutual funds, which have the highest net present value of their assets, which are invested in shares and currencies. Particularly they are: Concord Asset Management (CONCORD), Elana Asset Management(ELANA), Profit Asset Management (PROFIT), Texim (TEXIM), Central Cooperative Bank Lider (LIDER), Asset Management UBB Patrimonium (PATRIM), Asset Management UBB Premium (PREMIUM), Asset Management DSK Growth (GROWTH). The data about each mutual fund have been obtained from legal internet sources as Bulgarian Association of Asset Management Companies (www.baud.bg), Government Financial Supervision Commission (www.fsc.bg) and from the internet site of each mutual fund. The initial data, which are used in this example concern the mean return values per month for 2018. Having these data, the yearly mean returns, volatilities and covarition matrix are calculated, Fig. 5.

	A	B	C	D	E	F	G	H	I
1	month 2018	CONCORD	ELANA	PROFIT	TEXIM	LIDER	PATRIM	PREMIUM	GROWTH
2	January	0.7298	0.312	0.132	0.197	-0.054	-0.021	0.361	0.791
3	February	-0.2581	-0.414	0.134	0.181	-0.244	-0.924	-0.953	-0.917
4	March	-0.411	-0.197	-0.206	-0.304	-0.494	-0.899	-0.606	-0.448
5	April	0.0193	0.159	0.147	0.256	0.107	0.174	-0.091	-0.063
6	May	-0.4491	-0.362	0.046	-0.136	-0.024	0.416	-0.462	-0.299
7	June	-0.062	0.173	-0.062	-0.134	-0.246	-0.041	-0.37	-0.429
8	July	-0.0869	0.045	0.033	0.108	0.086	0.205	-0.211	0.473
9	August	0.0446	-0.055	-0.03542	-0.251	-0.151	0.176	-0.151	-0.079
10	Septembe	0.2544	0.174	0.105	0.264	-0.135	-0.339	-0.029	-0.322
11	October	-0.0728	0.148	-0.026	-0.003	-0.121	-0.553	-0.661	-0.661
12	Novembe	-0.5829	-0.525	-0.378	-0.393	0.109	0.379	-0.748	0.099
13	Decembe	0.7244	0.111	0.58	0.758	-0.308	-1.021	0.663	-0.038
14	wcapit	0.155	0.159	0.139	0.137	0.12	0.096	0.11	0.084
15	E	-0.012525	-0.03592	0.039132	0.04525	-0.12292	-0.204	-0.2715	-0.15775
16	STD	0.417379	0.273612	0.229691	0.318346	0.18279	0.526381	0.46554	0.469805

COV	CONCORD	ELANA	PROFIT	TEXIM	LIDER	PATRIM	PREMIUM	GROWTH
CONCORD	0.159688	0.08155	0.066914	0.094008	-0.00751	-0.05295	0.159582	0.080864
ELANA	0.08155	0.068625	0.063314	0.040608	-0.00075	-0.00987	0.080044	0.040523
PROFIT	0.066914	0.063314	0.048362	0.063314	-0.0061	-0.04621	0.070033	0.007817
TEXIM	0.094008	0.040608	0.063314	0.092899	-0.00322	-0.06888	0.093779	0.017825
LIDER	-0.00751	-0.00075	-0.0061	-0.00322	0.030628	0.070669	0.000594	0.039516
PATRIM	-0.05295	-0.00987	-0.04621	-0.06888	0.070669	0.253988	-0.00788	0.113633
PREMIUM	0.159582	0.080044	0.070033	0.093779	0.000594	-0.00788	0.198667	0.123228
GROWTH	0.080864	0.040523	0.007817	0.017825	0.039516	0.113633	0.123228	0.202324

Fig. 5. Initial historical data of mean returns and the covariation matrix

These data has been introduced in MATLAB environment for the evaluations, according to BL model. Without the usage of such mathware environment, the evaluations cannot be performed, due to the complex formalism of the portfolio theory and the usage of big amount of initial and sequential evaluated data. The block-diagram of the evaluations performed by the corresponding program suit is given in Fig. 6.

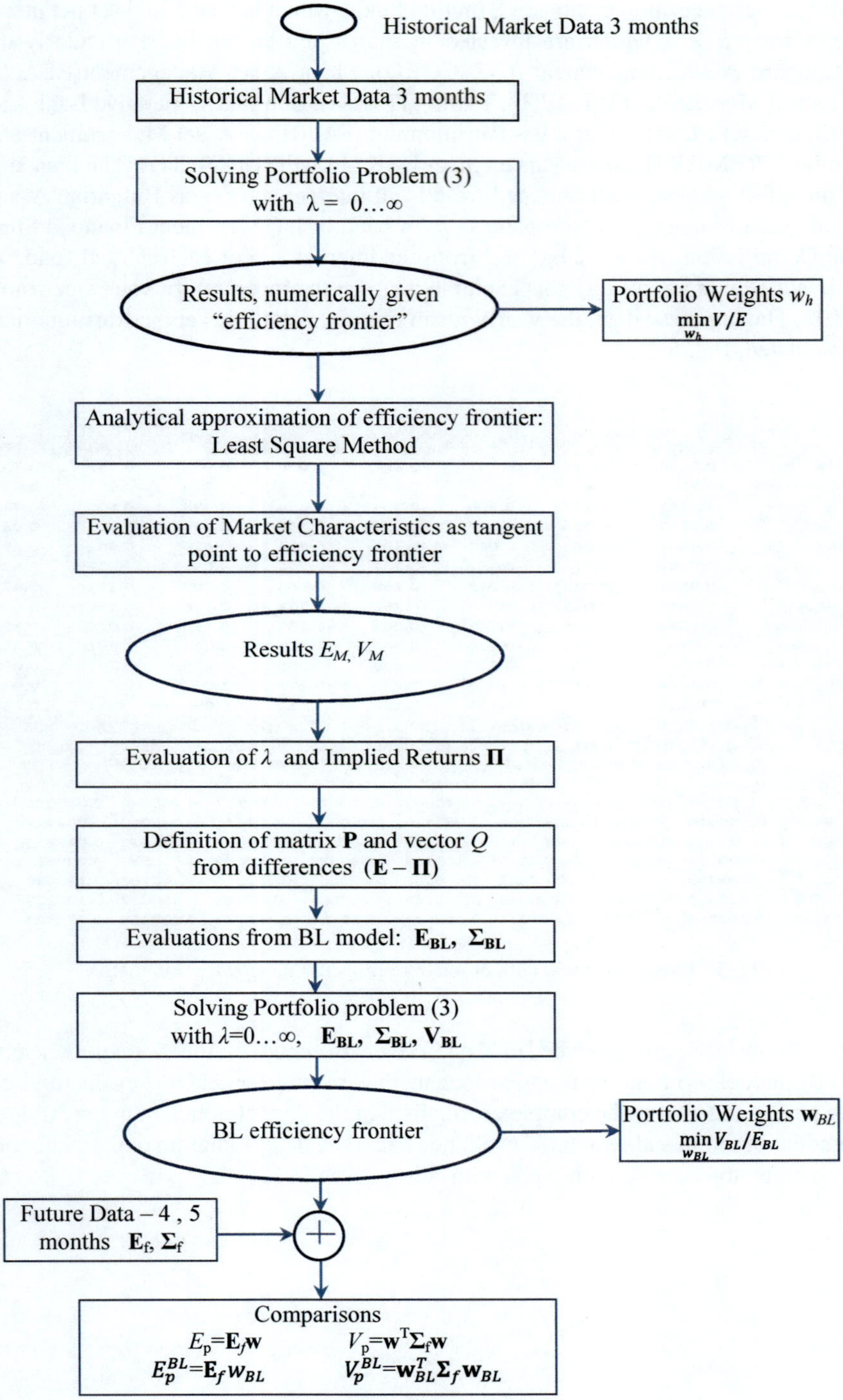

Fig. 6. Block-diagram of Mathlab program suit

For the assessment of the classical portfolio optimization and this one, performed by the BL model it has been used a moving strategy for the portfolio management. The moving strategy applies the procedure to make portfolio calculations for the first 8 months from 2018 as historical period. Then portfolio weights are evaluated twice: using historical data and with the modified data by BL model (11). These two sets of portfolio weights are assumed as investor decisions. The results of these decisions are calculated as portfolio return and risk, but with the mean return and risk of the next month, following the current historical period of 8 months, Fig. 7. Thus, the portfolio characteristics can be compared for efficiency, according to the applied portfolio model: classical, based on historical data or the modified BL one, described here. Then the historical period from 8 months move ahead with one month and the evaluations and comparisons for the next month are performed again, Fig. 7. Thus, for the overall 2018, the 8 months historical period can be moved four times by one month and the corresponding sets of portfolios to be assessed in the plain risk/return.

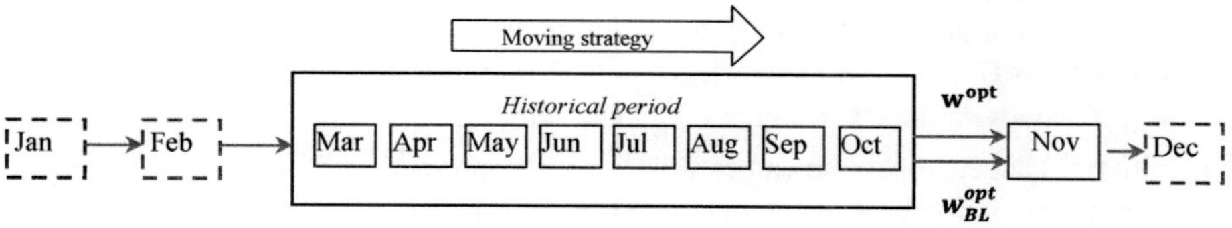

Fig. 7. Moving strategy of portfolio evaluations and assessment

The evaluations start with the historical data from the first eight months, calculating the 8 months historical average vector of returns **E** and the covariance matrix **Σ**. The Matlab functions *mean*() and *cov*() have been used.

The portfolio results are assets weights **w**, obtained by solving the portfolio problem (2), first with historical data **E** and **Σ** and second series of calculations with the BL data $\mathbf{E}_{BL}$ and $\mathbf{\Sigma}_{BL}$, which will give the weights $\mathbf{w}_{BL}$. The weights **w** are found from the efficient frontier evaluated with historical data **E** and **Σ** by solving (2) with a set of values of $\lambda = (0, \infty)$. The Matlab function *quadprog*() is used to solve the set of optimization problems. The optimal portfolio weights $\mathbf{w}^{opt}$ are chosen from the efficient frontier for the portfolio, which characteristics give minimal value of the relation Risk/Return (Information ratio):

$$\mathbf{w}^{opt} = arg\left\{ \min_{w} \left[\frac{\mathbf{w}^T \Sigma \mathbf{w}}{\mathbf{E}^T \mathbf{w}} \right] \right\}.$$

Because the efficient frontier is obtained numerically as a set of points, this research applies its analytical approximation to a quadratic curve. For this case the Matlab function *polyfit*() is used. Having this analytical description of the efficient frontier, the point of market portfolio, return $\mathbf{E_M}$ and risk $\mathbf{V_M}$ are analytically evaluated as a tangent point between the analytical approximation of the "efficient frontier" and Capital Market Line, which pass through the riskless point $(r_f, 0)$. Having values of $\mathbf{E_M}$ and $\mathbf{V_M}$, the implied returns Π are evaluated from (6). The weighted forms of **P** and **Q** are evaluated, according to (8, 9, 10, and 11). The BL asset parameters $\mathbf{E_{BL}}$ and $\mathbf{\Sigma_{BL}}$ are evaluated according to (11). With these new BL assets characteristics a new "efficient frontier" is numerically evaluated solving problem (2) with a set of values for $\lambda = (0,\infty)$. The optimal BL portfolio weights $\mathbf{w_{BL}^{opt}}$ are found, which minimize the Information ratio

$$w_{BL}^{opt} = arg\{ \min_{w} \left[\frac{\mathbf{w^T \Sigma_{BL} w}}{\mathbf{E_{BL}^T w}} \right]$$

These two sets of weights, $\mathbf{w^{opt}}$ and w_{BL}^{opt} are assessed by calculating the portfolio Return and Risk with data of mean return and risk but for the month, following the historical 8-th month period. Thus the investment decision, based on historical 8-th months data and BL model are applied for the next, future period of one month. Then these decisions are assessed with the real data, concerning mean return and risk of this future month. The notation for the assets returns for the future month are the mean values $\mathbf{fE}$ and the covariance matrix $\mathbf{f\Sigma}$. The last is calculated for 8 months period, but taking into consideration the current month with data $\mathbf{fE}$. Comparisons are made between the portfolio returns $\mathbf{fE^T}\, \mathbf{w}^{\mathrm{opt}}$ with historical evaluated weights $\mathbf{w^{opt}}$ and the portfolio return $\mathbf{fE^T}\, \mathbf{w_{BL}^{opt}}$ found with the BL weights $\mathbf{w_{BL}^{opt}}$. The same calculations are performed for the portfolio risks according to the values $\mathbf{w^{optT} f\Sigma w^{opt}}$ and $\mathbf{w_{BL}^{optT}}\; \mathbf{f\Sigma w_{BL}^{opt}}$. The meaning of this comparison is to check if the classical portfolio problem (2) gives better or worst results in comparison with the application of BL model, modified by the weighted form of expert views, derived in this research.

On Fig. 8 a graphical presentation is given for the analytical approximation of the efficient frontier and the tangent line, which passes through the riskless point $(r_f,0)$. The

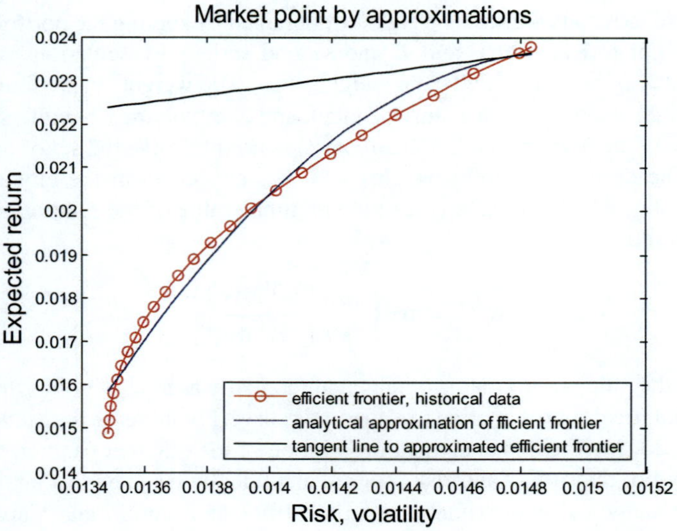

Fig. 8. Evaluation of the market return E_M and risk σ_M^2 as a tangent point

tangent point gives the market characteristics of return E_M and risk (volatility) σ_M^2, The evaluations are performed for value of the riskless asset $r_f = 0.1$

The evaluation of the BL portfolio characteristics $\mathbf{E}_{\text{BL}}$ and $\boldsymbol{\Sigma}_{BL}$ is performed with the coefficient $\tau = 0.5$. The graphical presentation of the efficient frontiers evaluated with historical data **E**, and the BL data $\mathbf{E}_{\text{BL}}$ and $\boldsymbol{\Sigma}_{BL}$ are given on Fig. 9.

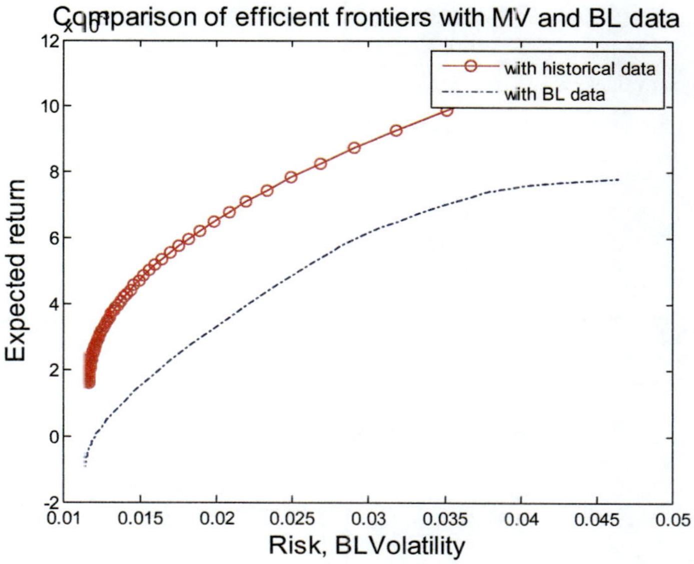

Fig. 9. Comparison of the efficient frontiers with historical (MV) and BL data

Using 8 months as a historical period, applying the evaluated portfolio weights $\mathbf{w}^{\text{opt}}$, $\mathbf{w}_{\text{BL}}^{\text{opt}}$ for the next (future) month and moving the historical period sequentially four times, the resulting portfolio characteristics are given in Table 1. These results assess the benefit of the application of the appropriate portfolio model for the investment.

The portfolio characteristics are presented graphically on the plain Risk/Return, Fig. 10. According to the portfolio theory, the point which is situated farther in Nord-West direction, is the preferable portfolio. The Nord-West situated portfolios have bigger return and less risk in comparison with the portfolios bellow them.

The results on Fig. 10 illustrate that the portfolios, obtained by the BL model are situated on the Nord-West part of the plain in comparison with the classical portfolio model. This give evidences that the modified BL model, given in this research, provides benefits in comparison to the classical portfolio model.

Table 1. Comparisons of the numerically evaluated investment decisions

Classical optimization with historical data	Portfolio Return $\mathbf{fE^T}\ w^{opt}$	Portfolio Risk w^{optT} $\mathbf{f\Sigma w_{BL}^{opt}}$
	0.1059 −0.0262 −0.5245 −1.0220	0.0137 0.0126 0.0672 0.2311
Historical mean values	−0.3667	0.0812
BL optimization	BL portfolio Return $\mathbf{fE^T\ w_{BL}^{opt}}$	BL portfolio Risk $\mathbf{w_{BL}^{optT}}$ $\mathbf{f\Sigma w_{BL}^{opt}}$
	0.1059 −0.027 0.1091 −0.3085	0.0137 0.0465 0.0385 0.0239
BL mean values	−0.0241	0.0307

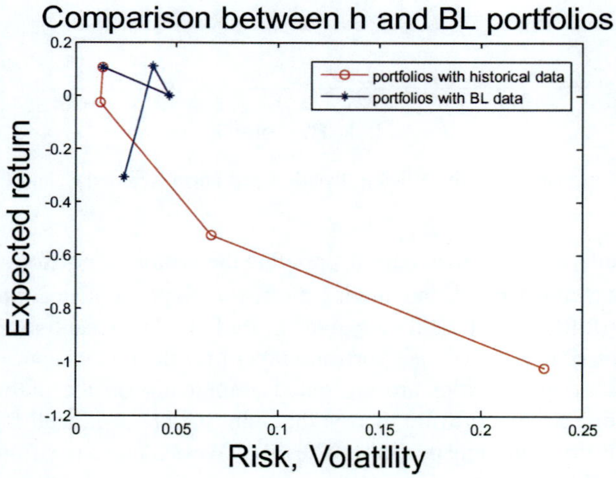

Fig. 10. Graphical presentation of the comparisons between classical and BL model

8 Conclusions

This research makes an important modification of the Black-Litterman model for evaluation of the assets characteristics. The modification is formalized by developing and applying new form of the expert views. Such a modification allows being used additional information, which is not subjectively given by the expert. In this work it was chosen to be used the difference between the historical mean values of the assets returns and their implied returns. Thus, for the definition and modification of the assets characteristics is applied objective data. This new modification allows the BL model to be compared with the classical Markowitz portfolio model. Numerical example, based on real and actual data of the Bulgarian market of Mutual Funds illustrates the benefit of this modification of the BL model in the part of the definition the expert views. Because the evaluations and portfolio analysis insist a lot of calculations and use initial big amount of data, as a mathware has been used Matlab environment in this research. The paper illustrates the utilization of specific Matlab functions, which decrease the need to invest programming efforts for the portfolio optimization. This research deals with portfolio optimization, modeling of portfolio characteristics and provides data driven analysis to define expert views in the BL model.

Acknowledgement. This work has been partly supported by project H12/8, 14.07.2017 of the Bulgarian National Science fund: Integrated bi-level optimization in information service for portfolio optimization, contract ДН12/10, 20.12.2017.

References

1. Allaj, E.: The Black-Litterman model and views from a reverse optimization procedure: An out-of-sample performance evaluation. SSRN Electron. J. (2017). https://doi.org/10.2139/ssrn.2999335. https://www.researchgate.net/profile/Erindi_Allaj/publication/320566797_Slides_Presentation/data/59ef0f9d0f7e9b0f76a981c4/presentation-black.pdf
2. Bertsimas, D., Gupta, V., Paschalidis, I.C.: Inverse optimization: a new perspective on the Black-Litterman model. Oper. Res. **60**(6), 1389–1403 (2012). https://www.ncbi.nlm.nih.gov/pmc/articles/PMC4224190/
3. Black, F., Litterman, R.: Asset Allocation: combining investor views with market equilibrium. J. Fixed Income **1**(2), 7–18 (1991)
4. Calvo, C.I., Liern, V.: On the computation of the efficient frontier of the portfolio selection problem. J. Appl. Math., 25 pages (2012). Article ID 105616. https://doi.org/10.1155/2012/105616. https://www.hindawi.com/journals/jam/2012/105616/
5. Cheung, W.: The Black-Litterman model explained. J. Asset Manage. **11**(4), 229–243 (2011). https://doi.org/10.2139/ssrn.1312664. https://ssrn.com/abstract=1312664
6. Harris R.D.F., Stoja, E., Tan, L.: The dynamic Black–Litterman approach to asset allocation. Eur. J. Oper. Res. **259**, 1085–1096 (2017). https://doi.org/10.1016/j.ejor.2016.11.045. https://www.sciencedirect.com/science/article/pii/S0377221716309705
7. He, G., Litterman, R.: The intuition behind Black-Litterman model portfolios. Technical report, Fixed Income. https://doi.org/10.2139/ssrn.334304
8. Idzorek. T.M.: A step-by-step guide to the Black-Litterman model (2002). https://faculty.fuqua.duke.edu/~charvey/Teaching/BA453_2006/Idzorek_onBL.pdf
9. Markowitz, H.: Portfolio selection. J. Financ. **7**, 77–91 (1952)
10. Merton R.C.: An analytic derivation of the efficient portfolio frontier. J. Financ. Quant. Anal. **7**(4), pp. 1851–1872 (1972). https://doi.org/10.2307/2329621. https://www.jstor.org/stable/2329621?seq=1#page_scan_tab_contents, http://www.stat.ucla.edu/~nchristo/statistics_c183_c283/analytic_derivation_frontier.pdf
11. Michaud R.O., Esch, D.N., Michaud, R.O.: Deconstructing Black-Litterman: how to get the portfolio you already knew you wanted. SSRN Electron. J. **11**(1), 6–20 (2013). https://doi.org/10.2139/ssrn.2641893 https://www.researchgate.net/publication/260362580_Deconstructing_Black-Litterman_How_to_Get_the_Portfolio_You_Already_Knew_You_Wanted

12. Satchell, S., Scowcroft, A.: A demystification of the Black–Litterman model: Managing quantitative and traditional portfolio construction. J. Asset Manage. **1**(2), 138–150 (2000). https://link.springer.com/chapter/10.1007%2F978-3-319-30794-7_3
13. Sharpe, W.: Portfolio Theory and Capital Market, 316 p. McGraw Hill (2000)
14. Vladimirov, M., Stoilov, T., Stoilova, K.: New formal description of expert views of Black-Litterman asset allocation model. J. Cybern. Inf. Technol. **17**(4), 87–98 (2017). ISSN: ISSN Print: 1311-9702, ISSN Online: 1314-408. https://doi.org/10.1515/cait-2017-0043
15. Walters, J.: Reconstructing the Black-Litterman Model (2014). https://doi.org/10.2139/ssrn.2330678. https://ssrn.com/abstract=2330678

Contributed Papers

Large-Scale Molecular Dynamics Simulations on Modular Supercomputer Architecture with Gromacs

S. Markov[1], P. Petkov[1,2(✉)], and V. Pavlov[1]

[1] NCSA, Sofia, Bulgaria
Stoyan.Markov@cern.ch, x.pavlov@gmail.com
[2] University of Sofia, Sofia, Bulgaria
peicho@phys.uni-sofia.bg

Abstract. Modular Supercomputer Architecture is an innovative idea to bring advantages of different HPC architectures currently available together and make it possible for the users to benefit from it by running different parts of their applications on the most appropriate hardware to reach the best performance with high energy efficiency. DEEP-EST project aims not just at building MSA prototype, but rather co-design it by a collaboration of hardware and application software experts. The HPC applications encompass scientific fields like neuroscience, molecular dynamics, radio astronomy, space weather, data analytics in earth science and high energy physics. Here, we present our research on running large-scale molecular dynamics simulations with one the worlds' fastest MD software - Gromacs on two modules (Booster - KNL based HPC system and Cluster - Haswell based HPC system) of JURECA supercomputing system at Juelich Supercomputer Centre. We tested the performance scalability of the simulation of 20 million atoms biomolecular system running Gromacs in Cluster-Booster configuration. The long-range electrostatics calculations were conducted on the JURECA Cluster nodes, while all of the rest calculations - on JURECA Booster. Through source code profiling and Gromacs internal performance counters analysis, we investigated the influence of the performance scalability the different application's parts on the overall application performance scalability. We optimized the job parameters like Cluster to Booster number of cores ration and the corresponding number of module-interconnecting gateways. Our results prove the applicability of the MSA concept in the field of Molecular Dynamics.

1 Introduction

Modular Supercomputer Architecture is an innovative idea to bring advantages of different HPC architectures currently available together and make it possible

P. Petkov - for DEEP-EST Project.

I. Dimov and S. Fidanova (Eds.): HPC 2019, SCI 902, pp. 359–367, 2021.
https://doi.org/10.1007/978-3-030-55347-0_30

for the users to benefit from it by running different parts of their applications on the most appropriate hardware to reach the best performance with high energy efficiency [1,2]. Molecular dynamics (MD) [3] is a theoretical approach widely used in the fields of material sciences, life sciences, chemistry and physics for studies of processes and phenomena on spatial and temporal scales unreachable by experimental techniques nowadays. MD consists of solving the classical equations of motion numerically to calculate the time evolution of a system of particles (atoms, molecules or generalized particle-like objects) and is very computationally demanding.

We report tests of running MD simulations on JURECA supercomputer at JSC mimicking Modular Supercomputing architecture approach. We investigated two cases of running Gromacs [http://www.gromacs.org] in Cluster-Booster configuration: when the Booster module is based on MIC architecture and when Booster nodes are equipped with GPU accelerators.

2 Test Platforms and MD Software

Modular Supercomputing Approach and JURECA system at JSC are well described in [2].

2.1 JURECA Supercomputing System

JURECA BOOSTER has 1640 compute nodes based on Intel Xeon Phi 7250-F - 7250- F, 68 cores, 1.7 GHz, quad mode, 96 GB main memory, + 16 GB MCDRAM. The compute nodes are configured currently in hybrid50 mode. The nodes interconnecting network is of Intel Omni-Path Architecture with Full fat-tree topology and bandwidth of 100 Gb/s. JURECA booster's peak performance is about 5 PF/s. **JURECA CLUSTER** (peak performance 2.2 PF/s) has 1882 compute nodes based on dual-socket Intel Xeon Haswell - 2680v3, 24 cores, 2.5 GHz, connected by Mellanox InfiniBand EDR100 Gb/s network with Full fat-tree topology. Most of the computing nodes have 128 GB main memory, and there are 128 nodes with 256 GB, 64 nodes with 512 GB and 75 nodes equipped with 2×Nvidia K80 GPUs.

JURECA Cluster-Booster Configuration: JURECA Booster and JURECA cluster interconnecting networks connected by 196 bridging nodes each with 1×[EDR InfiniBand HCA] and 1×[Omni-Path Architecture HFI] cards, ensuring common MPI environment.

2.2 Gromacs Software Package

Gromacs [4–9] is one of the fastest molecular dynamics simulators in the world. It is used mainly for soft matter molecular dynamics (MD) simulations with implementation in life sciences. Gromacs tracks the trajectories of a system of particles (atoms) that evolves in time by solving differential equations of motion numerically. The coordinates and velocities of the particles are calculated by using the coordinates and velocities from the previous time step. In each time step, it calculates the forces acting on each atom, which is indeed the most time-consuming operation. From the computational point of view, the force acting on a particle is a result of a summation over:

- pairs of particles connected by "bonds" – bonded interactions,
- pairs of particles satisfying some distance criteria – short-range nonbonded interactions,
- long-range corrections including calculations where data of all the simulated particles are needed – long-range interactions.

The pairs of atoms are defined in a predefined cut-off radius calculating short-range interactions [8], while the long-range interactions are calculated using FFT based algorithm, namely Particle Mesh Ewald (PME) [10].

Gromacs can run on almost every modern computing machine architecture [9]. The simulation program uses multi-level parallelism to utilize the computational power offered. Through domain decomposition, the calculations are spread over distributed memory computational resources – compute nodes. On the domain decomposition level, MPI is used. At MPI rank level, shared memory model is implemented with both OpenMP and GPU accelerators. Finally, SIMD registers are used to vectorize the calculations in each CPU core.

The PME algorithm uses FFT to solve long-range electrostatic contributions to real-space direct Coulomb sums. The specific implementation involving All-to-All MPI communications causes the performance scalability to drop. Overlapping real-space calculations and Fourier-space calculations can solve the latter. In Gromacs the MPI ranks are divided into two groups (see Fig. 1): one for real-space calculations (PP nodes) and the rest being dedicated to PME calculations (PME nodes) [7]. The resulting flowchart (Fig. 3 in [7]) in an MD step can be described in the following manner. Each PP node has a corresponding PME node. At the beginning of the time step, each PP node sends coordinates and charges to its corresponding PME node, and once the PME calculations are completed, each PME node sends the resulting forces back to the corresponding PP node. Meanwhile, all collective communications proceed only between PME nodes as well as only between PP nodes and overlapping the FFT All-to-All communications (exchanged between PME nodes only) with real-space calculations. Consequently, one must optimize the number of PP and PME nodes in such a way that PME nodes need to send the forces they have calculated in the exact moment when the PP nodes need Fourier-space forces, energies and virial.

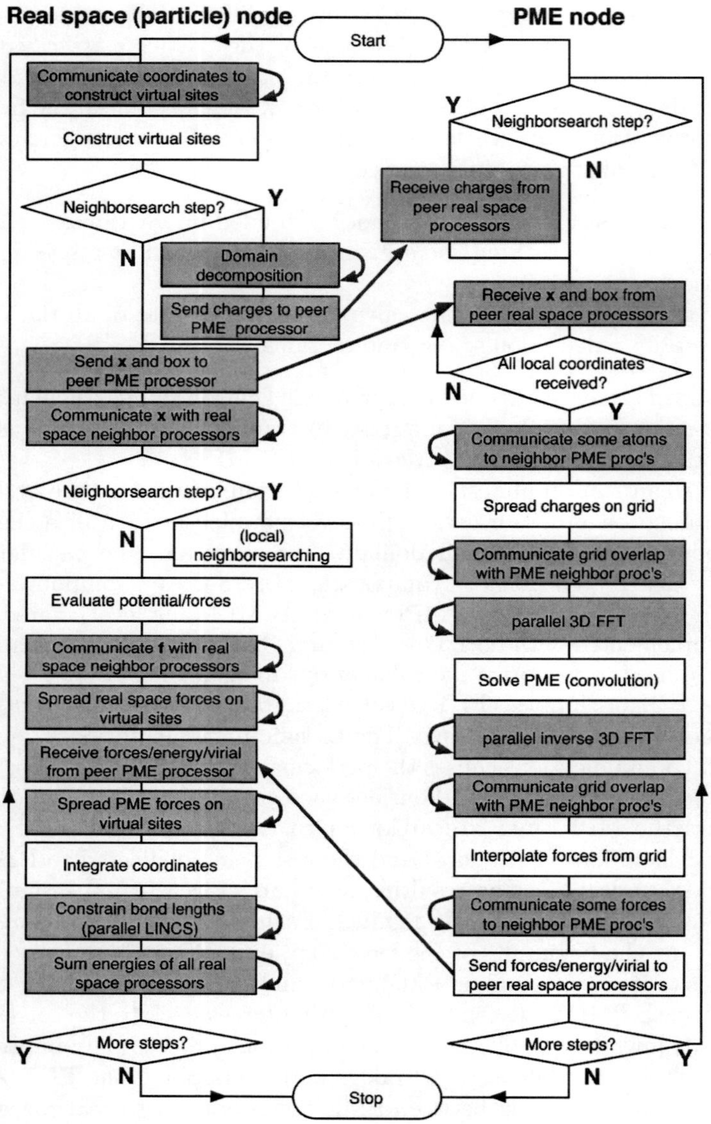

Fig. 1. Flowchart for a typical simulation step for both PP and PME nodes. Shaded boxes involve communication, with gray arrows indicating whether the communication only involves similar types of nodes or synchronization between the two domains [7].

2.3 MSA Mapping

The two groups of nodes introduced above could be mapped to two modules of Modular Supercomputing Architecture (as depicted in Fig. 2):

- The PP nodes, which calculate pair forces over neighbour particles. Gromacs has SIMD kernels for efficient utilization of vector operations offered by the JURECA Booster. MPI parallelization would be used between processes running on CPU cores and groups of them. Finally, OpenMP parallelization could also be used.
- A small number of PME nodes, is preferable in order to reduce All-to-All MPI communications. Therefore, this part of the application should be run on a module which 5 offers relatively higher performance per CPU core, i.e. JURECA Cluster. An optimal number of PME ranks would balance the calculation time of both PP and PME nodes. For this, hybrid MPI/OpenMP parallelization is envisaged.

The hardware resources required to run a specific simulation should guarantee similar computation and communication times within groups of PP and PME nodes as well as between the groups.

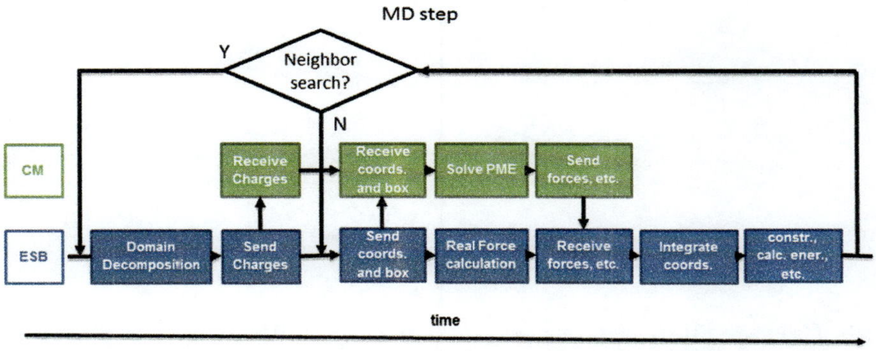

Fig. 2. Mapping the PP and PME nodes to a cluster module (CM) and a booster module (ESB), respectively.

3 Results and Discussion

The results shown below are obtained running medium-size, big and large MD simulations with 325k, 2.2M and 20M atoms (64 times bigger than the first one), respectively. We tested Cluster-Booster configuration in two cases: JURECA BOOSTER-JURECA CLUSTER and JURECA GPUS partition-JURECA CLUSTER. JURECA GPUS partition is a subset on JURECA CLUSTER consisting of nodes with the same architecture but equipped with 4 NVidia K80 GPUs per node.

The Gromacs performance scalability plot for the large MD simulation in Cluster-Booster configuration is shown in Fig. 3. The marker labels represent the number of cluster cores used in the test. As seen in the plot, the slope changes in the middle of the investigated number of booster nodes range. After the inspection of the timing of different application parts vs the number of bridging gateway nodes, we concluded that the reason for the latter is the connecting network bandwidth limitations. Those are due to the insufficient number of gateway nodes that ensure module interconnection.

To investigate the benefits of using Cluster-Booster configuration, we did tests with a smaller system which has performance scalability saturation region starting at 16 nodes on JURECA BOOSTER (blue bars in Fig. 4). We added MPI processes (cluster cores) to the Booster computing resources to run PME part of the application there, as shown in Table 1. The resulting Cluster-Booster performance depicted with red bars in Fig. 4 is better due to the additional computational power for 1 and 4 nodes, but for a bigger number of nodes improves the scalability bringing back the Booster performance to normal by offloading the communication demanding PME part to the Cluster module.

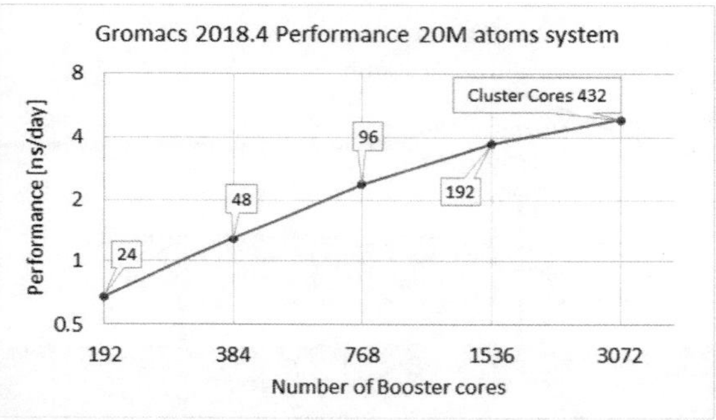

Fig. 3. Gromacs 2018.4 performance vs. the number of JURECA BOOSTER cores. Data points' labels show the number of JURECA Cluster cores used in the simulations for PME calculations. 64 cores per Booster node, 1 MPI process per core and 2 OpenMP threads per MPI process. 24 cores per Cluster node, 1 MPI process per node and 2 OpenMP thread per MPI process.

Table 1. Number of MPI processes per module used in Booster only and Cluster-Booster configuration for data shown on Fig. 3

Number of Booster nodes	Booster only	Cluster-Booster
8	64 MPI KNL (16 PME procs.)	64 MPI KNL + 8 MPI Haswell
16	256 MPI KNL (72 PME procs.)	256 MPI KNL + 48 MPI Haswell
32	2048 MPI KNL (1024 PME procs.)	2048 MPI KNL + 384 MPI Haswell
64	4096 MPI KNL (2048 PME procs.)	4096 MPI KNL + 768 MPI Haswell

We run 2.2M atoms MD simulation on GPUS partition of the JURECA CLUSTER to investigate the case when Booster module is a highly scalable

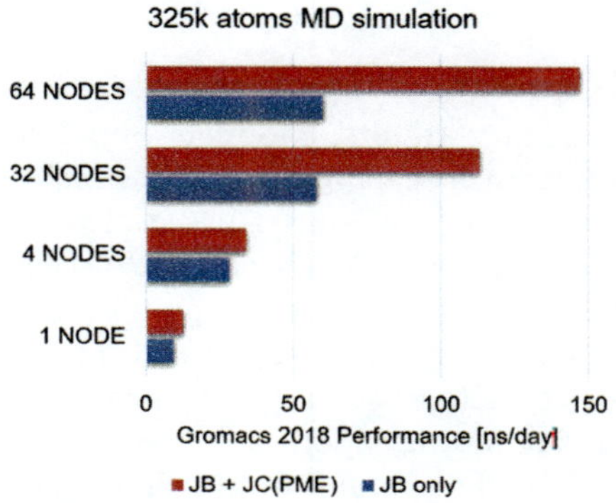

Fig. 4. Gromacs 2018.4 performance vs. the number of JURECA BOOSTER nodes. 64 cores per Booster node, 1 MPI process per core and 2 OpenMP threads per MPI process. 24 cores per Cluster node, 1 MPI process per node and 2 OpenMP thread per MPI process. JURECA BOOSTER only – blue. JURECA BOOSTER + JURECA CLUSTER – red.

GPU-based HPC system, which was very important for the design of the DEEP-EST prototype. The resulting performance is plotted in blue in Fig. 5. The scalability significantly deviates from the ideal above 8 nodes. When we offloaded the PME calculations to additional CPU-only nodes in a similar manner as described above, the performance scalability gets closer to the ideal (red line in Fig. 5).

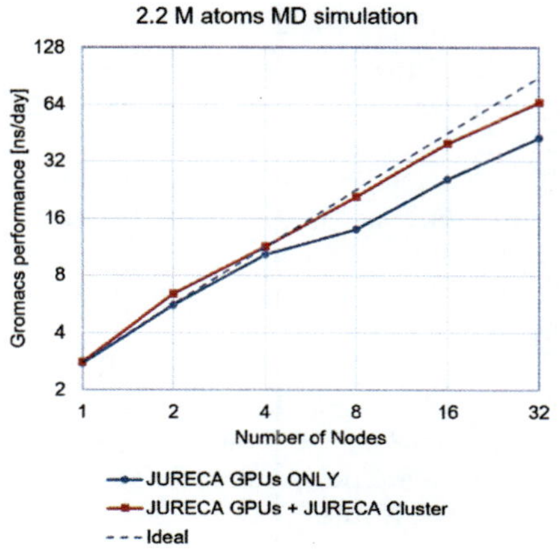

Fig. 5. Gromacs 2019.2 performance vs. the number of JURECA GPU nodes. 4 GPUs per node. GPUS partition only - blue. GPUS + Cluster nodes - red.

4 Conclusion

According to our preliminary results, Modular Supercomputing architecture could improve the performance scalability of MD simulations with long-range electrostatic interactions treated by Fourier-based algorithms. We showed that hybrid HPC systems comprising of MIC- or GPU-accelerated nodes together with high-performance CPU nodes, possess a promising potential for conducting large-scale MD simulations. In such architectures, the accelerated nodes can be beneficially used for the intensive short-range calculations, while the high-performance CPU nodes are best used for the long-range calculations. The latter relies heavily on collective MPI communications. Thus, good CPU-node interconnect network is a necessary condition for good performance scalability. Booster modules usually are designed with energy efficiency better than that of the ordinary clusters. That makes the MD simulations on MSA more energy efficient. The module interconnecting network is crucial for the MD simulations' performance, and the results of the current study are taken into account for the design of the DEEP-EST prototype machine.

Acknowledgements. The DEEP projects have received funding from the European Union's Seventh Framework Programme (FP7) for research, technological development and demonstration and the Horion2020 (H2020) funding framework under grant agreement no. FP7-ICT-287530 (DEEP), FP7-ICT-610476 (DEEP-ER) and H2020-FETHPC-754304 (DEEP-EST).

References

1. Eicker, N., Lippert, T., Moschny, T., Suarez, E. (for the DEEP project): Concurrency and computation: practice and experience **28**(8), 2394 (2016). https://doi.org/10.1002/cpe.3562
2. Suarez, E., Eicker, N., Lippert, T.: Modular Supercomputing Architecture: From Idea to Production; 3rd edn., vol. 3, pp. 223–251. CRC Press, FL, USA (2019). http://juser.fz-juelich.de/record/862856
3. Tuckerman, M.E., Martyna, G.J.: J. Phys. Chem. B **104**(2), 159 (2000). https://doi.org/10.1021/jp992433y
4. Berendsen, H., van der Spoel, D., van Drunen, R.: Comput. Phys. Commun. **91**(1), 43 (1995). https://doi.org/10.1016/0010-4655(95)00042-E. http://www.sciencedirect.com/science/article/pii/001046559500042E
5. Lindahl, E., Hess, B., van der Spoel, D.: Mol. Model. Annu. **7**(8), 306 (2001). https://doi.org/10.1007/s008940100045
6. Van Der Spoel, D., Lindahl, E., Hess, B., Groenhof, G., Mark, A.E., Berendsen, H.J.C.: J. Comput. Chem. **26**(16), 1701 (2005). https://doi.org/10.1002/jcc.20291
7. Hess, B., Kutzner, C., van der Spoel, D., Lindahl, E.: J. Chem. Theor. Comput. **4**(3), 435 (2008). PMID: 26620784. https://doi.org/10.1021/ct700301q
8. Pronk, S., Páll, S., Schulz, R., Larsson, P., Bjelkmar, P., Apostolov, R., Shirts, M.R., Smith, J.C., Kasson, P.M., van der Spoel, D., Hess, B.: Bioinformatics **29**(7), 845 (2013). https://doi.org/10.1093/bioinformatics/btt055

9. Abraham, M.J., Murtola, T., Schulz, R., Páll, S., Smith, J.C., Hess, B., Lindahl, E.: SoftwareX **1–2**, 19 (2015). https://doi.org/10.1016/j.softx.2015.06.001. http://www.sciencedirect.com/science/article/pii/S2352711015000059
10. Essmann, U., Perera, L., Berkowitz, M.L., Darden, T., Lee, H., Pedersen, L.G.: J. Chem. Phys. **103**(19), 8577 (1995). https://doi.org/10.1063/1.470117

Finite Element Approximation for the Sturm-Liouville Problem with Quadratic Eigenvalue Parameter

A. B. Andreev[1,2](✉) and M. R. Racheva[2]

[1] Institute of Information and Communication Technologies, BAS, Sofia, Bulgaria
[2] Technical University of Gabrovo, Gabrovo, Bulgaria
andreev@tugab.bg

Abstract. The eigenvalue problem considered in this paper contains both spectral parameter λ as well as λ^2. A method for linearization is presented and finite element approximation is used. Error estimate of eigenpairs is proved. Computational aspects for more general problems in this type are discussed. Finally, for purpose of illustration numerical implementation is given.

1 Introduction

The paper deals with the Sturm-Liouville problem with quadratic dependence on the spectral parameter [1,6]:

$$-(p(x)y'(x))' + q(x)y(x) - \lambda r(x)y(x) = \lambda^2 y(x), \quad 0 < x < 1, \tag{1}$$

with homogeneous conditions $y(0) = y(1) = 0$ or $y(0) = y'(1) = 0$.

Here, p, q, r are real-valued functions satisfying $p > 0,\ q \geq 0,\ r > 0$ and $q,\ 1/p \in L_1(0,1)$ and $r \in L_\infty(0,1)$.

Moreover, for the eigenvalue problem

$$Ay - \lambda^2 By - \lambda y = 0,$$

where A and B are positive definite compact self-adjoint operators, Turner [8] showed the following: The spectrum consists of two sequences of real eigenvalues, the positive eigenvalues converge to zero and the negative ones tending to minus infinity.

Nonlinear Sturm-Liouville problems are subject of numerous investigations in the theory of hydrodynamics, acoustics, quantum mechanics and other branches of natural sciences (see [1,4]).

Our main goal here is to avoid the presence of λ^2 into the Eq. (1). For this purpose, we introduce a new unknown function in order to obtain a system of

I. Dimov and S. Fidanova (Eds.): HPC 2019, SCI 902, pp. 368–375, 2021.
https://doi.org/10.1007/978-3-030-55347-0_31

two equations in which λ takes part linearly [9]. A similar idea is used by the authors for a fourth-order quadratic eigenvalue problem [2].

Let V be a closed subspace of $H^1(0,1)$, $H_0^1(0,1) \subseteq V \subseteq H^1(0,1)$, $H^m(0,1)$ and $H_0^m(0,1)$ being the usual $m-$th order Sobolev spaces on $(0,1)$ with norm $\|\cdot\|_{m,(0,1)}$. The letter C will denote a generic positive constant which may vary with context.

The Eq. (1) could be written as (we drop the argument x):

$$Ly = \lambda^2 y + \lambda r y, \quad 0 < x < 1, \tag{2}$$

where $Ly = -(py')' + qy$ is a second-order self-adjoint positive definite operator in V.

Multiplying Ly by the function $z(x) \in H_0^1(0,1)$ and integrating by parts, we easily obtain the following bilinear form:

$$\begin{aligned} a(y,z) &= \int_0^1 Lyz\,dx = -py'z\,|_0^1 + \int_0^1 py'z'\,dx + \int_0^1 qyz\,dx \\ &= \int_0^1 (py'z' + qyz)\,dx. \end{aligned}$$

Obviously, the $a-$form is symmetric and coercive on $V \subset H^1(0,1)$, i.e.

$$a(y,z) = a(z,y), \quad \rho_1\|y\|_{1,(0,1)}^2 \leq a(y,y), \quad \forall y,z \in V,$$

for some constant $\rho_1 > 0$.

Moreover, the bilinear form $b(y,z) = \displaystyle\int_0^1 ryz\,dx$ is symmetric and coercive on $L_2(0,1)$, i.e.

$$b(y,z) = b(z,y), \quad \rho_0\|y\|_{0,(0,1)}^2 \leq b(y,y), \quad \forall y,z \in V,$$

for some constant $\rho_0 > 0$.

Thus, from (2) follows a variational formulation of the Sturm-Liouville Problem (1): find a number $\lambda \in R$ and a function $y \in V$, $y \neq 0$ such that

$$a(y,z) = \lambda^2(y,z) + \lambda b(y,z), \quad \forall z \in V. \tag{3}$$

Using the properties of the $a-$ and $b-$form we can conclude that the Problem (3) has a countable set of real eigenvalues λ_i, $i = 1,2,\ldots$ having finite multiplicity. The corresponding eigenfunctions y_i, $i = 1,2,\ldots$ are normalized $\|y_i\|_{0,(0,1)} = 1$.

2 Main Results

To start with, let us introduce the function $\sigma = \lambda y$. So, $\sigma \in V$ and we can write from (2) the following system:

$$\left| \begin{aligned} & L\sigma = \lambda L y \\ & Ly - r\sigma = \lambda \sigma. \end{aligned} \right. \tag{4}$$

In this system the eigenvalue parameter λ explicitly participates linearly.

Lemma 1. *The Sturm-Liouville Problem (3) admits a variational formulation with linear eigenvalue parameter.*

Proof. Consider a couple of functions $(y_1, \sigma_1) \in V \times V$. Multiplying the first equation of (4) by y_1 and the second one by σ_1 and integrating on the interval $(0,1)$ we get:

$$\left| \begin{aligned} & (L\sigma, y_1) = \lambda (Ly, y_1) \\ & (Ly, \sigma_1) - (r\sigma, \sigma_1) = \lambda(\sigma, \sigma_1), \quad \forall (y_1, \sigma_1) \in V \times V). \end{aligned} \right.$$

From here we easily obtain

$$a(\sigma, y_1) + a(y, \sigma_1) - b(\sigma, \sigma_1) = \lambda \left(a(y, y_1) + (\sigma, \sigma_1) \right), \quad \forall (y_1, \sigma_1) \in V \times V. \tag{5}$$

Although this equation is linear with respect to λ, it contains two unknown functions. This disadvantage could be overcome by using that the two unknowns are connected. On the other hand, after discretization of (5) by finite element method (FEM), the corresponding matrix equation allows us to find the first (essential) eigenpairs.

Consider a family of uniform finite element partitions τ_h of the interval $[0,1]$, where h is the length of any subinterval.

With a partition τ_h we associate a finite-dimensional subspace V_h of V such that the restriction of every function $z_h \in V_h$ over every finite element $e \in \tau_h$ is a polynomial of degree n at most.

We assume that V_h verifies the following approximation property for second-order differential equation (see [5]): For every $g \in H^m(0,1)$, $m > 0$, there exists a function $z_h \in V_h$ such that

$$\|g - z_h\|_{s,(0,1)} \leq C h^{\mu} \|g\|_{m,(0,1)}, \quad s = 0; 1,$$

where $\mu = \min\{n + 1 - s, m - s\}$.

Using FEM, we determine the approximate eigencouples $(\lambda_h, (y_h, \sigma_h))$ in the mixed form:

$$\left| \begin{aligned} & a(\sigma_h, y_1) = \lambda_h a(y_h, y_1), \quad \forall y_1 \in V_h, \\ & a(y_h, \sigma_1) - b(\sigma_h, \sigma_1) = \lambda_h (\sigma_h, \sigma_1), \quad \forall \sigma_1 \in V_h. \end{aligned} \right. \tag{6}$$

From Lemma 1 we consider both equations from (6) coupled:

$$a(\sigma_h, y_1) + a(y_h, \sigma_1) - b(\sigma_h, \sigma_1) = \lambda_h \left(a(y_h, y_1) + (\sigma_h, \sigma_1)\right), \quad \forall (y_1, \sigma_1) \in V_h \times V_h. \tag{7}$$

Theorem 1. *Let (λ, y) be an exact eigenpair of (1). Then (λ_h, y_h) is the corresponding approximate eigenpair using finite element space V_h of piecewise polynomials of degree n.*

Supposing also all bilinear forms in the system (6) (or Eq. (7)) are coercive, bounded and continuous. Let the coefficient functions p, q and r be sufficiently smooth such that $y(x) \in H^{n+1}(0,1)$.

Then

$$|\lambda - \lambda_h| \le C(\lambda) h^{2n} \|y\|_{n+1,(0,1)}. \tag{8}$$

Proof. To estimate the error of the approximate eigenpairs $(\lambda_h, (y_h, \sigma_h))$ we consider the associated source and approximate source problems. Let G be a functional space.

Then the associated source problem is: Given $g \in G$, find $(y, \sigma) \in V \times V$ satisfying

$$a(\sigma, y_1) - a(g, y_1) = 0 \quad \forall y_1 \in V,$$
$$a(y, \sigma_1) - b(\sigma, \sigma_1) = (g, \sigma_1) \quad \forall \sigma_1 \in V.$$

Correspondingly, the approximate source problem is: Given $g \in G$, find $(y_h, \sigma_h) \in V_h \times V_h$ satisfying

$$a(\sigma_h, y_1) - a(g, y_1) = 0 \quad \forall y_1 \in V_h,$$
$$a(y_h, \sigma_1) - b(\sigma_h, \sigma_1) = (g, \sigma_1) \quad \forall \sigma_1 \in V_h.$$

So, we can define the corresponding component solution operators:

$$S : G \to V, \quad Sg = \sigma,$$
$$S_h : G \to V_h, \quad S_h g = \sigma_h,$$
$$T : G \to V, \quad Tg = y,$$
$$T_h : G \to V_h, \quad T_h g = y_h.$$

Since the Sturm-Liouville differential operator has compact inverse, our approximation results are based on the properties of compact operators [3].

Let λ be a simple eigenvalue. Consider the solution compact operator and the family of operators $\{T_h\}$ on the functional space G. We assume

$$\|T - T_h\|_{GG} \to 0 \quad \text{as} \quad h \to 0.$$

If $(\lambda, (y, \sigma))$ is an eigenpair, then $\lambda T y = y$.

Now, we apply the approximation properties of T and T_h (see [3], Theorem 7.3 and Theorem 7.4). For any fixed λ, let $\mathcal{P}_h$ be the spectral projection operator. The following estimate is valid (see, e.g. [6,7]):

$$\|y - \mathcal{P}_h y\|_{s,(0,1)} \leq C h^{n+1-s} \lambda^{(n+1)/2}, \quad s = 0; 1. \tag{9}$$

We denote by M the range of the spectral projection associated with T and λ^{-1}. Under the assumptions made in this theorem, there is a constant C such that (see [3], Theorem 11.2):

$$\|y - y_h\|_{1,(0,1)} \leq C \|(T - T_h)_{|M}\|_{GG}.$$

This inequality in conformity with (9) gives order of convergence

$$\|y - y_h\|_{1,(0,1)} = \mathcal{O}(h^n).$$

Finally, in order to prove the estimate (8), we have to use the general result concerning the compact solution operators ([3], Theorem 11.1). More precisely,

$$|\lambda - \lambda_h| \leq \left\{ \|(S - S_h)_{|M}\|^2 + \|(S - S_h)_{|M}\| \; \|(T - T_h)_{|M}\| + \|(T - T_h)_{|M}\|^2 \right\}.$$

Consequently, the order of convergence of $|\lambda - \lambda_h|$ is $\mathcal{O}(h^{2n})$.

Remark 1. If we let

$$A\left((y, \sigma), (y_1, \sigma_1)\right) = a(\sigma, y_1) + a(y, \sigma_1) - b(\sigma, \sigma_1)$$

and

$$B\left((y, \sigma), (y_1, \sigma_1)\right) = a(y, y_1) + (\sigma, \sigma_1),$$

then (5) can be written as

$$(y, \sigma) \in V \times V, \quad (y, \sigma) \neq (0, 0)$$

$$A\left((y, \sigma), (y_1, \sigma_1)\right) = \lambda B\left((y, \sigma), (y_1, \sigma_1)\right).$$

Remark 2. Our method of linearization ensures symmetric presentation with respect to the couples of functions (y, σ) and (y_1, σ_1) (see [2,3,5]). Thus, the spectrum of the Sturm-Liouville problem under consideration is real. The appearance of b–form in $A(\cdot, \cdot)$ makes some of the eigenvalues to be negative.

3 Matrix Representation of the Mixed Variational Equation

Let $\varphi_i(x)$, $i = 1, 2, \ldots, n_1$, $n_1 \in \mathbf{N}$, be the shape functions of the finite element space V_h. Then y_h and σ_h have the representation

$$y_h(x) = \sum_{i=1}^{n_1} y_i \varphi_i(x) \quad \text{and} \quad \sigma_h(x) = \sum_{j=1}^{n_1} \sigma_j \varphi_j(x), \quad x \in [0,1].$$

Finding the solution $(\lambda_h, (y_h, \sigma_h)) \in \mathbf{R} \times V_h \times V_h$ of (7) consists of resulting matrix equation solving.

If we denote $\mathcal{Y} = (y_1, y_2, \ldots, y_{n_1})^T$ and $\mathcal{S} = (\sigma_1, \sigma_2, \ldots, \sigma_{n_1})^T$, then from (7) we obtain matrix equation which could be written either as

$$\begin{pmatrix} -\mathcal{B} & \mathcal{A}^T \\ \mathcal{A} & \mathcal{O} \end{pmatrix} \begin{pmatrix} \mathcal{S} \\ \mathcal{Y} \end{pmatrix} = \lambda \begin{pmatrix} \mathcal{C} & \mathcal{O} \\ \mathcal{O} & \mathcal{A} \end{pmatrix} \begin{pmatrix} \mathcal{S} \\ \mathcal{Y} \end{pmatrix}, \tag{10}$$

or into the form

$$\begin{pmatrix} \mathcal{O} & \mathcal{A} \\ \mathcal{A}^T & -\mathcal{B} \end{pmatrix} \begin{pmatrix} \mathcal{Y} \\ \mathcal{S} \end{pmatrix} = \lambda \begin{pmatrix} \mathcal{A} & \mathcal{O} \\ \mathcal{O} & \mathcal{C} \end{pmatrix} \begin{pmatrix} \mathcal{Y} \\ \mathcal{S} \end{pmatrix}, \tag{11}$$

where the matrices $\mathcal{A}$ and $\mathcal{B}$ correspond to the bilinear forms $a(\cdot,\cdot)$ and $b(\cdot,\cdot)$, respectively, and $\mathcal{C}$ corresponds to the scalar product $(\cdot,\cdot)$.

More precisely,

$$\mathcal{A} = \left(p\ \varphi_i' \varphi_j'\right)_{i,j=1}^{n_1} + \left(q\ \varphi_i \varphi_j\right)_{i,j=1}^{n_1};$$

$$\mathcal{B} = \left(r\ \varphi_i \varphi_j\right)_{i,j=1}^{n_1}; \quad \mathcal{C} = \left(\varphi_i \varphi_j\right)_{i,j=1}^{n_1}$$

and $\mathcal{O}$ is square null matrix with dimensions $n_1 \times n_1$. It is to be noted that the matrices $\mathcal{A}, \mathcal{B}$ and $\mathcal{C}$ are symmetric. Thus the global block matrices in (10) and (11) are symmetric matrices of type $2n_1 \times 2n_1$.

4 Numerical Example

Consider the differential equation

$$-y''(x) - \lambda y(x) = \lambda^2 y(x),$$

$$0 < x < 1,$$

with boundary conditions $y(0) = y(1) = 0$, i.e. into the Eq. (1) we take $p(x) \equiv 1$; $q(x) \equiv 0$; $r(x) \equiv 1$.

This choice of coefficient is not likely to be realistic, but it serves as an confirmation of the proposed method and an illustration how does the method works.

In the case under consideration the exact eigenfunctions are computed as $y(x) = \sin k\pi x,\ k = 1, 2, \ldots$ and the exact eigenvalues are

$$\lambda^{+,-} = \frac{-1 \pm \sqrt{1 + 4k^2\pi^2}}{2},$$

so that

$$\lambda_1^+ = 2.6811326; \qquad \lambda_2^+ = 5.8030483; \qquad \lambda_3^+ = 8.9380316; \ldots$$

$$\lambda_1^- = -3.6811326; \quad \lambda_2^- = -6.8030483; \quad \lambda_3^- = -9.9380316; \ldots$$

The interval $(0, 1)$ is divided into N subintervals, $N = 8;\ 16;\ 32;\ 64;\ 128;\ 256;\ 512$, so that the mesh parameter h is equal to $1/N$.

We solve the Eq. (7) using the matrix representation (10) by means of linear finite elements. The motivation of this choice is to illustrate that on the base of the proposed method eigenvalue problem in which the eigenvalue parameter λ appears quadratically as well as linearly could be successfully solved even if simplest finite elements are used.

In Table 1 approximations of $\lambda_1^{\pm}$, $\lambda_2^{\pm}$ and λ_3^+ by means of the proposed method solving are presented. Also, the trends are given and, as it is seen, they depend on the sign of the approximated eigenvalue. It is well-known that standard conforming finite element methods give approximations of the eigenvalues from above, while for any finite element method of mixed type this is not clear in general. From Table 1 it is seen that for the concrete numerical example under consideration the demonstrated method in fact approximates the absolute values of the exact eigenvalues assymptotically from above.

The numerical results confirm the error estimate from Theorem 1 and illustrate that the method works even in case of course meshes.

Table 1. Approximations $\lambda_{1,h}^+, \lambda_{1,h}^-, \lambda_{2,h}^+, \lambda_{2,h}^-, \lambda_{3,h}^+$ of the exact eigenvalues $\lambda_1^+, \lambda_1^-, \lambda_2^+, \lambda_2^-, \lambda_3^+$ obtained after finite element implementation of the proposed method by means of linear FEs

N	$\lambda_{1,h}^+$	$\lambda_{1,h}^-$	$\lambda_{2,h}^+$	$\lambda_{2,h}^-$	$\lambda_{3,h}^+$
8	2.7011062	−3.7011062	5.9650265	−6.9650265	9.4868931
16	2.6861189	−3.6823787	5.8433684	−6.8433675	9.0345715
32	2.6823787	−3.6823787	5.8131145	−6.8131145	8.9682250
64	2.6814441	−3.6814441	5.8055640	−6.8055639	8.9465378
128	2.6812104	−3.6812104	5.8036771	−6.8036771	8.9401578
256	2.6811520	−3.6811520	5.8032055	−6.8032055	8.9385631
512	2.6811374	−3.6811374	5.8030876	−6.8030876	8.9380186
Trend	↘	↗	↘	↗	↘
Exact	2.6811326	−3.6811326	5.8030483	−6.8030483	8.9380316

5 Conclusion

In this work a new approach for numerical solving of Sturm-Liouville problem in which eigenvalue parameter appears quadratically as well as linearly is presented. This method is analyzed and an error estimate is proved.

References

1. Al-Gwaiz, M.A.: Sturm-Liouville Theory and Its Applications, vol. 7. Springer, Berlin (2008)
2. Andreev, A.B., Racheva, M.R.: A method for linearization of a beam problem. In: International Conference on Numerical Analysis and Its Applications, pp. 180–186. Springer, Cham (2016)
3. Babuška, I., Osborn, J.: Eigenvalue problems. In: Lions, P.G., Ciarlet P.G. (eds.) Handbook of Numerical Analysis, vol. II, Finite Element Methods (Part 1), North-Holland, Amsterdam, pp. 641–787 (1991)
4. Barkwell, L., Lancaster, P., Markus, A.S.: Gyroscopically stabilized systems: a class of quadratic eigenvalue problems with real spectrum. Can. J. Math. **44**(1), 42–53 (1991)
5. Brenner, S., Scott, R.: The Mathematical Theory of Finite Element Methods, vol. 15. Springer Science & Business Media (2007)
6. Solov'ev, S.I.: Approximation of differential eigenvalue problems with a nonlinear dependence on the parameter. Differ. Equ. **50**(7), 947–954 (2014)
7. Solov'ev, S.I.: Finite element approximation with numerical integration for differential eigenvalue problems. Appl. Numer. Math. **93**, 206–214 (2015)
8. Turner, R.E.: Some variational principles for a nonlinear eigenvalue problem. J. Math. Anal. Appl. **17**(1), 151–160 (1967)
9. Weinberger, H.F.: On a nonlinear eigenvalue problem. J. Math. Anal. Appl. **21**, 506–509 (1968)

A Survey of Optimal Control Problems for PDEs

Owe Axelsson[1,2](✉)

[1] Institute of Geonics, The Czech Academy of Sciences, Ostrava, Czech Republic
[2] Department of Information Technology, Uppsala University, Uppsala, Sweden
owe.axelsson@it.uu.se

Abstract. Optimal control problems constrained by partial differential equations arise in a multitude of important applications, such as in Engineering, Medical and Financial research. They arise also in microstructure analyses.

They lead mostly to the solution of very large scale algebraic systems to be solved. It is then important to formulate the problems so that these systems can be solved fast and robustly, which requires construction of a very efficient preconditioner. Furthermore the acceleration method should be both efficient and cheap, which requires that sharp and tight eigenvalue bounds for the preconditioned matrix are available. Some types of optimal control problems where the above hold, are presented.

With the use of a proper iterative aceleration method which is inner product free, the methods can be used efficiently on both homogeneous and heterogeneous computer architectures, enabling very fast solutions.

1 Introduction

As is widely accepted, analyses and solutions of partial differential equations are merely just part of a more general solution process that includes some kind of optimization and sensitivity analyses where the PDE equation acts as a constraint. For example, one may want to control an equipment to have a, as close as possible desired, i.e. target behaviour. In other applications one must identify some coefficient, i.e. material properties or boundary values at an inaccessible part of the boundary of the domain, which is important to enable to control if various safety requirements are satisfied.

Such problems lead to an optimal control framework and are of inverse type, i.e. ill-posed and must be regularized. This is normally done by a Tichonov regularization, where a control cost term is added to the functional to be minimized. The function is a Lagrange functional involving the state solution, the control variable and an adjoint variable that acts as multiplier function for the PDE constraints, with the control variable acting as (part of the) source function.

The minimization leads to a three-by-three block system, which is discretized, normally by use of a finite element method. For some basic types of problems,

I. Dimov and S. Fidanova (Eds.): HPC 2019, SCI 902, pp. 376–390, 2021.
https://doi.org/10.1007/978-3-030-55347-0_32

this system can be reduced to a two-by-two block matrix system for which a very efficient preconditioner can be readily constructed. More involved problems require however more specialized treatments. As iterative acceleration method one can use a conjugate gradient method or, more generally, a minimal residual generalized conjugate gradient method, preferably modified for use of variable preconditioners, see e.g. [1,2] or a flexible method, see [3,4].

Such a flexible or variable version is advisable to use since it enables use of variable tolerances in the solution of the inner systems that arise in the application of the preconditoiner, which can save much computations if used efficiently.

In many problems eigenvalue bounds for the eigenvalues of the preconditoned matrix are known which enables use of an inner product free acceleration method, such as the second order Chebyshev iteration method, which is computationally stable.

Contrary to Krylov subspace methods here there is no need to orthogonalize the search vectors and since the condition number is known, an upper bound of the rate of convergence is also known, which implies that it is not even necessary to compute the, say L_2-norm, of the current residual, implying that the method is fully inner product free. This saves computations and limits spread of rounding errors and saves computational time. This occurs in particular when the method is implemented on a parallel computer where communication costs of data is expensive, but which is avoided since there are no parts of inner products to be communicated between the computer processors. It is also shortly shown how one can solve the arising inner linear systems efficiently on parallel computers using a special version of a domain decomposition method.

After a presentation of the above basic principles for optimal control, OPT-PDE problems and basic problem types, some more involved problems, such as with bound constrained controls and boundary control inverse problems, will be commented on. We show also how it can be efficient to use an exact factorization of the arising block matrices. As mentioned above the use of a Chebyshev iteration method enables one to avoid computations of costly inner products. It is also shown that by use of a polynomial preconditioner this method can be applied efficiently even for problems with outlier eigenvalues.

2 Basic Applications of OPT-PDE

2.1 Distributed Control

For basic optimal control problems one can use a very efficient preconditioner, named preconditioned square block (PRESB) method, see [5–9]. This method arose as a simple method to avoid complex arithmetics when solving complex valued systems, see [10]. To illustrate this, consider

$$(A + iB)(x + iy) = f + ig,$$

where A, B etc are real valued where we assume that $A + B$ is nonsingular. It can be rewritten in real valued form

$$\begin{bmatrix} A & -B \\ B & A \end{bmatrix} \begin{bmatrix} x \\ y \end{bmatrix} = \begin{bmatrix} f \\ g \end{bmatrix}.$$

We show that it can be solved easily and efficiently by the use of a PRESB preconditioned iteration method. This will be shown for more general problems of type

$$\mathscr{A}\begin{bmatrix} x \\ y \end{bmatrix} = \begin{bmatrix} A & -B^* \\ B & A \end{bmatrix}\begin{bmatrix} x \\ y \end{bmatrix} = \begin{bmatrix} f \\ g \end{bmatrix}$$

where B^* denotes the complex conjugate of B. We assume that A is symmetric and positive semidefinite, that $B + B^*$ is positive definite and that $\mathscr{N}(A) \cap \mathscr{N}(B + B^*) = \{0\}$.

Proposition 1. *Under the above conditions, $\mathscr{A}$ is nonsingular.*

Proof. If

$$\mathscr{A}\begin{bmatrix} x \\ y \end{bmatrix} = \begin{bmatrix} 0 \\ 0 \end{bmatrix},$$

then $Ax = B^*y$ and $Bx + Ay = 0$. Hence $x^*Ax + y^*Ay = 0$, that is, $x = 0$, $y = 0$, that is, since A is semidefinite, x, y belongs to $\mathscr{N}(A)$, which implies that x, y belongs to $\mathscr{N}(B)$, that is, $x = y = 0$. □

As preconditoner to $\mathscr{A}$ we take the PRESB matrix,

$$\mathscr{B} = \begin{bmatrix} A & -B^* \\ B & A + B + B^* \end{bmatrix}.$$

It is readily seen that $\mathscr{B}$ can be factorized as

$$\begin{bmatrix} I & -I \\ 0 & I \end{bmatrix}\begin{bmatrix} A + B & 0 \\ B & A + B^* \end{bmatrix}\begin{bmatrix} I & I \\ 0 & I \end{bmatrix}, \tag{1}$$

which shows that an action of $\mathscr{B}^{-1}$, besides a matrix vector multiplication with B and some vector additions, involves solving a linear system with matrix $A+B$ and with $A + B^*$.

In many problems there exists efficient solution methods for such systems, such as based on algebraic multigrid [11,12], or modified incomplete factorization, [13] or for very large problems, use of a domain decomposition method, such as in [14].

To find the eigenvalues λ of the preconditioned matrix $\mathscr{B}^{-1}\mathscr{A}$ we consider the eigenvalue problem

$$(1 - \lambda)\mathscr{B}\begin{bmatrix} \xi \\ \eta \end{bmatrix} = (\mathscr{B} - \mathscr{A})\begin{bmatrix} \xi \\ \eta \end{bmatrix} = \begin{bmatrix} 0 \\ (B + B^*)\eta \end{bmatrix}. \tag{2}$$

Proposition 2. *Under the stated conditions, the eigenvalues λ of $\mathscr{B}^{-1}\mathscr{A}$ are contained in the interval $\frac{1}{2} \le 1 - \varrho(D_0) \le \lambda \le 1$, where*

$$D_0 = ((A + B)A^{\dagger}(A + B^*))^{-1}(B + B^*).$$

If $B^* = B$ and B is spsd, then

$$\frac{1}{2} \le \lambda(\mathscr{B}^{-1}\mathscr{A}) \le \frac{1}{2}\left(1 + \max_{\mu(A+B)^{-1}B}(1 - 2\mu)^2\right) \le 1.$$

Proof. Clearly $\lambda = 1$ if and only if $\eta \in \mathcal{N}(B+B^*)$, any ξ. To find the eigenvalues $\lambda \neq 1$ it is convenient to first use a congruence transformation by $\begin{bmatrix} A^{-1/2} & 0 \\ 0 & A^{-1/2} \end{bmatrix}$ on both sides of (2). We assume then that A is perturbed by εI, $\varepsilon > 0$ to make A positive definite or we can use a generalized inverse of A. At the end of the derivation we will see that the eigenvalue bounds hold uniformly with respect to ε so we can let $\varepsilon \to 0$. From (2) follows then

$$(1-\lambda)\widehat{\mathscr{B}} \begin{bmatrix} \hat{\xi} \\ \hat{\eta} \end{bmatrix} = (1-\lambda) \begin{bmatrix} I & \widehat{B}^* \\ \widehat{B} & I + \widehat{B} + \widehat{B}^* \end{bmatrix} \begin{bmatrix} \hat{\xi} \\ \hat{\eta} \end{bmatrix} = \begin{bmatrix} 0 \\ (\widehat{B} + \widehat{B}^*)\hat{\eta} \end{bmatrix},$$

where $\widehat{B} = A^{-1/2} B A^{-1/2}$, $\hat{\xi} = A^{1/2}\xi$, $\hat{\eta} = A^{1/2}\eta$.

It follows from (1) that

$$\widehat{\mathscr{B}}^{-1} = \begin{bmatrix} I & -I \\ 0 & I \end{bmatrix} \begin{bmatrix} I + \widehat{B} & 0 \\ \widehat{B} & I + \widehat{B}^* \end{bmatrix}^{-1} \begin{bmatrix} I & I \\ 0 & I \end{bmatrix},$$

so it holds

$$\widehat{\mathscr{B}}^{-1} \begin{bmatrix} 0 & 0 \\ 0 & \widehat{B} + \widehat{B}^* \end{bmatrix} = \begin{bmatrix} I & -I \\ 0 & I \end{bmatrix} \begin{bmatrix} (I + \widehat{B})^{-1}(\widehat{B} + \widehat{B}^*) \\ (I + \widehat{B}^*)^{-1}(I + \widehat{B})^{-1}(\widehat{B} + \widehat{B}^*) \end{bmatrix} =$$

$$= \begin{bmatrix} 0 & B^*(I + \widehat{B})^{-1}(\widehat{B} + \widehat{B}^*) \\ 0 & ((I + \widehat{B})(I + \widehat{B}^*))^{-1}(\widehat{B} + \widehat{B}^*) \end{bmatrix}.$$

Hence the eigenvalues $1 - \lambda$ equal the eigenvalues of matrix

$$D_0 = ((I + \widehat{B})(I + \widehat{B}^*))^{-1}(\widehat{B} + \widehat{B}^*). \tag{3}$$

Since

$$0 \leq (I - \widehat{B})(I - \widehat{B}^*) = I + \widehat{B}\widehat{B}^* - (\widehat{B} + \widehat{B}^*)$$

it follows that $0 \leq \widehat{B} + \widehat{B}^* \leq I + \widehat{B}\widehat{B}^*$, so $0 \leq 1 - \lambda \leq \frac{1}{2}$, that is $\frac{1}{2} \leq \lambda \leq 1$. More precisely, $\lambda \geq 1 - \varrho(D_0)$, where $\varrho(D_0) \leq \frac{1}{2}$.

If $B^* = B$, then it follows from (3) that

$$(1 - \lambda)(\mathscr{B}^{-1}\mathscr{A})\eta = ((A + B)A^{-1}(A + B))^{-1} 2B\eta.$$

If $\mu(A + B)\xi = B\xi$, then with

$$(A+B)A^{-1}(A+B) = (A+B)(A+B-B)^{-1}(A+B) = (A+B)(I-(A+B)^{-1}B)^{-1}$$

and

$$((A + B)A^{-1}(A + B))^{-1} 2B\xi = 2(I - \mu)\mu\xi = 2\left(\frac{1}{2} + \frac{1}{2} - \mu\right)\left(\frac{1}{2} - \left(\frac{1}{2} - \mu\right)\right)\xi$$

$$= 2\left(\frac{1}{4} - \left(\frac{1}{2} - \mu\right)^2\right)\xi = \frac{1}{2}(1 - (1 - 2\mu)^2)\xi$$

so

$$\lambda(\mathscr{B}^{-1}\mathscr{A}) \geq \frac{1}{2}(1 + (1 - 2\mu)^2).$$

□

Remark 1. If $0 < \mu < \frac{1}{2}$, then the lower eigenvalue bound, $\lambda(\mathscr{B}^{-1}\mathscr{A}) > \frac{1}{2}$.

These eigenvalues bounds hold uniformly with respect to ε so we take $\varepsilon = 0$.

Corollary 1. *Let A be spd, $A+B$ nonsingular and assume that $Re(\mu) \geq 0$ where $\mu Ax = Bx$, $\|x\| \neq 0$. Then the eigenvalues of $\mathscr{B}^{-1}\mathscr{A}$ satisfy $1 \geq \lambda \geq \frac{1}{1+\alpha}$, where $\alpha = \max\limits_{\mu} \frac{Re(\mu)}{1+|\mu|^2}$.*

Proof. From (3) follows $1 - \lambda = \frac{2\mathrm{Re}(\mu)}{1+|\mu|^2+2\mathrm{Re}(\mu)}$. Hence $\lambda = \frac{1+|\mu|^2}{1+|\mu|^2+2\mathrm{Re}(\mu)} \geq \frac{1}{1+\alpha}$. □

Corollary 2. *The preconditioned matrix $\mathscr{B}^{-1}\mathscr{A}$ has the block matrix form*

$$\mathscr{B}^{-1}\mathscr{A} = \begin{bmatrix} I & E \\ 0 & I + D_0 \end{bmatrix},$$

where D_0 is defined in (3). Hence as D_0, it has a full eigenvector space, that is, is normal and can be transformed to a block diagonal matrix. It $\mathscr{A}$ has order $n \times n$, the number of unit eigenvalues equal $n + n_0$, where n_0 is the dimension of the null space of D_0.

Proof. Since D_0 has a complete eigenvector space, this follows directly from the proof of Proposition 2. □

It follows that the preconditioned iteration method converges fast and, since the eigenvalue bounds are known, it is even efficient to apply the Chebyshev iteration method, see further next section. As follows from [13–15], the rate of convergence factor is the bounded above by $\frac{\sqrt{2}-1}{\sqrt{2}+1} = \frac{1}{3+2\sqrt{2}} \approx \frac{1}{6}$.

As we shall see in Sect. 3, for time-harmonic eddy-current problems, the ratio $2\mathrm{Re}(\mu)/(1 + |\mu|^2)$ becomes very small for large values of the frequency ω, i.e. where $|\mu|$ is large, which implies that the eigenvalues cluster at unity, and implies a superlinear rate of convergence.

2.2 Control Restricted to a Subdomain

Consider now a problem where the control is restricted to a subdomain, see Sect. 4 for further details. In such a case the first order necessary conditions lead to a matrix of the form

$$\mathscr{A} = \begin{bmatrix} A & -B^* \\ B & A_1 \end{bmatrix},$$

where A and A_1 are symmetric. A is positive definite but A_1 is singular and

$$x^T A_1 x \leq x^T A x.$$

In typical applications, A is a mass matrix for the whole domain of definition and A_1 is a mass matrix restricted to a subdomain. If

$$\mathscr{A}\begin{bmatrix} x \\ y \end{bmatrix} = \begin{bmatrix} 0 \\ 0 \end{bmatrix},$$

then $Ax = B^*y$ i.e. $x = A^{-1}B^*y$ and $Bx + A_1y = 0$ implies that

$$y^*BA^{-1}B^*y + y^TA_1y = 0.$$

It follows that if $y \in \mathscr{N}(A_1)$ then also $y \in \mathscr{N}(BA^{-1}B^*)$, that is, $y \in \mathscr{N}(B)$. Hence, to get a nonsingular matrix $\mathscr{A}$ we must assume that $\mathscr{N}(A_1) \cap \mathscr{N}(B) = \{0\}$. Even if A_1 is nonsingular, the matrix $\mathscr{A}$ can be ill-conditioned and it is advisable to regularized. Therefore we consider $\widetilde{\mathscr{A}} = \begin{bmatrix} A & -B^* \\ B & A_1 + \delta A \end{bmatrix}$, where $\delta > 0$, but must be small. The preconditioner

$$\mathscr{B} = \begin{bmatrix} A & -B^* \\ B & A + B + B^* \end{bmatrix}$$

leads then to the generalized eigenvalue problem,

$$(1-\lambda)\mathscr{B}\begin{bmatrix} x \\ y \end{bmatrix} = \begin{bmatrix} B & -A \end{bmatrix}\begin{bmatrix} x \\ y \end{bmatrix} = \begin{bmatrix} 0 \\ ((1-\delta)A - A_1 + B + B^*)y \end{bmatrix}.$$

Here $\lambda = 1$ if and only if $y \in \mathscr{N}((1-\delta)A - A_1 + B + B^*)$, any x. For $\lambda \neq 1$ it holds $Ax = B^*y$, i.e. $x = A^{-1}B^*y$ and

$$(1-\lambda)[(BA^{-1}B^* + A + B + B^*)y] = ((1-\delta)A - A_1 + B + B^*)y.$$

Hence

$$\begin{aligned} &y^*(A + BA^{-1}B^* + B + B^*)y = \\ &y^*((1-\delta)A + B + B^* - A_1)y + +\lambda y^*(A + BA^{-1}B^* + B + B^*)y, \end{aligned}$$

that is,

$$y^*(\delta A + A_1 + BA^{-1}B^*)y = \lambda y^*(A + BA^{-1}B^* + B + B^*)y.$$

Here we use a congruence transformation on both sides with $A^{-1/2}$ to get

$$\tilde{y}^*(\delta I + \widehat{A}_1 + \widehat{B}\widehat{B}^*)\tilde{y} = \lambda\tilde{y}^*(I + \widehat{B}\widehat{B}^* + \widehat{B} + \widehat{B}^*)\tilde{y},$$

where $\widehat{B} = A^{-1/2}BA^{-1/2}$, $\widehat{A}_1 = A^{-1/2}A_1A^{-1/2}$, $\tilde{y} = A^{1/2}y$.

Let, $\gamma = y^*\widehat{B}y = y^*\widehat{B}^*y$, $\|y\| = 1$. It follows then

$$\frac{\delta}{(1+\delta)^2} \le \min_{\delta} \frac{\delta + |\gamma|^2}{(1+\gamma)(1+\gamma^*)} \le \lambda = \frac{\tilde{y}(\delta I + \widehat{A}_1 + \widehat{B}\widehat{B}^*)\tilde{y}}{\tilde{y}^*((I+\widehat{B})(I+\widehat{B}^*))\tilde{y}} \le 1 + \delta,$$

where the minimal value is taken for $\gamma = \delta$ and maximal value taken for $\widehat{A}_1 = I$.

Hence the eigenvalues are contained in the interval $\left[\frac{\delta}{(1+\delta)^2}, 1+\delta\right]$. Note that δ must be small in order not to permute the solution too much. Since the spectral condition number increases as $O(\delta^{-1})$ this unfortunately means that there will occur more outer iterations.

2.3 Eigenvalue Bounds for OPT-PDE Problem with an Indefinite Pivot Block Matrix

As has been shown in [8,16], it is possible to extend the above preconditioning method and the Chebyshev iteration method to cases where the pivot block matrix A is indefinite.

When it is negative definite it is then assumed that the spectral radius, $\varrho((I+\widehat{B}\widehat{B}^*)^{-1}(\widehat{B}+\widehat{B}^*)) \leq \gamma < 1$, where $\widehat{B} = (-A)^{-1/2}B(-A)^{-1/2}$. As shown in [16], the preconditioned matrix has then eigenvalues in the interval $[1, 1/(1-\gamma)]$.

3 HPC Efficient Inner Product Free Iteration Methods for Parallel Computers

Usually the preconditioned basic iteration method is accelerated by some Krylov subspace iteration method. To orthogonalize search direction vectors such methods are based on orthogonal or oblique projections, which imply heavy use of inner products of globally defined vectors.

This can be a bottleneck when the methods are implemented on parallel computers, because all locally computed parts of an inner product on the processors must be communicated to all other processors which needs much global communication time and synchronization time. Hence for a high performance computation it is important to use an inner product free acceleration method. A very efficient such method is based on the second order Chebyshev iteration method.

However, this method requires a priori knowledge of lower and upper eigenvalue bounds. There can be severe penalties for over-or-under estimating these bounds. However, as we have seen in the previous section, there exists important applications where such bounds, and even sharp bounds are given. In the next section we shall also see that even if the preconditioned matrix is not block diagonal, it may be diagonalizable and is hence a normal matrix, for which convergence is fully determined by the individual eigenvalues and the corresponding minimal polynomial. It also leads to a parameter independent convergence rate.

At each iteration typically a linear system $A+B$ and $A+B^*$ must be solved. In many problems A is a mass matrix and B is multiplied by a small factor, the square root of the control cost regularization parameter.

If the systems are complex valued we rewrite them on real valued form which leads to solving $M+K$ and $M+K^T$. When they are of large scale it is advisable to implement the method on a parallel computer and use a domain decomposition method, see e.g. [17]. Recently a very efficient domain decomposition method has been devised, see [13]. It is based on dividing the given domain in layers and solving double pair neighboring layers in parallel, say for even ordered followed by odd ordered domains, for which the Schwarz iteration method is used. As is wellknown, such Schwarz methods do not approximate smother eigenvalue nodes of the solution and its takes typically m^2 iterations where m is the number of layers, for the method to converge.

Therefore it must be stabilized, that is, after each, say second iteration a stabilization method based on the coarse mesh, or as has been shown in [13], a coarse-fine mesh method is applied. This leads to very few iterations. Clearly each subdomain system can also be solved in parallel. For three space dimensional problems, it is advisable to use a doubly layered subdomain division, that is, each 3D subdomain is itself solved by a Schwarz stabilized domain decomposition method. For numerical illustrations for 2D problems, see [13].

4 Optimal Control Problems

4.1 A Basic OPT-PDE Problem

Consider the optimal control problem with a distributed control, that is minimize

$$J(u, v) = \frac{1}{2}\|u - u_d\|_{\Omega}^2 + \frac{1}{2}\beta\|v\|_{\Omega}^2,$$

where u is the state solution, u_d is its target solution, v is the control variable and $\beta > 0$ is a control cost regularization parameter. This is subject to the constraint,

$$\mathscr{L}u = f + v,$$

where $\mathscr{L}$ is a given elliptic operator and f is a given source function. The corresponding Lagrange functional takes then the form

$$F(u, v, w) = J(u, v) + \int_{\Omega} w(\mathscr{L}u - f - v)$$

where w is the Lagrange multiplier, i.e. the adjoint variable corresponding to the constraint. This is rewritten by partial integration in a weak variational form, in this way giving corresponding equations for both the state and the adjoint variables.

If $\beta = 0$, this is an ill-posed problem, but for $\beta > 0$ there exists a unique solution of the first order necessary conditions, $\nabla F(u, v, w) = 0$, that is, after discretization,

$$\begin{bmatrix} M & 0 & K^* \\ 0 & \beta M & -M \\ K & -M & 0 \end{bmatrix} \begin{bmatrix} \mathbf{u} \\ \mathbf{v} \\ \mathbf{w} \end{bmatrix} = \begin{bmatrix} \mathbf{f} + M\mathbf{u}_d \\ 0 \\ 0 \end{bmatrix}.$$

where M is the mass matrix corresponding to the whole domain and K corresponds to the given elliptic operator $\mathscr{L}$. Since $v = \frac{1}{\beta}w$, it can be reduced to

$$\begin{bmatrix} M & K^* \\ K & -\frac{1}{\beta}M \end{bmatrix} \begin{bmatrix} \mathbf{u} \\ \mathbf{w} \end{bmatrix} = \begin{bmatrix} \mathbf{f} + M\mathbf{u}_d \\ 0 \end{bmatrix}$$

or after scaling with the factor $\sqrt{\beta}$ and a change of sign, to

$$\begin{bmatrix} M & -\sqrt{\beta}K^* \\ \sqrt{\beta}K & M \end{bmatrix} \begin{bmatrix} \mathbf{u} \\ \widetilde{\mathbf{w}} \end{bmatrix} = \begin{bmatrix} \mathbf{f} + M\mathbf{u}_d \\ 0 \end{bmatrix},$$

where $\widetilde{\mathbf{w}} = -\frac{1}{\sqrt{\beta}}\mathbf{w}$.

As described in the previous section, using the PRESB preconditioner, this system is readily solved, whereby each iteration involves solving systems with matrices $M + \sqrt{\beta}K$ and $M + \sqrt{\beta}K^*$. These matrices get increasingly better conditioned when β decreases. They can be readily solved by use of an algebraic multilevel iteration method, see e.g. [11], or by the modified incomplete factorization method.

4.2 Subdomain Controls

Consider now the optimal control problem when the control is restricted to a subdomain, $\Omega_1 \subset \Omega$. Further we consider a time-dependent optimal control problems which is solved by use of a harmonic expansion method. Such problems can be solved in parallel for different frequencies and it suffices here to consider only one frequency ω. A shown in [7,8], this leads to an elliptic operator $K = K_0 + i\omega M$ where K, M are real valued. The optimality condition matrix becomes now

$$\widetilde{A} = \begin{bmatrix} M & 0 & -K^* \\ 0 & \beta M & -N^T \\ -K & -N & 0 \end{bmatrix},$$

where M is the mass matrix corresponding to the subdomain Ω_1 and $N = [N_{ij}]$, $N_{ij} = \int_\Omega \varphi_i \psi_j$, $i = 1, \cdots, n$, $j = 1, \cdots, m$ is a rectangular "mass matrix" of order $n = \dim\ \Omega_h$, $m = \dim\ \Omega_{1,h}$. Here $\{\varphi_i\}_1^n$, $\{\psi_j\}_1^m$ are the corresponding finite element basis functions used for the whole domain Ω respectively for the subdomain Ω_1 and h denotes a discretization parameter and Ω_h, $\Omega_{1,h}$ the finite element meshes corresponding to the triangular meshes for Ω respectively Ω_1. Here we reduce the system matrix to

$$\mathscr{A} = \begin{bmatrix} M & -K^* \\ K & \frac{1}{\beta}\widehat{M_1} \end{bmatrix}$$

where $\widehat{M_1} = N M_1^{-1} N^T$.

We form a preconditioner $\mathscr{B}$ to $\mathscr{A}$ be replacing $\widehat{M_1}$ with M, that is,

$$\mathscr{B} = \begin{bmatrix} M & -K^* \\ K & \frac{1}{\beta}M \end{bmatrix}.$$

Here there are two choices:

(i) Replace $\mathscr{A}$ and $\mathscr{B}$ with the scaled versions

$$\widetilde{\mathscr{A}} = \begin{bmatrix} M & -\sqrt{\beta}K^* \\ \sqrt{\beta}K & M_1 \end{bmatrix} \quad \text{respectively} \quad \widetilde{\mathscr{B}} = \begin{bmatrix} M & -\sqrt{\beta}K^* \\ \sqrt{\beta}K & M \end{bmatrix}$$

and form the PRESB preconditioner to $\widetilde{\mathscr{A}}$,

$$\widehat{\mathscr{B}} = \begin{bmatrix} M & -\sqrt{\beta}K^* \\ \sqrt{\beta}K & M + \sqrt{\beta}(K + K^*) \end{bmatrix}.$$

As has been shown in [7,8], the eigenvalues of $\widehat{\mathscr{B}}^{-1}\widetilde{\mathscr{B}}$ are contained in the interval $[\frac{1}{2}, 1]$ which holds uniformly with respect to β and other parameters. Hence the efficiency of this approach depends on the spectrum for $\widehat{\mathscr{B}}\widetilde{\mathscr{A}}^{-1}$. This depends much on how close M_1 is to M. Hence the method may not be fully robust. Due to this we take another approach.

(ii) Factorize $\mathscr{A}$ and $\mathscr{B}$ as

$$\mathscr{A} = \begin{bmatrix} M & 0 \\ K & \frac{1}{\beta}(M_1 + \beta K M^{-1} K^* \end{bmatrix} \begin{bmatrix} I & -M^{-1}K^* \\ 0 & I \end{bmatrix},$$

$$\mathscr{B} = \begin{bmatrix} M & 0 \\ K & \frac{1}{\beta}(M + \beta K M^{-1} K^*) \end{bmatrix} \begin{bmatrix} I & -M^{-1}K^* \\ 0 & I \end{bmatrix}. \tag{4}$$

Computations show that the corresponding Schur complement matrices for $\mathscr{A}$ and $\mathscr{B}$ equal

$$\mathscr{S}_1 = \frac{1}{\beta}\widehat{M_1} + KM^{-1}K^* = \frac{1}{\beta}M_1 + (K_0 + i\omega M)M^{-1}(K_0 - i\omega M) =$$
$$= \frac{1}{\beta}\widehat{M_1} + \omega^2 M + K_0 M^{-1} K_0,$$

respectively

$$\mathscr{S} = \frac{1}{\beta}M + KM^{-1}K^* = \frac{1}{\beta}(1 + \beta\omega^2)M + K_0 M^{-1} K_0.$$

We factorize $\mathscr{S}$ exactly in complex factors as

$$\mathscr{S} = (\alpha M + iK_0)M^{-1}(\alpha M - iK_0),$$

where $\alpha = \sqrt{\frac{1+\beta\omega^2}{\beta}}$ and use this to replace the matrix in position 2, 2 in (4) and form

$$\mathscr{B} = \begin{bmatrix} M & 0 \\ K & \alpha M + iK_0 \end{bmatrix} \begin{bmatrix} I & -M^{-1}K^* \\ 0 & M^{-1}(\alpha M - iK_0) \end{bmatrix}.$$

Clearly, besides some matrix-vector multiplications and vector additions, an action of $\mathscr{B}^{-1}$ involves solving a system with the complex valued matrices $\alpha M + iK_0$ and $\alpha M - iK_0$. To avoid complex arithmetics, for these solution we can use the PRESB method in Sect. 2, that is

$$(\alpha M + iK_0)(x + iy) = f + ig$$

is rewritten as

$$\begin{bmatrix} \alpha M & -K_0 \\ K_0 & \alpha M \end{bmatrix} \begin{bmatrix} x \\ y \end{bmatrix} = \begin{bmatrix} f \\ g \end{bmatrix}$$

and correspondingly for $\alpha M - iK_0$.

In practice, to avoid having to repeat solving systems with $\mathscr{B}$, these inner systems should be solved to a fairly high accuracy. It remains to derive eigenvalue bounds for $\mathscr{B}^{-1}\mathscr{A}$. These bounds equal those for the Schur complement matrices $\mathscr{S}$ and $\mathscr{S}_1$, where

$$\mathscr{S}_1 = \frac{1}{\beta}(M - \widehat{M_1}).$$

Proposition 3. *Let $\mathscr{A}$, $\mathscr{B}$ be defined in (4). The eigenvalues of $\mathscr{B}^{-1}\mathscr{A}$ are contained in the interval $\left[\frac{\beta\omega^2}{1+\beta\omega^2}\right]$.*

Proof. It holds

$$1 \geq \frac{\eta^T \mathscr{S}_1 \eta}{\eta^T \mathscr{S} \eta} = 1 - \frac{\eta^T (M - \widehat{M_1})\eta}{(1+\beta\omega^2)\eta^T M \eta + \eta^T K_0 M^{-1} K_0 \eta} \geq 1 - \frac{1}{1+\beta\omega^2} = \frac{\beta\omega^2}{1+\beta\omega^2}.$$

□

Remark 2. In practical problems ω is normally fairly large which means that even for small values of the regularization parameter β, so the product $\beta\omega^2$ can take values close to unity so the eigenvalue interval can be fairly tight, leading to a fast convergence. Furthermore, the lower bound and values close to it, are taken for a very small subset of vectors, i.e. $\eta \in \mathscr{N}(M_1)$ and $\mathscr{N}(K_0)$. Besides these small eigenvalues the term $\eta^T K_0 M^{-1} K_0 \eta$ becomes large for most vectors η, so most eigenvalues are clustered near unity, leading to a superlinear rate of convergence.

4.3 Parameter Identification Problems

A very important type of optimal control problems are problems where one must identify some unknown boundary conditions at part of the boundary, for instance where a part of some equipment is not accessible for measurements. Another important class of problems arise if one needs to identify some material dependent coefficients in a differential equation. Such problems are more complicated than the problem dealt with here and they need a separate treatment. For unknown boundary conditions one can overimpose boundary values at an accessible part of the domain and let the control variable be defined to be the missing boundary data, see e.g. [19] for a novel approach.

5 Extended Applications of the Chebyshev Iteration Method

5.1 The Classical Second Order Chebyshev Iteration Method

Inner product free iteration methods such as the Chebyshev method can outperform more commonly used iteration methods based on Krylov subspaces, such as the classical conjugate gradient (CG) method and the generalized minimum residual (GMRES) method, at least when implemented on parallel computer platforms. This is due to the need to compute inner products in CG and GMRES, mainly for the orthogonalization of a new search direction with respect to previous search directions. Thereby each processor must send its local part of the inner product to all other processors or to a master processor, to enable the computation of the global inner product. This requires start up times for synchronization and global communication times. Since computer chips on modern computers now have reached nearly their physical limit of speed, this type of overhead will be even more dominating for future generations of parallel computer processors, where the only way to decrease the computer time is to use more parallel processors.

The Chebyshev second order semi-iteration method, see e.g. [13], is defined by the three term recursion,

$$\begin{aligned} x^{(k+1)} &= \alpha_k x^{(k)} + (1-\alpha_k)x^{(k-1)} - \beta_k r^{(k)}, \quad k = 1,2,\cdots \\ x^{(1)} &= x^{(0)} - \frac{1}{2}\beta_0 r^{(0)}, \end{aligned} \tag{5}$$

$\beta_0 = 4/(a+b)$, where $r^{(k)} = \mathscr{A}x^{(k)} - f$ is residual.

Given an eigenvalue interval $[a,b]$, it follows that

$$\alpha_k = \frac{a+b}{2}\beta_k, \quad \beta_k^{-1} = \frac{a+b}{2} - \left(\frac{b-a}{4}\right)^2 \beta_{k-1}, \quad k = 1,2,\cdots$$

where $\alpha_0 = \frac{4}{a+b}\beta_0$. Note that $\alpha_k > 1$, $k \geq 1$. The recursion (5) can alternatively be written in the form

$$x^{(k+1)} = \alpha_k\Big(x^{(k)} - x^{(k-1)} - \frac{2}{a+b}r^{(k)}\Big) + x^{(k-1)}, \quad k = 1,2,\cdots.$$

It can be seen that the sequence $\{\beta_k\}$ decreases monotonically and converges to

$$\tilde{\beta} = \frac{4}{(\sqrt{a}+\sqrt{b})^2}, \quad k \to \infty$$

and $\{\alpha_k\}$ to

$$\tilde{\alpha} = 2\frac{a+b}{a+b+2\sqrt{ab}}, \quad k \to \infty.$$

The classical conjugate gradient method contains also a three-term recursion and has been extended to the solution of nonsymmetric matrices, for example by the GMRES method. Here the search vectors must be orthogonalized which implies the need for heavy inner product computations.

The Chebyshev iteration method has the following advantages over CG type methods.

(i) On a parallel processor machine there are no start-up times, synchronization and global communication of local parts of inner products computed on individual processors. This can save much computer times, in particular on modern computer where one has nearly reached the physical limits of communication speed, that is, Moore's now 50 year old prediction of exponential decay of communication time, has leveled out.
(ii) There are fewer arithmetic computations since none or hardly any inner products need to be computed.

(iii) The methods are insensitive to round-off errors, while CG type methods are more sensitive since exact orthogonality among search direction vectors is lost.

The price to be paid is that one needs sufficiently accurate estimates of the bounds of the eigenvalues of the preconditioned matrix, otherwise the Chebyshev iteration method will become less competitive. In numerous applications of OPT-PDE problems it has been shown that such accurate eigenvalue bounds can be estimated.

5.2 Modification of the Chebyshev Iteration Method to Handle Outlier Eigenvalues

Assume first that there is just one outlier eigenvalue, $\lambda_{\max}$, which is assumed to be known. This can be 'killed' by use of a corrected right hand side vector,

$$\tilde{b} = \left(I - \frac{1}{\lambda_{\max}} \mathscr{A}\mathscr{B}^{-1}\right) b. \tag{6}$$

Then

$$\mathscr{B}^{-1}\tilde{b} = \left(I - \frac{1}{\lambda_{\max}} \mathscr{B}^{-1}\mathscr{A}\right)\mathscr{B}^{-1}b$$

and one solves

$$\mathscr{B}^{-1}\mathscr{A}\tilde{x} = \mathscr{B}^{-1}\tilde{b},$$

by use of the Chebyshev method with eigenvalue bounds corresponding to the remaining eigenvalues. Then, since

$$\mathscr{A}x = b = \mathscr{A}\tilde{x} + \frac{1}{\lambda_{\max}}\mathscr{A}\mathscr{B}^{-1}b$$

one can compute the full solution,

$$x = \tilde{x} + \frac{1}{\lambda_{\max}} \mathscr{B}^{-1}b.$$

Since, due to rounding and some errors in the approximate eigenvalues used, the Chebyshev method makes the dominating eigenvalue component 'awake' again, so only very few steps should be taken. This can be compensated for by repetition of the iteration method, but for the new residual. The resulting Algorithm is then:

Algorithm; The reduced condition number Chebyshev method:
For a current approximate solution vector x, until convergence, do:

1. Compute $r = b - \mathscr{A}x$
2. Compute $\hat{r} = \mathscr{B}^{-1}r$
3. Compute $q = (I - \frac{1}{\lambda_{\max}}\mathscr{B}^{-1}\mathscr{A})\hat{r}$

4. Solve $\mathscr{B}^{-1}\mathscr{A}\tilde{x} = q$, by the Chebyshev method with reduced condition number.
5. Compute $x = \tilde{x} + \frac{1}{\lambda_{\max}} q$
6. Repeat

Numerical tests, see [8], for an equation of motion (see e.g. [18]) with a frequency $\omega = 2\pi$ showed that the method needed 3 times repetitions of a third order Chebyshev method, i.e. in total 9 preconditioning steps, while the Chebyshev method for the full interval $[\frac{1}{2}, \lambda_{\max}(= 2.326)]$ needed 15 preconditioning steps.

For the case of more outlier eigenvalues, say an eigenvalue λ_2, one can use an additional matrix factor to (6),

$$\left(I - \frac{1}{\lambda_2}\mathscr{B}^{-1}\mathscr{A}\right)\mathscr{B}^{-1}r.$$

One can also add some eigenvalues larger but close to unity to the interval $[1/2, 1]$, used for the Chebyshev method, which will make the interval slightly larger but still preserves its rapid convergence.

For $\omega = 3\pi$ there are three outlier eigenvalues, $\lambda_{\max} = 21.02$, $\lambda_2 = 3.42$ and $\lambda_3 = 1.25$. Here only the first two outlier eigenvalues will be 'killed' and a 2'nd degree polynomially preconditioned Chebyshev method was used. Being close to the interval $[\frac{1}{2}, 1]$, the eigenvalue 1.25 was added to it. It needed 8 repetitions, i.e. in total 16 preconditioning steps, while the Chebyshev method for the full interval, $[\frac{1}{2}, \lambda_{\max}]$ needed 57 preconditioning steps, hence our method had a huge saving in computational costs. The above eigenvalues were computed by use of the well-known eigenvalues and eigenvectors of the model Laplacian operator on a unit square. For a more general problem one can estimate the largest eigenvalues by use of a power iteration method, at least if there are only few of them.

6 Concluding Remarks

The PRESB preconditioner has been used in several OPT-PDE problems, compared analytically and numerically with many other methods and found to outperform them in number of iterations and CPU times. It has also been shown that the Chebyshev method is a very efficient acceleration method for this method. The earlier publications have dealt with symmetric and positive semidefinite matrix blocks. In later papers it has been shown how the methods can be extended to handle indefinite matrix blocks and how to modify the Chebyshev method to handle outlier eigenvalues.

Acknowledgement. The work of Owe Axelsson is supported by The Ministry of Education, Youth and Sports of the Czech Republic from the National Programme of Sustainability (NPU II), project "IT4Innovations excellence in science - LQ1602" and by the project BAS-17-08, "Microstructure analysis and numerical upscaling using parallel numerical methods, algorithms for heterogeneous computer architectures and hi-tech measuring devices".

References

1. Axelsson, O., Vassilevski, P.S.: Algebraic multilevel preconditioning methods. I. Numer. Math. **56**, 157–177 (1989)
2. Vassilevski, P.S.: Multilevel Block Factorization Preconditioners. Springer, New York (2008)
3. Saad, Y., Schultz, M.H.: GMRES: a generalized minimal residual algorithm for solving nonsymmetric linear systems. SIAM J. Sci. Stat. Comput. **7**, 856–869 (1986)
4. Saad, Y.: A flexible inner-outer preconditioned GMRES algorithm. SIAM J. Sci. Comput. **14**, 461–469 (1993)
5. Axelsson, O., Farouq, S., Neytcheva, M.: Comparison of preconditioned Krylov subspace iteration methods for PDE-constrained optimization problems. Poisson and convection-diffusion control. Numer. Algorithms **73**, 631–663 (2016)
6. Liang, Z.-Z., Axelsson, O., Neytcheva, M.: A robust structured preconditioner for time-harmonic parabolic optimal control problems. Numer. Algorithms **79**, 575–596 (2018)
7. Axelsson, O., Lukáš, D.: Preconditioning methods for eddy current optimally controlled time-harmonic electromagnetic problems. J. Numer. Math. **27**, 1–22 (2019)
8. Axelsson, O., Salkuyeh, D.K.: A new version of a preconditioning method for certain two-by-two block matrices with square blocks. BIT Numer. Math. **59**(2), 321–342 (2018). https://doi.org/10.1007/s10543-018-0741-x
9. Axelsson, O., Liang, Z.-Z.: A note on preconditioning methods for time-periodic eddy current optimal control problems. J. Comput. Appl. Math. **352**, 262–277 (2019)
10. Axelsson, O., Neytcheva, M., Ahmad, B.: A comparison of iterative methods to solve complex valued linear algebraic systems. Numer. Algorithms **66**, 811–841 (2014)
11. Notay, Y.: AGMG software and documentation (2015). http://homepages.ulb.ac.be/~ynotay/
12. Axelsson, O.: Iterative Solution Methods. Cambridge University Press, Cambridge (1994)
13. Axelsson, O., Gustafsson, I.: A coarse-fine-mesh stabilization for an alternating Schwarz domain decomposition method. Numer. Linear Algebra Appl. **26**, 1–19 (2019)
14. Varga, R.S.: Matrix Iterative Analysis. Springer, Heidelberg (2000). Originally published by Prentice Hall, 1962 edition
15. Greenbaum, A.: Iterative methods for solving linear systems. Front. Appl. Math. **17** (1997). SIAM, Philadelphia, PA
16. Axelsson, O.: An inner product free solution method for an equation of motion with indefinite matrices. Studies in Computational Intelligence (2019)
17. Axelsson, O., Liang, Z.-Z.: An optimal control framework for the iterative solution of an incomplete boundary data inverse problem (in preparation)
18. Toselli, A., Widlund, O.: Domain Decomposition Methods-Algorithms and Theory. Springer, Heidelberg (2005)
19. Feriani, A., Perotti, F., Simoncini, V.: Iterative system solvers for the frequency analysis of linear mechanical systems. Comput. Meth. Appl. Mech. Eng. **190**, 1719–1739 (2000)

Method for Evaluating the Vulnerability of Random Number Generators for Cryptographic Protection in Information Systems

Ivan Blagoev(✉)

Institute of Information and Communication Technologies – BAS, Sofia, Bulgaria
i.blagoev@iit.bas.bg

Abstract. A random number generator (RNG) and cryptographic algorithm is the base of each encryption system. For this, it can be considered that, no matter how complex cryptographic algorithms are applied, they become as vulnerable as the random number generator which is at the root of this system. The theme of RNG and PRNG deserves particular attention, especially because cryptographic protection in information systems relies on it. It can also be said that this is also the basis for cyber security as a whole. For this, it is extremely important to be sure of the quality of a system or mechanism for generating random numbers. In our study, as a means of assessing reliability, we rely on the mathematics of time series. The results of the proposed method are discussed and possibilities for ensuring cryptographic protection in information systems are shown.

Keywords: Random number generator · Cyber security · Time series

1 Introduction

Two kinds of random number generators are used for the contemporary needs of cryptography - Random Number Generator (RNG) and Pseudo Random Number Generator (PRNG). The efficiency of random number generators is measured by the degree of entropy to generate random numbers.

- **Random Number Generator (RNG):** Apply at any given time to the RNG to generate values that must be unique and should not be repeated on subsequent calls to the RNG [1]. The numbers obtained with this type of RNG are applied to operations that require unique/unobservable numerical values generated over time. An example of such a situation is generating a cryptographic key for encoding/decoding data, initializing vectors, initial numerical values for controlled RNG, etc.
- **Pseudo Random Number Generator (PRNG):** this generator uses the SEED initial number [2]. From this value, all randomly generated consecutive numbers come from the algorithm. These values in order of their sequence are re-reproducible. The only

I. Dimov and S. Fidanova (Eds.): HPC 2019, SCI 902, pp. 391–397, 2021.
https://doi.org/10.1007/978-3-030-55347-0_33

unexpected and secret value that should be as unpredictable as possible is the SEED number, which is the root of the base of this numerical sequence and the basis for generating the entire numerical array. This technology is used most often for One Time Password (OTP) authentication and the generation of cryptographic keys derived from the Master Root Key, which is used to compile wallets in Block Chain - distributed ledger technology, HMAC authentication, etc.

But how predictive/vulnerable is a given random number generator?

This is an issue that has excited the community for a long time. Even if we use a good mix of cryptographic algorithms designed to protect the performance of an information system, but our PRNG is not good, then the whole cryptographic solution may be vulnerable [3]. Encryption algorithms are in themselves reproducible, and the simplest way is for an attacker to try to predict values from the random numbers coming from our PRNG that feed as input to cryptographic algorithms, and compare the output with the captured encrypted stream. With a correct prognosis and match of an attacker's RNG value, the security of the entire cryptographic solution will fail. And unlike brute force attacks that require a great deal of time and resources for which they are considered ineffective when combining reliable cryptographic algorithms. Intrusion detection attacks and predictability in a poor random number generator require much less time and resources that make this type of attack possible [4]. We, as a victim in turn, may feel fraudulently secure because we believe that the system relies on a good combination of cryptographic algorithms without suspecting the weakness of our RNG.

How is the reliability of an RNG measured?

The effectiveness of RNG is measured by the degree of entropy for the generation of random numbers. For example, we take a binary bit and it can have a value of 0 or 1. If we have no idea what the value is, we have entropy 1 bit (i.e. coin throwing with 2 possible outcomes). If the generated value is always 1 and we know this, we have entropy 0 bits. The predictability is opposed to unpredictability. If the binary bit 1 is falling in 99% of all cases, entropy may only be a fraction over 0 bits. In the area of cryptography, the more unpredictable bits we would obtain so much the better.

In other areas, such as statistics, signal processing, econometrics, and mathematical finance, to find cycle times and predict future values, time series are used [5]. Since a time series is a sequence of data points typically measured at successive time points located at unified time intervals, it is also possible to apply this approach to the quality analysis of a random number generation system. Regardless of which random number generator is more often used (uncontrolled or controlled), the overall success of the system is based on the quality of the random numbers produced.

The complexity of analysing a given RNG is a function of the quality of its entropy, such as seasonality and collision tendencies, or the creation of repetitive patterns. These are the moments when the RNG will generate a value that is cyclical or a range of values that lead to a repeat of an already output result or the generation of a new but expected value.

2 Architecture for Generating Random Numbers in PHP

As an example the architecture for generating random numbers in PHP programming language technology in different operating systems has been examined in the paper.

With reference to what has been written here, let's take a look at the architecture for generating random numbers in PHP programming language technology. Here the main means for random numbers are PRNGs:

1. Linear Congruential Generator (LCG), e.g. `lcg_value()`
2. The Marsenne-Twister algorithm, e.g. `mt_rand()`
3. Locally supported C function, i.e. `rand()`

They are reused internally also for functions like `array_rand()` and `uniqid()` in PHP. The downside of entropy and random number generators of the above described functions lies in the easy to predict future PRNG values. Insufficient entropy originates from the fact that the initial internal states or SEEDs of PRNG are limited and the output of values is in insufficient range and it is predictable from readily available modern computational resources.

Consider the following example:

```
<?php
mt_srand(3231153718);
for ($i = 1; Si < 15; $i ++) {
echo mt_rand(), PHP_EOL;
}
```

By performing the above script we get the following sequence of 15 pseudo-random numbers:

```
818549686, 167964715, 1051297630,
679916735, 46602984, 1183523200, 591658099, 619786907, 50517956,
1766971703, 380190080, 1985369304, 1829774512, 859734050
```

No matter how many times we run this script, we'll get the same sequence of numbers. We received the value of SEED = 3231153718 using the function `rand()`. In this case, if an attacker receives the value of SEED used in Mersenne Twister PRNGs of PHP, s/he will be able to predict the output of `mt_rand()`. For this, as a matter of protection, it is of principal importance in this case to protect the value of the SEED number.

To get a value for SEED in PHP, developers often get value for SEED by using above script, or they can use `mt_rand()` to give it automatically. This is particularly likely to be true when using older PHP-based systems and even with higher versions of PHP used in modern conditions. Therefore, there is a risk that recovery of SEED by an attacker is possible. Despite its passive nature, it is actually a real vulnerability. Even if information is missed at certain points in the system, it allows the attacker to make further attacks, which is a breach of the principle of in-depth protection.

Let's imagine working with an application that uses the following source code to generate token for different application purposes:

```
$newtoken = hash('sha512', mt_rand());
```

Surely there are more complex ways to generate a token, but it's also a nice example of just one call to `mt_rand()` that has been hashed with SHA512. In fact, if a programmer assumes that PHP's random value functions are "random enough," he would be much more inclined to embed a simple model of use (similar to the one presented above) as long as it does not include the word cryptography. SHA512 exudes programmers' awe because this is the largest value algorithm in the SHA-2 family. However, the method used above for generating tokens suffers from one drawback - random values are limited to digits (i.e. its uncertainty or entropy is close to negligible). If we check the output of `mt_getrandmax()`, we will find that the maximum random number `mt_rand()` can generate only 2.147 billion. This limited number of options makes it vulnerable to a brute force attack.

If we have a fairly good video card from the past few generations, things get even easier. In this case, the `hashcat` application can be used as an example, being one of the brute force software tools. If everything is configured and works properly, the `hashcat` will find the value of the hashed random number within a few minutes. This demonstrates in practice how entropy works that we describe here. The `mt_rand()` function is limited to so few options that the SHA512 hashes of all possible values can be calculated for a short time. Hence, using hash to hide the output of `mt_rand()` is useless.

To protect this type of system, higher quality random values should be generated. For use on non-trivial tasks, PHP requires high-end entropy sources that can be provided by the operating system. Linux is usually used with `/dev/urandom`, unless devices with even higher entropy are installed.

PHP can be accessed through the OpenSSL external library reuse functions, using `openssl_pseudo_random_bytes()` and `mcrypt_create_iv()`. These two functions are also optimized to use a cryptographically protected pseudo-random generator. In Windows, a source of such entropy is CSPRNG, but for now PHP does not have direct access to it without special extensions to provide it. In fact, we need to make sure that the PHP version that is used on our servers has enabled OpenSSL for Linux or Mcrypt for Windows.

In Linux, with the correct setup, a regular random number generator, which is of the PRNG type (which is a pseudo random number generator), is often recharged from a source of high entropy `/dev/random` which makes it resistant to attacks. Typically, direct reading from `/dev/random` is avoided, because it is with limited capacity, and can lead to blocking all attempts to be available. Addressing to `/dev/random` occurs only in critical cases when it is needed.

3 Investigating Random Numbers Generators with Time Series

In our study, as a means of assessing reliability, we rely on the mathematics of time series.

To write the computer program for generating and writing time series from PRNG, we used C# and the open-source Mono platform for Linux with the target framework

.NET Framework 4.7, it is equivalent in Windows to the Microsoft .NET Framework 4.7. So the compiled program or program code transfer is compatible with both Linux and Windows operating systems.

The .NET platform has a sufficiently reliable specialized random number generation module: `RNGCryptoServiceProvider`, which is FIPS-certified, it is located in `System.Cryptography` and should be used for cryptographic tasks. But in our case we will use `System.Random`, which has much weaker entropy and will be suitable for analysis from the computational power we have at our disposal.

The resulting raw data file from PRNG in our case worked with a program we created on Python version 3. It is easy to visualize the data in a linear graphic version with a two-dimensional coordinate system (Fig. 1). The graph resembles a common noise generated by random numbers. In the next interpretation of the same data in the graph, as points of values on a two-dimensional coordinate field, the problems begin to become more obvious (Fig. 2).

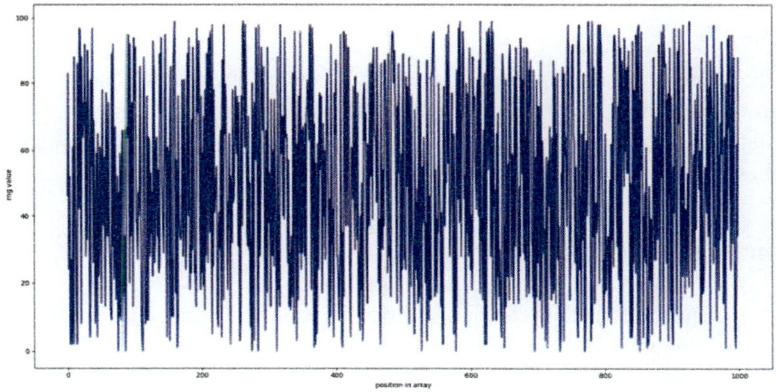

Fig. 1. Raw data file from PRNG

The graph in Fig. 2 demonstrates the weaknesses of the processed results. There are the patterns of repetitions that occur periodically over time. These cases are colored in red by our program, using predefined forecasting models.

If such a random number generator is used in cryptography, the SEED values produced by it could be attacked successfully.

Cryptographic algorithms used worldwide are generally available and reproducible. Their foundation is based on the fact that they must be combined with reliable sources of random numbers. For example, by predicting a next SEED value or by monitoring transmitted encrypted data, the values at the base of the encryption system may be perceived at some point.

In other cases, if an attacker has more information about a system, or about the encoded data it may be able to overcome cryptographic protection and even more rapidly. Finding repeat patterns also allows automated search by program and predicting subsequent PRNG values.

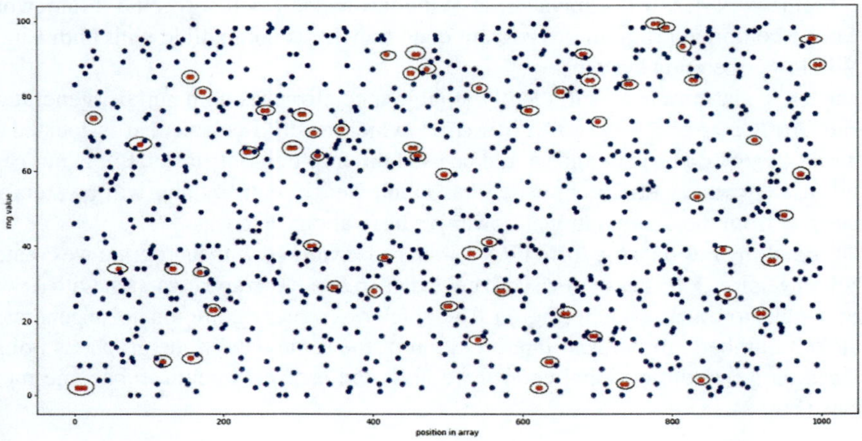

Fig. 2. Patterns of reappearances that occur periodically over time

It is therefore of the utmost importance that the same data to break and show through the prism of more than one system of presentation and analysis. This can also be said for the methods of assessing the quality of the PRNG. In any case, combining different approaches will lead to more precise end results and conclusions.

4 Conclusion

The theme of RNG and PRNG deserves particular attention, especially because cryptographic protection in information systems relies on it. It can also be said that this is also the basis for cyber security as a whole. For this, it is extremely important to be sure of the quality of a system or mechanism for generating random numbers. Therefore, it is desirable to evaluate such a hardware or software generator with more than one method of analysis.

From the studies presented up to now, it can be concluded that this approach would help in the study of the quality of random number generators. It would be even better to combine with other methods of PRNG analysis. The spectrum of application of such solutions is very wide. Since not only in cryptography, but also in other technology areas [6–8] good PRNG solutions are needed. Such solutions would therefore also help in the creation and refinement of the PRNG, as well as for more precise determination of the spectrum of tasks that the generator can perform well.

Acknowledgement. Research described in this article was partially supported by the National Scientific Program "Information and Communication Technologies for a Single Digital Market in Science, Education and Security (ICTinSES)", financed by the Ministry of Education and Science.

References

1. Random Number Service. https://www.random.org
2. Pseudo-Random Number Generators. https://crypto.stanford.edu/pbc/notes/crypto/prng.html
3. Lacharme, P., Röck, A., Strubel, V., Videau, M.: The Linux Pseudorandom Number Generator Revisited. Cryptology ePrint Archive, 251 (2012)
4. Dörre, F., Klebanov, V.: Practical detection of entropy loss in pseudo-random number generators. In: CCS 2016, Vienna, Austria, pp. 1–12 (2016)
5. Atanasova, T., Barova, M.: Exploratory analysis of Time Series for hypothesizes feature values. In: International Scientific Conference UniTech 2017, vol. II, pp. 399–403. University Publishing House V, Aprilov, Gabrovo (2017)
6. Tashev, T., Monov, V.: Large-scale simulation of uniform load traffic for modeling of throughput on a crossbar switch node. In: 8-th International Conference "Large-Scale Scientific Computations", Sozopol, Bulgaria, 6–10 June 2011. LNCS, vol. 7116, pp. 630–637. Springer (2012)
7. Balabanov, T., Zankinski, I., Shumanov, B.: Slot machines RTP optimization with generic algorithms. In: Numerical Methods and Applications. LNCS, vol. 9374, pp. 210–217. Springer, Switzerland (2015)
8. Dineva, K., Atanasova, T.: OSEMN process for working over data acquired by IoT devices mounted in beehives. Curr. Trends Nat. Sci. **7**(13), 47–53 (2018)

ETCCDI Climate Indices for Assessment of the Recent Climate over Southeast Europe

Hristo Chervenkov(✉) and Kiril Slavov

National Institute of Meteorology and Hydrology,
Tsarigradsko Shose blvd. 66, 1784 Sofia, Bulgaria
{hristo.tchervenkov,kiril.slavov}@meteo.bg

Abstract. An important task in the modern climatology is to assess extreme climate and weather events alongside various aspects of the mean climate. The analysis of such extreme events has become increasingly important due to their significant impact on society and natural systems. The climate indices are relatively simple, but statistically consistent, quantitative indicators of the climate extremes. The modern sets of such indices, among which the most widely used is the ETCCDI collection, are statistically robust, cover a wide range of climate conditions, and have a high signal-to-noise ratio. The World Meteorological Organization encourages the application of indices, stating that by using the same definitions of extremes and analyzing the data in a standardized way, it is possible to compare results from different places and to obtain coherent pictures of change around the world. The present study is based on source information from five free available datasets of ETCCDI climate indices. Five thermal and five precipitation-based indices, originating from different datasets, are analyzed and compared for their spatial structure over SE Europe and temporal evolution in the period 1979–2010. The study reveals overall similarity of the distribution pattern, at the same time substantial discrepancies for certain indices and years. The trend analysis of centenial-long time series of thermal indices, indicates clear and statistically significant warming in the twentieth century.

Keywords: ETCCDI climate indices · ClimData · Gridded datasets · Climate assessment · Trend analysis

1 Introduction

The oncoming climate changes will have an impact on the ecosystems, on all branches of the international economy, on the human health and quality of life. In the past, most analyses of long-term global climate change using observational temperature and precipitation data are focused on changes in mean values. One important task in the modern climatology is to assess not only the various aspects of the mean climate, but also extreme climate and weather events. Over

I. Dimov and S. Fidanova (Eds.): HPC 2019, SCI 902, pp. 398–412, 2021.
https://doi.org/10.1007/978-3-030-55347-0_34

the last two decades, the analysis of such extreme events has become increasingly important due to their significant impacts on society and natural systems [17]. The extreme weather phenomena are discussed in all reports of the Intergovernmental Panel on Climate Change (see, for example, [15]). There are various methods to investigate extreme events, but the computation and analysis of climate indices (CIs) derived from daily data is probably the most widely used non-parametric approach. In order to detect changes in climate extremes, it is important to develop a set of indices that are statistically robust, cover a wide range of climate conditions, and have a high signal-to-noise ratio [4]. The indices could be divided roughly into four categories: (1) absolute indices, which describe, for example, the hottest or coldest day of a year; (2) threshold indices, which count the number of days when fixed temperature or precipitation threshold is exceeded; (3) duration indices, which describe the length of some event; and (4) percentile-based threshold indices, which describe the exceedance rates above or below a threshold, which is defined as fixed percentile derived from the base period, most frequently—1961–1990 [17,22]. Hence they are used in several projects on climate change with focus on the present climate in different spatial scales, there is a lot of literature (see, for example, references in [4]). Other group of studies are dedicated primarily on the assessment of the present climate [1,14,17] and/or in future climate projections [18,19,21] using CIs, derived from climate models (CMs) output. Many of these studies (e.g. [5,19]) use partially pre-existing datasets of CIs (DCIs), the availability of such databases is from crucial importance for any research, which relies more or less on CIs-based analysis of the present and/or projected future climate. The present study is based entirely of already existing and free available DCIs. Our main aim is to assess the spatial and inter annual distribution of selected CIs, more especially the discrepancies, originating from the different nature, as explained in Sect. 2, of the corresponding sources. The considered domain is southeast (SE) Europe and the period of interest—the common time intersection of the used five DCIs, namely the 32 years long interval 1979–2010. Due the relevance of assessing trends in climate extremes, the time series of the areal averaged (AA) values of the selected indices are tested for presence of trend and its statistical significance.

The paper is structured as follows: The selected climate indices and the used DCIs are described concisely in Sect. 2. The main part of the study is placed in Sect. 3, dedicated on the performed calculations and the performed results. The brief concluding remarks, comments as well as outlook for possible continuation, are in Sect. 4.

2 Selected Climate Indices and Datasets

2.1 ETCCDI Climate Indices

The climate indices of the Expert Team on Climate Change, Detection, Monitoring and Indices (ETCCDMI), (http://www.clivar.org/organization/etccdi/etccdi.php) were constructed to sample a wide variety of climates and are calculated with a well-elaborated methodology [17,22]. The total number of

ETCCDI-indices is 29 (descriptive list at http://etccdi.pacificclimate.org/list_27_indices.shtml), among them 18 are temperature- (i.e. thermal) indices, the rest—precipitation-based. Most of the recent studies are limited to significantly smaller subset and, depending of the concrete task, various arguments for selection of key parameters could be outlined. We have considered 5 thermal- and 5 precipitation-based indices, which are listed in Table 1. TNn and TXx describe the coldest and hottest day of a year and are typical absolute indices. They are in focus of recent studies as, for example [17,19]. In contrast, TN10p and TX90p, which are often called "cold nights" and "warm days" respectively, are percentile-based indicators and are also subject of interest [5,13,14]. It have to be emphasized, motivating the selection of the indices in Table 1, that some widely used threshold indices are with limited applicability for the selected domain. Thus, for example, the tropical night index is substantial only in low-elevation areas (below 800 m), i.e. not over big areas of the Balkan Peninsula. Similarly, the number of ice/frost days index have little meaning along the Aegean, Ionic, Adriatic and Black Sea costs [5]. The GSL is the most valuable agrometeorological indicator among the ETCCDI-CIs, relevant for cultivated plants phenology and active growth of crops [5]. The index R10MM is usually referred as the number of the heavy precipitation episodes during the year [8,19]. As stated in [19], CDD is the only index that refers to the dry part of the year and can indicate regions vulnerable to droughts. RX5day can be used as a flood indicator because severe floods, on the space scales considered here, are generally not caused by a single heavy thunderstorm event, but more likely, by long-lasting heavy precipitation events that are extended over a large region [19]. R95p, called "Very wet days", characterizes the upper tail of the precipitation distribution and PRCPTOT is most widely recognizable precipitation indicator. Is it worthy of analyzing PRCPTOT and RCP95p in parallel: If different trends are observed for PRCPTOT and RCP95p, this would reflect changes in the character of precipitation. For example, a positive trend in RCP95p and a zero PRCPTOT trend would imply that precipitation has become more intense, but also less frequent as the total amount is not changed [14].

2.2 Datasets of Indices

The existence and barrier free availability of CIDs is of key importance for the present study and thus this subsection is dedicated on their concise description. HadEX2, the successor of HadEX, is a result of common international effort, leaded from the University of New South Wales Climate Change Research Centre (http://www.climdex.org) [8]. The gridded dataset is observation-based, is with global land coverage with $3.75° \times 2.5°$ longitude-latitude grid and time span 1901–2010. High-quality in situ observations from over 7000 temperature and 11000 precipitation meteorological stations across the globe were obtained to calculate the CIs over the period of record available for each station and then interpolated to the global grid. The spatial coverage of HadEX2 depends on the data availability and the search radius used in the interpolation procedure, which is determined by spatial correlation. Because station-data availabil-

Table 1. List of the considered ETCCDI climate indices

ID	Name	Definition	Units
TNn	Monthly minimum value of daily minimum temperature	Let TNn be the daily minimum temperature in month k in period j. The minimum daily minimum temperature each month is then $TNn_{kj} = \min(TNn_{kj})$	°C
TXx	Monthly maximum value of daily maximum temperature	Let TXx be the daily minimum temperature in month k in period j. The maximum daily maximum temperature each month is then $TXx_{kj} = \max(TXx_{kj})$	°C
TN10p	Percentage of days when TN < 10th percentile	Let TN_{ij} be the daily minimum temperature on day i in period j and let $TN_{in}10$ be the calendar day 10th percentile centered on a 5-day window for the base period 1961–1990. The percentage of time for the base period is determined where: $TN_{ij} < TN_{in}10$	%
TX90p	Percentage of days when TX > 90th percentile	Let TX_{ij} be the daily maximum temperature on day i in period j and let $TX_{in}90$ be the calendar day 90th percentile centered on a 5-day window for the base period 1961–1990. The percentage of time for the base period is determined where: $TX_{ij} > TX_{in}90$	%
GSL	Growing season length	Let td_{ij} be the mean temperature on day i in period j. Count the number of days between the first occurrence of at least six consecutive days with $td_{ij}>5\,°C$ and the first Occurrence after 1st July (NH) or 1st January (SH) of at least six consecutive days with $td_{ij}<5\,°C$	Days
R10mm	Annual count of days when PRCP ≥ 10 mm	Let RR_{ij} be the daily precipitation amount on day i in period j. Count the number of days where: $RR_{ij} \geq 10$ mm	Days
CDD	Maximum length of dry spell, max. number of consecutive days with RR < 1 mm	Let RR_{ij} be the daily precipitation amount on day i in period j. Count the largest number of consecutive days where: $RR_{ij} < 1$ mm	Days
Rx5day	Monthly maximum consecutive 5-day precipitation	Let RR_{kj} be the precipitation amount for the 5-day interval ending k, period j. Then maximum 5-day values for period j are: $Rx5day_j = \max(RR_{kj})$	mm
R95pTOT	Annual total PRCP when RR > 95p	Let RR_{wj} be the daily precipitation amount on a wet day w (RR ≥ 1.0 mm) in period i and let $RR_{wn}95$ be the 95th percentile of precipitation on wet days in the 1961–1990 period. If W represents the number of wet days in the period, then: $R95_{pj} = \sum_{w=1}^{W} RR_{wj}$ where $RR_{wj} > RR_{wn}95$	mm
PRCPTOT	Annual total precipitation in wet days	Let RR_{ij} be the daily precipitation amount on day i in period j. If I represents the number of days in j, then: $PRCPTOT_j = \sum_{i=1}^{I} RR_{ij}$	mm

ity and search radius differ from one index to another, different indices have different temporal and spatial coverage, as will be shown in Sect. 2. The next three out of considered five CIDs are prepared from the Canadian Centre for Climate Modelling and Analysis (CCCMA) and downloaded from the corresponding ETCCDI indices archive (EIA) website (http://www.cccma.ec.gc.ca/data/climdex/climdex.shtml) [17,18]. They are computed from the output of three widely used reanalyses: ERA-Interim of the European Centre for Medium-Range Weather Forecasts (ECMWF) [7], National Centers for Environmental Prediction/National Center for Atmospheric Research (NCEP/NCAR) Reanalysis 1 (NCEP1) [10,12] and NCEP-DOE Reanalysis 2 (NCEP2) [11]. These CIDs are noted further for conciseness as ERAI-EIA, NCEP1-EIA and NCEP2-EIA, correspondingly. The ERA-Interim is available on a $1.5^{\circ} \times 1.5^{\circ}$ regular longitude-latitude grid for the period from 1979 to 2010. Both NCEP1 and NCEP2 reanalysis data sets are available on a 192×94 Gaussian grid. The NCEP1-CIA and NCEP2-CIA are computed for the years 1948–2010 and 1979–2010, respectively. The fifth dataset is part of the created by the authors free accessible CID named ClimData https://repo.vi-seem.eu/handle/21.15102/VISEEM-343 [4,5]. It is based on the well known and widely used in the meteorological community E-OBS (https://www.ecad.eu/download/ensembles/download.php) of the European Climate Assessment & Dataset (ECA&D) project [9]. E-OBS is updated periodically and version 16.0, spanning from 1950 till the end of 2016, for domain, covering whole Europe with 0.25° horizontal spacing, is selected for the computation of this CID, noted further as EOBS-CD. The most significant information for the described above CIDs is briefly summarized in Table 2. As far as the nature of the input data for the computation of the listed in Table 2 is substantially different, it is reasonable to expect also nonnegligible differences in the output CID. The main advantage of the reanalysis, used for ERAI-EIA, NCEP1-EIA and NCEP2-EIA is obvious: it provides spatially and temporally continuous best estimate picture of the atmosphere in the form of gridded digital map. The physical consistency of all parameters is ensured with the incorporation of numerical weather prediction model. This makes the reanalysis data a potentially useful source of information for monitoring long-term changes in extremes. The principal applicability of reanalysis for deriving CIs is studied by many authors and the common conclusion is that the temperature data are generally reliable compared to data from representative meteorological stations [13]. Although reanalyses are essentially observationally constrained model output, variables that are directly assimilated in the reanalysis forecast model are typically closer to observations. Parameters such as precipitation that are primarily determined by the forecast model are poorly constrained by the assimilated observations and others, like the used for the computation of the thermal CIs, near-surface temperature fields, are strongly influenced from the model [17]. The rest two CIDs, HadEX2 and EOBS-CD, are entirely based on pre-processed near-surface observations. For the European region, HaDEX2 relies on the ECA&D data, on which EOBS-CD is also based. Due to the substantially different gridding procedure, however, it is reasonable to respect differences in the subsequent CIDs. The principal difference

Table 2. Summary of the considered climate indices datasets

CID	Source, Reference(s)	Input data	Coverage, spacing	Time span
ERAI-EIA	EIA [17,18]	ERA-I, ECMWF	Global, 1.5° × 1.5°	1979–2012
NCEP1-EIA		NCEP, NCAR		1948–2011
NCEP2-EIA		NCEP, DOE		1979–2011
HadEX2	HadEX2 project [8]	Station Observ.	Global, land 3.75° × 2.5°	1901–2010
EOBS-CD	ClimData [4,5]	E-OBS v16.0	EU domain, 0.25° × 0.25°	1950–2016

between HadEX2 and EOBS-CD is that, in the first case, the CID consist of gridded CIs, calculated from point measurements, and in the latter—CID, which are obtained from gridded in advance near-surface temperatures and daily precipitation sums. As stated in [17], this results in a spatial-scale mismatch, because the latter represents area (grid box) averages rather than point values. Other relevant source of potential discrepancies between the considered CIDs is the up to an order different (HadEX2 from one site and EOBS-CD from other) horizontal spacing. All of them, except EOBS-CD, are typical planetary application and thus any aspects, specific for the regional climate of the domain, are not resolvable, by means of these CIDs. Due to this reason, it is methodologically correct to deal with spatially generalized, for example AA values of the considered CIs, as shown in the next Sect. 3.

3 Calculations and Results

3.1 Mean Values and Cases

First, in order to facilitate the next computations, are prepared subsets with spatial extend over the domain of interest and containing the listed in Table 1 variables only. All netCDF file manipulations are performed with the Climate Data Operators (https://code.zmaw.de/projects/cdo) collection. The NCEP1-EIA and NCEP2-EIA are regrided from the original Gaussian to the regular grid of ERAI-EIA, using a first-order conservative remapping procedure as in [17]. The common time intersection of the used five DCIs, namely the 32 years long interval 1979–2010, is close to the interval 1981–2010, which has been increasingly used in reports and evaluation studies as reference of modern climate. Figure 1 shows the spatial distribution of the multi-year means of the thermal indices for the common time intersection. As first and foremost conclusion from the analyses of Fig. 1 is the overall coherency of the mean fields of the thermal CIs. The differences between the CIDs with the biggest discrepancy in the spacing—HadEX2 an EOBS-CD, seems most apparent. To check this coherency also for concrete years, two limit cases are selected—the relatively cold year 2006 and the exceptionally hot year 2000, as depicted on Fig. 2. As expected, the differences among the fields of the same index from different CIDs is more apparent for single years rather than the multi-year means. Although some general similarities still exists, the discrepancies, especially for some CIs and CIDs,

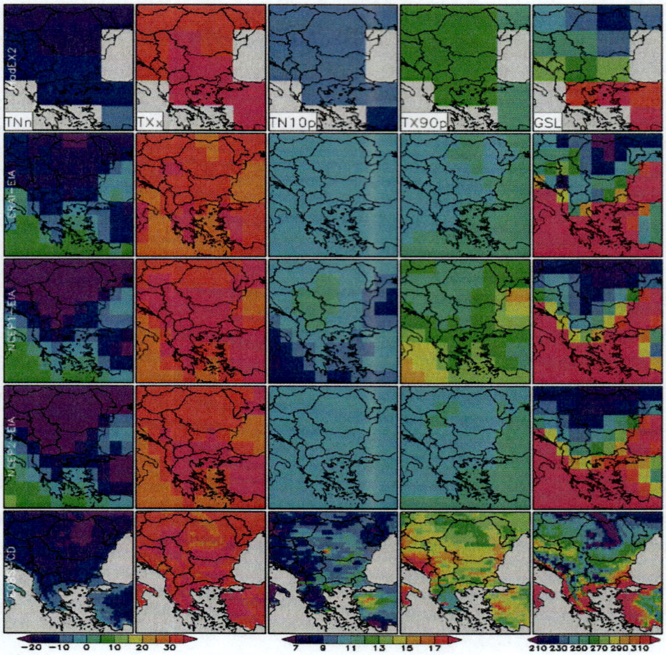

Fig. 1. Multi-year means of the thermal indices for 1979–2010. The units are according to Table 1.

for example the warm days in HadEX2 and ERAI-CIA, are significant. It is worth comparing the time series of the AA values of the thermal CIs from the considered CIDs. To be such comparison methodologically correct, however, the reanalyses-based CIDs, namely ERAI-CIA, NCEP1-CIA and NCEP2-CIA, have to be constrained only over the land. This is done applying the ERA-Interim land-sea mask with the same grid spacing. The evolution of the AA values of the thermal indices is shown on Fig. 3. Again, as for the spatial distributions shown on Fig. 1, the overall impression is for acceptable coherency between all CIDs in the inter-annual dynamics. The range of the discrepancies in the ensemble remains limited: for the coldest/hottest day is, roughly speaking, below 5 °C, and for the cold nights/warm days remains below 5%. Interesting is to check the time series of the ERAI-CIA: It produces the warmest AA coldest nights among all CIDs and vice versa for the hottest day. It produces also relatively (in comparison with the others) big values for the AA of TN10p and TX90p. The ERA-Interim output can be considered to be of very high quality, particularly in the Northern Hemisphere extratropical areas where reanalysis uncertainty is negligible [2]. The reason could be rooted in the fact, that the land-sea mask can not exclude completely the values over the water bodies in the domain: the analysis of the map shows, that if even small part of any gridcell is occupied from land, the whole gridcell is accepted as land. Thus, for example, due to the

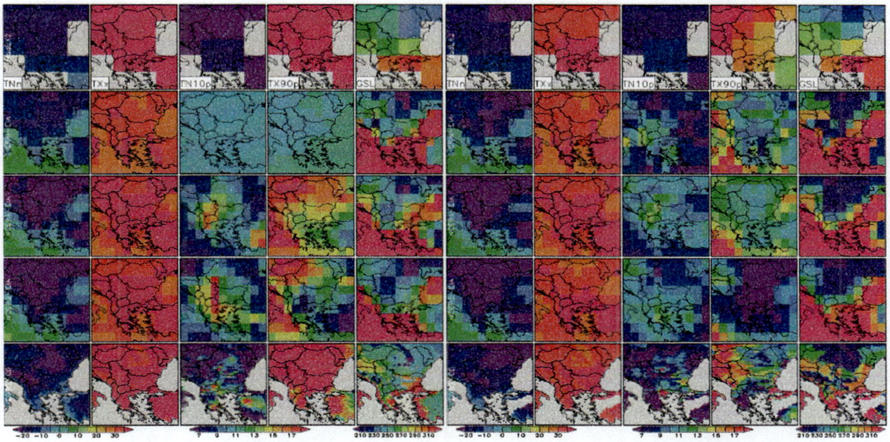

Fig. 2. Same as Fig. 1, but for the year 2000 (on left) and 2006 (on right).

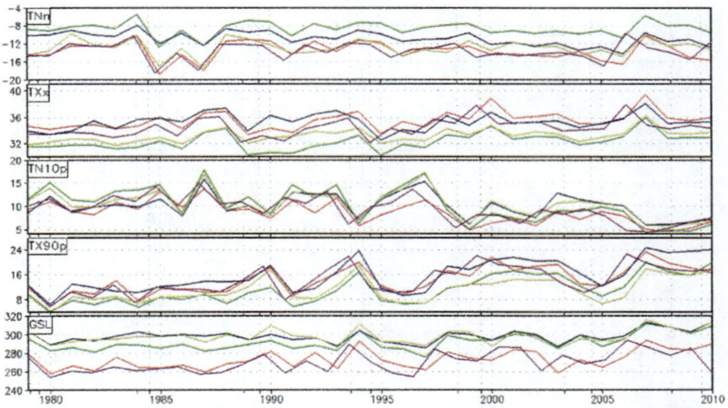

Fig. 3. Time series of the AA values of the thermal indices for 1979–2010 from HadEX2 (in red), ERAI-CIA (in green), NCEP1-EIA (in blue), NCEP2-EIA (in yellow) and EOBS-CIA (in purple). The units are according to Table 1.

presence of lot of islands in the Aegean Sea, the whole sea is treated (in this grid resolution) as continuous land. The different share of land and sea areas over fixed domain in two CIDs could produce substantial differences in the thermal indices due to the not equal seasonal warming/cooling of the land and sea. The warming influence of the sea, deep inside the land, on the spatial structure of some cold-related ETCCDI indices, is emphasized in [13] studying the winter climate conditions over the Balkan Peninsula. The question, why this is not an issue for the other reanalyses-based CIDs, for which the same mask is applied, however, remains.

Similarly to the analysis in the previous paragraph, the results for the precipitation-related indices, listed in Table 1, will be presented. Figure 4 shows the spatial distribution of the multi-year means. As expected, the spatial heterogeneity of the precipitation-related indices is bigger then this for the thermal ones. The general impression is for the overall similarity in the distribution patterns. Hence the NCEP-DOE Reanalysis 2 is essentially successor of NCEP/NCAR Reanalysis 1, it is worth to emphasize the differences in NCEP1-EIA and NCEP2-EIA in Fig. 4. Obviously these differences are consequence of the improved description skill of the precipitation in NCEP-DOE Reanalysis 2 [11].

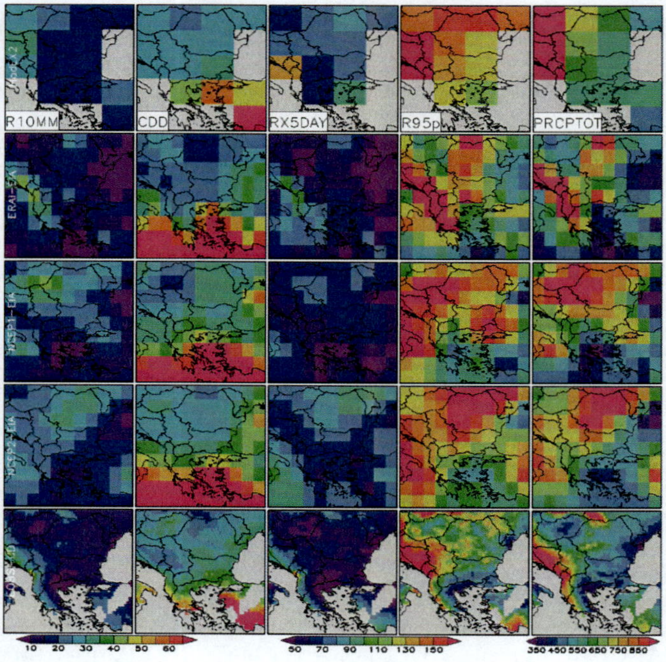

Fig. 4. Multi-year means of the precipitation-related indices for 1979–2010. The units are according to Table 1.

Drought is among the most complex climatic phenomena, which poses significant problems around the world, and places huge demands on rural and urban water resources as well as enormous burdens on agricultural and energy production. Due to its advantages, the Standardized Precipitation Index (SPI) has been selected by the WMO as a key indicator for monitoring drought and is among the two most frequently used in Europe [3]. The SPI is not based on daily data and subsequently is not in the ETCCDI-CI collection. To illustrate the drought severity for one particulate year, which is selected as drought case, the SPI for 6 months window (SPI-6), based on the precipitation inventory of the WMO GPCC, is depicted on Fig. 5. It is remarkable that the drought spreads

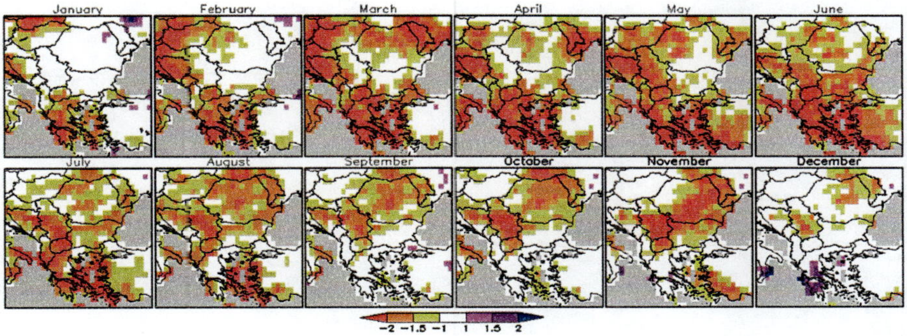

Fig. 5. Monthly values of the SPI-6 over the model domain for the year 1990

over the bigger part of the domain and the year; its severity is quantified with values, falling in the severe, even extreme drought bin (i.e. bellow −1.5 and bellow −2.0, correspondingly). The underling digital map, together with other three centennial-long global gridded datasets SPI, forms free accessible SPI data base (ftp://xeo.cfd.meteo.bg/SPI, see [3] for details). The field for the considered precipitation-based indices for this year is shown on Fig. 6. Figure 6 shows overall

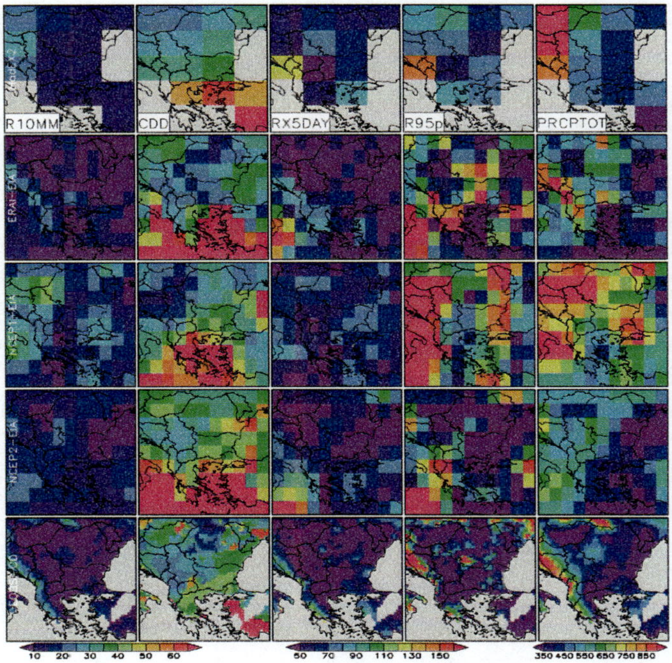

Fig. 6. Same as Fig. 4, but for the year 1990

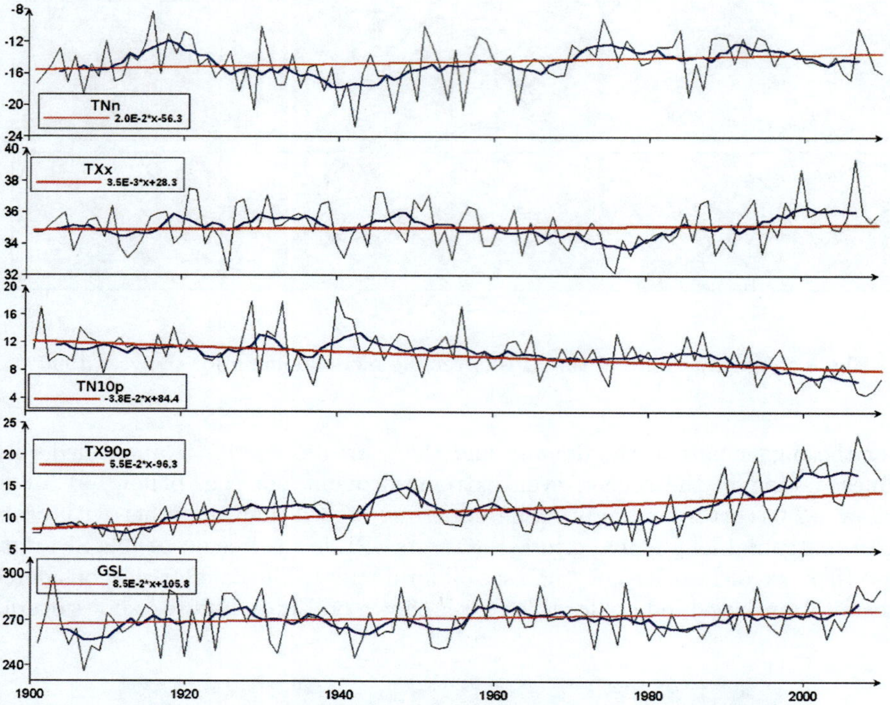

Fig. 7. Time series of the AA values of the thermal CIs (in black), running 7-year mean (in blue) and TSE trend line (in red, equation in legend).

similarity in the common distribution patterns, in the same time nonnegligible differences from CID to CID and index to index. Thus, for example, it should be outlined the discrepancy between NCEP1-CIA and NCEP2-CIA for R95p and PRCPTOT. Figures 4 and 6 demonstrates also one specific feature of HadEX2, already mentioned in Subsect. 2.2: different indices have different temporal and spatial coverage. It can be seen that RX5DAY and R95p, unlike R10MM, CDD and PRCPTOT, has no coverage over Asia Minor. Thus can lead to undesired effect calculating the AA values. This effect is strengthened when the domain is covered by relatively small number of gridcells, as in our case. For this reason and for sake of brevity we will not perform comparison of the time series, analogous to this on Fig. 3.

3.2 Trend Analysis

The importance of assessing trends in weather extremes is often emphasized. The main reason is that extreme weather conditions related to temperature, precipitation, storms or other aspects of climate, can cause loss of life, severe damage and large economic and societal losses (see [14] and references therein). One of

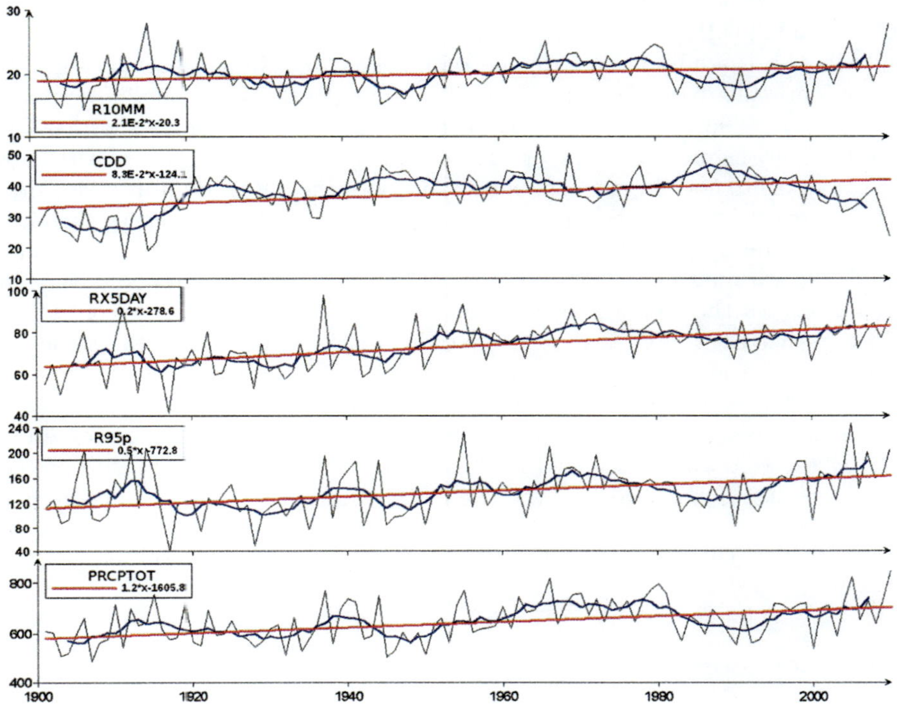

Fig. 8. Same as Fig. 7, but for the precipitation-based indices

the general conclusion in [5] is the confirmed warming signal over Europe, which is spatially dominating. This conclusion is based on analysis of the contained in ClimData data for linear trend estimation and its statistical significance of the CIs. The trend estimation in this study is performed by means of the Theil–Sen Estimator (TSE) [16,20], which is increasingly used as superior replacement of the Ordinary Least Squares (LSE) [14]. The main reason is that, despite its popularity, the LSE, in contrast to the TSE, is a non-robust technique. Comparing TSE and LSE according their trend estimation skills [6] reveals that both regression methods produce different, in the common case nonnegligible, results concerning the slope of the line. In some special cases, even if few outliers in the analyzed time series are present, LSE can lead to completely wrong results. Hence the HadEX2 time coverage is significantly longer (see Table 1) than the others, the time series of the AA values from this CID are analyzed. Figure 7 shows the outcomes for the thermal indices. The temporal cooling in the 1940s, mentioned in other studies, for example in [8], is notable in the dynamics of the running 7-year mean on Fig. 7. All indices, except TN10p, shows positive trend. The slope of the line for TNn, TXx, TX90p and GSL is 2.0 °C/100 yr, 0.4 °C/100 yr 5.5%/100 yr and 8.5 days/100 yr, correspondingly. The slope of the trend of TN10p is −3.8%/100 yr. All trends are tested for statistical significance with

Mann-Kendall (MK) test [5], which is widely used in the climatological studies. The MK test is a non-parametric and rank-based procedure, especially suitable for non-normally distributed data, data containing outliers and nonlinear trends [5,8,14]. The result of the MK test reveals that the trend of all thermal CIs is statistically significant with confidence more than 99% ($p < 0.01$). Although these results are not directly commensurable with the outcomes from other similar researches for the region (see, for example, [5] and the references therein), there is general agreement with their conclusions: There is clear and statistically significant warming signal. The comparison of the slopes of TNn and TXx is in coherence with the modern concept that the (European) climate becomes less colder, rather then more hotter [14], although the comparison of the slopes of TN10p and TX90p contradicts to this conclusion.

Figure 8 is equivalent of Fig. 7 for the precipitation-based indices. We would like to emphasize, that due to specific issues of the HadEX2, concerning especially these indices, as explained in Subsect. 3.1, the results should be interpreted with big caution. For example, the rise of the PRCPTOT with 12 mm/decade appears as exceptionally suspicious. This formal result is caused most probably by the internal inhomogeneities and/or gaps of the HadEX2-subset over the domain.

4 Conclusion

The database ClimData and especially the ETCCDI archive of the CCCMA are useful and reliable source of gridded CIDs, organized as continuous digital maps, both in space and time. The free accessibility makes these resources convenient versatile for broad range of experts—meteorologists, climatologists, hydrologist, which scientific research includes CIs-based analysis.

The comparison of the spatial distribution patterns as well as the temporal evolution of the areal average values, of the considered CIs from every CIDs with their counterparts form the other CIDs, reveals that, despite the presence of general similarity, certain, nonnegligible differences exist. Such conclusion, although in global context, is outlined also in [17]. These differences varies from one index to another and are stronger for one year, rather for the time mean in the period 1979–2010. The trend analysis reveals statistically significant warming trend over the domain. Although this result agrees with the outcomes of recent studies, it have to be confirmed with subsequent analysis, including, in particular, estimation of the magnitude of the warming (i.e. the lapse rate) by means of other methods and datasets. The results of the trend analysis of the precipitation-related indices, which, as explained, have to be treated with great caution, indicates the possible shortcomings when using problematic data and/or not proper analytic technique. This principle have to be kept always in mind, accounting for the high public significance and sensitivity to the results of the climatological studies.

Our study could be extended further, investigating, for instance, the seasonal aspects of the regional climate of the considered domain. Hence some of the ETCCDI CIs are also available on a monthly basis, this analysis can be performed in a similar way as in this study, relying solely on ClimData and EIA. The logical next step, which could be also based on EIA, is to complement this research with evaluation of the projected future climate over the region.

Acknowledgement. The authors would like to express their deep gratitude to the CCCMA, HadEX2 project, ClimData, MPI-M, UNI-DATA for the providing free of charge software and data. Personal thanks to I. Tsonevsky from ECMWF for the cooperation. The study is funded by the Bulgarian National Science Fund (DN–14/3 13.12.2017). We acknowledge the provided access to the e-infrastructure of the Centre for Advanced Computing and Data Processing, with the financial support by the Grant No BG05M2OP001–1.001–0003, financed by the Science and Education for Smart Growth Operational Program (2014–2020) and co-financed by the European Union through the European structural and Investment funds.

References

1. Alexander, L.V., Arblaster, J.M.: Assessing trends in observed and modelled climate extremes over Australia in relation to future projections. Int. J. Climatol. **29**, 417–435 (2009). https://doi.org/10.1002/joc.1730
2. Brands, S., Gutiérrez, J.M., Herrera, S., Cofiño, A.S.: On the use of reanalysis data for downscaling. J. Climate **25**, 2517–2526 (2012). https://doi.org/10.1175/JCLI-D-11-00251.1
3. Chervenkov, H., Tsonevsky, I., Slavov, K.: Presentation of four centennial-long global gridded datasets of the standardized precipitation index. Int. J. Environ. Agric. Res. (IJOEAR) **2**(3), 93–110 (2016). ISSN 2454-1850
4. Chervenkov, H., Slavov, K., Ivanov, V.: STARDEX and ETCCDI climate indices based on E-OBS and CARPATCLIM. Part one: general description. In: Nikolov, G., et al. (eds.) NMA 2018. LNCS, vol. 11189, pp. 360–367. Springer, Cham (2019). https://doi.org/10.1007/978-3-030-10692-8_40
5. Chervenkov, H., Slavov, K.: STARDEX and ETCCDI climate indices based on E-OBS and CARPATCLIM. Part two: ClimData in use. In: Nikolov, G., et al. (eds.) NMA 2018. LNCS, vol. 11189, pp. 368–374. Springer, Cham (2019). https://doi.org/10.1007/978-3-030-10692-8_41
6. Chervenkov, H., Slavov, K.: Theil-Sen estimator vs. ordinary least squares—trend analysis for selected ETCCDI climate indices. C. R. Acad. Bulg. Sci. **72**(1), pp. 47–54 (2019). https://doi.org/10.7546/CRABS.2019.01.06
7. Dee, D.P., et al.: The ERA-interim reanalysis: configuration and performance of the data assimilation system. Q. J. Roy. Meteor. Soc. **137**, 553–597 (2011). https://doi.org/10.1002/qj.828
8. Donat, M.G., et al.: Updated analyses of temperature and precipitation extreme indices since the beginning of the twentieth century: the HadEX2 dataset. J. Geophys. Res. Atmos. **118**, 2098–2118 (2013). https://doi.org/10.1002/jgrd.50150
9. Haylock, M.R., Hofstra, N., Klein Tank, A.M.G., Klok, E.J., Jones, P.D., New, M.: A European daily high-resolution gridded dataset of surface temperature and precipitation. J. Geophys. Res (Atmos.) **113**, D20119 (2008). https://doi.org/10.1029/2008JD10201
10. Kalnay, E., et al.: The NCEP/NCAR 40-year reanalysis project. Bull. Am. Meteorol. Soc. **77**, 437–471 (1996)

11. Kanamitsu, M., Ebisuzaki, W., Woollen, J., Yang, S.K., Hnilo, J., Fiorino, M., Potter, G.L.: NCEP-DOE AMIP-II reanalysis (R-2). Bull. Am. Meteorol. Soc. **83**, 1631–1643 (2002)
12. Kistler, R., et al.: The NCEP-NCAR 50-year reanalysis: monthly means CD-ROM and documentation. Bull. Am. Meteorol. Soc. **82**, 247–267 (2001)
13. Malcheva, K., Chervenkov, H., Marinova, T.: Winter severity assessment on the basis of measured and reanalysis data. In: Proceedings of the 16th International Multidisciplinary Scientific GeoConference, SGEM 2016, Albena, Bulgaria, 28 June–7 July, Book 4, vol. 1, pp. 719–726 (2016). ISSN 1314-2704
14. Moberg, A., et al.: Indices for daily temperature and precipitation extremes in Europe analyzed for the period 1901–2000. J. Geophys. Res. **111**, D22106 (2006). https://doi.org/10.1029/2006JD007103
15. Pachauri, R.K., et al.: Climate change 2014: synthesis report. In: Pachauri, R., Meyer, L. (eds.) Contribution of Working Groups I, II and III to the Fifth Assessment Report of the Intergovernmental Panel on Climate Change, IPCC, Geneva, Switzerland, 151 p. (2014). ISBN 978-92-9169-143-2
16. Sen, P.K.: Estimates of the regression coefficient based on Kendall's tau. J. Am. Stat. Assoc. **63**, 1379–1389 (1968)
17. Sillmann, J., Kharin, V.V., Zhang, X., Zwiers, F.W., Bronaugh, D.: Climate extremes indices in the CMIP5 multimodel ensemble: part 1. Model evaluation in the present climate. J. Geophys. Res. Atmos. **118**, 1716–1733 (2013). https://doi.org/10.1002/jgrd.50203
18. Sillmann, J., Kharin, V.V., Zwiers, F.W., Zhang, X., Bronaugh, D.: Climate extremes indices in the CMIP5 multimodel ensemble: part 2. Future climate projections. J. Geophys. Res. Atmos. **118**, 2473–2493 (2013). https://doi.org/10.1002/jgrd.50188
19. Sillmann, J., Röckner, E.: Clim. Change **86**, 83 (2008). https://doi.org/10.1007/s10584-007-9308-6
20. Theil, H.: A rank-invariant method of linear and polynomial regression analysis. I, II, III, Nederl. Akad. Wetensch., Proc. **53**, 386–392, 521–525, 1397–1412 (1950)
21. Tebaldi, C., Hayhoe, K., Arblaster, J.M., et al.: Clim. Change **79**, 185 (2006). https://doi.org/10.1007/s10584-006-9051-4
22. Zhang, X., Alexander, L., Hegerl, G.C., Jones, P., Tank, A.K., Peterson, T.C., Trewin, B., Zwiers, F.W.: Indices for monitoring changes in extremes based on daily temperature and precipitation data. WIREs Clim. Change **2**, 851–870 (2011). https://doi.org/10.1002/wcc.147

Sensitivity of Selected ETCCDI Climate Indices from the Calculation Method for Projected Future Climate

Hristo Chervenkov(✉) and Valery Spiridonov

National Institute of Meteorology and Hydrology, Tsarigradsko Shose blvd. 66, 1784 Sofia, Bulgaria
{hristo.tchervenkov,valery.spiridonov}@meteo.bg

Abstract. Statistical bias correction of climate models' outputs is being increasingly used in the recent decade for assessing the impact of climate change on several sectors. In this study, output from simulations for the present-day (1961–1990) climate, as well as for the near future scenario (2021–2050), performed with ALADIN-Climate in the frame of the CECILIA project, are used for calculation of selected ETCCDI climate indices. The same subset, but based on observational database E-OBS, is taken from the open archive ClimData and used as reference. The results of the computations, performed over the interior of the Balkan Peninsula, demonstrates the possibilities of the delta-change approach and its modifications. The outcome from the study confirms the high sensitivity of the absolute-threshold indices from the applied bias corrections. Overall, the sign of the change caused by all of the applied methods, agrees with the consolidated outcome from the recent studies of the projected future climate in SE Europe. The general distribution pattern of the climate change signal, expressed with the thermal indices, is similar for all considered techniques. This signal shows, however, in the field of the precipitation-based indices, significant spatial irregularities and non-negligible differences depending on the method applied.

Keywords: ETCCDI climate indices · ALADIN-climate · Bias correction · E-OBS · ClimData

1 Introduction

The oncoming climate changes will have an impact on the ecosystems, on all branches of the international economy, and on the quality of life. Global Circulation Models (GCMs) are the most widespread and successful tools employed for both numerical weather forecast and climate research since the 1980s [6,23,25]. Regional Climate Models (RCMs) are important tools used for downscaling climate simulations from GCMs [2,24]. The RCMs also still have substantial errors, partly inherited from the driving GCMs [13,21]. A very important question is

I. Dimov and S. Fidanova (Eds.): HPC 2019, SCI 902, pp. 413–427, 2021.
https://doi.org/10.1007/978-3-030-55347-0_35

thus how to interpret the model simulation results, when the simulations for the present-day shows certain, in the common case non-negligible, deviations from the observed reality. It is methodologically correct to expect that the future scenario simulation also inherits such a feature. Intended to be used as an input for process-based impact models, output from climate models is usually post-processed to reduce their systematic biases (i.e., the errors compared to observations over a reference, present-day period). These methods become known as bias correction (BC). The primary goal of BC is to try to adjust selected statistics of a climate model simulation to better match observed statistics over the present-day reference period. In general, core of the BC is the so called 'transfer function' (TF), which transforms the probability density function (PDF) of the present-day model output of some variable (e.g., temperature) in new one that maximally mimics the PDF of the distribution of the reference dataset. Consequently, the obtained TF is applied on the future scenario model output. The basic assumption is that bias changes are negligible compared to climate change or, equivalently, that the bias itself is time-invariant [4,11]. Due to its relative simplicity and low computational demand along with growing databases of global and regional climate model simulations, bias correction (BC) has become very popular in climate impact research. Over recent years, many different methods have been developed and widely applied to post-process climate projections (see, for instance, [9,10,24] and references therein). In the BC-technique, proposed in [20], the TF is a predefined analytic function, with some degree of arbitrariness. The choice of a particular form of TF to fit is based on a trade-off between robustness, i.e. using a minimal number of parameters, and goodness-of-fit obtainable with the emerging perfect TF. [20] argues that the other, non-parametric estimation of the cumulative distribution function is computationally expensive, as it requires the fitting of individual observed and modeled empirical distribution functions. The implementation of more sophisticated methods increases also the risk of loss of climate change signal, especially when the reference dataset is based on a measurement network that is significantly coarser than the model grid. Despite the problems, there is broad consensus about the added value of the BC: The results of [8] show that the applied BC-technique performed satisfactorily not only for mean quantities but even for time dependent statistical properties, such as the number of consecutive dry days and the cumulative amount of rainfall for consecutive heavy precipitation days. In [10] is stated that, among other, that BC reduces the temperature climate change over most of southern Europe (with local differences up to more than 0.7 °C). The aim of this paper is to assess the sensitivity of a number of climate indices (CIs) from the Expert Team on Climate Change Detection and Indices (ETCCDI) suite from the calculation method for projected future climate, rather than to give comprehensive review of the various aspects of the BC. A concise description of the theoretical background and the conceptional aspects of the BC can be found in [17] and more detailed presentation of specific methods - in [8,9,24]. The selected CIs have been calculated for the present (1961–1990) and near future

(2021–2050) climate from the output of the ALADIN simulation and modified subsequently with the considered BC methods.

The paper is structured as follows. Concise description of the ALADIN simulation, as well presentation of its output, is in Sect. 2. The considered ETCCDI CIs are addressed shortly in Sect. 3. The main part of the article is in Sect. 5, where the performed calculations and obtained results are described. The question how the applied BC affects the climate change signal, in terms of CIs, is from primary importance from end-user point of view. Thus, the Sect. 6, is focused on this issue. Concise conclusion remarks are placed in Sect. 7.

2 ALADIN Simulation

The climate simulations with model system ALADIN-Climate with focus on average precipitation and 2 m temperature were performed in the frame of the EU Framework-supported CECILIA (Central and Eastern Europe Climate Change Impact and Vulnerability Assessment, http://www.cecilia-eu.org/, see [12] for details) project (2006–2009). The used version of ALADIN-Climate corresponds to cycle 24 of the ARPEGE/IFS code, which physical parameterization package is derived directly from the one used in GCM ARPEGE-CLIMAT 4 [2]. The model was driven by the boundary condition of 50 km horizontal resolution coming from a "stretch mesh" version of ARPEGE-CLIMAT 4 GCM. This version of the GCM has a variable horizontal resolution being around 50 km over Southern Europe and decreasing to approximately 300 km at the antipode. The runs are performed for three time slices: 1961–1990 (present climate control run, CR), 2021–2050 (near future run, NF), and 2071–2100 (far future run, FF) in grid spacing of 10 km. All future simulations were carried out using CO_2 concentrations as described by the IPCC AR4 SRES A1B scenario [19]. The model domain is centered over Bulgaria, but accommodates significant part of the Balkan Peninsula. The well-known and freely available for the research community database E-OBS version 16.0 of the European Climate Assessment & Dataset (ECA&D) project [15] is used as a reference. It is useful to demonstrate briefly the ability of ALADIN to reproduce the present-day climate before discussing how BC impacts the projected future climate. Figure 1 shows maps of the averaged in the period 1961–1990 (i.e. norms) daily thermal parameters as well as daily precipitation sum from E-OBS and ALADIN-CR. Figure 1 shows for all thermal parameters prevailing, especially over the mountains, cold bias where it exceeds 5 °C. The precipitation bias is predominantly negative except over some mountainous areas. The outcomes of many evaluation studies of single [5] or ensembles of RCMs (see [16] and references therein) becomes available in the recent decade. The direct comparison between them and the results above is hampered by many factors as different time-windows of consideration, regions, statistical metrics of evaluation and others. Most of the models show, however generally cold bias over South Europe, which are in principal agreement with the ALADIN-outcome. For the precipitation, where the uncertainties are bigger and, correspondingly, the model spread larger, there is a predominantly wet bias over the region. This result, however, is in some extent, strange compared with the ALADIN output.

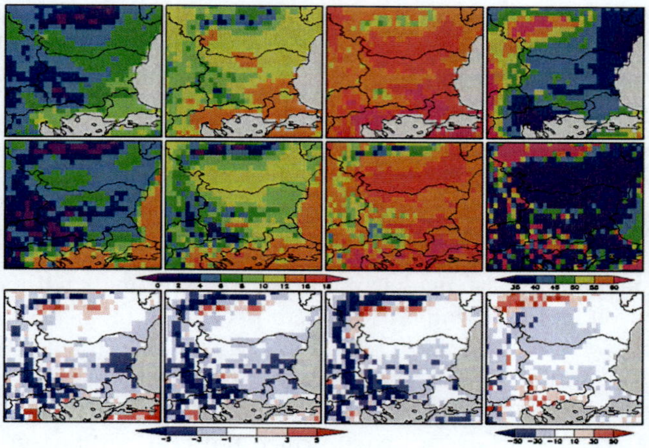

Fig. 1. Present-day norms of the minimum, mean and maximum temperature (in the first, second and thirth column correspondingly, °C) as well as monthly precipitation sum (fourth column, mm) from E-OBS (first row) and ALADIN-CR (second row). The bias ALADIN-CR–E-OBS is depicted in the last row.

3 Considered ETCCDI Indices

The climate analysis of extreme events for past, present and projected future climate, which is based on CIs, is one of the most widely used non-parametric approaches and consequently the literature is large (see [1,18] for details). It has been recognized in the last decades that it is important to document the exact formulation of an internationally agreed suite of indices of climate extremes from daily precipitation and temperature data. The ETCCDI defined a set of indicators that have subsequently become known as the ETCCDI-indices. These indices were chosen to sample a wide variety of climates and includes three main classes: absolute-threshold indices, percentile-based indices, and indices based on the duration of an event. From all 29 ETCCDI-indices (see full list with definitions on http://etccdi.pacificclimate.org/list_27_indices.shtml) we consider subset of the 19 absolute-thresholds and based on the duration of an event CIs, which are listed in Table 1. Depending of the method, these indices are calculated from the non-corrected or corrected ALADIN CR- and NF-output, as demonstrated in the next sections.

Table 1. List of the considered ETCCDI climate indices

ID	Name	Definition	Units
FD	Frost days	Let $\mathrm{tn}_{\mathrm{ij}}$ be the daily minimum temperature on day i in period j. Count the number of days where $\mathrm{tn}_{\mathrm{ij}} < 0\,^{\circ}\mathrm{C}$	Days
ID	Icing days	Let $\mathrm{tx}_{\mathrm{ij}}$ be the daily maximum temperature on day i in period j. Count the number of days where $\mathrm{tx}_{\mathrm{ij}} < 0\,^{\circ}\mathrm{C}$	Days
SU	Summer days	Let $\mathrm{tx}_{\mathrm{ij}}$ be the daily maximum temperature on day i in period j. Count the number of days where $\mathrm{tx}_{\mathrm{ij}} > 25\,^{\circ}\mathrm{C}$	Days
TR	Tropical nights	Let $\mathrm{tn}_{\mathrm{ij}}$ be the daily minimum temperature on day i in period j. Count the number of days where $\mathrm{tn}_{\mathrm{ij}} > 20\,^{\circ}\mathrm{C}$	Days
GSL	Growing season length	Let $\mathrm{td}_{\mathrm{ij}}$ be the mean temperature on day i in period j. Count the number of days between the first occurrence of at least six consecutive days with $\mathrm{td}_{\mathrm{ij}} > 5\,^{\circ}\mathrm{C}$ and the first Occurrence after 1st July (NH) or 1st January (SH) of at least six consecutive days with $\mathrm{td}_{\mathrm{ij}} < 5\,^{\circ}\mathrm{C}$	Days
Rx1day	Monthly max. 1-day precipitation	Let $\mathrm{RR}_{\mathrm{ij}}$ be the daily precipitation amount on day i in period j. The maximum 1-day value for period j are: $\mathrm{R} \times 1\mathrm{day}_{\mathrm{j}} = \max(\mathrm{RR}_{\mathrm{ij}})$	mm
Rx5day	Monthly max. consecutive 5-day precipitation	Let $\mathrm{RR}_{\mathrm{kj}}$ be the precipitation amount for the 5-day interval ending k, period j. Then maximum 5-day values for period j are: $\mathrm{R} \times 5\mathrm{day}_{\mathrm{j}} = \max(\mathrm{RR}_{\mathrm{kj}})$	mm
R10mm	Annual count of days when $\mathrm{PRCP} \geq 10\,\mathrm{mm}$	Let $\mathrm{RR}_{\mathrm{ij}}$ be the daily precipitation amount on day i in period j. Count the number of days where: $\mathrm{RR}_{\mathrm{ij}} \geq 10\,\mathrm{mm}$	Days
CDD	Max. length of dry spell, max. number of consecutive days with $\mathrm{RR} < 1\,\mathrm{mm}$	Let $\mathrm{RR}_{\mathrm{ij}}$ be the daily precipitation amount on day i in period j. Count the largest number of consecutive days where: $\mathrm{RR}_{\mathrm{ij}} < 1\,\mathrm{mm}$	Days
PRCPTOT	Annual total precipitation in wet days	Let $\mathrm{RR}_{\mathrm{ij}}$ be the daily precipitation amount on day i in period j. If I represents the number of days in j, then $PRCPTOT_j = \sum_{i=1}^{I} RR_{ij}$	mm

4 Methodology

Following the notation in [17], the simulated present-day model (predictor) time series of length N of chosen variable will be denoted as x_i^p, the corresponding reference (predict and) time series as y_i^p. The mean of the uncorrected model over the considered present period (i.e. 1961–1990) μ_{raw}^p can be estimated as $\hat{\mu}_{raw}^p = \overline{x_i^p}$ (where traditionally the hat denotes the estimator, the bar averaging in time), the corresponding real mean μ_{real}^p as $\hat{\mu}_{real}^p = \overline{y_i^p}$.

The most simple approach used for BC is the so-called delta change approach. It is widely used in climate impact research [17]. In its most basic application, performed in the present work also, a time series of future climate is generated as:

$$x^f_{i,corr} = y^p_i + (\overline{x^f_i} - \overline{x^p_i}). \tag{1}$$

The quantity $\overline{x^f_i} - \overline{x^p_i}$ is actually the model derived climate change signal. Thus, Eq. 1 could be treated as observed time series corrected with the simulated climate change. The method based on Eq. 1 is called correspondingly additive delta change method. Such approach is suitable for parameters such as the daily minimum (tn), mean (td) and maximum (tx) temperatures. They are core climatic parameters particularly involved in determining climate change impacts on society and ecosystems [3] and are typical input for calculation of CIs (in case of ETCCDI—tn and tx only). For variables such as precipitation, which assume positive values only, one would typically consider relative changes, i.e.:

$$x^f_{i,corr} = y^p_i \times \frac{\overline{x^f_i}}{\overline{x^p_i}}, \tag{2}$$

called further proportional delta change. Obviously this approach conserves the observed weather sequence and with it the linear spatial-, temporal- and multi-variable dependence structure. The delta change approach is therefore sensible only when these aspects may be assumed unchanged for the considered future climate.

The method of the relative change has one positive feature, that has to be outlined. Let's assume that the (main part of the) error of the model is proportional to the magnitude of its non-biased value or $x^p_{err} = k\overline{x^p_{i,nb}}$ for the present-day norm and $x^f_{err} = k\overline{x^f_{i,nb}}$ for the future scenario, where k is the coefficient of proportion and the subscript nb marks the non-biased values. Keeping in mind that

$$\overline{x^p_i} = \overline{x^p_{i,nb}} + x^p_{err} \text{ and } \overline{x^f_i} = \overline{x^f_{i,nb}} + x^f_{err} \tag{3}$$

the fraction in the RHS of Eq. 2 can be rewritten as:

$$\frac{\overline{x^f_i}}{\overline{x^p_i}} = \frac{\overline{x^f_{i,nb}} + k\overline{x^f_{i,nb}}}{\overline{x^p_{i,nb}} + k\overline{x^p_{i,nb}}} = \frac{(1+k)\,\overline{x^f_{i,nb}}}{(1+k)\,\overline{x^p_{i,nb}}} = \frac{\overline{x^f_{i,nb}}}{\overline{x^p_{i,nb}}}. \tag{4}$$

Thus, the implementation of the method of the relative change leads to the elimination of the errors, proportional to the magnitude of the non-biased value of the considered variable.

Figure 2 illustrates the impact of the delta change of the thermal parameters and the precipitation - the time series (running 7-days mean and only 3-years long subset for clearness hence the whole period is too long) of the spatial averages are shown. Figure 2 shows, first, that the performed with Eq. 1 BC leads to overall increase of tx and, second, absence of well expressed systematic shift in tn. Overall, the BC of Eq. 2 do not causes apparent systematic shift of the precipitation also.

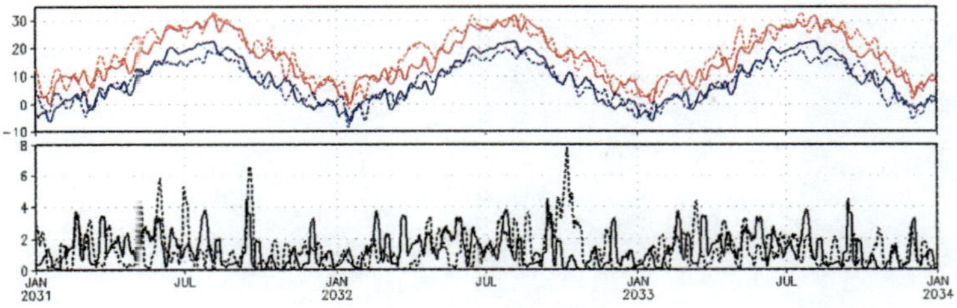

Fig. 2. Upper panel: Time series of the running 7-days spatial averages *tn* (in blue) and *tx* (in red) (continuous lines, °C) of the row ALADIN-NF output as well as modified with the delta change approach values (dotted lines, °C). Lower panel: Same as the upper panel but for the precipitation (mm).

5 Performed Calculations and Obtained Results

The E-OBS-based indices are part of the free-available ClimData database (https://repo.vi-seem.eu/handle/21.15102/VISEEM-343) and are used as a reference. The CIs are calculated by means of the ETCCDI-supplied software, integrated in a program written specifically for this study. The calculations are performed for each year of the both simulation periods, on every grid cell of the domain and are subsequently time averaged. For sake of brevity, however, we have considered for sensitivity analysis only part of the all 19 CIs listed in Table 1. The delta change approach presented briefly in Sect. 4, could be applied on the input, necessary for the calculation of the CIs, parameters or, alternatively, directly on the CR- and NR- CIs calculated with ALADIN-output. In the first case, noted further for brevity as "method 1", the *tn* and *tx* as well as the daily precipitation sum, are modified with the delta change approach, according to Eqs. 1 and 2 for the temperatures and the precipitation correspondingly and the CIs are calculated subsequently. In the second approach the ALADIN-output is not altered before the calculation of the CIs; the delta change is applied as a second step, on the CIs calculated with the row ALADIN CR- and NF-output. In this case, unlike "method 1", the additive variant of the delta-change BC (Eq. 1), noted as "method 3", as well as the proportional variant of the delta change BC (Eq. 2), noted as "method 2", are principally suitable for the precipitation-related and thermal CIs correspondingly. CIs, applied without any BC, are most frequently used in studies for projected future climate ([14,22] and others). The present climate CIs as well as their sensitivity from the calculation method are analyzed in the following Subsects. 5.1 and 5.2. The indices for the future scenario, obtained with the three methods, together with these, calculated directly from the ALADIN-NF output, are intercompared as well.

5.1 Thermal Indices

First, it is necessary to put together the reference CIs with these from ALADIN-CR, similar to the performed in Sect. 2 comparison, as shown on Fig. 3. It is

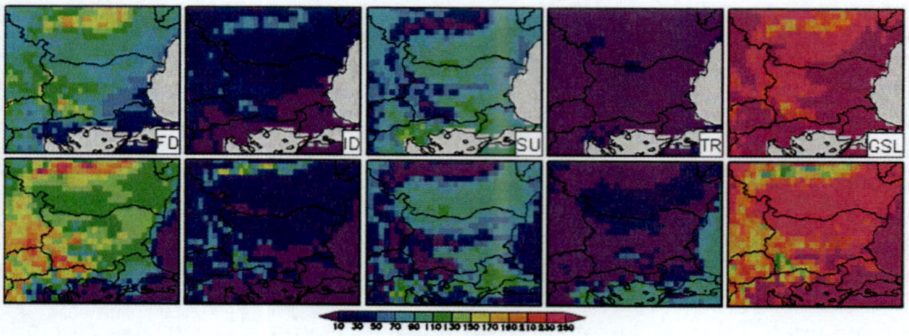

Fig. 3. Present climate thermal CIs from E-OBS (first row) and ALADIN-CR (second row).

worth emphasizing that the distribution of the ALADIN-CR CIs are rather adequate to their reference counterparts. The prevailing positive bias for the "cold" indices (i.e. FD and ID) and negative for the "warm" ones (SU, TR and GSL) is direct consequence of the cold bias in the model discussed in Sect. 2. Small exceptions can be detected over limited areas, for example over the Thracian valley for the TR, where the sign of the bias of the *tn* is opposite. Figure 4 shows the distribution of the thermal CIs for the near future scenario, calculated directly from the ALADIN-NF output and with the aforementioned three methods. Overall, the distribution patterns of the corresponding CIs are very similar. The corrected "cold" indices are generally smaller; the "hot" indices are bigger except TR, where the differences between the methods seems most significant. The subplots for SU and TR on Fig. 4, demonstrates also the obvious weakness of method 2 dealing with quantities, close to zero in the denominator of Eq. 2 - the corrected CIs becomes unrealistically high. In [3] it is stated, that the changes in the occurrence of TR are substantial only in low-elevation areas (below 800 m), located outside the Carpathian Mountains range, which are particularly exposed to persistent and intense warm spells in summer. Generally, TR are not characteristic to the mountain climate of the considered domain. Over many pixels there CI and SU are constantly zero and method 2 is not applicable.

5.2 Precipitation-Related Indices

Similarly to Subsect. 5.1, the analysis begins with comparison of the reference and ALADIN-CR CIs, as shown on Fig. 5. The significantly higher spatial inhomogeneity in ALADIN-CR appears as a main difference between the distributions of the both groups of CIs. The reason could be rooted in the fact that E-OBS, due to the embedded interpolation procedure, oversmooths the precipitation field [15]. The lower values of PRCPTOT over the flat regions in ALADIN-CR are due to the dry bias in the model mentioned in Sect. 2. RX1DAY and R10MM, especially RX5DAY, are often accepted as indicators for the precipitation extremes. It is frequently emphasized (see references in

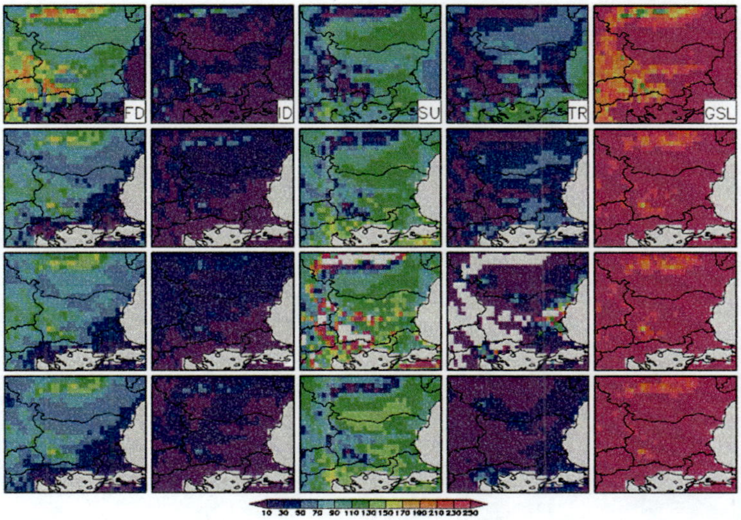

Fig. 4. Thermal CIs for the NF, calculated directly from the ALADIN-NF output (first row), method 3 (second row), method 2 (third row) and method 1 (fourth row). The units are according Table 1.

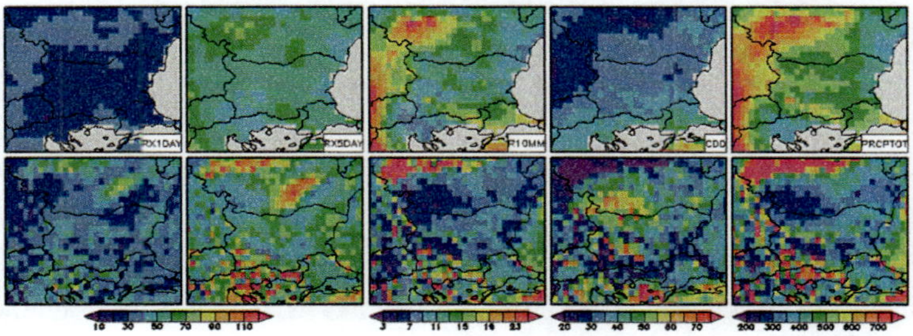

Fig. 5. Same as Fig. 3, but for the precipitation-related indices.

[1]) that the increased number of precipitation extremes are possible in certain cases in generally dryer climate (real or simulated). Thus, for these three Cis the ALADIN-CR-based values are greater than the reference for some pixels. The CDD-distribution reveals positive bias over the flat regions and negative one over the mountains. Figure 6 shows the distribution of the precipitation-related CIs for the near future scenario, calculated directly from the ALADIN-NF output, and the three methods. The general distribution pattern is very similar, as in the case of the thermal CIs. Roughly speaking, the fields of the indices, corrected with methods 2 and 3, are practically identical without general disagreements with the non-corrected CIs. The involved in the BC E-OBS as row data (fourth row) or as reference CIs (second and third row), leads apparently to overall spatial smoothing. The impact of method 1 appears stronger than method 2

and method 3 - it causes significant reduction of RX1DAY, RX5DAY, R10MM and PRCTOT, especially for the highest values. In contrast, the CDD, which is indicator for the drought conditions, is increased.

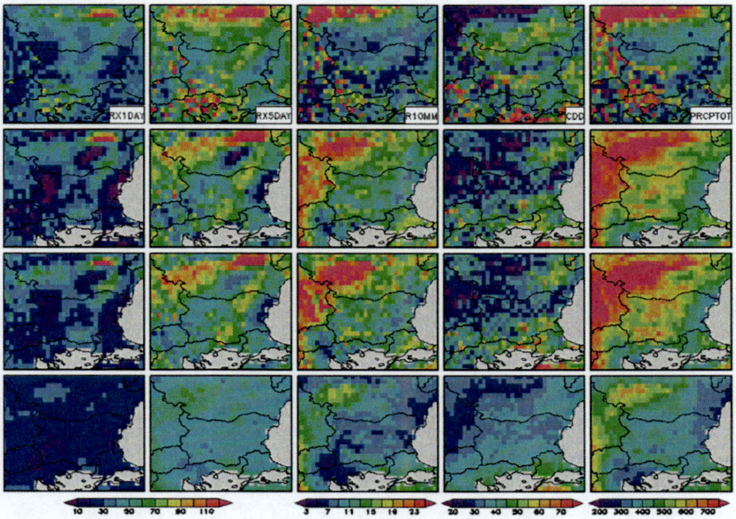

Fig. 6. Same as Fig. 4, but for the precipitation-related indices.

5.3 Analysis of the Sensitivity

The Analysis of the sensitivity could be performed by means of various methods, for example by the global sensitivity analysis (GSA) approach proposed by Sobol and studied in a number of research papers [7]. Although the GSA approach is more reliable since it takes into account the smoothness of the data and avoids high oscillations that may happen, we use the so-called local sensitivity analysis based on the variation of coefficients due to its simplicity. The coefficient of variation (CV), defined as $CV = \frac{\sigma}{\mu}$, could be applied as common quantitative estimator of the sensitivity. Here σ is the standard deviation and μ is the mean. The CV of the observations/simulations is used to describe the level of variability within the domain independently of their absolute values. If they differ markedly, or are of different variables (indices) then it is needed to use a standardized measure - such as the CV. If the dataset has a $CV < 1$, its distribution is considered of having low-variance, while if $CV > 1$, the distribution is considered as high-variance. The exponential distribution, for instance, where the mean equals the standard deviation, i.e. $CV = 1$, would be considered a medium-variance distribution. The mean, the standard deviation as well as the CV for all considered datasets are presented in Table 2. For sake of brevity, only the thermal

CIs are given. Similarly to the spatial distributions (see Fig. 4), the results for the FD, ID and especially for GSL, shown in Table 2, are very analogous. All methods produce for these indices commensurable mean values, which are light increased in comparison to the reference. The same is valid also generally for the CVs. For SU, method 1 and method 3 produce similar outcomes for the mean as well as for the CV, which are, in case of the mean values, reasonably greater than the reference. The differences in the results of the considered BC methods are most apparent in the field of the TR.

Table 2. Mean, Standard deviation and CV for the Thermal CIs

	×	Reference	CR	NF	Method 1	Method 2	Method 3
FD	μ	94.18	137.24	111.14	63.10	75.78	68.14
	σ	32.38	38.15	38.67	28.56	29.30	31.88
	CV	0.34	0.28	0.35	0.45	0.39	0.47
ID	μ	22.79	26.41	17.52	11.50	14.89	14.63
	σ	15.29	25.23	18.03	9.95	10.06	12.88
	CV	0.67	0.96	1.03	0.87	0.68	0.88
TR	μ	2.55	12.09	40.17	11.46	11.61	30.63
	σ	4.31	18.39	30.09	13.07	24.79	20.03
	CV	1.69	1.52	0.75	1.14	2.13	0.65
SU	μ	75.59	60.87	81.15	102.86	131.80	95.88
	σ	28.79	36.65	35.28	30.38	144.86	29.72
	CV	0.38	0.60	0.43	0.30	1.10	0.31
GSL	μ	259.96	223.33	245.83	289.10	286.35	282.46
	σ	32.41	38.02	42.44	32.56	38.65	36.88
	CV	0.12	0.17	0.17	0.11	0.13	0.13

6 Effect of BC on the Projections of the CIs

The effect of the three BC-methods, described in Sect. 4 as well as the non-corrected projections of the CIs, are estimated with the difference DIFF = Estimation–Reference for each of the considered indicators. The result for the thermal CIs is shown on Fig. 7. As shown in [2], ALADIN produces positive climate response for the NF (i.e. warming) with average values typically in the range of 2–4 °C for Bulgaria. These results are in principal agreement with the outcomes for the domain of the other models, participating in the CECILIA project. Thus, the results on Fig. 7 concerning the non-corrected FD and GSL (first row) for part of the domain as well as in some areas for ID and SU, are strange: it could be expected that the FD is reduced and the GSL is

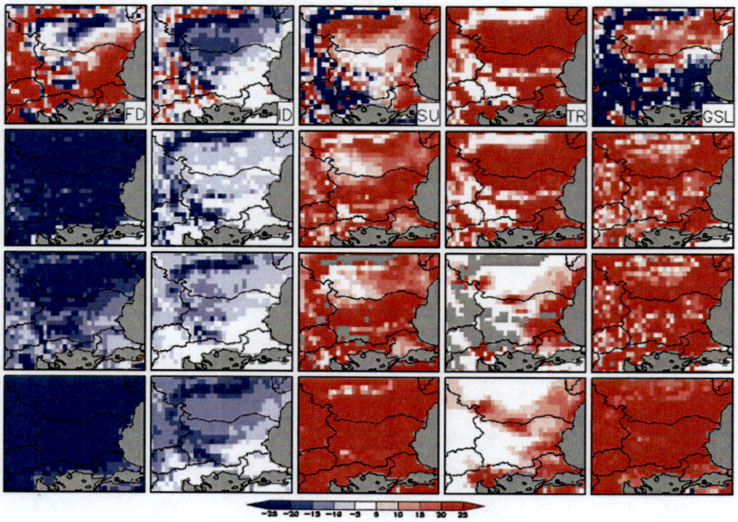

Fig. 7. Differences between the reference thermal CIs and the thermal CIs for the projected future climate. The order of the subplots and the units are as in Fig. 4.

increased in warmer climate. The reason is based partially in the cold bias of the model mentioned in Sect. 2. It has to be emphasized also that, according the common CECILIA methodology, the climate response in [2] is calculated as ALADIN-NF–ALADIN-CR and not ALADIN-NF–Reference. The former difference, as explained in [2] in contrast to the latter, reduces the systematic bias. The results in the second, third and fourth row on Fig. 7 show that the considered CIs are largely affected by the different BCs; For the TR and GSL even the sign of the climate response is changed. The strong sensitivity of the absolute-threshold indices is also one of the main conclusion of [10]. There is apparent overall agreement between the different methods. As expected, the climate response of method 2 and method 3 are practically identical. Method 1 changes the climate signal significantly for the TR and very strong for ID and GSL. It is worth emphasizing that this approach produces the most spatially consistent fields. The overall picture of the climate change signal in the field of the precipitation-related CIs, shown on Fig. 8, is much more complicated than this for the thermal indices. It can be seen, the non-corrected signal, as well as these from method 2 and method 3, is spatially very heterogeneous, with significant differences even in sign over neighboring regions. Practically only method 1 produces reasonable and spatially consistent distribution pattern. The reduction of RX1DAY, RX5DAY and R10MM, as well as the increase of CDD, agrees with the overall consensus in [2] in the projected future climate. However, the most prominent result is the reduction of PRCPTOT with more than 90 mm. This result, together with the dry bias shown in Sect. 2, fits well with to the outcomes in [2].

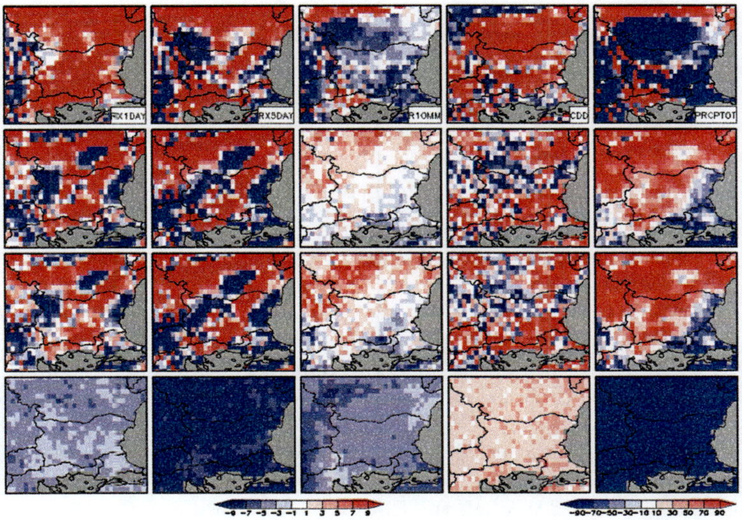

Fig. 8. Same as Fig. 7, but for the precipitation-related indices. RX1DAY, RX5DAY, R10MM and CDD shares one color bar, the color bar for PRCPTOT is placed bellow the last column.

7 Conclusion

It is known that the BC may alter the mean climate signal of the adjusted variable; however, the effect on the projected occurrence of climate extremes, as stated in [10], is less commonly investigated. The proposed two novel approaches, which are modifications of the delta change method, but handles directly the CIs, rather than the input parameters, seems promising. Generally, the investigated three methods of BC are similar. The direct methods, are, significantly easier for implementation than the others, and are applicable even if the input parameters are not available. The outcome from the study confirms the high sensitivity of the absolute-threshold indices, both thermal and precipitation-based, to the applied bias corrections. Most significant differences appears for TR and SU for the thermal CIs. They describe extreme and as a consequence rare events. Overall, the sign of the change caused by all of the applied methods, agrees with the consolidated outcome from the recent studies of the projected future climate in SE Europe. Despite some disagreement about the magnitude and the spatial and temporal distribution, common conclusion from all these studies is a expected reduction in precipitation totals and increase of mean temperatures [2]. The present study demonstrates, that the general distribution pattern of the climate change signal, expressed by means of the thermal indices, is similar for all considered techniques. This signal shows, however, for precipitation-based indices, significant spatial irregularities and non-negligible differences depending on the method applied. From a practical point of view, the possibility of using the methods, proposed in this study, has a definite advantage together with the computation simplicity. It is worth emphasizing, however, that these methods as any other BC, cannot correct for the model's own deficiencies (e.g., dynamics or physical parameterization). As such, the projections of a model that misrepresents some fundamental processes will remain unreliable, even after bias adjustment [10,17]. In any case it should be emphasized that the productive application of the proposed here or any other BC technique is based on a understanding not only of the implemented statistical method, but also the relevant climatic processes and their representation of the considered climate model [6].

Acknowledgements. The authors would like to express their deep gratitude to the organizations and institutes for the providing free of charge software and data. Personal thanks to I. Tsonevsky from ECMWF for the cooperation. The study is funded by the Bulgarian National Science Fund (DN–14/3 13.12.2017). We acknowledge the provided access to the e-infrastructure of the Centre for Advanced Computing and Data Processing, with the financial support by the Grant No BG05M2OP001–1.001–0003, financed by the Science and Education for Smart Growth Operational Program (2014–2020) and co-financed by the EU through the European structural and Investment funds.

References

1. Alexander, L.V., et al.: Global observed changes in daily climate extremes of temperature and precipitation. J. Geophys. Res. **111**, D05109 (2006). https://doi.org/10.1029/2005JD006290
2. Belda, M., Skalák, P., Farda, A., Halenka, T., Déqué, M., Csima, G., Bartholy, J., Torma, C., Boroneant, C., Caian, M., Spiridonov, V.: CECILIA regional climate simulations for future climate: analysis of climate change signal. Adv. Meteorol. **2015** (2015). https://doi.org/10.1155/2015/354727. 13 pages, Article ID 354727
3. Birsan, M.V., Dumitrescu, A., Micu, D.M., et al.: Nat. Hazards **74**, 1899 (2014). https://doi.org/10.1007/s11069-014-1290-5
4. Buser, C.M., Künsch, H.R., Lüthi, D., Wild, M., Schär, C.: Bayesian multi-model projection of climate: bias assumptions and interannual variability. Clim. Dyn. **33**, 849–868 (2009). https://doi.org/10.1007/s00382-009-0588-6
5. Chervenkov, H., Ivanov, V., Gadzev, G., Ganev, K.: Sensitivity study of different RegCM4.4 model set-ups–recent results from the TVRegCM experiment. Cybern. Inf. Technol. **17**(5) 17–26 (2018). https://doi.org/10.1515/cait-2017-0051
6. Dai, A.: Precipitation characteristics in eighteen coupled climate models. J. Climate **19**, 4605–4630 (2006). https://doi.org/10.1175/JCLI3884.1
7. Dimov, I., Georgieva, R., Ivanovska, S., Ostromsky, T., Zlatev, Z.: Studying the sensitivity of pollutants' concentrations caused by variations of chemical rates. J. Comput. Appl. Math. **235**, 391–402 (2010). https://doi.org/10.1016/j.cam.2010.05.041
8. Dosio, A., Paruolo, P.: Bias correction of the ENSEMBLES high-resolution climate change projections for use by impact models: evaluation on the present climate. J. Geophys. Res. **116**, D16106 (2011). https://doi.org/10.1029/2011JD015934
9. Dosio, A., Paruolo, P., Rojas, R.: Bias correction of the ENSEMBLES high resolution climate change projections for use by impact models: analysis of the climate change signal. J. Geophys. Res. **117**, D17110 (2012). https://doi.org/10.1029/2012JD017968

10. Dosio, A.: Projections of climate change indices of temperature and precipitation from an ensemble of bias-adjusted high-resolution EURO-CORDEX regional climate models. J. Geophys. Res. Atmos. **121**, 5488–5511 (2016). https://doi.org/10.1002/2015JD024411
11. Haerter, J.O., Hagemann, S., Moseley, C., Piani, C.: Climate model bias correction and the role of timescales. Hydrol. Earth Syst. Sci. **15**, 1065–1079 (2011). https://doi.org/10.5194/hess-15-1065-2011
12. Halenka, T.: Regional climate modeling activities in CECILIA project: introduction. Idöjárás **112**, III–IX (2008)
13. Hall, A.: Projecting regional change. Science **346**(6216), 1461–1462 (2014). https://doi.org/10.1126/science.aaa0629
14. Harding, A., Rivington, M.E., Mineter, M.J., Tett, S.F.B.: Agro-meteorological indices and climate model uncertainty over the UK. Clim. Change **128**, 113–126 (2015). https://doi.org/10.1007/s10584-014-1296-8
15. Haylock, M.R., Hofstra, N., Klein Tank, A.M.G., Klok, E.J., Jones, P.D., New, M.: A European daily high-resolution gridded dataset of surface temperature and precipitation. J. Geophys. Res. (Atmospheres) **113**, D20119 (2008). https://doi.org/10.1029/2008JD10201
16. Kotlarski, S., Keuler, K., Christensen, O.B., Colette, A., Déqué, M., Gobiet, A., Goergen, K., Jacob, D., Lüthi, D., van Meijgaard, E., Nikulin, G., Schär, C., Teichmann, C., Vautard, R., Warrach-Sagi, K., Wulfmeyer, V.: Regional climate modeling on European scales: a joint standard evaluation of the EURO-CORDEX RCM ensemble. Geosci. Model Dev. **7**, 1297–1333 (2014). https://doi.org/10.5194/gmd-7-1297-2014
17. Maraun, D.: Bias correcting climate change simulations - a critical review. Curr. Clim. Change Rep. **2**, 211–220 (2016). https://doi.org/10.1007/s40641-016-0050-x
18. Moberg, A., et al.: Indices for daily temperature and precipitation extremes in Europe analyzed for the period 1901–2000. J. Geophys. Res. **111**, D22106 (2006). https://doi.org/10.1029/2006JD007103
19. Nakicenovic, N.: Intergovernmental Panel on Climate Change: Emission Scenarios. Cambridge University Press, Cambridge (2000). A Special Report of Working Group III of the Intergovernmental Panel on Climate Change
20. Piani, C., Weedon, G.P., Best, M., Gomes, S.M., Viterbo, P., Hagemann, S., Haerter, J.O.: Statistical bias correction of global simulated daily precipitation and temperature for the application of hydrological models. J. Hydrol. **395**, 199–215 (2010). https://doi.org/10.1016/j.jhydrol.2010.10.024
21. Rummukainen, M.: State-of-the-art with regional climate models. WIREs Clim. Change **1**, 82–96 (2010). https://doi.org/10.1002/wcc.8
22. Sillmann, J., Röckner, E.: Indices for extreme events in projections of anthropogenic climate change. Clim. Change **86**, 83 (2008). https://doi.org/10.1007/s10584-007-9308-6
23. Spiridonov, V., Déqué, M., Somot, S.: ALADIN-CLIMATE: from the origins to present date. ALADIN Newslett. **29**, 2005 (2005)
24. Teutschbein, C., Wetterhall, F., Seibert, J.: Evaluation of different downscaling techniques. Dyn. Clim. (2011). https://doi.org/10.1007/s00382-010-0979-8
25. Xue, Y., Janjic, Z., Dudhia, J., Vasic, R., De Sales, F.: A review on regional dynamical downscaling in intraseasonal to seasonal simulation/prediction and major factors that affect downscaling ability. Atm. Res. **147–148**(2014), 68–85 (2014). https://doi.org/10.1016/j.atmosres.2014.05.001

Second Order Shifted Approximations for the First Derivative

Venelin Todorov[1,2], Yuri Dimitrov[1,3](✉), and Ivan Dimov[2]

[1] Department of Information Modeling, Institute of Mathematics and Informatics, Bulgarian Academy of Sciences, 1113 Sofia, Bulgaria
vtodorov@math.bas.bg, venelin@parallel.bas.bg

[2] Department of Parallel Algorithms, Institute of Information and Communication Technologies, Bulgarian Academy of Sciences, 1113 Sofia, Bulgaria
ivdimov@bas.bg

[3] Department of Mathematics and Physics, University of Forestry, 1756 Sofia, Bulgaria
yuri.dimitrov@ltu.bg

Abstract. In the present paper we construct second order shifted approximations for the first derivative which have exponential and logarithmic generating functions. Applications of the approximations for numerical solution of first order ordinary differential equations and the heat equation are studied in the paper.

1 Introduction

Approximations for fractional derivatives and integrals and the properties of their generating functions are studied in [2–5,7]. Let $\mathscr{A}_n[y(t)] = \tau^{-\alpha}\sum_{k=0}^{n} w_k y_{n-k}$ be an approximation for the fractional derivative of order α on the interval $[t_0, T]$, where $\tau = (T-t_0)/N$ and $t_n = t_0 + n\tau,\ y_n = y(t_n)$ for $0 \leq n \leq N$. Approximation $\mathscr{A}_n[y(t)]$ has a generating function $G(t) = \sum_{n=0}^{\infty} w_n t^n$. The values of α in the approximations for the first derivative and the definite integral are $\alpha = 1$ and $\alpha = -1$. In [2] we construct approximations for the definite integral of order 2, 3 and 4 which have generating functions $\bar{G}_1(t) = \pi \sec\left(\pi\sqrt{t}/2\right)/4$ and $\bar{G}_2(t) = \pi \tan\left(\pi\sqrt{t}/2\right)/(4\sqrt{t})$. The functions $\bar{G}_i(t)$ satisfy the condition $\lim_{t\to 1}(t-1)\bar{G}_i(t) = -1$ for $i = 1, 2$. The first order central difference approximation is a second order approximation for the first derivative with a shift value 0.5τ and a generating function $G_0(t) = 1 - t$

$$\frac{1}{\tau}(y_n - y_{n-1}) = y'_{n-0.5} + O\left(\tau^2\right). \tag{1}$$

In Sect. 2 we construct the second order shifted approximations for the first derivative which have exponential and logarithmic generating functions

$$G_1(t) = a - ae^{\frac{t-1}{a}}, \qquad G_2(t) = \frac{1}{b}\ln(b+1-bt),$$

I. Dimov and S. Fidanova (Eds.): HPC 2019, SCI 902, pp. 428–437, 2021.
https://doi.org/10.1007/978-3-030-55347-0_36

where $|a| > 1$ and $-0.5 < b < 1$ $(b \neq 0)$. The generating functions $G_i(t)$ of the approximations of the first derivative have properties $G_i(1) = 0$ and $G_i'(1) = -1$ for $i = 0, 1, 2$. Denote $H_i(t) = G_i(e^{-t})$. The functions $H_i(t)$ have third order Maclaurin series $H_i(t) = t - \gamma_i t^2 + O(t^3)$. From the Maclaurin series of the functions $G_i(t)$ and $H_i(t)$ we obtain the second order approximations $(i = 1, 2)$:

$$\frac{1}{\tau}\sum_{k=0}^{n-1} w_k^{(i)} y_{n-k} = y'_{n-\gamma_i} + O\left(\tau^2\right), \tag{2}$$

Where

$$w_0^{(1)} = a - ae^{-\frac{1}{a}}, \quad w_k^{(1)} = -\frac{e^{-\frac{1}{a}}}{k!a^{k-1}}, \quad (1 \leq k \leq n-1); \qquad \gamma_1 = \frac{a+1}{2a}, \tag{3}$$

$$w_0^{(2)} = \frac{\ln(1+b)}{b}, \quad w_k^{(2)} = -\frac{1}{bk}\left(\frac{b}{b+1}\right)^k, \ (k \geq 1); \qquad \gamma_2 = \frac{b+1}{2}. \tag{4}$$

Shifted approximations (2) have a second order accuracy when $y(t_0) = y'(t_0) = 0$. In Sect. 3 we derive a modification of the last three weights of approximations (2):

$$w_{n-2}^{(m,i)} = w_{n-2}^{(i)} + \frac{W_n^{(i)}}{2}, \quad w_{n-1}^{(m,i)} = w_{n-1}^{(i)} - 2W_n^{(i)}, \quad w_n^{(m,i)} = \frac{3W_n^{(i)}}{2} - \sum_{k=0}^{n-1} w_k^{(i)}, \tag{5}$$

Where $W_n^{(i)} = \sum_{k=0}^{n-1}(n-k)w_k^{(i)} - 1$. The modified weights (5) are computed with $O(N)$ operations. Approximations (2) with weights $w_{n-2}^{(m,i)}, w_{n-1}^{(m,i)}$ and $w_n^{(m,i)}$ have a second order accuracy for all functions $y \in C^2[t_0, T]$. In Sect. 3 we apply shifted approximations (1) and (2) for computing the numerical solutions of ordinary and partial differential equations. When $a > 1$ and $0 < b < 1$ the shift values and the weights of approximations (2) satisfy $1/2 < \gamma_i < 1$ and

$$w_0^{(i)} > 0, \quad w_1^{(i)} < w_2^{(i)} < \cdots < w_n^{(i)} < \cdots < 0. \tag{6}$$

The expansion formulas of the approximations of the fractional derivatives involve their integer-order derivatives [3]. Approximations (2) are suitable for construction of approximations for the fractional derivatives which have properties (6) of the weights. The first derivative and its approximations have applications in the sensitivity analysis of mathematical models in engineering, physics, telecommunications and human cyber-physical systems [6,8,9].

2 Second Order Approximations of the First Derivative

Let $G_i(t) = \sum_{n=0}^{\infty} w_n^{(i)} t^n$ and $H_i(t) = \sum_{n=0}^{\infty} c_n^{(i)} t^n$ be the Maclaurin series of the functions $G_i(t)$ and $H_i(t)$. The functions $H_i(t)$ satisfy $H_i(0) = c_0^{(i)} = 0,\ H_i'(0) =$

$c_1^{(i)} = 1,$

$$H_1(t) = G_1(e^{-t}) = a - ae^{\frac{e^{-t}-1}{a}}, \qquad H_1''(0) = -\frac{a+1}{a} = -2\gamma_1,$$
$$H_2(t) = G_2(e^{-t}) = \frac{1}{b}\ln\left(b+1-be^{-t}\right), \quad H_2''(0) = -b-1 = -2\gamma_2,$$

and have third order Mclaurin series expansions

$$H_1(t) = t - \frac{a+1}{2a}t^2 + O(t^3), \quad H_2(t) = t - \frac{b+1}{2}t^2 + O(t^3). \tag{7}$$

From (7) we obtain the approximations for the first derivative

$$\frac{1}{\tau}\sum_{k=0}^{n} w_k^{(i)} y(T-k\tau) = y'(T) - \gamma_i y''(T)\tau + O\left(\tau^2\right), \tag{8}$$

$$w_0^{(1)} = a - ae^{-\frac{1}{a}},\ w_k^{(1)} = -\frac{e^{-\frac{1}{a}}}{k!a^{k-1}}; \quad w_0^{(2)} = \frac{\ln(1+b)}{b},\ w_k^{(2)} = \frac{-1}{bk}\left(\frac{b}{b+1}\right)^k,$$

Where $k = 1, 2, \cdots$. Second order shifted approximations (2) are obtained from (8) and $y'(T-\gamma_i\tau) = y'(T) - \gamma_i y''(T)h + O\left(\tau^2\right)$. The shift value $\gamma_1 = \frac{a+1}{2a} \neq 1/2$ and satisfies $0 < \gamma_1 < 1$ when $|a| > 1$, where $1/2 < \gamma_1 < 1$ when $a > 1$. The weights $w_k^{(2)}$ converge to zero when $b > -1/2$. The shift value $\gamma_2 = \frac{b+1}{2} \neq 1/2$ and satisfies $0 < \gamma_2 < 1$ when $b < 1$, where $1/2 < \gamma_2 < 1$ when $0 < b < 1$. Now we prove that approximations (8) have a second order accuracy when the function y satisfies the condition $y^{(i)}(t_0) = 0$ for $i = 0, 1, 2$. The reminder term of the Taylor series of the function y is expressed in an integral form as

$$y(T) - \sum_{k=0}^{n-1} \frac{y^{(k)}(t_0)}{k!}(T-t_0)^k = \frac{1}{(n-1)!}\int_{t_0}^{T}(T-t)^{n-1}y^{(n)}(t)dt.$$

By using integration by parts we can show that

$$\frac{1}{m!}\int_{t_0}^{T}(T-t)^m y^{(n)}(t)dt = y^{(n-m-1)}(T), \tag{9}$$

When the function y satisfies $y^{(m)}(t_0) = 0$ for all $0 \leq m \leq n-1$. Denote by $P_m^{(i)}(t)$ the coefficients of z^m in the series expansion of the functions $e^{tz}H_i(z\tau)$

$$e^{tz}H_i(z\tau) = \sum_{m=0}^{\infty} P_m^{(i)}(t)z^m.$$

The polynomials $P_m^{(i)}(t)$ have degree $m-1$ and

$$P_m^{(i)}(t) = \sum_{k=0}^{m-1} \frac{c_{m-k}^{(i)}\tau^{m-k}}{k!}t^k.$$

From (9), when the function y satisfies $y^{(k)}(t_0) = 0$ for $0 \leq k \leq m-1$

$$\int_{t_0}^{T} P_{m-1}^{(i)}(T-t) y^{(m)}(t) dt = \sum_{k=1}^{m-1} c_k^{(i)} y^{(k)}(T) \tau^k. \tag{10}$$

The polynomial $P_2^{(i)}(t) = \tau t + c_2^{(i)} \tau^2$. From (10) with $m = 3$ we obtain

$$J_2^{(i)} y(t) = \int_{t_0}^{T} P_2^{(i)}(T-t) y'''(t) dt = y'(T)\tau + c_2^{(i)} y''(T)\tau^2. \tag{11}$$

The function $H_i(\tau z) e^{tz}$ satisfies

$$\begin{aligned} H_i(\tau z) e^{tz} &= e^{tz} G_i(e^{-z\tau}) = e^{tz} \sum_{m=0}^{\infty} w_m^{(i)} e^{-mz\tau} = \sum_{m=0}^{\infty} w_m^{(i)} e^{z(t-m\tau)} \\ &= \sum_{m=0}^{\infty} w_m^{(i)} \sum_{k=0}^{\infty} \frac{z^k (t-m\tau)^k}{k!} = \sum_{m=0}^{\infty} \frac{z^m}{m!} \sum_{k=0}^{\infty} w_k^{(i)} (t-k\tau)^m. \end{aligned}$$

The polynomials $P_m^{(i)}(t)$ are expressed with the coefficients $w_k^{(i)}$ as

$$P_m^{(i)}(t) = \frac{1}{m!} \sum_{k=0}^{\infty} w_k^{(i)} (t-k\tau)^m. \tag{12}$$

Theorem 1. *Let $y \in C^3[t_0, T]$ and $y^{(k)}(t_0) = 0$ for $k = 0, 1, 2$. Then*

$$\mathscr{A}_n^{(i)}[y(t)] = \frac{1}{\tau} \sum_{k=0}^{n} w_k^{(i)} y(T-k\tau) = y'(T) + c_2^{(i)} y''(T)\tau + O\left(\tau^2\right).$$

Proof. Denote $S_n^{(i)} y(t) = \tau \mathscr{A}_n^{(i)}[y(t)]$. From (9) with $m = 2$ and $n = 3$

$$S_n^{(i)} y(t) = \sum_{k=0}^{n-1} w_k^{(i)} y(T-k\tau) = \frac{1}{2} \sum_{k=0}^{n-1} w_k^{(i)} \int_{t_0}^{T-kh} (T-k\tau-t)^2 y'''(t) dt. \tag{13}$$

From (12)

$$J_2^{(i)} y(t) = \int_{t_0}^{T} P_2^{(i)}(T-t) y'''(t) dt = \frac{1}{2} \sum_{k=0}^{\infty} w_k^{(i)} \int_{t_0}^{T} (T-t-k\tau)^2 y'''(t) dt,$$

$$\begin{aligned} J_2^{(i)} y(t) &= \frac{1}{2} \sum_{k=0}^{n-1} w_k^{(i)} \int_{t_0}^{T-kh} (T-t-k\tau)^2 y'''(t) dt \\ &+ \frac{1}{2} \sum_{k=0}^{n-1} w_k^{(i)} \int_{T-kh}^{T} (T-t-k\tau)^2 y'''(t) dt + \frac{1}{2} \sum_{k=n}^{\infty} w_k^{(i)} \int_{t_0}^{T} (T-t-k\tau)^2 y'''(t) dt. \end{aligned} \tag{14}$$

Let $|y'''(t)| \le C$. Denote

$$Q_n^{(i)} y(t) = \frac{1}{2} \sum_{k=n}^{\infty} w_k^{(i)} \int_{t_0}^{T} (T - t - k\tau)^2 y'''(t) dt.$$

The numbers $Q_n^{(i)} y(t)$ satisfy

$$|Q_n^{(i)} y(t)| \le \frac{C}{2} \sum_{k=n}^{\infty} |w_k^{(i)}| \int_{t_0}^{T} (T - t - k\tau)^2 dt = \frac{Ch^3}{6} \sum_{k=n}^{\infty} |w_k^{(i)}| (k^3 - (k-n)^3),$$

$$|Q_n^{(i)} y(t)| \le \frac{Ch^3}{6} \sum_{k=n}^{\infty} |w_k^{(i)}| k^3.$$

From (13) and (14)

$$|J_2^{(i)} y(t) - S_n^{(i)} y(t)| \le \frac{1}{2} \sum_{k=0}^{n-1} |w_k^{(i)}| \int_{T-k\tau}^{T} (T - t - k\tau)^2 |f'''(t)| dt + |Q_n^{(i)} y(t)|$$

$$\le \frac{C}{2} \sum_{k=0}^{n-1} |w_k^{(i)}| \int_{T-k\tau}^{T} (T - t - k\tau)^2 dt + |Q_n^{(i)} y(t)| = \frac{Ch^3}{6} \sum_{k=0}^{n-1} |w_k^{(i)}| k^3 + |Q_n^{(i)} y(t)|,$$

Hence

$$|J_2^{(i)} y(t) - S_n^{(i)} y(t)| \le \frac{Ch^3}{6} \sum_{k=0}^{n-1} |w_k^{(i)}| k^3 + \frac{C\tau^3}{6} \sum_{k=n}^{\infty} |w_k^{(i)}| k^3 = \frac{Ch^3}{6} \sum_{k=0}^{\infty} |w_k^{(i)}| k^3.$$

The series $\sum_{k=1}^{\infty} |w_k^{(i)}| k^3$ converge when $|a| > 1$ and $-0.5 < b < 1$. From (11)

$$\mathscr{A}_n^{(i)} [y(t)] = \frac{1}{\tau} S_n^{(i)} y(t) = \frac{1}{\tau} J_2^{(i)} y(t) + O\left(\tau^2\right) = y'(T) + c_2^{(i)} y''(T) \tau + O\left(\tau^2\right).$$

□

3 Numerical Solutions of Differential Equations

In this section we compute the numerical solutions of the first order ODE

$$y'(t) + P(t) y(t) = F(t), \; y(0) = y_0, \quad t \in [0, T] \tag{15}$$

and the heat equation which use shifted approximations (1) and (2) of the first derivative. Let $\tau = T/N$ and $t_n = n\tau$, $y_n = y(t_n)$. By approximating $y'_{n-0.5}$ with (1) and $y_{n-0.5} = 0.5(y_n + y_{n-1}) + O\left(\tau^2\right)$ we obtain

$$\frac{1}{\tau}(y_n - y_{n-1}) + \frac{P_{n-0.5}}{2}(y_n + y_{n-1}) = F_{n-0.5} + O\left(\tau^2\right).$$

$$2(y_n - y_{n-1}) + \tau P_{n-0.5}(y_n + y_{n-1}) = 2\tau F_{n-0.5} + O\left(\tau^3\right).$$

The numerical solution $\{U_n\}_{n=0}^N$, where U_n is an approximation for the value of the solution y_n of Eq. (15) is computed with

$$U_n = \frac{1}{2+\tau P_{n-0.5}}\left((2-\tau P_{n-0.5})U_{n-1} + 2\tau F_{n-1/2}\right). \tag{NSCD1}$$

When $n=1$ the number $U_1 = ((2-\tau P_{0.5})U_0 + 2\tau F_{0.5})/(2+\tau P_{0.5})$ is a second order approximation for the value of the solution $y(\tau)$ of Eq. (15). Let

$$\frac{1}{\tau}\sum_{k=0}^{n} w_k y_{n-k} = y'_{n-\gamma} + O\left(\tau^2\right) \tag{*}$$

be a second order shifted approximation of the first derivative. By approximating $y'_{n-\gamma}$ with (*) and $y_{n-\gamma} = \gamma y_{n-1} + (1-\gamma)y_n + O\left(\tau^2\right)$ we obtain

$$\frac{1}{\tau}\sum_{k=0}^{n} w_k y_{n-k} + P_{n-\gamma}(\gamma y_{n-1} + (1-\gamma)y_n) = F_{n-\gamma} + O\left(\tau^2\right),$$

$$(w_0 + \tau(1-\gamma)P_{n-\gamma})y_n = \tau F_{n-\gamma} - \tau P_{n-\gamma}\gamma y_{n-1} - \sum_{k=1}^{n} w_k y_{n-k} + O\left(\tau^3\right).$$

The numerical solution $\{U_n\}_{n=0}^N$ of Eq. (15) is computed as

$$U_n = \frac{1}{w_0 + \tau(1-\gamma)P_{n-\gamma}}\left(\tau F_{n-\gamma} - hP_{n-\gamma}\gamma U_{n-1} - \sum_{k=1}^{n} w_k U_{n-k}\right). \tag{NS1(*)}$$

Example 1

$$y'(t) + (1+t^2)y(t) = t^3(1+\ln t(4+t+t^3)),\ y(0)=0, \quad t\in[0,1]. \tag{16}$$

The solution $y(t) = t^4 \ln t$ of Eq. (16) satisfies $y^{(k)}(0) = 0$ for $k = 0,1,2$. From Theorem 1 shifted approximations (2) of the first derivative of the function $t^4 \ln t$ have a second order accuracy. Denote by (2:3) and (2:4) approximations (2) with weights (3) and (4) respectively. The numerical results for the error and the order of numerical solutions NS1(2:3), NS1(2:4), NSCD1 and initial conditions $U_0 = U_1 = 0$ are presented in Table 1.

Now we use the method from [3,4] for modifying the last three weights of (2). The modified approximations (2) have a second order accuracy for all functions $y \in C^2[t_0, T]$. From Theorem 1

$$\mathscr{A}_n^{(i)}[y(t)] = \frac{1}{\tau}\sum_{k=0}^{n-1} w_k^{(i)} y_{n-k} = y'_n + c_2^{(i)} y''_n \tau + O\left(\tau^2\right)$$

when the function y satisfies $y(t_0) = y'(t_0) = 0$ and the accuracy of approximations (2) is exponential when $y(t) = (t-t_0)^2$. Let $z(t) = y(t) - y(t_0) - (t-t_0)y'(t_0)$. Then

$$y'(t) = z'(t) + y'(t_0),\ y''(t_0) = z''(t_0),$$

Table 1. Error and order of numerical solutions of Eq. (16) and $\gamma_1 = \gamma_2 = 0.45$.

τ	NS1(2 : 3), a = −10		NS1(2 : 4), b = −0.1		NSCD1	
	Error	*Order*	*Error*	*Order*	*Error*	*Order*
0.0125	4.677×10^{-5}	2.003	4.811×10^{-5}	2.002	6.736×10^{-5}	1.999
0.00625	1.167×10^{-5}	2.002	1.202×10^{-5}	2.001	1.685×10^{-5}	1.999
0.003125	2.916×10^{-6}	2.001	3.002×10^{-6}	2.001	4.212×10^{-6}	2.000

$$y'_n - y'_0 + c_2^{(i)} y''_n \tau = z'_n + c_2^{(i)} z''_n \tau$$

and

$$\begin{aligned}\mathscr{A}_n^{(i)}[y(t)] &= \mathscr{A}_n^{(i)}[z(t)] + y_0 \mathscr{A}_n^{(i)}[1] + y'_0 \mathscr{A}_n^{(i)}[t - t_0] \\ &= y'_n + c_2^{(i)} y''_n \tau + y_0 \mathscr{A}_n[1] + y'_0 (\mathscr{A}_n[t - t_0] - 1) + O\left(\tau^2\right),\end{aligned}$$

$$\mathscr{A}_n[y(t)] = y'_n + c_2^{(i)} y''_n \tau + \frac{y_0}{\tau} \sum_{k=0}^{n-1} w_k^{(i)} + y'_0 \left(\sum_{k=0}^{n-1} (n-k) w_k^{(i)} - 1 \right) + O\left(\tau^2\right). \tag{17}$$

Denote

$$W_n^{(i)} = \sum_{k=0}^{n-1} (n-k) w_k^{(i)} - 1.$$

By substituting $y'_0 = (-0.5y_2 + 2y_1 - 1.5y_0)/\tau + O\left(\tau^2\right)$ in (17) we obtain

$$\mathscr{A}_n[y(t)] = y'_n + c_2^{(i)} y''_n \tau + \frac{y_0}{\tau} \sum_{k=0}^{n-1} w_k^{(i)} + \left(-\frac{1}{2} y_2 + 2y_1 - \frac{3}{2} y_0 \right) \frac{W_n^{(i)}}{\tau} + O\left(\tau^2\right). \tag{18}$$

Let $\bar{\mathscr{A}}_n^{(i)}[y(t)] = \frac{1}{\tau} \sum_{k=0}^{n-1} \bar{w}_k^{(i)} y_{n-k}$, where $\bar{w}_k^{(i)} = w_k^{(i)}$ for $0 \leq k \leq n-3$ and

$$\bar{w}_{n-2}^{(i)} = w_{n-2}^{(i)} + \frac{W_n^{(i)}}{2}, \quad \bar{w}_{n-1}^{(i)} = w_{n-1}^{(i)} - 2W_n^{(i)}, \quad \bar{w}_n^{(i)} = \frac{3W_n^{(i)}}{2} - \sum_{k=0}^{n-1} w_k^{(i)}.$$

When $n = 1$ the shifted approximations of the first derivative have a second order accuracy only when the shift value is -0.5τ. From (18)

$$\bar{\mathscr{A}}_n[y(t)] = y'_n + c_2^{(i)} y''_n \tau + O\left(\tau^2\right)$$

for all functions $y \in C^2[t_0, T]$ and $n \geq 2$.

Example 2

$$y'(t) + \frac{y(t)}{1+t^2} = \frac{e^t(2+t^2)}{1+t^2}, \; y(0) = 1, \quad t \in [0, 1]. \tag{19}$$

Equation (19) has the solution $y(t) = e^t$. The numerical results for the error and the order of numerical solutions NS1(2:3), NS1(2:4) and NSCD1 of Eq. (19) and initial conditions $U_0 = 1$ and $U_1 = ((2 - \tau P_{0.5})U_0 + 2\tau F_{0.5})/(2 + \tau P_{0.5})$ are presented in Table 2.

Table 2. Error and order of numerical solutions of Eq. (19) and $\gamma_1 = \gamma_2 = 0.55$.

τ	NS1(2 : 3), a = 10		NS1(2 : 4), b = 0.1		NSCD1	
	Error	*Order*	*Error*	*Order*	*Error*	*Order*
0.0125	3.204×10^{-5}	1.998	3.244×10^{-5}	1.998	7.781×10^{-5}	1.994
0.00625	8.115×10^{-6}	1.999	8.115×10^{-6}	1.999	1.949×10^{-5}	1.997
0.003125	2.004×10^{-6}	2.000	2.029×10^{-6}	2.000	4.878×10^{-6}	1.999

Now we construct finite difference schemes for numerical solution of the heat equation which use approximations (1) and (2) for the first derivative.

$$\begin{cases} u_t(x,t) = Du_{xx}(x,t) + F(x,t), \quad (x,t) \in [0,X] \times [0,T], \\ u(x,0) = u_0(x),\ u(0,t) = u_1(t),\ u(1,t) = u_2(t). \end{cases} \tag{20}$$

Let $h = X/N, \tau = T/M$ and U_n^m be an approximation of the solution $u(nh, m\tau)$ of Eq. (20), where $0 \leq n \leq N$ and $0 \leq m \leq M$. The Crank–Nicolson method is a main second order method for numerical solution of the heat equation [1]:

$$\frac{U_n^m - U_n^{m-1}}{\tau} = \frac{D}{2h^2}\left(U_{n-1}^m - 2U_n^m + U_{n+1}^m + U_{n-1}^{m-1} - 2U_n^{m-1} + U_{n+1}^{m-1}\right) + F_n^{m-0.5},$$

$$-U_{n-1}^m + \left(2 + \frac{2h^2}{D\tau}\right)U_n^m - U_{n+1}^m = U_{n-1}^{m-1} + \left(\frac{2h^2}{D\tau} - 2\right)U_n^{m-1} + U_{n+1}^{m-1} + \frac{2h^2}{D}F_n^{m-0.5}. \tag{NSCD2}$$

By approximating $u_t(x_n, t_{m-\gamma})$ with (*) and $\gamma u_{xx}(x_n, t_{m-1}) + (1-\gamma)u_{xx}(x_n, t_m)$ using second order central differences we obtain

$$\frac{1}{\tau}\sum_{k=0}^{m} w_k U_n^{m-k} = \frac{D}{h^2}((1-\gamma)(U_{n-1}^m - 2U_n^m + U_{n+1}^m) + \gamma(U_{n-1}^{m-1} - 2U_n^{m-1} + U_{n+1}^{m-1})) + F_n^{m-\gamma},$$

$$-U_{n-1}^m + \left(2 + \frac{h^2 w_0}{(1-\gamma)\tau D}\right)U_n^m - U_{n+1}^m = \frac{h^2}{(1-\gamma)D}F_n^{m-\gamma} - \frac{h^2}{\tau(1-\gamma)D}\sum_{k=1}^{m} w_k U_n^{m-k} + \frac{\gamma}{1-\gamma}(U_{n-1}^{m-1} - 2U_n^{m-1} + U_{n+1}^{m-1}), \tag{NS2(*)}$$

where $2 \leq m \leq M$. When $m = 1$ finite difference scheme NS2(*) of the heat equation uses the linear system from NSCD2. Finite difference schemes NSCD2 and NS2(*) have boundary conditions $U_0^m = u_1(t_m)$, $U_N^m = u_2(t_m)$ for $1 \leq n \leq N-1$ at time $m\tau$.

Example 3

$$\begin{cases} u_t(x,t) = 2u_{xx}(x,t) - e^{x+t}, & (x,t) \in [0,1] \times [0,1], \\ u(x,0) = e^x, \; u(0,t) = e^t, \; u(1,t) = e^{1+t}. \end{cases} \tag{21}$$

Equation (21) has the solution $u(x,t) = e^{x+t}$. The numerical results for the error and the order of numerical solutions NS2(2:3), NS2(2:4) and NSCD2 of heat Eq. (21) are presented in Table 3.

Table 3. Error and order of numerical solutions of Eq. (19) and $\gamma_1 = \gamma_2 = 0.505$.

$h = \tau$	$NS2(2:3),\ a = 100$		$NS2(2:4),\ b = 0.01$		NSCD2	
	Error	*Order*	*Error*	*Order*	*Error*	*Order*
0.0125	1.593×10^{-5}	1.999	1.593×10^{-5}	1.999	1.604×10^{-5}	1.999
0.00625	3.984×10^{-6}	1.999	3.984×10^{-6}	1.999	4.011×10^{-6}	1.999
0.003125	9.960×10^{-7}	2.000	9.959×10^{-7}	2.000	1.003×10^{-6}	2.000

4 Conclusion

In the present paper we construct second order shifted approximations (2) of the first derivative. An advantage of approximations (2) is that they can be applied for construction of approximations of the Caputo derivative which have properties (6) of the weights. The method for construction of approximations of the first derivative which have exponential and logarithmic generating functions can be extended to approximations of the second and higher order derivatives. In future work we will prove that the numerical solutions studied in the paper have a second order accuracy and we will apply the approximations for construction of approximations of the Caputo derivative.

Acknowledgements. The present study was initiated and implemented with the support of the National Science Program "Young Scientists and Postdoctoral Students". We express our gratitude for the concern for the young researchers, scientists and professors of Bulgaria. The first author Venelin Todorov is supported by Task 1.2.5 "Prediction and Guarantee of the Quality of Service in Human-Cyber-Physical Systems" of the Bulgarian National Scientific Program "Information and Communication Technologies for a Single Digital Market in Science, Education and Security" (ICT in SES), financed by the Ministry of Education and Science and by the National Scientific Program "Young scientists and Postdoctoral candidates" 2020–2021.

References

1. Burden, R.L., Faires, J.D.: Numerical Analysis, 9th edn. Brookscole, Boston (2011)
2. Dimitrov, Y., Miryanov, R., Todorov, V.: Quadrature formulas and Taylor series of secant and tangent. Econ. Comput. Sci. **4**, 23–40 (2017)
3. Dimitrov, Y.: Approximations for the Caputo derivative (I). J. Fractional Calc. Appl. **9**(1), 35–63 (2018)
4. Dimitrov, Y., Miryanov, R., Todorov, V.: Asymptotic expansions and approximations of the Caputo derivative. Comput. Appl. Math. **37**(4), 5476–5499 (2018)
5. Ding, H., Li, C.: High-order numerical algorithms for Riesz derivatives via constructing new generating functions. J. Sci. Comput. **71**(2), 759–784 (2017)
6. Epstein, M.: Partial Differential Equations: Mathematical Techniques for Engineers. Springer, Cham (2017)
7. Lubich, C.: Discretized fractional calculus. SIAM J. Math. Anal. **17**(3), 704–719 (1986)
8. Poryazov S., Saranova E., Ganchev I.: conceptual and analytical models for predicting the quality of service of overall telecommunication systems. In: Ganchev, I., van der Mei, R., van den Berg, H., (eds) Autonomous Control for a Reliable Internet of Services. State-of-the-Art Survey, Lecture Notes in Computer Science, vol. 10768, pp. 151–181. Springer, Cham (2018)
9. Poryazov, S.A.: A suitable unit of sensitivity in telecommunications. In: International Conference Telecom, Sofia, Bulgaria, pp. 165-172 (2011). ISSN 1314-2690

Parallel Correction for Hierarchical Re-Distancing Using the Fast Marching Method

Michael Quell[1(✉)], Georgios Diamantopoulos[1], Andreas Hössinger[2], Siegfried Selberherr[3], and Josef Weinbub[1]

[1] Christian Doppler Laboratory for High Performance TCAD, Institute for Microelectronics, TU Wien, Vienna, Austria
{quell,diamantopoulos,weinbub}@iue.tuwien.ac.at
[2] Silvaco Europe Ltd., Saint Ives, UK
andreas.hoessinger@silvaco.com
[3] Institute for Microelectronics, TU Wien, Vienna, Austria
selberherr@iue.tuwien.ac.at

Abstract. Topography simulation is typically implemented with the level-set method which uses the level-set function to represent the interface. The signed-distance property, capturing the distances from the entire simulation domain towards the interface, has to be regularly restored during a simulation in order to update the distances relative to the evolution of the interface. The restoring process is called 're-distancing' and the most established algorithm is the Fast Marching Method. For Cartesian grids, which are commonly used, high-performance applications require an adaptive resolution for geometric features, such as narrow trenches or sharp corners. Among the most important challenges is the need to utilize the solution in higher resolved regions to correct the solution in the embedding coarser regions. We present a parallelized bottom-up (i.e. from finest to coarsest resolution levels) correction technique for hierarchical re-distancing using the Fast Marching Method, which increases the accuracy of the discretized level-set function on the coarser grids. The coarser grids are corrected by interpolation on grid points covered by finer regions and a partial restart of the Fast Marching Method for the remaining grid points, thus minimizing the computational effort. This parallel correction step has been integrated into a recently developed parallel re-distancing algorithm, is implemented in C++ using OpenMP, and is evaluated for different geometries. The correction step significantly reduces the error in the signed-distance function, introducing a performance penalty of less than 10%.

1 Introduction

The level-set method [1] is used in topography simulations to track deforming volumes $\Omega \subset \mathbb{R}^d$. This can be achieved by implicitly representing the boundary of

I. Dimov and S. Fidanova (Eds.): HPC 2019, SCI 902, pp. 438–451, 2021.
https://doi.org/10.1007/978-3-030-55347-0_37

the volume as the zero level-set of a level-set function $\phi(\mathbf{x}, t)$. The time evolution of Ω is given by

$$\frac{\partial \phi}{\partial t} - V(\mathbf{x}, t) \cdot \nabla \phi = 0, \quad (1)$$

with V being the deforming velocity of Ω. A recent review on the level-set method and its applications, such as computational fluid dynamics, minimal surfaces, and epitaxial growth, is given in [2].

The signed-distance property ($|\nabla \phi(\mathbf{x})| = 1$) of the level-set function has to be regularly restored, since solving the level-set Eq. (1) does not generally preserve it [3]. This restoration procedure, called *re-distancing*, modifies a given level-set function ϕ so that the signed-distance property is fulfilled without changing the position of the zero level-set [4,5]. This is equivalent to solving the Eikonal equation

$$\begin{aligned} |\nabla \phi(\mathbf{x})| &= f(\mathbf{x}) \qquad \mathbf{x} \in \mathbb{R}^n, \\ \phi(\mathbf{x}) &= g(\mathbf{x}) \qquad \mathbf{x} \in \partial\Omega, \end{aligned} \quad (2)$$

for $f(\mathbf{x}) = 1$ and $g(\mathbf{x}) = 0$.

The Fast Marching Method (FMM) [1] is a widely used method to solve (2). The FMM is a one-pass algorithm using upwind differences. However, the use of a priority queue in the algorithm prohibits straightforward, efficient parallelization. This gave rise to other algorithms, most importantly the Fast Sweeping Method (FSM) [6] and the Fast Iterative Method (FIM) [7]. Both methods have been shown to be not competitive regarding accuracy due to their iterative nature [8]. Additionally, applications of the level-set method often employ narrow-band re-distancing[1] to reduce computational costs, which is not possible for the FSM [9]. Recently, a new approach using the Hopf-Lax formula, which is also iterative, has been developed showing a good parallel speed-up. However, comparisons to the FMM have yet to be made [9].

In another approach, a parallel version of the FMM via a distributed-memory domain decomposition approach has been developed, offering excellent accuracy [10]. Based on this approach a shared-memory parallelized FMM for multiple meshes within a hierarchical grid was developed [11], which was later extended to handle multiple meshes within a hierarchical grid [12,13].

Hierarchical grids enable covering geometrically challenging regions (e.g., corners, narrow trenches) with higher resolution (or *finer*) meshes on a separate, higher resolution grid level [14]. This method, therefore, supports higher accuracy in particular regions of interest without the need to globally increase the mesh resolution, thereby balancing the number of mesh elements with the accuracy. This is also known as adaptive mesh refinement.

As another consequence and the main context of this work, the accuracy of the solution in coarser grid levels can be increased by incorporating the more accurate solutions from the relevant finer levels, thereby correcting the local solution. In this work, such a correction algorithm tailored to the level-set method,

---

[1] The distance is only computed up to a given threshold value around the zero level-set [1].

is introduced and directly integrated into a FMM based re-distancing step with manageable computational overhead.

This work is structured as follows. Section 2 provides necessary background on the FMM. Section 3 introduces details of the employed hierarchy of grids and the scheme for re-distancing this hierarchy. Section 4 provides the details of the correction step. Section 5 evaluates the presented correction step with respect to improved accuracy using two-dimensional test cases. Additionally, computational benchmarks are presented to assess the parallel performance using a three-dimensional test case inspired by simulating a typical fabrication process step of a microelectronic structure.

2 Fast Marching Method

The FMM is based on the idea behind Dijkstra's algorithm [15] for finding the shortest path in a graph. The FMM hast two stages: initialization and marching. The initialization sets for all grid points the flag `Far` and their distance to infinity. Next, the flag `KnownFix` is assigned to the set of *given* grid points (i.e., points for which the distances are known in advance) and their pre-defined distances are assigned accordingly. Their neighboring grid points with flag `Far` are assigned the flag `Band` and the distance is calculated based on the upwind finite difference discretization

$$\begin{bmatrix} \max\left(D^{-x}_{ijk}\phi, -D^{+x}_{ijk}\phi, 0\right)^2 + \\ \max\left(D^{-y}_{ijk}\phi, -D^{+y}_{ijk}\phi, 0\right)^2 + \\ \max\left(D^{-z}_{ijk}\phi, -D^{+z}_{ijk}\phi, 0\right)^2 \end{bmatrix}^{\frac{1}{2}} = 1 \tag{3}$$

with D^-_{ijk} and D^+_{ijk} as the first order backward and forward difference operators [1].

For the marching stage, the grid point with the smallest distance to the interface and marked `Band` is set to `Known`. Subsequently, all its neighbors' distances are recalculated. If the newly calculated distance for a neighbor is smaller than the current distance, the new distance is assigned and the neighbor is marked `Band`. This is repeated until all grid points are flagged `Known` or `KnownFix`. In case of narrow-band re-distancing, the marching procedure is stopped, if the grid point with the smallest distance to the interface and marked `Band` is beyond the narrow-band threshold value.

3 Hierarchical Re-Distancing

Critical geometrical features of the interface, such as narrow trenches and sharp corners, require high spatial resolution to be appropriately represented [2]. As previously indicated, a hierarchical grid enables capturing of regions of interest by introducing additional hierarchy levels, offering locally defined meshes with increased resolution and thus avoiding a globally increased mesh resolution. In the following an overview of hierarchical grids and hierarchical re-distancing is given.

The here considered hierarchical grid consists of several levels. Level 1 is the coarsest grid. A subsequential Level N is refined by a factor $\mathbf{n_{N-1}}$ compared to Level $N-1$. Different resolutions for each spatial dimensions are possible. Every level holds a set of meshes, to represent the regions of interest which require a higher mesh resolution. Data is stored in a cell-centered manner.

We consider the following hierarchical grid properties: On Level 1 there is always only one mesh covering the whole domain. The meshes on the finer levels must conform to the following: (1) Meshes must not overlap, (2) a mesh must have exactly one parent mesh in the next coarser grid, (3) the meshes are aligned to the parent grid points, and (4) all the grid points in the parent mesh have neighboring grid points on the same refinement level. Figure 1 shows a two-dimensional example.

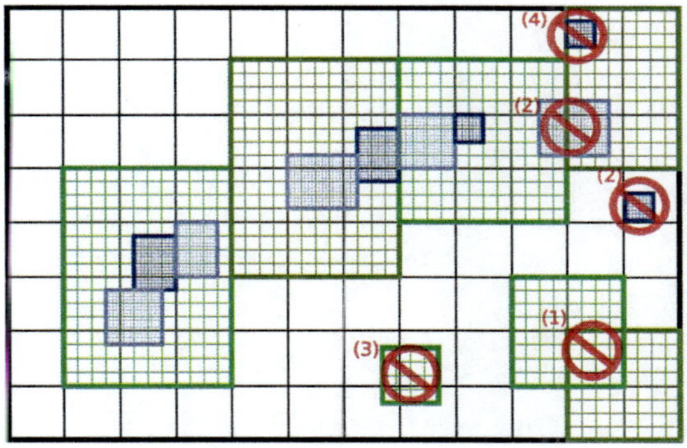

Fig. 1. Two-dimensional hierarchical grid with depth 3 and refinement factor 4 for all spatial directions and levels. Invalid configurations are highlighted in red with the number corresponding to the violated rules (1)–(4).

Hierarchical Re-Distancing

The re-distancing begins on Level 1, as this is the only mesh for which all the boundary conditions, i.e. the domain boundaries, are set. The FMM is initialized with the grid points set by the domain boundaries and the grid points which have a neighboring grid point with the opposite sign, since they are next to the interface and must not be modified. After the FMM finished processing a level, the boundaries of the meshes on the next finer level are set by interpolation and re-distancing is performed on it. This procedure is repeated for every level until the finest level is calculated.

The processing of a grid level has been parallelized by a coarse grained approach treating the individual meshes in parallel and enforcing an explicit data exchange on shared mesh boundaries via ghost layers [13]. This is implemented using OpenMP[2] task directives, where each mesh is processed by a separate task.

4 Correction Step

The developed correction step is based on the previously described hierarchical re-distancing step (i.e., the flags and distances are set according to the re-distancing algorithm [13]). The bottom-up correction begins on the finest level. For notational coherence as the correction spans over two levels, meshes on the finer level will be called child meshes and the meshes on the next coarser level are called parent meshes. Algorithm 1 shows the pseudocode of the correction step algorithm.

Algorithm 1: Correction Step

```
// From finest to coarsest level
ChildLevel ← FinestLevel
ParentLevel ← FinestLevel.parentLevel()
while ChildLevel != CoarsestLevel do
    // Initialization step
    for every Parent mesh in parallel do
        for every Child mesh in parallel do
            ChildList ← interpolateParentGridPoints()
        mergeChildLists()
        initializeFMMCorrection(ChildList)
    Barrier
    // Marching step (identical to the original FMM)
    for every Parent mesh in parallel do
        parallelFastMarchingMethod()
    Barrier
    // Move one level up
    ChildLevel ← ParentLevel
    ParentLevel ← ParentLevel.parentLevel()
```

Parent grid points covered by child grid points are corrected by evaluating the level-set function ϕ at the position of the parent grid point on the child level. ϕ is evaluated by linear interpolation.

Figure 2 provides a schematic representation of the different interpolation cases: There, the extracted position of the interface on the parent and child level is shown by the green lines, thick for the parent mesh and thin for the child mesh. Red dots are used for grid points located on one side of the interface and

[2] https://www.openmp.org/.

blue dots for the other side of the interface. All parent grid points are labeled with a letter from A to I. Not all child grid points are marked by a dot, because some are not computed by the FMM on the child level, due to narrow-band re-distancing.

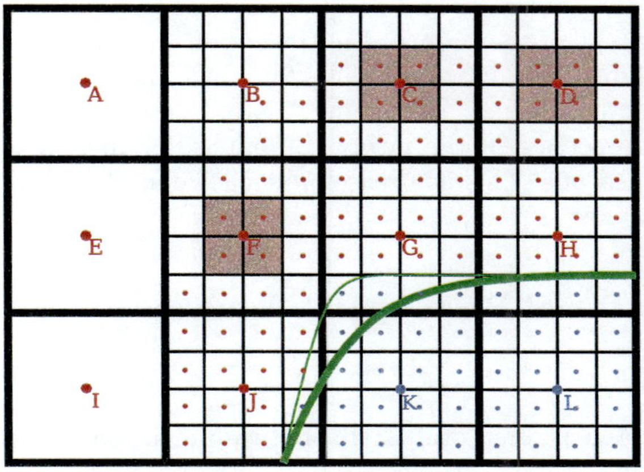

Fig. 2. Schematic representation of different cases for the interpolation of the parent level grid points.

The parent grid points G, H, J, K, and L have a neighbor with an opposite sign. They are marked `KnownFix` by the FMM and, therefore, must not be modified. The parent grid points A, E, and I are not covered by a fine mesh and thus cannot be interpolated. The parent grid point B is also not interpolated, because the narrow-band re-distancing on the child level did not compute all the necessary points for the interpolation. Finally, parent grid points C, D, and F are interpolated. For each of those grid points the linear interpolation is based on the four closest child grid points (marked by a red background).

The interpolated grid points are stored in separate *lists* for each child mesh. Subsequently, the lists are used for initializing the FMM. The mesh based parallelization approach presented in Sect. 3 is employed. An OpenMP task is created for all child meshes, since a parent grid point is refined by at most one child mesh (cf. Algorithm 1, Line 6).

The previously created lists for every child mesh with a common parent mesh are merged together. The lists are then used to initialize the FMM on the parent grid level. The actual initialization step, however, is modified compared to the original FMM. The grid points keep their distance and their flag from the previous re-distancing step. If a parent grid point is marked `KnownFix`, the correction is not applied, in order to avoid modifying the zero level-set (cf. Algorithm 1, Line 8). This is performed in parallel for all parent meshes with OpenMP tasks.

Then, the unmodified marching of the FMM is performed, correcting the solution in regions not covered by child meshes (cf. Algorithm 1, Line 10). The parallelization of the marching step is the same as for the regular re-distancing discussed in Sect. 3. After the FMM has finished processing, it continues by moving up in the hierarchy until Level 1 has been corrected.

The proposed initialization of the FMM avoids the re-computation of all grid points. The original FMM initializations (cf. Sect. 2) would trigger a re-computation of all grid points and therefore double the runtime on each level.

The disadvantage of the proposed initialization is, that during the marching step (cf. Algorithm 1, Line 11) grid points, for which the distance to the interface was under-estimated are not corrected. This is inherent to the FMM, as grid points are only processed if the newly computed distance is smaller than the previously assigned one. This is a reasonable choice as the FMM tends to over-estimate the distance [16].

5 Computational Results and Analyses

5.1 Test Cases

The correction step is first evaluated on two-dimensional examples: *Corner*, *Sharp Corner*, and *Narrow Trench* (Fig. 3a, Fig. 3b, and Fig. 4, respectively). The error norms for the different levels are compared in Table 1, Table 2, and Table 3. Additionally, a *Three-Dimensional Trench* example inspired by a fabrication simulation in the field of microelectronics is given (cf. Fig. 5). The error norms, as the simulations compare to the exact solution, are shown in Table 4. Due to the computationally significant load induced by the three spatial dimensions, the parallel speed up is investigated.

The domain for all test cases is $[-1, 1]^d$, with $d \in \{2, 3\}$ and symmetric boundary conditions. The two-dimensional test cases are discretized by 40 grid points in each direction on the coarsest level. The three-dimensional test case is discretized using 50 grid points in each direction. In all test cases, two levels of refinement with a constant factor 4 in all spatial directions are used. In total, there are three levels. The initialization and the exact solution are geometrically computed from the triangle ($d = 3$) and line ($d = 2$) representations of the investigated geometries based on the Computational Geometry Algorithms Library (CGAL)[3].

Benchmarking Platform

The benchmarking results have been obtained by using a single node of the Vienna Scientific Cluster 3 (VSC3)[4]. The node is equipped with two 8-core Intel Xeon E5-2650v2 processors, running at 2.6 GHz and 64 GB of DDR3 memory. The benchmark implementation is written in C++11 and compiled in a GNU/Linux environment using GCC-7.3 with optimization flag `-O3`.

[3] https://www.cgal.org/.

[4] http://vsc.ac.at/.

5.1.1 Corner Examples

(a) *Corner* example (b) *Sharp Corner* example

Fig. 3. Isolines on Level 1 (coarsest grid): Black lines denote the solutions based on re-distancing without the correction step, red lines with the correction step, and white lines for the exact solution. The green and blue background colors give the distance to the interface. The yellow boxes show the outline of the meshes, there is only one mesh on each level.

The corner test cases have the refined levels around the corner near the center of the domain, as shown by the yellow boxes. The symmetric boundary condition causes additional corners at the domain boundary. For these corners no refinement is employed, so that a correction is not possible (cf. Fig. 3a, lower right and Fig. 3b, lower left). Figure 3 shows the isolines extracted from Level 1. Notably, the corner is not represented by a single point as the corner is purposely not grid aligned, as is the usual case in practical level-set simulations. The FMM over-estimates the distance for rarefaction waves (reflex angle side) and under-estimates the distance at shock waves. This can be seen on the obtuse angle side in Fig. 3a and on the, acute angel side in Fig. 3b.

The proposed correction step reduces the error (cf. Table 1 and Table 2) in regions covered by a finer mesh and for the rarefaction wave. In the *Corner* example the errors in L1-norm and L2-norm are reduced by a factor of 2.1 on Level 1 and by a factor of 1.9 on Level 2. For the *Sharp Corner* example the reduction is even higher, 2.7 on Level 1 and 2.1 on Level 2, as sharper corners benefit more from the correction. On Level 3 the correction is not possible, as it is the finest level.

Table 1. Error norms for the *Corner* example, with and without the correction step applied, and the corresponding factor by which the error norm is reduced.

Level	L1-norm	L1-reduc	L2-norm	L2-reduc	inf-norm	inf-reduc
1	5.437e−3		3.260e−4		4.785e−2	
1 corrected	2.491e−3	2.2	1.550e−4	2.1	3.079e−2	1.6
2	1.122e−3		5.101e−5		1.393e−2	
2 corrected	6.035e−4	1.9	2.792e−5	1.8	8.541e−3	1.6
3	5.126e−4		1.819e−5		3.757e−3	

Table 2. Error norms for the *Sharp Corner* example, with and without the correction step applied, and the corresponding factor by which the error norm is reduced.

Level	L1-norm	L1-reduc	L2-norm	L2-reduc	inf-norm	inf-reduc
1	9.110e−3		4.823e−4		6.212e−2	
1 corrected	3.388e−3	2.7	1.812e−4	2.7	2.707e−2	2.3
2	1.894e−3		7.264e−5		1.753e−2	
2 corrected	8.957e−4	2.1	3.484e−5	2.1	9.546e−3	1.8
3	8.866e−4		2.569e−5		4.661e−3	

5.1.2 Narrow Trench

This test case consists of a narrow trench on a flat surface. The trench has a width of 0.01 (i.e., 0.5% of the domain dimension) whereas the grid spacing on Level 1 is only 0.05, thus too coarse to resolve the trench, as no grid points with opposite signs exist along the trench. The re-distancing algorithm without the correction step does not result in a trench, yielding only a small dent in the surface (cf. Fig. 4a). The resolution on the finer levels allows to resolve the trench (cf. Fig. 4b), thus the correction step enables its representation also on Level 1. The errors in L1-norm and L2-norm are reduced by a factor of 15.3 and 14.4, respectively for Level 1 but only by a factor of 1.6 for Level 2 (cf. Table 3). The substantial difference in the improvement between Level 1 and Level 2 is due to the fact that the trench is already representable on Level 2. This can be seen by comparing the black isolines in Fig. 4 for Level 1 and Level 2.

5.1.3 Three-Dimensional Trench

The three-dimensional example combines corners and a trench. There is a step and a trench with a width smaller than the grid resolution of Level 1 (cf. Fig. 5). The trench has a slight dent, so that it is not aligned with the grid. Geometrical configurations like this often appear in microelectronic devices [17], in particular, high aspect ratios are common [18].

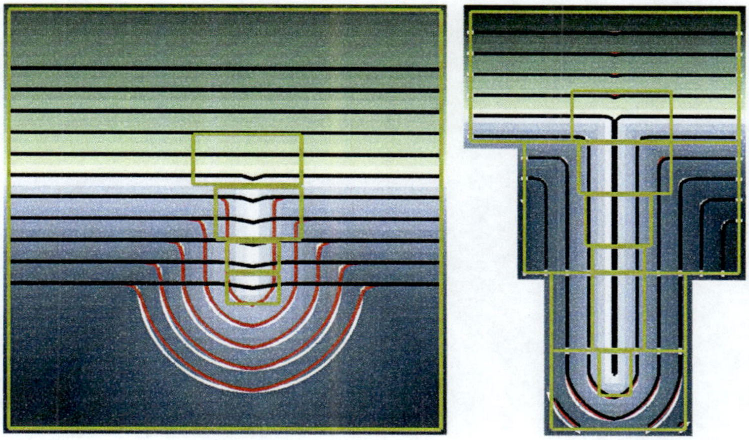

(a) Level 1 (b) Level 2

Fig. 4. Isolines for Level 1 and Level 2 for the *Narrow Trench* example. Black lines show the results based on re-distancing without the correction step, red lines with the correction step, and white lines for the exact solution. The green and blue background colors give the distance to the interface. The yellow boxes show the mesh configuration on Level 1 and Level 2 in (a) and Level 2 and Level 3 in (b).

Table 3. Error norms for the *Narrow Trench* example, with and without the correction step applied, and the corresponding factor by which the error norm is reduced.

Level	L1-norm	L1-reduc	L2-norm	L2-reduc	inf-norm	inf-reduc
1	9.101e−2		4.539e−3		4.839e−1	
1 corrected	5.941e−3	15.3	3.148e−4	14.4	4.005e−2	12.1
2	6.732e−4		5.222e−5		1.262e−2	
2 corrected	4.129e−4	1.6	3.339e−5	1.6	8.398e−3	1.5
3	6.709e−5		4.236e−6		3.400e−3	

The given mesh placement is visualized in Fig. 6 for the different levels. There are 107 meshes on Level 2 containing 674 496 grid points in total and on Level 3 there are 591 meshes with a total of 6 146 688 grid points.

In this test case the errors in the L1-norm and L2-norm for the corrected levels are smaller by a factor of at least 4.2 on Level 1 and 1.5 on Level 2. The reasons are similar to the ones presented in the *Narrow Trench* example. The inf-norm for the error on Level 2 has not decreased, since the grid point causing it is not covered by a refined mesh and is also located on a shock wave. The FMM cannot correct such errors, though a skeleton[5] aware modification

[5] A skeleton is the union of all points with more than one interface point closest to them [19].

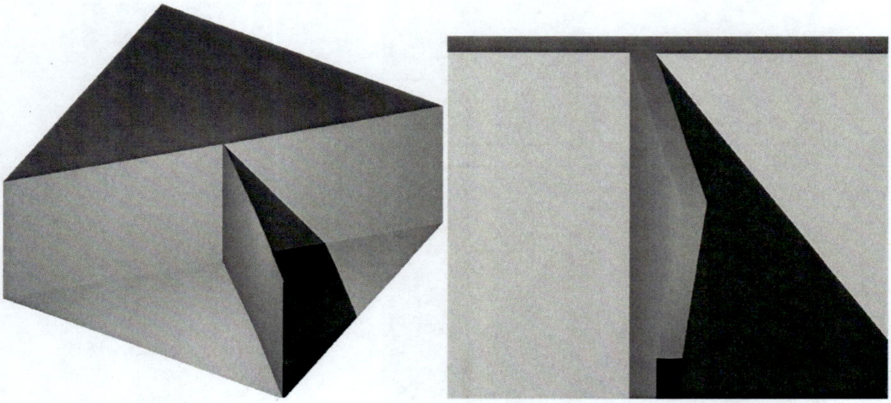

(a) Full view (b) Closeup of the dent

Fig. 5. Rendering of the *Three-Dimensional Trench* example (view from the bottom). The dent is only visible from a particular angle.

Table 4. Error norms for the *Three-Dimensional Trench* example, with and without the correction step applied, and the corresponding factor by which the error norm is reduced.

Level	L1-norm	L1-reduc	L2-norm	L2-reduc	inf-norm	inf-reduc
1	1.853e−2		1.356e−4		4.593e−1	
1 corrected	4.422e−3	4.2	2.644e−5	5.1	3.478e−2	13.2
2	2.426e−4		1.259e−6		1.418e−2	
2 corrected	1.588e−4	1.5	8.429e−7	1.5	1.418e−2	1.0
3	3.165e−5		7.340e−8		3.017e−3	

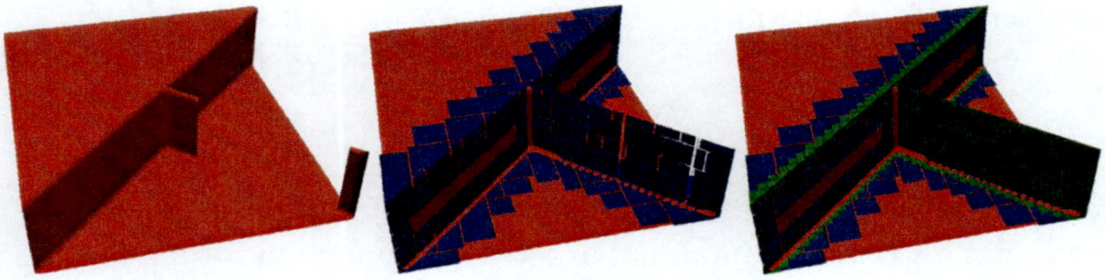

(a) Level 1 (b) Level 2 (c) Level 3

Fig. 6. Interface representation on the different mesh levels (view from the bottom). On Level 1 (red) the trench is too thin to be resolved by the spatial discretization. Level 2 (blue) and Level 3 (green) show the placement of the different meshes (small gaps in between).

of the FMM could potentially improve those grid points, by considering only grid points, which are not separated by the skeleton, for the calculation of the distance.

In Fig. 7a the runtime for re-distancing and the correction steps, together with the total runtime are shown. The correction step extends the re-distancing runtime by 4% for one thread and up to 10% for 16 threads. The increased runtime contribution for higher thread counts is caused by the limited parallel speed-up (cf. Fig. 7b) of the correction step.

A speed-up for the re-distancing step with correction of 9.3 has been achieved for 16 threads (cf. Fig. 7b), which corresponds to a parallel efficiency of 58%. The parallel speed-up for the correction step alone is inferior, because the non-parallelized Level 1 contributes more than half of the runtime for higher thread counts (cf. Fig. 8a). For a fair comparison (cf. Fig. 8b) the speed-up of the correction step is compared to the re-distancing on a per-level-basis for all levels. The speed-up strongly depends on the level, because the number and computational load of the meshes varies considerably among the levels.

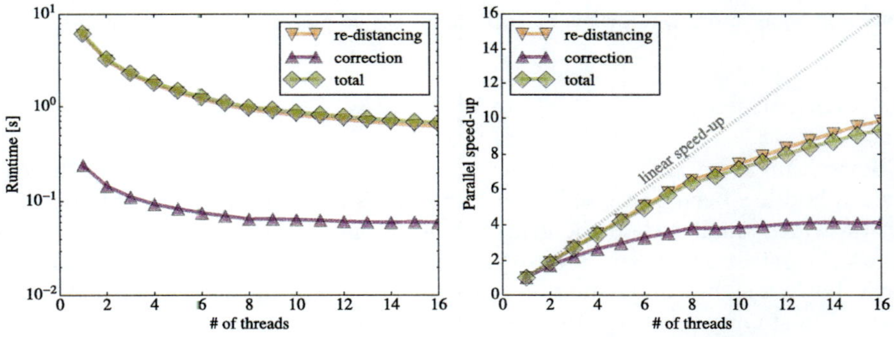

(a) Runtime (b) Parallel speed-up

Fig. 7. Runtime and speed-up (all levels combined) for re-distancing and the correction step for up to 16 threads.

On Level 2 the correction step shows a speed-up of 7.5, which is nearly the same as for the re-distancing step (speed-up of 7.8). The speed-up is reduced because there is less computational load and, therefore, synchronization tasks take up proportionally more time. The second reason is that the initialization does not show any speed-up for more than 8 threads, due to non-uniform memory access (NUMA) effects.

On Level 1 there is practically no speed-up, because the mesh based parallelism cannot be applied for a single mesh. The shortest runtime for the initialization on Level 1 is achieved with 8 threads. With more threads the initialization step is limited by NUMA effects.

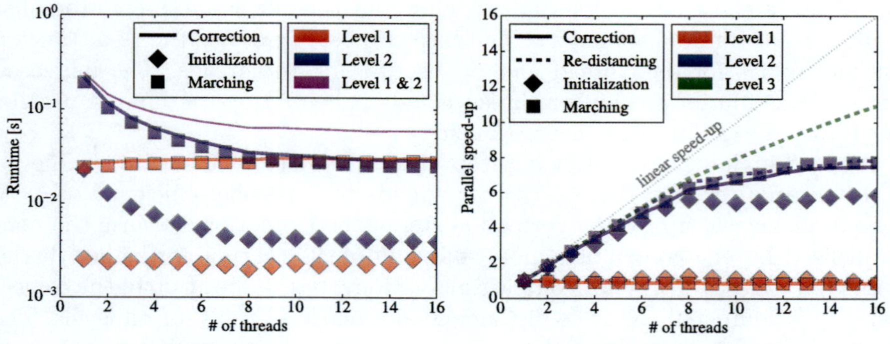

(a) Runtime

(b) Parallel speed-up

Fig. 8. Runtime and speed-up per level up to 16 threads. The speed-up on the same level is similar between the re-distancing and correction steps. The initialization and marching times are given for the correction step. The marching time dominates.

6 Conclusion

A bottom-up correction step for hierarchical re-distancing using the fast marching method has been presented. The correction step allows to represent trenches with a width smaller than the grid resolution on coarse grids and higher accuracy on rarefaction waves. Several test examples in two and three dimensions have shown a significant reduction of the error. The performance penalty of the correction step ranges from 4 to 10% of the original runtime. The total parallel speed-up of re-distancing with the correction step is 9.3 for 16 threads. Finally, we show that the level-by-level speed-up is similar to re-distancing without correction.

Acknowledgments. The financial support by the *Austrian Federal Ministry for Digital and Economic Affairs* and the *National Foundation for Research, Technology and Development* is gratefully acknowledged. The computational results presented have been achieved using the Vienna Scientific Cluster (VSC).

References

1. Sethian, J.A.: A fast marching level set method for monotonically advancing fronts. Proc. Natl. Acad. Sci. **93**, 1591–1595 (1996)
2. Gibou, F., Fedkiw, R., Osher, S.: A review of level-set methods and some recent applications. J. Comput. Phys. **353**, 82–109 (2018)
3. Cheng, L.-T., Tsai, Y.-H.: Redistancing by flow of time dependent Eikonal equation. J. Comput. Phys. **227**, 4002–4017 (2008)
4. Adalsteinsson, D., Sethian, J.A.: A fast level set method for propagating interfaces. J. Comput. Phys. **118**, 269–277 (1995)
5. Russo, G., Smereka, P.: A remark on computing distance functions. J. Comput. Phys. **163**, 51–67 (2000)

6. Detrixhe, M., Gibou, F., Min, C.: A parallel fast sweeping method for the Eikonal equation. J. Comput. Phys. **237**, 46–55 (2013)
7. Jeong, W.-K., Whitaker, R.T.: A fast iterative method for Eikonal equations. SIAM J. Sci. Comput. **30**, 2512–2534 (2008)
8. Weinbub, J., Hössinger, A.: Comparison of the parallel fast marching method, the fast iterative method, and the parallel semi-ordered fast iterative method. Procedia Comput. Sci. **80**, 2271–2275 (2016)
9. Royston, M., Pradhana, A., Lee, B., Chow, Y.T., Yin, W., Teran, J., Osher, S.: Parallel redistancing using the Hopf-Lax formula. J. Comput. Phys. **365**, 7–17 (2018)
10. Yang, J., Stern, F.: A highly scalable massively parallel fast marching method for the Eikonal equation. J. Comput. Phys. **332**, 333–362 (2017)
11. Weinbub, J., Hössinger, A.: Shared-memory parallelization of the fast marching method using an overlapping domain-decomposition approach. In: Proceedings of the 24th High Performance Computing Symposium, pp. 1–8 (2016)
12. Diamantopoulos, G., Weinbub, J., Selberherr, S., Hössinger, A.: Evaluation of the shared-memory parallel fast marching method for re-distancing problems. In: Proceedings of the 17th International Conference on Computational Science and Its Applications, pp. 1–8 (2017)
13. Diamantopoulos, G., Hössinger, A., Selberherr, S., Weinbub, J.: A shared memory parallel multi-mesh fast marching method for re-distancing. In: Advances in Computational Mathematics, pp. 1–17 (2019). https://doi.org/10.1007/s10444-019-09683-z
14. Joppich, W., Mijalković, S.: Multigrid Methods for Process Simulation. Computational Microelectronics. Springer, Vienna (1993)
15. Dijkstra, E.W.: A note on two problems in connexion with graphs. Numer. Math. **1**, 269–271 (1959)
16. Popovici, A.M., Sethian, J.A.: 3-D imaging using higher order fast marching traveltimes. Geophysics **67**, 604–609 (2002)
17. Radjenović, B., Lee, J.K., Radmilović-Radjenović, M.: Sparse field level set method for non-convex hamiltonians in 3D plasma etching profile simulations. Comput. Phys. Commun. **174**, 127–132 (2006)
18. Liu, P., Zhang, D., Guo, J., Wang, W., Yang, F.: Optimization of photoresist development and DRIE processes to fabricate high aspect ratio Si structure in 5 nm scale. J. Micromech. Microeng. **29**, 035006 (2019)
19. Cornea, N., Silver, D., Min, P.: Curve-skeleton applications. In: Proceedings of IEEE Visualization, pp. 95–102 (2005)

A Convergence Result of a Linear SUSHI Scheme Using Characteristics Method for a Semi-linear Parabolic Equation

Abdallah Bradji[1(✉)] and Moussa Ziggaf[2,3,4]

[1] LMA (Laboratoire de Mathématiques Appliquées), Faculty of Sciences, University of Annaba, Annaba, Algeria
abdallah.bradji@gmail.com, abdallah.bradji@etu.univ-amu.fr, bradji@math.univ-paris13.fr

[2] Université Sorbonne Paris Nord, LAGA, CNRS, UMR 7539, 93430 Villetaneuse, France
ziggaf@math.univ-paris13.fr

[3] CSEHS, Mohammed VI Polytechnic University, Lot 660, 43150 Ben Guerir, Morocco
Moussa.ZIGGAF@um6p.ma

[4] ENSAO, LM2N, Complexe Universitaire, B.P. 669, 60000 Oujda, Morocco
https://www.i2m.univ-amu.fr/perso/abdallah.bradji/, https://www.math.univ-paris13.fr/~bradji/

Abstract. This work is an extension and improvement of [1] which dealt with a convergence analysis of a FVS (Finite Volume Scheme) using the Characteristic method for non-stationary LINEAR advection-diffusion equations. In this note, we address the case of non-stationary SEMILINEAR advection-diffusion equations. We establish two FVSs, one is linear and the other is nonlinear, which uses the discrete gradient developed in [5] and an approximation of the equation using the Characteristic method. For the sake of simplicity of the present note, we only focus on the linear scheme and we prove its convergence. The convergence analysis relies mainly on a well developed new discrete *a prior estimate.*

This work is a continuation of the previous one [2] in which we derived directly a finite volume scheme for the semilinear heat equation along with a convergence analysis.

Keywords: Nonlinear parabolic equation · Finite volume methods · Characteristics method · Linear scheme · Nonlinear scheme

1 Aim of This Note

We are concerned with the following SLUADP (Semilinear Unsteady Advection-Diffusion Problem):

$$u_t(\boldsymbol{x},t) - \Delta u(\boldsymbol{x},t) + \operatorname{div}(\mathbf{v}u)(\boldsymbol{x},t) + b(\boldsymbol{x})u(\boldsymbol{x},t) = f(u)(\boldsymbol{x},t),\ \boldsymbol{x} \in \Omega,\ t \in (0,T), \quad (1)$$

where Ω is an open bounded connected subset of $\mathbb{R}^d$ ($d \in \mathbb{N}^\star$), $T > 0$, f is a given function, and $\mathbf{v} = \mathbf{v}(\boldsymbol{x},t) \in \mathbb{R}^d$ is a vector field. We assume here that b is only depending on $\boldsymbol{x}$.

I. Dimov and S. Fidanova (Eds.): HPC 2019, SCI 902, pp. 452–462, 2021.
https://doi.org/10.1007/978-3-030-55347-0_38

Initial conditions are given by:

$$u(0) = u^0. \tag{2}$$

where u^0 is a given functions defined on Ω. Dirichlet boundary conditions are given by

$$u(\boldsymbol{x},t) = 0, \ (\boldsymbol{x},t) \in \partial\Omega \times (0,T). \tag{3}$$

For the sake of simplicity, we assume that the following hypotheses are satisfied:

Assumption 1 (Assumptions on *v*, *b*, and *f*). *We assume that the functions* $\boldsymbol{v}$, b, *and* f *are satisfying:*

$$\boldsymbol{v} \in \mathscr{C}^1(\overline{\Omega} \times [0,T]), \ \ \boldsymbol{v}(\boldsymbol{x},t) = 0, \ \ \forall (\boldsymbol{x},t) \in \partial\Omega \times [0,T], \tag{4}$$

$$b \in \mathscr{C}^1(\overline{\Omega}), \ \ \text{and} \ \ b(\boldsymbol{x}) + \operatorname{div}\boldsymbol{v}(\boldsymbol{x},t) \geq 0, \ \ \forall (\boldsymbol{x},t) \in \Omega \times [0,T], \tag{5}$$

and $f \in \mathscr{C}^1(\mathbb{R})$ *such that, for a given* $\kappa > 0$

$$|f'(s)| \leq \kappa, \ \ \forall s \in \mathbb{R}. \tag{6}$$

In this note, we consider SUSHI (Scheme Using Stabilization and Hybrid Interfaces) developed in [5] to approximate SLUADP (1)–(3) combined with characteristics method. Characteristics method is the replacement of the advective part of the equation by total differentiation along characteristics. Such method is well developed in the context of FEM (Finite Element Method), see [7, Subsection 12.5, pp. 423–427]. However, to the best of our knowledge, the characteristics method has not been attracted the attention it merits in FV methods for parabolic equations. This work is an extension and improvement of our previous note [1] in which we applied SUSHI combined with the Characteristic method for the linear case. It is also a continuation of our previous works [2–4] in which we derived directly schemes for the linear heat equation $u_t - \Delta u = f$ and the semilinear heat equation $u_t - \Delta u = f(u)$ along with convergence analysis.

This work and our recent note [1] are initiation in the area of the application of FV methods combined with the characteristics methods.

2 Space and Time Discretizations

We consider as discretization in space the mesh of [5] in which a SUSHI is developed. In brief, such mesh is defined as the triplet $\mathscr{D} = (\mathscr{M}, \mathscr{E}, \mathscr{P})$ where $\mathscr{M}$ is the set of cells, $\mathscr{E}$ is the set of edges, and $\mathscr{P}$ is a set of points $\boldsymbol{x}_K$ in each cell K. We assume that, for all $K \in \mathscr{M}$, there exists a subset $\mathscr{E}_K$ of $\mathscr{E}$ such that $\partial K = \cup_{\sigma\in\mathscr{E}_K}\overline{\sigma}$. For any $\sigma \in \mathscr{E}$, we denote by $\mathscr{M}_\sigma = \{K, \sigma \in \mathscr{E}_K\}$. We then assume that, for any $\sigma \in \mathscr{E}$, either $\mathscr{M}_\sigma$ has exactly one element and then $\sigma \subset \partial\Omega$ (the set of these interfaces, called boundary interfaces, denoted by $\mathscr{E}_{\text{ext}}$) or $\mathscr{M}_\sigma$ has exactly two elements (the set of these interfaces, called interior interfaces, denoted by $\mathscr{E}_{\text{int}}$). For all $\sigma \in \mathscr{E}$, we denote by $\boldsymbol{x}_\sigma$ the barycentre of σ. For all $K \in \mathscr{M}$ and $\sigma \in \mathscr{E}$, we denote by $\mathbf{n}_{K,\sigma}$ the unit vector normal to σ outward to K. Denoting by $d_{K,\sigma}$ the Euclidean distance between $\boldsymbol{x}_K$ and the hyperplane including σ, one assumes that $d_{K,\sigma} > 0$. We then denote by $\mathscr{D}_{K,\sigma}$ the cone with vertex $\boldsymbol{x}_K$ and basis

σ. Also, h_K is used to denote the diameter of K. For more details on the mesh, we refer to [5, Definition 2.1, p. 1012].

We define the discrete space $\mathscr{X}_{\mathscr{D},0}$ as the set of all $v = \left((v_K)_{K\in\mathscr{M}}, (v_\sigma)_{\sigma\in\mathscr{E}}\right)$, where $v_K, v_\sigma \in \mathrm{I\!R}$ and $v_\sigma = 0$ for all $\sigma \in \mathscr{E}_{\text{ext}}$. Let $H_{\mathscr{M}}(\Omega) \subset L^2(\Omega)$ be the space of functions which are constant on each control volume K of the mesh $\mathscr{M}$. For all $v \in \mathscr{X}_{\mathscr{D}}$, we denote by $\Pi_{\mathscr{M}} v \in H_{\mathscr{M}}(\Omega)$ the function defined by $\Pi_{\mathscr{M}} v(\boldsymbol{x}) = v_K$, for a.e. $\boldsymbol{x} \in K$, for all $K \in \mathscr{M}$. In order to analyze the convergence, we need to consider the size of the discretization $\mathscr{D}$ defined by $h_{\mathscr{D}} = \sup\{\text{diam(K)}, K \in \mathscr{M}\}$ and the regularity of the mesh given by $\theta_{\mathscr{D}} = \max\left(\max_{\sigma\in\mathscr{E}_{\text{int}},K,L\in\mathscr{M}} \frac{d_{K,\sigma}}{d_{L,\sigma}}, \max_{K\in\mathscr{M},\sigma\in\mathscr{E}_K} \frac{h_K}{d_{K,\sigma}}\right)$.

The formulation of the scheme we want to consider involves the discrete gradient, denoted by $\nabla_{\mathscr{D}}$, developed in [5]. The value of $\nabla_{\mathscr{D}} u$, where $u \in \mathscr{X}_{\mathscr{D},0}$, is defined by, for all $K \in \mathscr{M}$, for a.e. $\boldsymbol{x} \in \mathscr{D}_{K,\sigma}$

$$\nabla_{\mathscr{D}} u(\boldsymbol{x}) = \frac{1}{\mathrm{m}(K)} \sum_{\sigma\in\mathscr{E}_K} \mathrm{m}(\sigma)(u_\sigma - u_K)\mathbf{n}_{K,\sigma} + \left(\frac{\sqrt{d}}{d_{K,\sigma}}(u_\sigma - u_K - \nabla_K u \cdot (\boldsymbol{x}_\sigma - \boldsymbol{x}_K))\right)\mathbf{n}_{K,\sigma}. \tag{7}$$

We introduce now the inner product defined on $\mathscr{X}_{\mathscr{D},0} \times \mathscr{X}_{\mathscr{D},0}$ and given by

$$\langle u, v\rangle_F = \int_\Omega \nabla_{\mathscr{D}} u(\boldsymbol{x}) \cdot \nabla_{\mathscr{D}} v(\boldsymbol{x}) d\boldsymbol{x}.$$

The time discretization is performed with a constant time step $k = \dfrac{T}{N+1}$, where $N \in \mathrm{I\!N}^\star$, and we shall denote $t_n = nk$, for $n \in [\![0, N+1]\!]$. We denote by ∂^1 the discrete first time derivative given by

$$\partial^1 v^{j+1} = \frac{v^{j+1} - v^j}{k}.$$

Throughout this paper, the letter C stands for a generic positive constant independent of the parameters of the space and time discretizations.

3 Definition of Characteristics

We quote in this section some results of the method of characteristics. These results can be found for instance in [7, Section 12.5, pp. 423–427]. For any $s \in [0, T]$ and $\boldsymbol{x} \in \Omega$, we define the characteristic lines associated to $\mathbf{v} = \mathbf{v}(\boldsymbol{x}, t)$ as the vector functions $\boldsymbol{\Phi} = \boldsymbol{\Phi}(t; \boldsymbol{x}, s) : [0, T] \longrightarrow \Omega$ satisfying the following differential equation:

$$\begin{cases} \dfrac{d\boldsymbol{\Phi}}{dt}(t; \boldsymbol{x}, s) = \mathbf{v}(\boldsymbol{\Phi}(t; \boldsymbol{x}, s), t), \quad t \in (0, T) \\ \boldsymbol{\Phi}(s; \boldsymbol{x}, s) = \boldsymbol{x}. \end{cases} \tag{8}$$

Let us quote some properties concerning the characteristic lines associated to a function $\mathbf{v} = \mathbf{v}(\boldsymbol{x}, t)$:

- The existence and uniqueness of the characteristic lines for each choice of s and $\boldsymbol{x}$ hold under suitable assumptions on $\mathbf{v}$, for instance $\mathbf{v}$ continuous in $\overline{\Omega} \times [0,T]$ and Lipschitz continuous in $\overline{\Omega}$, uniformly with respect to $t \in [0,T]$. These conditions are satisfied under Assumption 1.
- The uniqueness stated in the previous item implies that

$$\boldsymbol{\Phi}(t;\boldsymbol{\Phi}(s;\boldsymbol{x},\tau),s) = \boldsymbol{\Phi}(t;\boldsymbol{x},\tau). \tag{9}$$

Indeed

$$\frac{d\boldsymbol{\Phi}(t;\boldsymbol{\Phi}(s;\boldsymbol{x},\tau),s)}{dt} = \mathbf{v}(\boldsymbol{\Phi}(t;\boldsymbol{\Phi}(s;\boldsymbol{x},\tau),s),t) \quad \text{and} \quad \frac{d\boldsymbol{\Phi}(t;\boldsymbol{x},\tau)}{dt} = \mathbf{v}(\boldsymbol{\Phi}(t;\boldsymbol{x},\tau),t).$$

In addition to this

$$\boldsymbol{\Phi}(s;\boldsymbol{\Phi}(s;\boldsymbol{x},\tau),s) = \boldsymbol{\Phi}(s;\boldsymbol{x},\tau).$$

- Taking $t = \tau$ in (9) yields

$$\boldsymbol{\Phi}(\tau;\boldsymbol{\Phi}(s;\boldsymbol{x},\tau),s) = \boldsymbol{\Phi}(\tau;\boldsymbol{x},\tau) = \boldsymbol{x}. \tag{10}$$

- For any t and s, the inverse of the function $\boldsymbol{x} \mapsto \boldsymbol{\Phi}(t;\boldsymbol{x},s)$ is $\boldsymbol{x} \mapsto \boldsymbol{\Phi}(s;\boldsymbol{x},t)$.

Let us define

$$\overline{u}(\boldsymbol{x},t) = u(\boldsymbol{\Phi}(t;\boldsymbol{x},0),t). \tag{11}$$

Replacing $\boldsymbol{x}$ by $\boldsymbol{\Phi}(0;\boldsymbol{x},t)$ in (11) yields

$$u(\boldsymbol{\Phi}(t;\boldsymbol{\Phi}(0;\boldsymbol{x},t),0),t) = \overline{u}(\boldsymbol{\Phi}(0;\boldsymbol{x},t),t). \tag{12}$$

Using (10), (12) implies that

$$u(\boldsymbol{x},t) = \overline{u}(\boldsymbol{\Phi}(0;\boldsymbol{x},t),t). \tag{13}$$

We have, using the definition (11)

$$\begin{aligned}\frac{\partial \overline{u}}{\partial t}(\boldsymbol{x},t) &= \frac{\partial u}{\partial t}(\boldsymbol{\Phi}(t;\boldsymbol{x},0),t) + \sum_{i=1}^{d} \frac{\partial u}{\partial \boldsymbol{x}_i}(\boldsymbol{\Phi}(t;\boldsymbol{x},0),t)\frac{\partial \boldsymbol{\Phi}_i}{\partial t}(t;\boldsymbol{x},0) \\ &= \frac{\partial u}{\partial t}(\boldsymbol{\Phi}(t;\boldsymbol{x},0),t) + \nabla u(\boldsymbol{\Phi}(t;\boldsymbol{x},0),t)\cdot \mathbf{v}(\boldsymbol{\Phi}(t;\boldsymbol{x},0),t).\end{aligned} \tag{14}$$

Replacing x by $\boldsymbol{\Phi}(0;\boldsymbol{x},t)$ in (1) and using (14) yield

$$\begin{aligned}&\frac{\partial \overline{u}}{\partial t}(\boldsymbol{x},t) - \nabla u(\boldsymbol{\Phi}(t;\boldsymbol{x},0),t)\cdot \mathbf{v}(\boldsymbol{\Phi}(t;\boldsymbol{x},0),t) - \overline{\Delta u}(\boldsymbol{x},t) \\ &+ \mathrm{div}(\mathbf{v}u)(\boldsymbol{\Phi}(0;\boldsymbol{x},t),t) + \overline{b}(\boldsymbol{x})\overline{u}(\boldsymbol{x},t) = f(\overline{u}(t)).\end{aligned}$$

This is equivalent to

$$\frac{\partial \overline{u}}{\partial t} - \overline{\Delta u} + \left(\overline{b} + \overline{\mathrm{div}\mathbf{v}}\right)\overline{u} = f(\overline{u}(t)). \tag{15}$$

Taking $t = t_{n+1}$ as argument in Eq. (15) leads to

$$\frac{\partial \overline{u}}{\partial t}(t_{n+1}) - \overline{\Delta u}(t_{n+1}) + \left(\overline{b} + \overline{\mathrm{div}\mathbf{v}}(t_{n+1})\right)\overline{u}(t_{n+1}) = f(\overline{u}(t_{n+1})). \tag{16}$$

Let us consider the first order approximation

$$\frac{\partial \overline{u}}{\partial t}(\boldsymbol{x}, t_{n+1}) \approx \frac{\overline{u}(\boldsymbol{x}, t_{n+1}) - \overline{u}(\boldsymbol{x}, t_n)}{k}.$$

This gives, thanks to (11)

$$\frac{\partial \overline{u}}{\partial t}(\boldsymbol{x}, t_{n+1}) \approx \frac{u(\boldsymbol{\Phi}(t_{n+1};\boldsymbol{x},0), t_{n+1}) - u(\boldsymbol{\Phi}(t_n;\boldsymbol{x},0), t_n)}{k}.$$

Using (10) (with $\tau = t_{n+1}$ and $s = 0$) and (9) (with $t = t_n$, $s = 0$, and $\tau = t_{n+1}$), we get the following approximation for $\frac{\partial \overline{u}}{\partial t}(\boldsymbol{\Phi}(0;\boldsymbol{x},t_{n+1}), t_{n+1})$

$$\frac{u(\boldsymbol{\Phi}(t_{n+1};\boldsymbol{\Phi}(0;\boldsymbol{x},t_{n+1}),0), t_{n+1}) - u(\boldsymbol{\Phi}(t_n;\boldsymbol{\Phi}(0;\boldsymbol{x},t_{n+1}),0), t_n)}{k} = \frac{u(\boldsymbol{x}, t_{n+1}) - u(\boldsymbol{\Phi}(t_n;\boldsymbol{x},t_{n+1}), t_n)}{k}.$$

Let us set

$$\begin{aligned}
&\frac{\partial \overline{u}}{\partial t}(\boldsymbol{\Phi}(0;\boldsymbol{x},t_{n+1}), t_{n+1}) \\
&= \frac{u(\boldsymbol{\Phi}(t_{n+1};\boldsymbol{\Phi}(0;\boldsymbol{x},t_{n+1}),0), t_{n+1}) - u(\boldsymbol{\Phi}(t_n;\boldsymbol{\Phi}(0;\boldsymbol{x},t_{n+1}),0), t_n)}{k} + \mathbb{T}_1^{n+1}(\boldsymbol{x}) \\
&= \frac{u(\boldsymbol{x}, t_{n+1}) - u(\boldsymbol{\Phi}(t_n;\boldsymbol{x},t_{n+1}), t_n)}{k} + \mathbb{T}_1^{n+1}(\boldsymbol{x}).
\end{aligned} \tag{17}$$

Using a Taylor expansion to get

$$|\mathbb{T}_1^{n+1}| \le Ck\|u\|_{\mathscr{C}^1([0,T];\mathscr{C}(\overline{\Omega}))}.$$

Taking $\boldsymbol{x} = \boldsymbol{\Phi}(0;\boldsymbol{x},t_{n+1})$ in (16), using (13), and gathering the result with (17) yield

$$\begin{aligned}
&\frac{u(t_{n+1}) - u(t_n)(\boldsymbol{\Phi}(t_n;\boldsymbol{x},t_{n+1}))}{k} - \Delta u(t_{n+1}) + (b + \mathrm{div}\mathbf{v}(t_{n+1}))\, u(t_{n+1}) \\
&= f(u(t_{n+1})) - \mathbb{T}_1^{n+1}(\boldsymbol{x}).
\end{aligned} \tag{18}$$

Let us now look for a suitable approximation for $\boldsymbol{\Phi}(t_n;\boldsymbol{x},t_{n+1})$. Taking $s = t_{n+1}$ in (8) leads to

$$\frac{d\boldsymbol{\Phi}}{dt}(t;\boldsymbol{x},t_{n+1}) = \mathbf{v}(\boldsymbol{\Phi}(t;\boldsymbol{x},t_{n+1}), t) \;\text{ and }\; \boldsymbol{\Phi}(t_{n+1};\boldsymbol{x},t_{n+1}) = \boldsymbol{x}. \tag{19}$$

Integrating (19) over $t \in (t_n, t_{n+1})$ yields

$$\boldsymbol{x} - \boldsymbol{\Phi}(t_n;\boldsymbol{x},t_{n+1}) = \int_{t_n}^{t_{n+1}} \mathbf{v}(\boldsymbol{\Phi}(t;\boldsymbol{x},t_{n+1}), t)dt.$$

This gives

$$\boldsymbol{\Phi}(t_n;\boldsymbol{x},t_{n+1}) = \boldsymbol{x} - \int_{t_n}^{t_{n+1}} \mathbf{v}(\boldsymbol{\Phi}(t;\boldsymbol{x},t_{n+1}), t)dt \approx \boldsymbol{x} - k\mathbf{v}(\boldsymbol{\Phi}(t_{n+1};\boldsymbol{x},t_{n+1}), t_{n+1}) = \omega^{n+1}(\boldsymbol{x}), \tag{20}$$

where, using (10),

$$\omega^{n+1}(\boldsymbol{x}) = \boldsymbol{x} - k\mathbf{v}(\boldsymbol{x}, t_{n+1}). \tag{21}$$

Let us set

$$\begin{aligned}\Phi(t_n;\boldsymbol{x},t_{n+1}) &= \omega^{n+1}(\boldsymbol{x}) + \mathbb{T}_2^{n+1}(\boldsymbol{x}) \ \text{ and } \ u(t_n)(\Phi(t_n;\boldsymbol{x},t_{n+1})) \\ &= u(t_n)(\omega^{n+1}(\boldsymbol{x})) + \mathbb{T}_3^{n+1}(\boldsymbol{x}).\end{aligned} \tag{22}$$

We can check that ω^{n+1} is a second order accurate approximation for $\Phi(t_n;\cdot,t_{n+1})$. This implies that $\mathbb{T}_3^{n+1}$ is of order two, i.e.

$$|\mathbb{T}_3^{n+1}| \le Ck^2 \|u\|_{\mathscr{C}^1([0,T];\mathscr{C}(\overline{\Omega}))}.$$

Under hypothesis (4) of Assumption 1, we have $\mathbf{v}(\boldsymbol{x}, t_{n+1}) = 0$ for all $\boldsymbol{x} \in \partial\Omega$. Let $\boldsymbol{x}^\star(\boldsymbol{x})$ be the point of $\partial\Omega$ which has minimal distance from $\boldsymbol{x}$. We have then, using (20)

$$\begin{aligned}&\left|\omega^{n+1}(\boldsymbol{x}) - \boldsymbol{x}\right| = k\left|\mathbf{v}(\boldsymbol{x},t_{n+1})\right| = k\left|\mathbf{v}(\boldsymbol{x},t_{n+1}) - \mathbf{v}(\boldsymbol{x}^\star(\boldsymbol{x}),t_{n+1})\right| \\ &\le k|\boldsymbol{x} - \boldsymbol{x}^\star(\boldsymbol{x})| \sup_{(\boldsymbol{x},t)\in\overline{\Omega}\times[0,T]} |\nabla\mathbf{v}(\boldsymbol{x},t)|,\end{aligned} \tag{23}$$

where $|\cdot|$ denotes the Euclidean norm in $\mathbb{R}^d$. Let us choose k sufficiently small such that

$$k \sup_{(\boldsymbol{x},t)\in\overline{\Omega}\times[0,T]} |\nabla\mathbf{v}(\boldsymbol{x},t)| < 1. \tag{24}$$

Gathering this fact with (23) implies that $\left|\omega^{n+1}(\boldsymbol{x}) - \boldsymbol{x}\right| < |\boldsymbol{x} - \boldsymbol{x}^\star(\boldsymbol{x})|$. This yields $\omega^{n+1}(\boldsymbol{x}) \in \Omega$.

From (18) and (22), we deduce that

$$\begin{aligned}\Delta u(t_{n+1}) + f(u(t_{n+1})) &= \frac{u(t_{n+1}) - u(t_n)(\omega^{n+1})}{k} + (b + \mathrm{div}\mathbf{v}(t_{n+1}))\, u(t_{n+1}) \\ &+ \mathbb{T}_1^{n+1}(\boldsymbol{x}) - \frac{\mathbb{T}_3^{n+1}(\boldsymbol{x})}{k}.\end{aligned} \tag{25}$$

This also implies that, using a Taylor expansion and Assumption 1

$$\begin{aligned}\Delta u(t_{n+1}) + f(u(t_n)) &= \frac{u(t_{n+1}) - u(t_n)(\omega^{n+1})}{k} \\ &+ (b + \mathrm{div}\mathbf{v}(t_{n+1}))\, u(t_{n+1}) + \mathbb{T}_4^{n+1}(\boldsymbol{x}),\end{aligned} \tag{26}$$

where

$$|\mathbb{T}_4^{n+1}| \le Ck\|u\|_{\mathscr{C}^1([0,T];\mathscr{C}(\overline{\Omega}))}. \tag{27}$$

4 Formulation of Two FV Schemes and Statement of the Convergence of the Linear Scheme

In this section, we apply the SUSHI to SLUADP (1)–(3). We establish two schemes. One is nonlinear and it is based on the approximation (25) of Eq. (1) and the second one is based on (26).

Definition 1 (Formulation of a nonlinear FV scheme for (1)–(3) using characteristics). *Let ω^{n+1} be the function given by* (21)*. We define the following FV scheme as an approximation for the SLUADP* (1)–(3)*: Find $u^0_{\mathscr{D}} \in \mathscr{X}_{\mathscr{D},0}$ such that*

$$\langle u^0_{\mathscr{D}}, v\rangle_F = -\left(\Delta u^0, \Pi_{\mathscr{M}} v\right)_{L^2(\Omega)}, \; \forall v \in \mathscr{X}_{\mathscr{D},0}, \tag{28}$$

and for any $n \in [\![0,N]\!]$, find $u^{n+1}_{\mathscr{D}} \in \mathscr{X}_{\mathscr{D},0}$ such that, for all $v \in \mathscr{X}_{\mathscr{D},0}$

$$\frac{1}{k}\left(\Pi_{\mathscr{M}}\left(u^{n+1}_{\mathscr{D}} - u^n_{\mathscr{D}}(\omega^{n+1})\right), \Pi_{\mathscr{M}} v\right)_{L^2(\Omega)} + \langle u^{n+1}_{\mathscr{D}}, v\rangle_F$$
$$+\left((b+\mathrm{div}\mathbf{v}(t_{n+1}))\,\Pi_{\mathscr{M}} u^{n+1}_{\mathscr{D}}, \Pi_{\mathscr{M}} v\right)_{L^2(\Omega)} = \left(f(\Pi_{\mathscr{M}} u^{n+1}_{\mathscr{D}}), \Pi_{\mathscr{M}} v\right)_{L^2(\Omega)}. \tag{29}$$

Definition 2 (Formulation of a linear FV scheme for (1)–(3) using characteristics). *Let ω^{n+1} be the function given by* (21)*. We define the following FV scheme as an approximation for the SLUADP* (1)–(3)*: Find $u^0_{\mathscr{D}} \in \mathscr{X}_{\mathscr{D},0}$ such that*

$$\langle u^0_{\mathscr{D}}, v\rangle_F = -\left(\Delta u^0, \Pi_{\mathscr{M}} v\right)_{L^2(\Omega)}, \; \forall v \in \mathscr{X}_{\mathscr{D},0}, \tag{30}$$

and for any $n \in [\![0,N]\!]$, find $u^{n+1}_{\mathscr{D}} \in \mathscr{X}_{\mathscr{D},0}$ such that, for all $v \in \mathscr{X}_{\mathscr{D},0}$

$$\frac{1}{k}\left(\Pi_{\mathscr{M}}\left(u^{n+1}_{\mathscr{D}} - u^n_{\mathscr{D}}(\omega^{n+1})\right), \Pi_{\mathscr{M}} v\right)_{L^2(\Omega)} + \langle u^{n+1}_{\mathscr{D}}, v\rangle_F$$
$$+\left((b+\mathrm{div}\mathbf{v}(t_{n+1}))\,\Pi_{\mathscr{M}} u^{n+1}_{\mathscr{D}}, \Pi_{\mathscr{M}} v\right)_{L^2(\Omega)} = (f(\Pi_{\mathscr{M}} u^n_{\mathscr{D}}), \Pi_{\mathscr{M}} v)_{L^2(\Omega)}. \tag{31}$$

For the sake of simplicity, we will only study the linear scheme (30)–(31) of Definition 2. We will focus on the nonlinear scheme (28)–(29) of Definition 1 in a future work. We state now one of the main results of this work, that is the convergence of the linear FV scheme given by (30)–(31).

Theorem 1 (Existence, uniqueness, and an $L^\infty(L^2)$-error estimate for scheme (30)–(31)). *Let Ω be a polyhedral open bounded subset of $\mathbb{R}^d$, where $d \in \mathbb{N} \setminus \{0\}$. Assume that the solution of* (1)–(3) *satisfies $u \in \mathscr{C}^1([0,T]; \mathscr{C}^2(\overline{\Omega}))$ and $\theta_{\mathscr{D}}$ satisfies $\theta \geq \theta_{\mathscr{D}}$. Let $\nabla_{\mathscr{D}}$ be the discrete gradient defined as in* (7)*. Let $k = \frac{T}{N+1}$, with $N \in \mathbb{N}^\star$, and denote by $t_n = nk$, for $n \in [\![0, N+1]\!]$. For any $n \in [\![0,N]\!]$, let ω^{n+1} be the function defined on Ω and given by* (21)*. Assume in addition that the step k is sufficiently small in the sense of* (24)*.*

Then, under Assumption 1, there exists a unique solution $(u^n_{\mathscr{D}})_{n=0}^{N+1} \in \mathscr{X}^{N+2}_{\mathscr{D},0}$ for scheme (30)–(31) *and the following $L^\infty(L^2)$-error estimate holds:*

$$\max_{n=0}^{N+1} \|\Pi_{\mathscr{M}} u^n_{\mathscr{D}} - u(t_n)\|_{L^2(\Omega)} \leq C\left(k + h_{\mathscr{D}} + \frac{h_{\mathscr{D}}}{k}\right) \|u\|_{\mathscr{C}^1([0,T];\mathscr{C}^2(\overline{\Omega}))}. \tag{32}$$

If we assume in addition that for some given positive δ, we have $h_{\mathscr{D}} \leq Ck^{1+\delta}$, then the error estimate (32) *becomes as*

$$\max_{n=0}^{N+1} \|\Pi_{\mathscr{M}} u^n_{\mathscr{D}} - u(t_n)\|_{L^2(\Omega)} \leq C\left(k + k^\delta + h_{\mathscr{D}}\right) \|u\|_{\mathscr{C}^1([0,T];\mathscr{C}^2(\overline{\Omega}))}. \tag{33}$$

To prove Theorem 1, we use the following new a priori estimate which can be viewed as a generalization and an improvement of the one proved in the recent work [1], see the Remark 1.

Theorem 2 (A priori estimate for the discrete problem). *Under the same hypotheses of Theorem 1, we assume that there exists* $(\eta^n)_{n=0}^{N+1} \in (\mathscr{X}_{\mathscr{D},0})^{N+2}$ *such that* $\eta^0_{\mathscr{D}} = 0$ *and for all* $n \in [\![0,N]\!]$

$$\frac{1}{k}\left(\Pi_{\mathscr{M}}\left(\eta^{n+1}_{\mathscr{D}} - \eta^n_{\mathscr{D}}(\omega^{n+1})\right), \Pi_{\mathscr{M}} v\right)_{L^2(\Omega)} + \langle \eta^{n+1}_{\mathscr{D}}, v\rangle_F + \left((b+\mathrm{div}\boldsymbol{v}(t_{n+1}))\,\Pi_{\mathscr{M}}\eta^{n+1}_{\mathscr{D}}, \Pi_{\mathscr{M}} v\right)_{L^2(\Omega)}$$
$$\leq \left(\mathscr{S}^{n+1}, \Pi_{\mathscr{M}} v\right)_{L^2(\Omega)} + C\|\Pi_{\mathscr{M}}\eta^n_{\mathscr{D}}\|_{L^2(\Omega)}\|\Pi_{\mathscr{M}} v\|_{L^2(\Omega)}, \tag{34}$$

where $\mathscr{S}^{n+1} \in L^2(\Omega)$*, for all* $n \in [\![0,N]\!]$*. Then, the following estimate holds:*

$$\max_{n=0}^{N+1} \|\Pi_{\mathscr{M}}\eta^n_{\mathscr{D}}\|_{L^2(\Omega)} \leq C\mathscr{S}. \tag{35}$$

where $\mathscr{S} = \max_{n=0}^{N} \|\mathscr{S}^{n+1}\|_{L^2(\Omega)}$.

An overview on the proof of Theorem 2. Taking $v = \eta^{n+1}_{\mathscr{D}}$ in (34) and using hypothesis (5) of Assumption 1 and the Cauchy Schwarz inequality yield

$$\|\Pi_{\mathscr{M}}\eta^{n+1}_{\mathscr{D}}\|^2_{L^2(\Omega)} \leq \left(k\mathscr{S} + \|\Pi_{\mathscr{M}}\eta^n_{\mathscr{D}}(\omega^{n+1})\|_{L^2(\Omega)} + Ck\|\Pi_{\mathscr{M}}\eta^n_{\mathscr{D}}\|_{L^2(\Omega)}\right)\|\Pi_{\mathscr{M}}\eta^{n+1}_{\mathscr{D}}\|_{L^2(\Omega)}.$$

This implies that

$$\|\Pi_{\mathscr{M}}\eta^{n+1}_{\mathscr{D}}\|_{L^2(\Omega)} \leq k\mathscr{S} + \|\Pi_{\mathscr{M}}\eta^n_{\mathscr{D}}(\omega^{n+1})\|_{L^2(\Omega)} + Ck\|\Pi_{\mathscr{M}}\eta^n_{\mathscr{D}}\|_{L^2(\Omega)}. \tag{36}$$

We have, thanks to the definition of $\Pi_{\mathscr{M}}$, $\Pi_{\mathscr{M}}\eta^n_{\mathscr{D}}(\omega^{n+1})(\boldsymbol{x}) = \eta^n_K$ if $\omega^{n+1}(\boldsymbol{x}) \in K$. Therefore

$$\|\Pi_{\mathscr{M}}\eta^n_{\mathscr{D}}(\omega^n)\|^2_{L^2(\Omega)} = \sum_{K\in\mathscr{M}} \mathrm{m}\left(\{\boldsymbol{x} \in \Omega : \omega^{n+1}(\boldsymbol{x}) \in K\}\right)(\eta^n_K)^2.$$

Using a change of variable yields $\mathrm{m}\left(\{\boldsymbol{x} \in \Omega : \omega^{n+1}(\boldsymbol{x}) \in K\}\right) = \int_{(\omega^{n+1})^{-1}(K)} d\boldsymbol{x} \leq (1+Ck)\mathrm{m}(K)$ (Note that $(\omega^{n+1})^{-1}$ exists since it is injective thanks to (24)). This gives

$$\|\Pi_{\mathscr{M}}\eta^n_{\mathscr{D}}(\omega^{n+1})\|^2_{L^2(\Omega)} \leq (1+Ck)\|\Pi_{\mathscr{M}}\eta^n_{\mathscr{D}}\|^2_{L^2(\Omega)}. \tag{37}$$

This gives $\|\Pi_{\mathscr{M}}\eta^n_{\mathscr{D}}(\omega^{n+1})\|_{L^2(\Omega)} \leq (1+Ck)^{\frac{1}{2}}\|\Pi_{\mathscr{M}}\eta^n_{\mathscr{D}}\|_{L^2(\Omega)}$. This with (36) imply that

$$\begin{aligned}
\|\Pi_{\mathscr{M}}\eta^{n+1}_{\mathscr{D}}\|_{L^2(\Omega)} &\leq k\mathscr{S} + \left((1+Ck)^{\frac{1}{2}} + Ck\right)\|\Pi_{\mathscr{M}}\eta^n_{\mathscr{D}}\|_{L^2(\Omega)}\\
&\leq k\mathscr{S} + \sqrt{1+Ck}\left(1 + C\frac{k}{\sqrt{1+Ck}}\right)\|\Pi_{\mathscr{M}}\eta^n_{\mathscr{D}}\|_{L^2(\Omega)}\\
&\leq k\mathscr{S} + \sqrt{1+Ck}\,(1+Ck)\,\|\Pi_{\mathscr{M}}\eta^n_{\mathscr{D}}\|_{L^2(\Omega)}.
\end{aligned}$$

This implies using the induction on n and that fact that $\eta_{\mathscr{D}}^0 = 0$

$$\|\Pi_{\mathscr{M}}\eta_{\mathscr{D}}^{n+1}\|_{L^2(\Omega)} \le k\mathscr{S}\frac{(\sqrt{1+Ck})^{N+1}(1+Ck)^{N+1}-1}{\sqrt{1+Ck}(1+Ck)-1}$$
$$\le C\mathscr{S}\left((\sqrt{1+Ck})^{N+1}(1+Ck)^{N+1}\right). \tag{38}$$

Let us bound the right hand side of (38). We have, since $k = T/(N+1)$

$$(\sqrt{1+Ck})^{N+1} = \sqrt{G(s)},$$

with $G(s) = (1+\frac{r}{s})^s$, $s = N+1$, and $r = CT$. We are able to check that $G'(s) >$, for all $s > 0$, and consequently G is an increasing function. This implies that, for all $s > 0$, $G(s) \le \lim_{s\to+\infty} G(s) = \exp(r)$. Therefore $(\sqrt{1+Ck})^{N+1} \le \exp\left(\frac{CT}{2}\right)$. Using the same reasoning to get $(1+Ck)^{N+1} \le \exp(CT)$. These two estimates on $(\sqrt{1+Ck})^{N+1}$ and $(1+Ck)^{N+1}$ together with (38) imply estimate (35). □

Remark 1 *(Comments on the a prior estimate of Theorem* 2 *and the one presented in* [1] *)***.** In the a prior estimate of our previous work [1], we considered the following hypothesis instead of (34)

$$\frac{1}{k}\left(\Pi_{\mathscr{M}}(\eta_{\mathscr{D}}^{n+1}-\eta_{\mathscr{D}}^{n}(\omega^{n+1})),\Pi_{\mathscr{M}}v\right)_{L^2(\Omega)} + \langle\eta_{\mathscr{D}}^{n+1},v\rangle_F$$
$$+\left((b+\mathrm{div}\mathbf{v}(t_{n+1}))\Pi_{\mathscr{M}}\eta_{\mathscr{D}}^{n+1},\Pi_{\mathscr{M}}v\right)_{L^2(\Omega)} = \left(\mathscr{S}^{n+1},\Pi_{\mathscr{M}}v\right)_{L^2(\Omega)}. \tag{39}$$

Therefore the hypothesis (34) is more general than that assumed in the discrete a priori estimate of [1]. Two additional facts in (34): the first one is the additional second term in the right hand side and the second fact is that (34) is an inequality whereas (39) is an equality. These facts allow us to deal with the non-linearity caused by the presence of $f(u)$ in (1).

Sketch of the Proof of Theorem 1

The uniqueness and existence for scheme (30)–(32) are consequences of the facts that $\|\nabla_{\mathscr{D}}\cdot\|_{L^2(\Omega)^d}$ is a norm on $\mathscr{X}_{\mathscr{D},0}$, a priori estimate (35), and the matrices involved are square. To prove estimate (32), we compare (30)–(31) with the following scheme: For all $n \in [\![0, N+1]\!]$, find $\bar{u}_{\mathscr{D}}^n \in \mathscr{X}_{\mathscr{D},0}$ such that

$$\langle\bar{u}_{\mathscr{D}}^n, v\rangle_F = -\left(\Delta u(t_n), \Pi_{\mathscr{M}}v\right)_{L^2(\Omega)}, \quad \forall v \in \mathscr{X}_{\mathscr{D},0}. \tag{40}$$

Taking $n = 0$ in (40), using the fact that $u(0) = u^0$ (subject of (2)), and comparing with scheme (30), we get $\eta_{\mathscr{D}}^0 = 0$, where, for all $n \in [\![0, N+1]\!]$, $\eta_{\mathscr{D}}^n \in \mathscr{X}_{\mathscr{D},0}$ is given by $\eta_{\mathscr{D}}^n = u_{\mathscr{D}}^n - \bar{u}_{\mathscr{D}}^n$.

First step: Comparison between u and $\bar{u}_{\mathscr{D}}^n$. We have (see [5])

$$\|u(t_n) - \Pi_{\mathscr{M}}\bar{u}_{\mathscr{D}}^n\|_{L^2(\Omega)} \le Ch_{\mathscr{D}}\|u\|_{\mathscr{C}(0,T;\mathscr{C}^2(\overline{\Omega}))}. \tag{41}$$

Second step: Comparison between $\bar{u}^n_{\mathscr{D}}$ and $u^n_{\mathscr{D}}$. Writing (40) in the level $n+1$ and subtracting the result from (31) yield

$$\frac{1}{k}\left(\Pi_{\mathscr{M}}(u^{n+1}_{\mathscr{D}} - u^n_{\mathscr{D}}(\omega^{n+1})), \Pi_{\mathscr{M}} v\right)_{L^2(\Omega)} + \langle \eta^{n+1}_{\mathscr{D}}, v\rangle_F$$
$$+ \left((b+\operatorname{div}\mathbf{v}(t_{n+1}))\, \Pi_{\mathscr{M}} u^{n+1}_{\mathscr{D}}, \Pi_{\mathscr{M}} v\right)_{L^2(\Omega)}$$
$$= (f(\Pi_{\mathscr{M}} u^n_{\mathscr{D}}) + \Delta u(t_{n+1}), \Pi_{\mathscr{M}} v)_{L^2(\Omega)}. \tag{42}$$

Replacing $\Delta u(t_{n+1}) + f(u(t_n))$ by its value of (26) in (42) yields

$$\frac{1}{k}\left(\Pi_{\mathscr{M}}(\eta^{n+1}_{\mathscr{D}} - \eta^n_{\mathscr{D}}(\omega^{n+1})), \Pi_{\mathscr{M}} v\right)_{L^2(\Omega)} + \langle \eta^{n+1}_{\mathscr{D}}, v\rangle_F + \left((b+\operatorname{div}\mathbf{v}(t_{n+1}))\, \Pi_{\mathscr{M}} \eta^{n+1}_{\mathscr{D}}, \Pi_{\mathscr{M}} v\right)_{L^2(\Omega)}$$
$$= (\mathscr{S}^{n+1}, \Pi_{\mathscr{M}} v)_{L^2(\Omega)} + (f(\Pi_{\mathscr{M}} u^n_{\mathscr{D}}) - f(\Pi_{\mathscr{M}} \bar{u}^n_{\mathscr{D}}), \Pi_{\mathscr{M}} v)_{L^2(\Omega)}, \tag{43}$$

where

$$\mathscr{S}^{n+1} = \frac{u(t_{n+1}) - u(t_n)(\omega^{n+1}(\boldsymbol{x}))}{k} - \frac{\Pi_{\mathscr{M}}\left(\bar{u}^{n+1}_{\mathscr{D}} - \bar{u}^n_{\mathscr{D}}(\omega^{n+1})\right)}{k}$$
$$+ (b+\operatorname{div}\mathbf{v}(t_{n+1}))\, \Pi_{\mathscr{M}}\left(u(t_{n+1}) - \bar{u}^{n+1}_{\mathscr{D}}\right)$$
$$+ f(\Pi_{\mathscr{M}} \bar{u}^n_{\mathscr{D}}) - f(u(t_n)) + \mathbb{T}^{n+1}_4. \tag{44}$$

We have using the hypothesis (6) of Assumption 1 and the Cauchy Schwarz inequality

$$(f(\Pi_{\mathscr{M}} u^n_{\mathscr{D}}) - f(\Pi_{\mathscr{M}} \bar{u}^n_{\mathscr{D}}), \Pi_{\mathscr{M}} v)_{L^2(\Omega)} \le \kappa \|\Pi_{\mathscr{M}} \eta^n_{\mathscr{D}}\|_{L^2(\Omega)} \|\Pi_{\mathscr{M}} v\|_{L^2(\Omega)}. \tag{45}$$

From (43) and (45) and the fact that $\eta^0_{\mathscr{D}} = 0$, we deduce that $(\eta^n)_{n=0}^{N+1} \in (\mathscr{X}_{\mathscr{D},0})^{N+2}$ is satisfying the hypotheses of Theorem 2. This allows to apply a priori estimate (35) to get $\max_{n=0}^{N+1} \|\Pi_{\mathscr{M}} \eta^n_{\mathscr{D}}\|_{L^2(\Omega)} \le C\mathscr{S}$, where $\mathscr{S} = \max_{n=0}^{N} \|\mathscr{S}^{n+1}\|_{L^2(\Omega)}$, and $\mathscr{S}^{n+1}$ is given by (44). Using estimate (41), (37), (27), and Assumption 1 imply that $\max_{n=0}^{N+1} \|\Pi_{\mathscr{M}} \eta^n_{\mathscr{D}}\|_{L^2(\Omega)} \le C\left(k + h_{\mathscr{D}} + \frac{h_{\mathscr{D}}}{k}\right) \|u\|_{\mathscr{C}^1([0,T];\mathscr{C}^2(\overline{\Omega}))}$. This with (41) and triangle inequality imply the desired estimate (32). Estimate (33) is a straightforward consequence of the proved estimate (32). This completes the proof of Theorem 1. □

5 Some Concluding Remarks and Perspectives

We applied SUSHI combined with characteristics method to approximate SLUADP (1)–(3). This model is more general than that we considered in [2] in which we derived schemes directly along with a convergence analysis for the semilinear heat equation $u_t - \Delta u = f(u)$. We established two finite volume schemes: one is linear (the subject of (30)–(31) of Definition 2) and the other is nonlinear (subject of (28)–(29) of Definition 1). For the sake of simplicity and clarity, we focused on the linear scheme. We proved the $L^\infty(L^2)$–error estimate (32) which is a conditional convergence. This error estimate is proved thanks to the well developed new a prior estimate of Theorem 2. Scheme

(30)–(31), Theorem 2, and convergence order (32) are new. The convergence order $k+h_{\mathscr{D}}+\frac{h_{\mathscr{D}}}{k}$ of (32) is similar to the one obtained in our previous work [1] and to the result obtained in the context of FEM for linear UADP, that is $k+h+\frac{h^2}{k}$, see [7, p. 427] but it is different because of the presence of $\frac{h_{\mathscr{D}}}{k}$ instead of $\frac{h^2}{k}$. This difference stems from the first order $L^\infty(L^2)$–error estimate (41) which is second order in FEM.

For the nonlinear scheme (28)–(29) of Definition 1, we are able to prove the existence and uniqueness using the method of contractive mapping (see for instance [6, p. 56]). We will address this subject together with other ones, e.g. Crank-Nicolson method and its linearized version, in future works.

As perspective, we will try also to prove an unconditional convergence instead of the conditional one stated in (32). The proof of a convergence in the discrete energy norm of $L^\infty(H^1)$ is also interesting task to work on.

References

1. Benkhaldoun, F., Bradji, A.: Convergence analysis of a finite volume gradient scheme for a linear parabolic equation using characteristic methods. In: Lirkov, I., Margenov, S. (eds.) Large-Scale Scientific Computing. LSSC 2019. LNCS, vol. 11958. Springer, Cham (2020). https://doi.org/10.1007/978-3-030-41032-2_65
2. Bradji, A.: An analysis for the convergence order of gradient schemes for semilinear parabolic equations. Comput. Math. Appl. **72**(5), 1287–1304 (2016)
3. Bradji, A., Fuhrmann, J.: Error estimates of the discretization of linear parabolic equations on general nonconforming spatial grids. C. R. Math. Acad. Sci. Paris **348**(19–20), 1119–1122 (2010)
4. Bradji, A., Fuhrmann, J.: Some abstract error estimates of a finite volume scheme for a non-stationary heat equation on general nonconforming multidimensional spatial meshes. Appl. Math. **58**(1), 1–38 (2013)
5. Eymard, R., Gallouët, T., Herbin, R.: Discretization of heterogeneous and anisotropic diffusion problems on general nonconforming meshes. IMA J. Numer. Anal. **30**(4), 1009–1043 (2010)
6. Feistauer, M., Felcman, J., Straskraba, I.: Mathematical and Computational Methods for Compressible Flow. Numerical Mathematics and Scientific Computation. Oxford University Press, Oxford (2003)
7. Quarteroni, A., Valli, A.: Numerical Approximation of Partial Differential Equations. Springer Series in Computational Mathematics, vol. 23. Springer, Berlin (2008)

Author Index

I. Dimov and S. Fidanova (Eds.): HPC 2019, SCI 902, pp. 463–464, 2021.
https://doi.org/10.1007/978-3-030-55347-0